SILURIAN

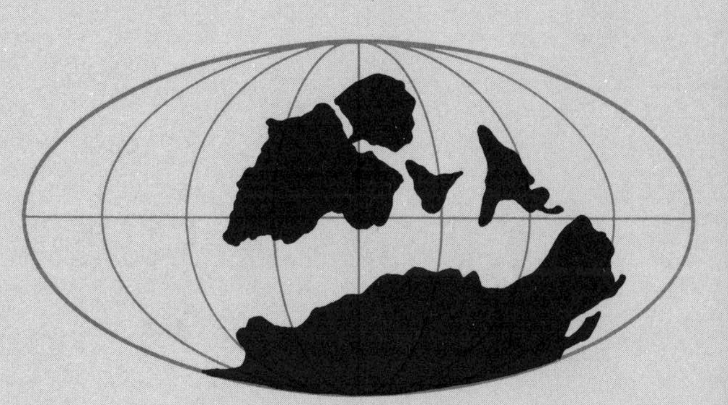

DEVONIAN

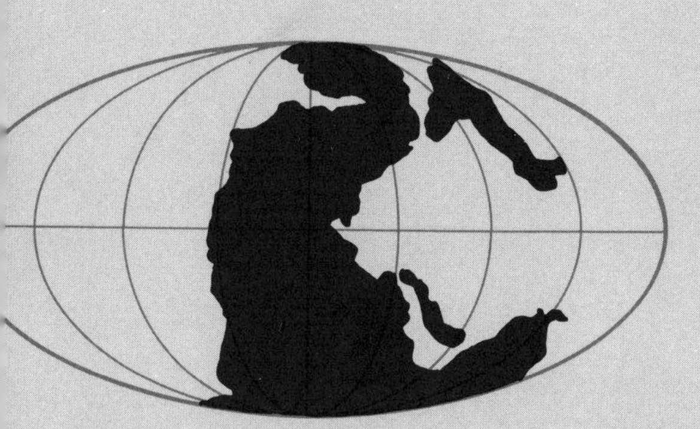

PERMIAN

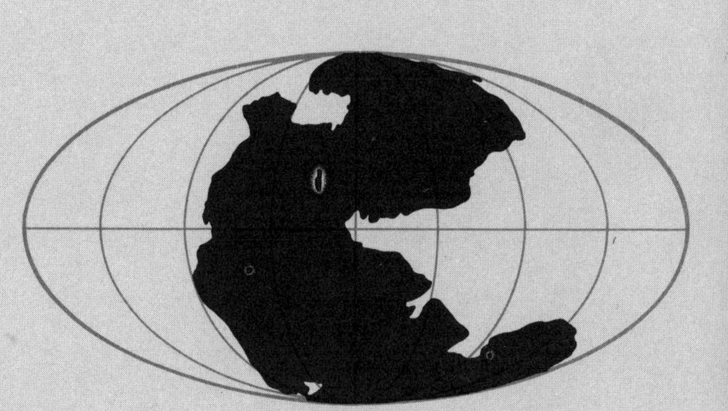

TRIASSIC

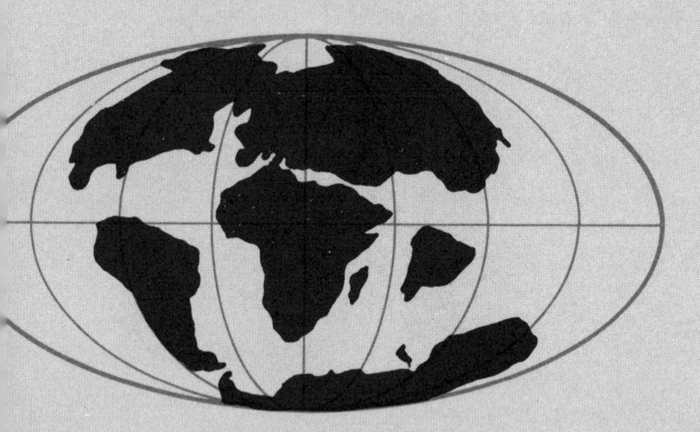

PALEOGENE

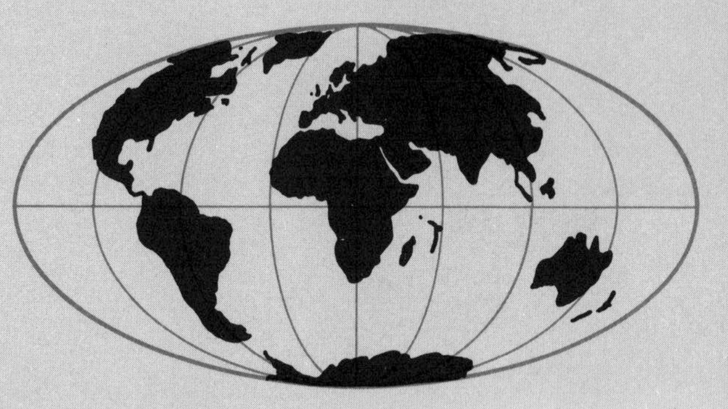

TODAY

HISTORY OF THE EARTH

HISTORY OF THE EARTH

Don L. Eicher
University of Colorado
A. Lee McAlester
Southern Methodist University

Prentice-Hall, Inc.
Englewood Cliffs, New Jersey 07632

Don L. Eicher and A. Lee McAlester, **History of the Earth**

Printed in the United States of America

10 9 8 7 6 5 4 3 2 1

Library of Congress Cataloging in Publication Data
Eicher, Don L.
History of the earth.
Includes bibliographies and index.
1. Historical geology.
I. McAlester, Arcie Lee, joint author.
II. Title.
QE28.3.E52 551.7 79-20857
ISBN 0-13-390047-9

Editorial supervision by Serena Hoffman
Design and layout by Rita Kaye Schwartz
Illustrations by Vantage Art
Cover photograph of Capitol Reef, Utah by Logan M. Campbell

Chapter Opening Photographs:
Title Page, Logan M. Campbell; Chapter 1, NASA; Chapter 2, Don L. Eicher; Chapter 3, W. A. Braddock; Chapter 4, Don L. Eicher; Chapter 5, The Smithsonian Institution; Chapter 6, U.S. Navy; Chapter 7, U.S. Navy; Chapter 8, Logan M. Campbell; Chapter 9, NASA; Chapter 10, Rudolph Zallinger mural, Yale Peabody Museum: Chapter 11, U.S. Coast Guard; Chapter 12, James Hutton lost drawings, Scottish Academic Press Limited.

Prentice-Hall International, Inc., *London*
Prentice-Hall of Australia Pty. Limited, *Sydney*
Prentice-Hall of Canada, Ltd., *Toronto*
Prentice-Hall of India Private Limited, *New Delhi*
Prentice-Hall of Japan, Inc., *Tokyo*
Prentice-Hall of Southeast Asia Pte. Ltd., *Singapore*
Whitehall Books Limited, *Wellington, New Zealand*

OUTLINE CONTENTS

COMPLETE CONTENTS

2 EARTH CHRONOLOGY

THE PRECAMBRIAN EARTH

4 THE EXPANSION OF LIFE

5 THE PHANEROZOIC RECORD

PLATE TECTONICS

7 CAMBRIAN TO DEVONIAN TIME: The Marine Realm Dominant

8 MISSISSIPPIAN TO TRIASSIC TIME: The Expanding Terrestrial World

9 THE BREAKUP OF PANGAEA AND EARTH REORGANIZATION

10

JURASSIC TO PLIOCENE TIME: Progressive Modernization of the Earth

11

THE PLEISTOCENE EARTH AND MAN

12
EPILOGUE: The History of Earth History

PREFACE

Teachers of historical geology face two related yet often conflicting tasks: on the one hand, they must distill a wealth of factual knowledge about the earth's past into a brief chronology of events; on the other, they must present enough of the underlying principles so that students will appreciate just *how* that hard-won wealth was obtained. Until the 1960s most textbooks on the subject ignored the second task and presented elaborate encyclopedias of facts about the earth's past. In recent years this tradition has been reversed. Most contemporary textbooks devote about half their content to stratigraphic, tectonic, and evolutionary principles. Typically the first half of such a book emphasizes background material, while the second half treats factual chronology. This scheme has the advantage of laying a foundation of principles before erecting an edifice of chronology upon it. The disadvantage is that many students have difficulty relating the principles to the concrete events that follow. What does the structure of the mantle have to do with Paleozoic geosynclines many chapters later? Or fruit-fly mutations with dinosaur evolution? An alternative might be to mix chronology with relevant background throughout. For example, radiometric dating might be discussed along with Precambrian history, and plate tectonic theory with the Mesozoic breakup of Pangaea. Unfortunately, such a scheme, while logically appealing, would be both repetitious and inflexible, requiring each user to design a course to follow the sequence presented.

This book represents a compromise between the extremes of separation and integration. Individual chapters emphasize either factual chronology (Chapters 1, 3, 7, 8, 10, 11) or background principles (Chapters, 2, 4, 5, 6, 9). We have arranged the chapters in what we have found to be a natural sequence, with principles being introduced near the most relevant descriptive chapters.

Each chapter is, however, an independent unit that can be assigned *in any order* to accommodate differing course organizations.

In the belief that students shouldn't wait until midterm to be introduced to earth chronology, we treat the cosmic origin and earliest history of the earth at the outset in Chapter 1. This material, based largely on planetary astronomy and seismology, requires little stratigraphic background. Chapter 2 then discusses those stratigraphic principles which do *not* depend upon fossils and are thus most relevant to the following chapter, which surveys Precambrian history. Chapter 4, which treats the origin and expansion of life, provides background for Chapter 5, on the principles of biostratigraphy, which, in turn, sets the stage for the four chapters (7, 8, 10, 11) on Phanerozoic history. Because of the central role of plate theory in an understanding of Phanerozoic history, two chapters are devoted to aspects of the subject: Chapter 6, introducing the Phanerozoic, provides essential background on the discovery and interpretation of plate motions; Chapter 9, which introduces the Mesozoic breakup of Pangaea, discusses in greater detail those plate motions which have shaped the modern earth.

A perennial question in teaching historical geology concerns the *level of detail* used in discussing Phanerozoic history. We believe that students are often discouraged by elaborate descriptions of local events, particularly those which cannot be reinforced by first-hand observation in the field. For this reason we have not attempted an exhaustive summary of even North American stratigraphy and tectonic history in Chapters 7, 8, 10, and 11. Instead, we have used a brief survey of events to lead into more general interpretive principles that are well illustrated by each span of Phanerozoic time. Accordingly, the Cambrian to Devonian chapter emphasizes the interpretation of marine sedimentary rocks; the Mississippian to Triassic chapter, terrestrial rocks; the Jurassic to Pliocene chapter, tectonic history; and the Pleistocene chapter, glacial geology and human evolution.

Some material dealing with stratigraphic principles and life history has been adapted from treatments of these subjects in the authors' earlier paperback books in the Prentice-Hall Foundations of Earth Science Series: *Geologic Time* and *The History of Life*. This material has been condensed and extensively recast to fit the special needs of this text.

We wish to express our appreciation for the guidance provided by those who reviewed early drafts of the manuscript: Robert L. Anstey, Michigan State University; Richard H. Bailey, Northeastern University; Peter Bretsky, State University of New York at Stony Brook; John Dewey, University of Calgary; Gerald A. Fowler, University of Wisconsin at Kenosha; Lee C. Gerhard, University of North Dakota; Robert E. Gernant, University of Wisconsin at Milwaukee; Brian F. Glenister, University of Iowa; Roger L. Kaesler, University of Kansas; and Ronald L. Parsley, Tulane University.

DON L. EICHER
A. LEE MCALESTER

INTRODUCTION

Less than 100 years ago, most people thought the earth was only 6,000 years old. A distinguished English student of earth history, Arthur Holmes (1890–1965), noted that, in the small community where he grew up at the turn of the century, no one questioned whether the earth might be older. "I can still remember the magic fascination of the date of Creation, 4004 B.C., which then appeared in the margin of the first page of the Book of Genesis. I was puzzled by the odd '4.' Why not a nice 4000 years? And why such a recent date? And how could anyone know?"* In the decades to follow, Holmes himself helped to show that the traditional date was about a million times too short!

Toward the end of the nineteenth century, even as the boy Holmes wondered about the assumed date of Creation, scientists had already accepted the idea that the earth was considerably more ancient. Dispute over the question of just *how* ancient, however, had divided them into two camps. Members of the first were impressed by the earth's elaborate record of fossil life; they believed that the development of today's living world from the multitudes of less diverse ancestors required several hundred million years. A second group, those generally opposed to organic evolution, were hoping for much less available time—a few million years at most. The matter became further confused in the late nineteenth century, when the British physicist Lord Kelvin proposed an impressively precise age based on the rate of heat loss from the earth's interior. Kelvin's calculations indicated a few tens of millions of years for the earth's age—too brief a time for organic evolution but still too long for the antievolutionists. (We now know that Kelvin's calculations were inac-

*K. Rankama (ed.), "Introduction," *The Precambrian* (New York: Interscience, 1963), pp. xv–xvi.

curate because he didn't take into account the heat continually generated within the earth by radioactivity.)

The discovery of radioactivity at the turn of the century finally provided a tool for dating rocks directly. Today many different methods of radiometric dating are in wide use that can give the precise ages of rocks ranging from the earth's oldest to the very youngest. Such techniques indicate that the earth has existed for 4.6 billion years—so great a time by human standards that we cannot readily comprehend it! The earliest writings date from only 5,000 years ago, or roughly 1/900,000 of the earth's duration—a very small fraction indeed.

To demonstrate, suppose you were to lay out a straight white thread 50 kilometers long. If that thread represents the age of the earth, a needle 5 centimeters long at the end of it represents recorded human history. Stick that needle through 20 pages of this book and that distance represents the duration of the industrial revolution, the time in which society harnessed the earth's energy and effectively conquered its surface.

The earth's solid crust and the water and air that cover it house a multitude of dynamic systems. Some of them, such as volcanoes and earthquakes, originate deep within the earth; others are external systems, such as atmospheric and oceanic circulation, erosion, sedimentation, and various chemical and biological processes. The interactions among these many systems make up the total panorama of earth history. In the chapters that follow, we shall survey these interactions and the ways in which they have shaped and modified our planet during its 4.6-billion-year existence.

ORIGIN OF THE EARTH

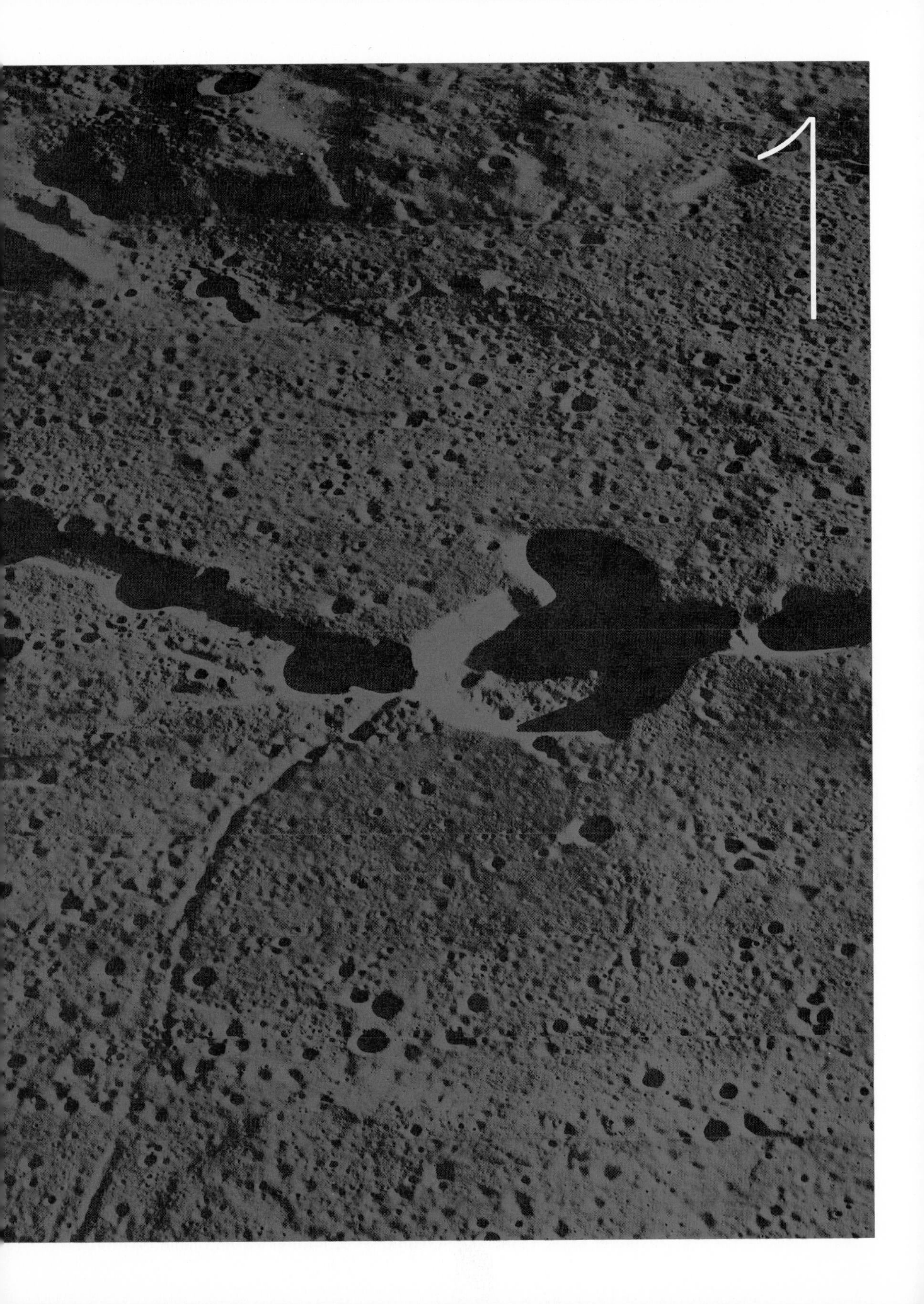
1

The physical substance of the earth was first produced by stars and shaped in interstellar space. Thus many questions concerning our planet's earliest history can be answered only through an understanding of the cosmic processes from which it arose. How did the earth originate as a solid body? And when? Is its composition the same as that of the universe as a whole? Or is it unique in some respects? Was it always composed of internal layers with an external ocean and atmosphere? Or have these evolved gradually over the long course of earth history?

These are difficult questions to answer because we have no direct evidence of the planet's remote beginnings. The oldest rocks we see postdate the origin of the earth by several hundred million years and show unmistakable signs that the solid earth, ocean, and atmosphere were already present when they formed. Therefore, we are forced to rely on indirect evidence to answer some of the most fundamental questions of the earth's past, those related to our planet's origin and early history. These inferences will be the subject of Chapter 1.

The Earth's Setting in Space

The Solar System

The earth, sun, moon, and their companion planets and satellites are known collectively as the solar system. The term is a good one because all nine planets and their satellites constitute a highly ordered system that moves about the sun with the regularity of an elaborate clock. The sun (an ordinary, average star) not only acts as an energy source, but, with 99.9 percent of the total mass, also overwhelmingly dominates the system. All the planets and most of the satellites orbit the sun in the same direction—counterclockwise as viewed from above the earth's northern hemisphere. Moreover, the orbits all lie very nearly in the same plane; that is, we can diagram them reasonably well using a single surface. Such regularity could not have arisen from chance events. It indicates a common origin from a single mass of matter with a well-defined rotational motion. Today astronomers are convinced that the planets all developed near their present orbits at about the same time. The origin of the earth is thus intimately linked to the origin of the entire solar system, whose main feature is a star called the sun. This provides the departure point for our investigation.

A great deal of empty space separates the sun and planets. The mean distance between the sun and the earth is about 150 million kilometers. The distance from the sun to the outermost planet (Pluto) is about 40 times that far. To place these distances in perspective, we can refer to the model shown in Figure 1.1. With the sun the size of a golf ball, the earth is a small grain of sand 4.6 meters away. Pea-sized Jupiter and Saturn are at a distance of 25 and 44 meters. Pluto, another small grain of sand, lies at a distance of 183 meters, or a couple of city blocks away. On the scale shown in Figure 1.1, the moon lies only 12 millimeters from the earth. The earth-moon system, including the total diameter of the moon's orbit, could be contained within the golf-ball sun.

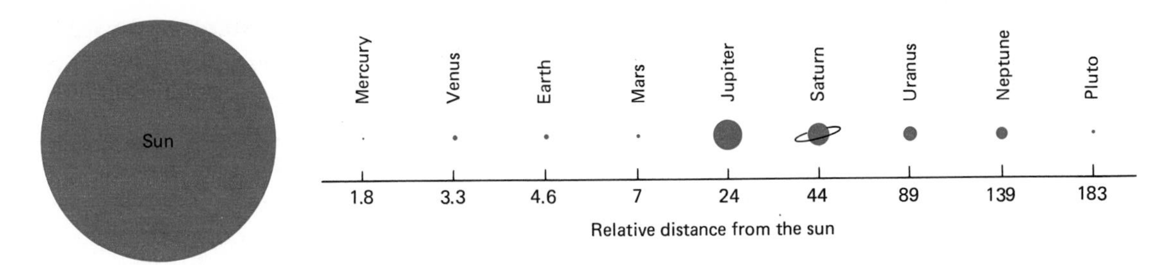

FIGURE 1.1

A solar system model. The sizes of the sun and planets are drawn to scale. The distances in meters each should be placed from the golf-ball-sized sun in order to make an accurate model are indicated, but not drawn to scale.

Stars and Galaxies

Distribution of Stars and Galaxies Dispersed as the solar system is, it appears relatively compact by contrast with the concentration of matter in surrounding space. The nearest star beyond the solar system is 4.3 light-years distant. (A light year is the distance a light beam travels in one year at its velocity of about 300,000 kilometers per second.) On the same scale indicated in Figure 1.1, in which the sun is a golf ball and the radius of the entire solar system is 183 meters, the nearest star would be represented by another golf ball at a distance of 1,240 kilometers. In all directions within a radius of 3,200 kilometers, there would be but nine additional golf balls. Thus it is easy to see why even the nearest stars appear only as points of light to our telescopes, and why other planetary systems can never be seen. Observing our planetary system from the nearest star would be comparable to observing the spots that represent the planets in Figure 1.1 from a distance of 1,240 kilometers!

Observations deeper into space reveal some regions in which stars are abundant and others where they are sparse. Inspection of the sky as a whole shows that stars are concentrated in the narrow band that constitutes the Milky Way. By actual count, 40 times as many stars are seen through large telescopes by looking directly at the Milky Way as are seen by looking at right angles to it. The Milky Way is, in fact, an enormous disc-shaped concentration of stars of which we are a part. Such concentrations are called *galaxies*.

The patchy appearance of the Milky Way is caused by huge clouds of interstellar gas and dust, which prohibit visual study of large portions of it. The advent of radio astronomy in recent decades, however, has greatly extended the distances we can explore within our galaxy. In addition to light waves, stars emit invisible radio waves that penetrate interstellar dust. Radio observations have made it possible to map most of the Milky Way galaxy and have shown clearly that it consists of a huge spiral structure. We can compare our galaxy to other great spiral galaxies that are visible in deep space, and we can assume that, viewed from a distance, it would appear very much like Figure 1.2. The Milky Way galaxy contains about 100 billion stars. Much of its mass is concentrated in a dense nucleus; its outer fringes are indistinct. Its maximum thickness is about 15,000 light years, and its diameter about 110,000 light years.

FIGURE 1.2 A spiral galaxy in the constellation of Andromeda of the same type as the Milky Way. (*Courtesy of Hale Observatories.*)

FIGURE 1.3 The distribution of matter in space: (*a*) Stars are not distributed evenly in space but occur in large concentrations called galaxies. Large galaxies, like the Milky Way galaxy in which our solar system is situated, measure about 100,000 light years across. (*b*) Similarly, galaxies are not distributed uniformly in space but occur in clusters that contain from a few to hundreds of galaxies. These are of the order of 10^6 to 10^7 light years across. (*c*) Clusters of galaxies are, in turn, concentrated in groups (superclusters), which occupy a region in space about 100×10^6 light years in diameter.

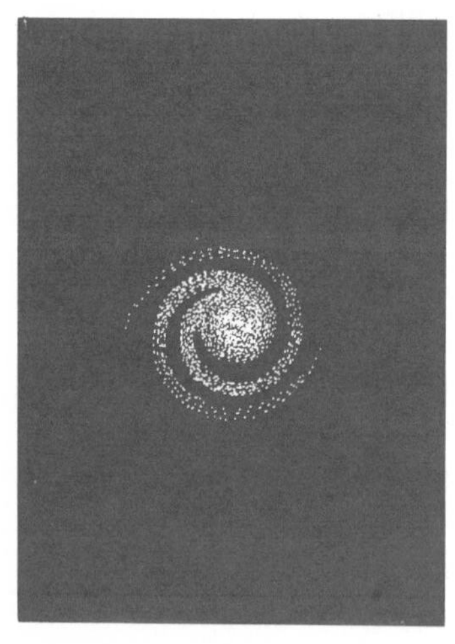

(*a*)

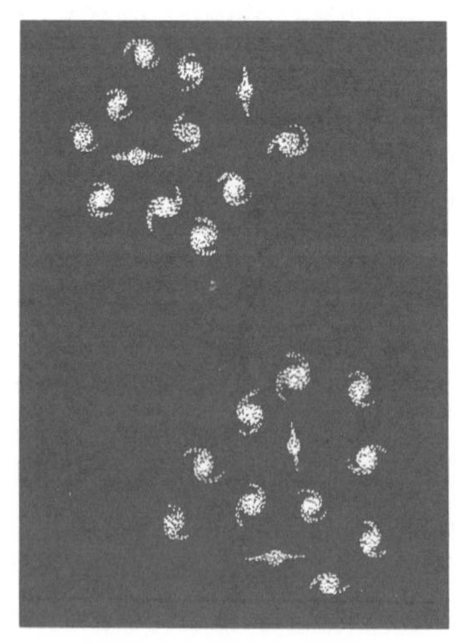

(*b*)

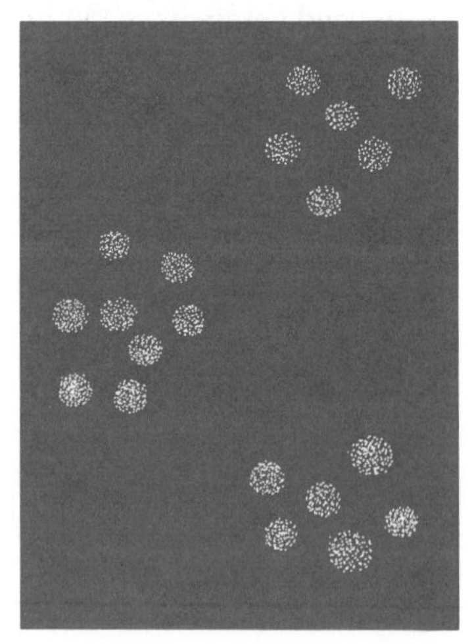

(*c*)

Our sun lies about 30,000 light years from the center, a bit more than halfway out in one of the spiral arms.

On the scale of Figure 1.1, in which the radius of the solar system is about 183 meters, the diameter of the Milky Way galaxy would be 320 million kilometers! Put another way, if the Milky Way galaxy were the size of the coterminous United States, our solar system would be about the size of this capital letter O.

Space beyond the Milky Way galaxy over huge distances appears to be empty, but tremendous numbers of stars are concentrated within other galaxies, the nearest of which are about a quarter of a million light-years distant. The nearest galaxies are visible to the unaided eye as faint, blurry objects. Telescopes not only resolve their individual stars, but also reveal different types of stars, such as exist within the Milky Way galaxy.

The total number of galaxies within reach of our most sensitive astronomical instruments may exceed 100 billion, or about the number of stars in the Milky Way. However, galaxies are not equally distributed in space. Our own Milky Way galaxy belongs to a relatively isolated cluster of 17 galaxies, whose diameter is 3 million light years. In this cluster, only one other galaxy—the Great Nebula in Andromeda, which is about 2 million light years away—is as large as the Milky Way galaxy (diameter about 110,000 light years). Other galaxies in the local cluster range in diameter from 3,000 light years to 30,000 light years.

Beyond our local cluster of galaxies we see similarly isolated clusters, the nearest of which is about 7 million light years distant. These neighboring clusters are, in turn, concentrated in a band of space, indicating that our local cluster of galaxies is part of a vast, flattened *supercluster* that consists of thousands of small clusters made up of tens of thousands of member galaxies. The supercluster appears to have a diameter of about 130 million light years and a thickness of about 30 million light years.

Beyond the limits of this supercluster, other clusters and superclusters can be observed. Present evidence is skimpy, but superclusters may be part of still higher-order groupings of matter. Clusters of superclusters having dimensions in the order of one billion light years may exist. The clustering of matter in an apparently open-ended series of ever greater scale indicates that *clumpiness* is a basic property of the distribution of matter in the universe (Figure 1.3).

Composition of Stars and Galaxies Virtually all that we know about the chemical composition of the stars and galaxies we have learned from the spectrograph, a device that breaks light up into its component colors and shows the exact wavelengths that the light is made of. When atoms of a particular element are excited by radiation, they *emit* additional radiation having particular wavelengths, some in the visible and some in the invisible parts of the spectrum, that are the characteristic "fingerprints" of the element. When atoms of the same element are present in a cool atmosphere around a star, they *absorb* from the background of white light the same characteristic wavelengths they emit when excited. These phenomena provide powerful tools for studying the composition of stars and galaxies and ultimately of the universe as a whole.

The spectrograph shows us that hydrogen is by far the most abundant element throughout the universe, constituting 93 out of every 100 atoms. This amounts to 76 percent of the total mass of the universe. Helium, next in abundance, accounts for 23 percent. Of the remaining 1 percent, the ten most common elements in decreasing order of abundance are oxygen, carbon, nitrogen, neon, magnesium, silicon, iron, sulfur, argon, and sodium. The abundance of all other remaining elements generally decreases with increasing atomic weight, the rarest elements being also the heaviest.

Motions of Stars and Galaxies The precise wavelengths of radiation that we receive from distant stars are governed not only by the composition of the star, but also by its relative motion. If a star is advancing toward the earth, the easily identified hydrogen wavelengths appear to become shorter, shifting toward the blue end of the spectrum; if the star is receding, they appear to become longer, shifting toward the red (Figure 1.4). This change in wavelength that results from the relative motion of the source is known as the **Doppler shift.** From the magnitude of the Doppler shift, we can easily calculate the relative velocity of stars and galaxies toward or away from the earth. Within the Milky Way galaxy, for example, stars in the region of the constellation Hercules are approaching at about 20 kilometers per second, and those in the constellation Columba are retreating at about the same velocity. Numerous similar observations have demonstrated the nature of the circulation of the galaxy. Our region of the galaxy makes a revolution about the galactic center once in about 200 million years. Nearer the center, the galaxy rotates faster, and nearer the edges, more slowly.

Similarly, galaxies belonging to the local cluster have varied motions relative to the earth; some are advancing, and others are retreating. The Great Nebula in Andromeda, for example, is moving toward us at about 50 kilometers per second. In contrast, light from more distant galaxies—in all directions in space without exception—is shifted toward the longer wavelengths, indicating

FIGURE 1.4

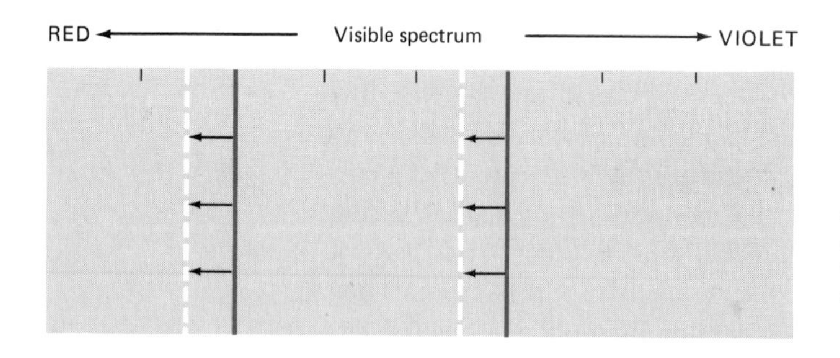

A rapidly receding light source will cause emission lines of an element to shift from their normal positions to new positions (dashed) that are always toward the red end of the spectrum.

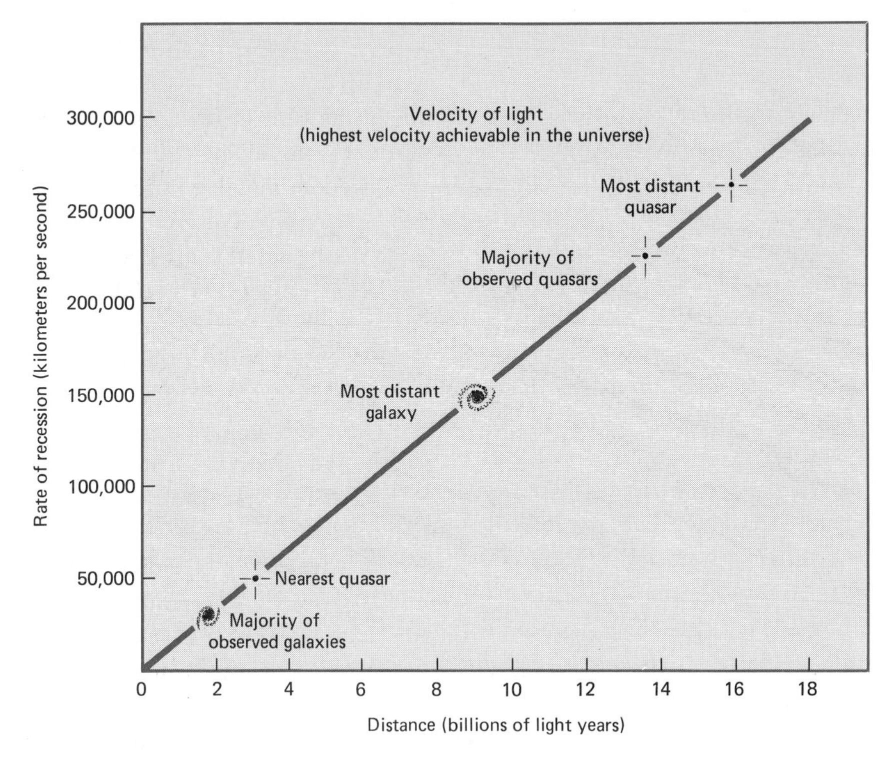

FIGURE 1.5

The increasing rate of recession of galaxies and quasars with increasing distance. The velocity of light provides an upper limit for the distance (about 18 billion light years) that an object can be from us, and hence for the time (about 18 billion years) since the universe began.

that all are *receding* from us. Moreover, the farther away a galaxy, the more rapidly it is receding (Figure 1.5). For every million light-years' distance from us, the galaxies recede about 16 kilometers per second faster.

The farthest galaxies observable are around 6 billion light years away, and they are receding at about half the speed of light. The term "are receding" is somewhat misleading in this context, because the light we see left these galaxies 6 billion years ago! All we can really say is that 6 billion years ago, the distant galaxies were 6 billion light years away and were then retreating from us with about half the speed of light. Where are they now? In such a framework, the term "now" loses significance. Light from our sun has not even reached these distant galaxies, because 6 billion years ago the sun did not yet exist! And all during the intervening 6 billion years, these distant galaxies have been retreating farther still. Hence, when we observe such distant objects, we cannot know their position now, or even if they exist now. This time lag prohibits us from observing the entire universe as it is today. But in a broader sense, the lag is extremely advantageous, because the more deeply we look in space, the further back in time we look. Studies of deep space, therefore, provide information on the early history and evolution of the universe.

An unusual kind of object revealed by studies of deep space is the **quasar.** Quasars are the most distant objects we can see. They are not huge galaxies but small, abnormally bright, starlike objects that may represent some early phase that matter went through when the universe was very young. Quasars have been discovered that are possibly as far away as 16 billion light years. In observing them, we are not only looking across vast distances, but are simultaneously looking far back into the abyss of time. Quasars were 50 times more plentiful 4.6 billion years ago, at the time the solar system formed, and were perhaps 1,000 times more plentiful around 13 billion years ago, when the universe was young. Observations of objects that far back in time indicate a much greater density of matter than exists now, a condition that is predicted by our present conception of the origin of the universe.

Origin of the Universe

Even though every distant galaxy and quasar seems to be retreating from our region of space, this does not indicate that we are near the center. Instead, all these distant objects are receding from one another in a universal expansion; the relative movement of galaxies would appear much the same from any one of them. What does this expansion indicate? Apparently all matter was once confined to an extremely small region of space. This concentration of matter then exploded violently to initiate the expansion that is still in progress today. This is known as the **big bang theory** of the origin of the universe.

How long ago did this big bang occur? Eighteen billion years would seem to be an absolute upper limit, based on the observation that, for every million light-years' distance, an object retreats 16 kilometers per second faster. At this rate, objects that are 18 billion light years away from us (although none at such a great distance has ever been observed) must be retreating at the speed of light. This velocity cannot be exceeded in the universe. No object can be farther away from us than this, because none can have been moving away from us at a higher velocity. The velocity of light thus puts an upper limit on the *distance* an object can be from us, but more important, it also puts a limit on the *time* that has elapsed since the big bang, that is, about 18 million years, as Figure 1.5 indicates. It is possible that the rate of expansion has, in fact, decreased throughout cosmic time. If so, the big bang occurred somewhat less than 18 billion years ago. Most recent estimates of the age of the universe lie between 12 billion and 18 billion years. The Milky Way galaxy, of which our sun is a part, is one of the fragments of that cosmic event.

Origin of the Solar System

The sun, like all stars, evolved from enormous clouds of interstellar gas and dust. Within galaxies, such clouds continuously condense into stars, burn their nuclear fuel, and then disperse as cold cosmic ashes, some of which may again be incorporated into newly condensing stars. The initial collapse of the sun's star-forming cloud—from a diameter of approximately one light year to a protostar with a diameter approximately the size of the orbit of Mercury—probably occurred very quickly, possibly within a few decades. Subsequent contraction to the sun's present size occurred much more slowly. Probably around 50 million years elapsed before the core of the gradually contracting mass attained the million-degree temperature necessary to initiate hydrogen fusion and to stabilize the sun as a normal star. The sun is slightly unusual in being a single star, because more than half the stars we can observe are members of binary star systems; yet the number of solitary stars is also huge, and being a solitary star may well be a prerequisite for the possession of planets. Stars the size of the sun require about 10 billion years to deplete their nuclear fuel and are therefore relatively permanent members of the universe. We shall see that the sun apparently formed about 5 billion years ago and thus has consumed only about half of its supply of hydrogen fuel.

Nebular Theory versus Encounter Theory

Two basic hypotheses have been advanced to explain how the planets may have been created around the early sun. The first assumes that they formed at the same time as the sun as a by-product of contraction of the interstellar cloud. The second assumes that they formed later, when the sun was violently disturbed in a near collision with another star. We can refer to these as the **nebular theory** and the **encounter theory**.

According to the nebular theory, the enormous cloud from which the sun and planets condensed exhibited some initial rotation. As the cloud contracted, its rate of rotation increased, much as a spinning skater moves faster with arms pulled inward. Simultaneously, its increasing rotation caused the cloud to flatten progressively into a disc. The massive central region of the disc ultimately became the sun, and the outer portion of the disc condensed into planets, satellites, asteroids, and comets. This idea was first put forth in the mid-eighteenth century; it was developed in some detail in the late eighteenth century by the French mathematician Laplace.

One important peculiarity of the solar system that is difficult to explain under this scheme is that, although the sun has more than 99 percent of the mass of the solar system, it has less than 1 percent of the angular momentum. (**Angular momentum** is a measure of the difficulty of stopping a body in rotation.) Far greater angular momentum is displayed by the giant outer planets. The sun, at the center of the solar system, should be spinning much more rapidly than it is, in order for its angular momentum to be proportional to its mass. Simple condensation from a contracting nebula cannot explain the excessive angular momentum of the planets. Beginning in the 1940s, astronomers began to formulate complex models that explain the transfer of angular momentum from the center to the outer parts of the disc through the eddy currents and twisted magnetic fields characteristic of a rotating mass of ionized gas. As magnetic lines of force become entwined in the spinning nebular disc, they would actually impart a rigidity to it (Figure 1.6). The effect of this would be to accelerate the outer parts of the disc and thus increase their angular momentum at the expense of the protosun. This and other refinements have answered most objections to the nebular theory.

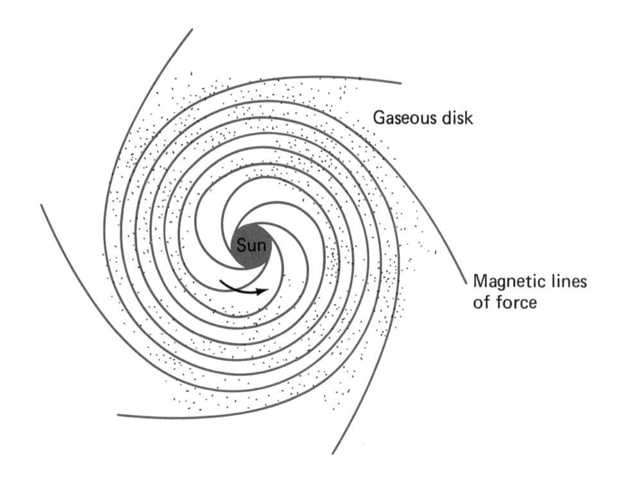

Twisted magnetic lines of force in the contracting solar nebula could have transferred angular momentum to the evolving planets.

FIGURE 1.6

Before the angular-momentum problem was solved, the encounter theory was proposed as an attractive substitute, and early in this century it was widely accepted. This theory held that the planets were formed, not at the same time as the sun, but from material that was torn from the sun some time later, during a near collision with another star. The gravitational forces between the sun and the passing star caused disruptive tides, and some of the sun's surface gases were flung far out into space. The gravitational attraction of the other star accelerated this material into orbit about the sun, thus supplying its great angular momentum. As this orbiting gaseous material cooled, it condensed into tiny solid particles and these, in turn, collected into larger bodies of rock that subsequently accumulated to form the planets.

What is the likelihood of such a near collision in space? If we compare the sun to a dot on this page, the nearest star, another dot, is 32 kilometers away, and each moves about half a meter annually. From this we can infer that the chances of collision or near collision are very small. Indeed, calculations indicate that no more than a few have probably ever occurred in the entire universe. Hence, the encounter theory attributes the formation of the solar system to an exceedingly *rare* event. In this respect it differs fundamentally from the nebular theory, which considers the creation of the solar system a normal, nonaccidental by-product of stellar evolution. If the solar system were the result of a near collision between two stars, it would probably be unique in the galaxy, and this alone is sufficient cause to view the encounter theory with suspicion. Scientists have learned to be wary of ideas that espouse uniqueness, particularly when they concern the position of human beings in the scheme of things.

Philosophical doubts about the validity of the encounter theory were supported by serious physical objections that began to emerge prior to the 1940s. Detailed analysis revealed that (1) there were no plausible circumstances by which a passing star could have given the planets the amount of angular momentum they actually possess, and (2) a mass of gas at the temperature of the sun suddenly released into space would be blown apart so violently by its own intense radiation that it could never condense into solid particles, however small. No revisions of the encounter theory seem capable of resolving these difficulties, and it has generally been abandoned.

Virtually all current work on the origin of the solar system consists of refinements of the nebular theory, in which the planets are derived from a flattened gaseous disc spinning about the contracting protosun. Although many of the exact circumstances surrounding the origin of our solar system will never be known, recent studies of the moon, planets, and meteorites have narrowed the range of possibilities. The general scheme is illustrated in Figure 1.7.

A possible way the solar system formed: (*a*) A huge dispersed nebula of interstellar gas and dust condenses under its own gravitational attraction. (*b*) Nebula contracts, begins to rotate perceptibly, and flattens. Most material falls into the rapidly accumulating central protosun. (*c*) Nebula rotates faster and flattens further. Protosun becomes distinct from surrounding disc. Particles condense and rapidly accrete in disc eddys. (*d*) Protosun heats further and becomes an infrared star. Disc still consists mostly of gases, but planets are largely formed. Warped magnetic fields transfer momentum to planets. (*e*) Contracting sun begins to shine visibly; intense solar wind at this stage drives off gases in the surrounding disc. (*f*) Sun begins burning hydrogen and is now stabilized. Planets and asteroids are virtually all that remain of disc.

FIGURE 1.7

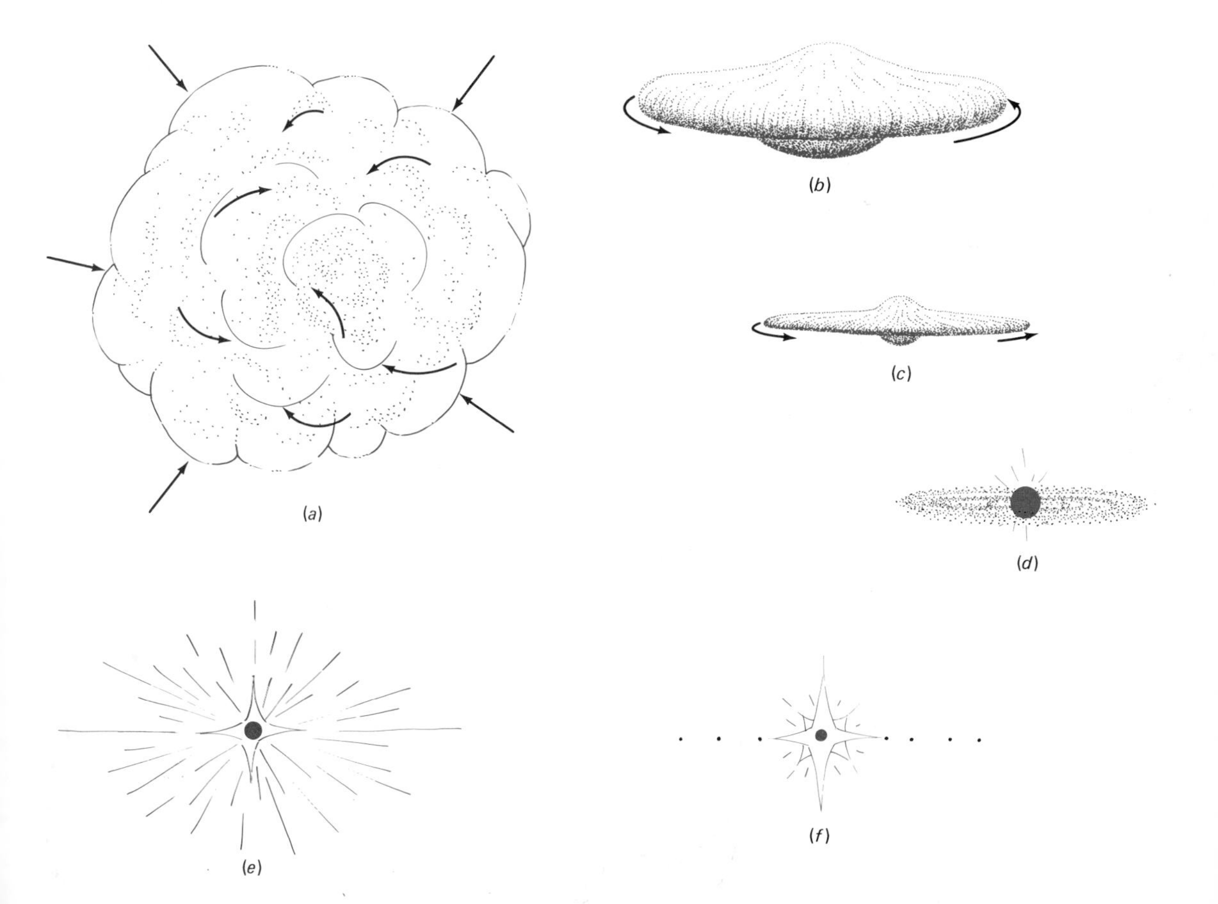

Accumulation of the Planets

During the early rapid collapse of the nebula, the flattening disc, as well as the central part of the mass that was to become the sun, heated up. The disc achieved a temperature of perhaps 1,700°C. Because of the high temperatures, the material within it remained in a gaseous state. Ultimately, when the contraction of the nebula began to slow down, the center continued to heat up, but the disc, where matter was less concentrated, began to cool, and as it cooled, the planets took shape.

Within the cooling disc, tiny particles of matter condensed like dew, first forming liquid droplets and then tiny crystals. The first materials to condense were probably high-temperature silicate minerals and rare, difficult-to-melt metals. Shortly after the temperature dropped below 1,200°C, an abundance of nickel-enriched iron must have begun to appear. As particles formed, they began to accumulate where eddys concentrated them in the swirling gaseous disc. In these regions the particles literally fell together under their own mutual gravitational attraction, and this created larger particles that, in turn, grew through collisions with others. Thus the earth and the other planets formed and were largely solid bodies from the beginning.

How much time was required for the planets to accumulate from the cooling disc? In one sense, accumulation continues even today, inasmuch as tiny meteors continually add to the mass of the earth and the other planets. But the present rate of accumulation is infinitesimally small. Astronomers now believe that once they began to form, the planets reached nearly their present mass very quickly, perhaps within 10,000 years.

As the earth accumulated in this fashion from tiny particles that were newly condensed from the nebula, the surrounding gases that were left behind became increasingly enriched in hydrogen compounds and also in the inert gases: neon, argon, krypton, xenon, and particularly helium. In the late stages, huge quantities of these must have surrounded the newly formed protoearth, and we might expect that they would have been trapped by gravity to form a primitive atmosphere. If these gases ever were trapped, however, they were subsequently removed, for today's atmosphere is severely depleted in all of them.

Hydrogen compounds and helium are similarly underrepresented in the other terrestrial planets (Mercury, Venus, and Mars). In the giant outer planets, however, they are abundant, and they contribute importantly to these planets' large sizes and low overall densities (Figure 1.8). This means that the giant planets are much closer in composition to the sun than to the terrestrial planets. Astronomers customarily distinguish between: (1) *earthy materials* (for example, iron, silicon, magnesium, and their oxides), which are solid on all the planets; (2) *icy materials* (compounds of hydrogen, carbon, oxygen, and nitrogen), which are abundant solids at the low temperatures that prevail in the portion of the solar system that lies beyond the earth; and (3) *gaseous materials* (chiefly elemental hydrogen and helium), which probably remain gaseous in all but perhaps the outermost reaches of the solar system.

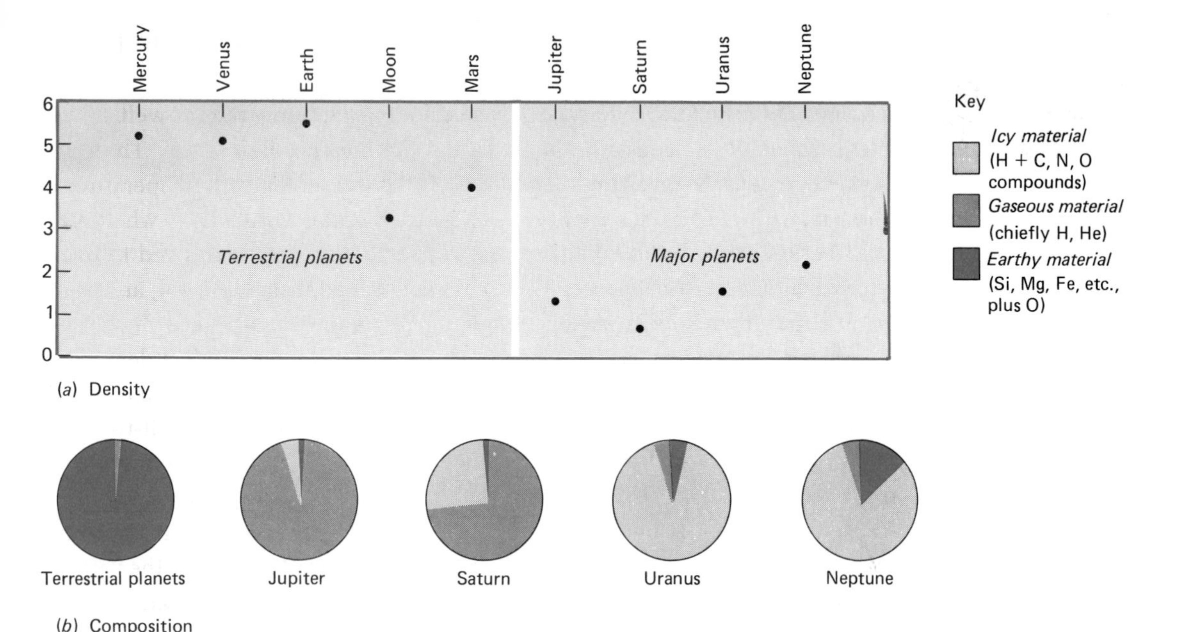

Major planets are much less dense than terrestrial planets because they contain more helium, hydrogen, and hydrogen compounds.

FIGURE 1.8

Assuming that the planets were derived from a nebula of about the same composition as the sun, the earth was concentrated from a portion of the nebula that was initially about 500 times more massive than the earth is today because of the abundance of the elements that have been lost. What is the cause of the great depletion of icy and gaseous materials in the earth and the other terrestrial planets? First, these planets are small, and their comparatively low gravitational attraction permits light gases to escape easily. Second, their proximity to the sun subjected them from the beginning to higher temperatures, which vaporize icy material and enhance the escape of gases. Finally, the inner planets have always been vulnerable to solar emissions. Most stars the size of the sun undergo a *variable phase* at about the time they become stabilized, during which they produce intense solar winds. Such solar winds could easily have driven away gaseous material from the inner planets, just as the present mild solar wind drives the material in a comet's tail away from its head.

In addition to undergoing an early variable phase, most stars of the sun's type have another peculiarity: They tend to rotate much more slowly than stars that are smaller or larger. Are these phenomena linked? During their stabilization, do stars about the size of the sun undergo a variable phase because they transfer so much of their angular momentum to the surrounding nebula? Is it even possible that planetary systems are somehow a cause or an integral by-product of these processes? If so, perhaps planets show a greater tendency to form around stars approximately the size of the sun.

How Did the Earth Get Its Moon?

The moon, the earth's closest partner in space, poses a separate problem all its own. Although it has now been studied at first hand, the question of how it formed remains unanswered. There are three basic theories: (1) the moon formed by the breaking up of the early earth; (2) it formed elsewhere in the solar system and was subsequently captured; or (3) it formed near the earth as a separate planetary body.

The low average density of the moon (3.3 grams per cubic centimeter, as compared to the earth's 5.5) indicates that it has little or no iron core and is made up largely of silicate minerals, as is the outer part of the earth. This observation lends support to the earth-breakup theory, which proposes that the moon was originally part of the earth, but that the earth-moon body was spinning so rapidly that it became unstable and broke apart (Figure 1.9*a*). The moon thus might consist simply of a severed portion of the earth's mantle.

One problem with this theory is that the angular momentum that the earth and moon now possess is inadequate to trigger their separation, if they were merged into the same body. Another objection is the major chemical differences between the earth and the moon. Samples collected by the Apollo missions indicate that, relative to the earth, the moon's surface is significantly depleted in elements such as potassium and rubidium, whose compounds become gaseous at comparatively low temperatures (approaching 2,000°C), and it is correspondingly enriched in heat-resistant elements like titanium and chromium. Both the angular momentum and the compositional objections might be accounted for if about half of the moon's mass had somehow been vaporized and lost by intense heating during the breakup. However, this would require a particularly complex sequence of events, and hence it is not a very satisfactory explanation.

It is possible to explain the considerably different composition of the moon by assuming that it formed in a far-removed region of the primordial nebula and was captured at a much later date (Figure 1.9*b*). Yet the moon's density of 3.3 grams per cubic centimeter is substantially below the density range (4 to 5.5) of the terrestrial planets, and hence the moon appears to be fundamentally different in its overall makeup and structure. Even if the moon formed elsewhere in the nebula, it is still difficult to see why its properties are not more like those of the terrestrial planets. In addition to these difficulties, most astronomers feel that the capture of the moon is a statistically improbable event.

Perhaps the simplest solution to the problem of lunar origin would be to assume that the moon formed in orbit close to the earth. Here it could have condensed independently in a nearby region of the nebular disc (Figure 1.9*c*). If this were the case, however, how could it come to differ so drastically in composition, being depleted in volatile elements and lacking an iron core?

A possible answer to this question is that the moon did not form from a separate nearby center of accumulation at all, but that it condensed instead out of a very hot, distended, disc-shaped atmosphere of vaporized material, chiefly silicates, that surrounded the accumulating earth (Figure 1.9*d*). In this view,

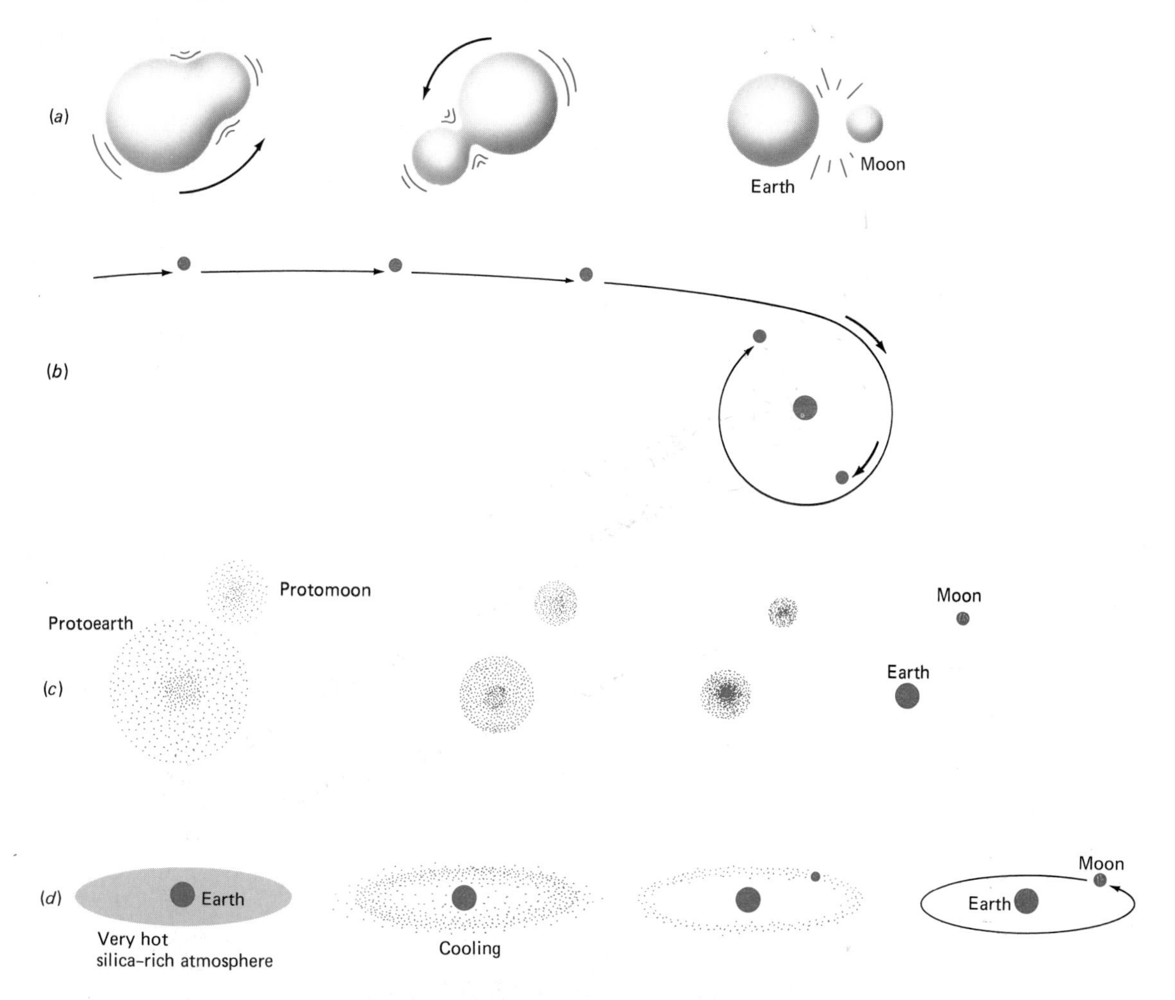

Possible ways in which the moon formed: (*a*) Fission of combined earth-moon body early in geologic history. (*b*) Capture of body that was formed earlier elsewhere in the primordial nebula. (*c*) Independent condensation from a source near the earth. (*d*) Rapid formation of the moon in a distended, silicate-rich atmospheric disc around a large, hot protoearth.

FIGURE 1.9

the silicate-rich atmosphere was vaporized from the earth's surface when it achieved extreme temperatures due to the high-energy impact of infalling particles during the late stages of accumulation.

At the 2,000-degree temperatures required for such vaporization, the more volatile materials in the white-hot atmosphere were particularly vulnerable to being swept deep into space by the intense solar wind that probably existed at that time. As the primitive atmosphere cooled, the less volatile silicate compounds that remained behind condensed into particles that formed a disc around the earth analogous to the rings of Saturn, only relatively more massive. The moon then rapidly condensed from the heat-resistant elements in the disc in as short a time as 100 years. Following the accumulation of the moon, both the earth and moon continued to cool, and the distance between them increased to its present size.

Regardless of how the moon came to be a satellite of the earth, satellites of other planets need not have originated in the same way. Some may have formed in orbit; others may have resulted from the breakup of planets; still others may have formed independently, only to be captured later. In any case, we are certain that all the planets and their satellites evolved from the same parent material at the same time. In the next chapter we shall examine evidence provided by radiometric dating showing that this event took place about 4.6 billion years ago.

Differentiation of the Early Earth

The materials making up the present-day solid earth are not mixed together in a homogeneous mass. Rather, they are arranged in a series of concentric layers of differing density. *The heaviest materials, molten metals, lie at the center of the earth and are overlain by progressively lighter layers of silicate minerals.* This differentiation by density is the most fundamental structural feature of the modern earth. The oldest known rocks, formed about 4 billion years ago, indicate that relatively low-density surface minerals, as well as a still lighter overlying ocean and atmosphere, were already in existence very early in earth history. For this reason, the fundamental differentiation by density of the earth must have developed still earlier—most probably along with, or soon after, the planet's initial accumulation.

Factual clues to the origin of this differentiation come from two principal sources: (1) study of the earth's present-day internal structure; and (2) similar studies of the earth's nearest planetary neighbors, particularly the moon, Mars, Venus, and those smaller samples of interplanetary matter that reach the earth as meteorites.

The Earth's Interior Today

The most direct clues to the differentiation of the solid earth come from studies of its present-day internal structure. Only the thin outer skin of the earth's rocks can be sampled directly at the surface and in mines or bore holes, the deepest of which extend downward about 10 kilometers, which is only 1/500 of the distance to the center of the earth (Figure 1.10). Many of the rocks exposed today at and near the earth's surface originated at deeper levels and were brought near the surface by processes of mountain building, but even these rocks apparently formed in only the outermost 5 percent of the earth's thickness. For knowledge of the materials and structure of the earth's deeper interior, we must rely on indirect evidence, by far the most important of which is provided by earthquake shock waves. Earthquake waves, called **seismic waves**, pass through the earth's interior and then emerge at the surface, where they can be measured by sensitive recording devices called **seismographs** (Figure 1.11). Hundreds of thousands of such measurements have shown that seismic waves do not travel at a uniform velocity through the earth's interior;

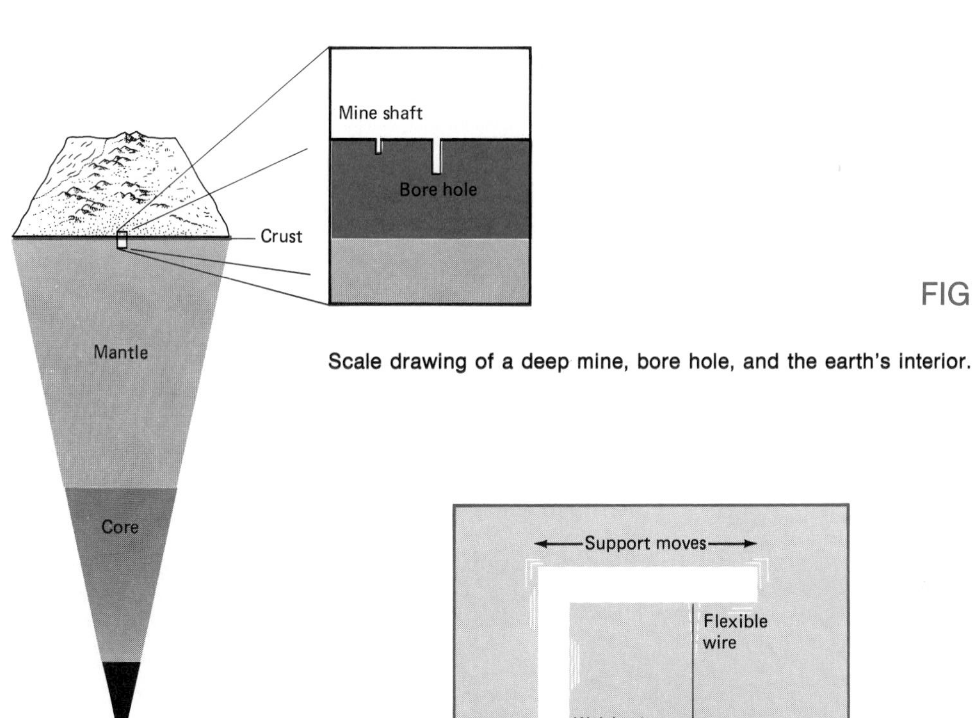

FIGURE 1.10

Scale drawing of a deep mine, bore hole, and the earth's interior.

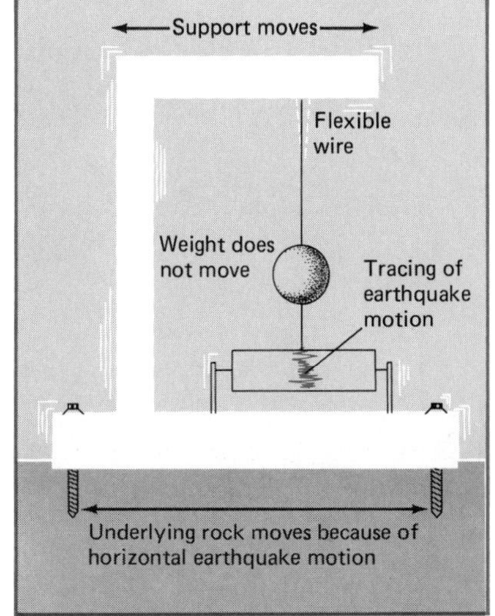

FIGURE 1.11

Schematic diagram of a seismograph. Earthquake waves cause movements of the underlying rock and attached support; the weight tends to remain stationary while the pen records the motion.

instead, the velocity of wave movement tends to increase with depth and shows particularly sharp changes at depths of about 40 and 2,900 kilometers (Figure 1.12). These **seismic discontinuities,** as they are called, indicate fundamental changes in the materials of the earth's interior, for the speed of wave travel can be shown in laboratory measurements to depend in large part on the *density* of the materials through which the waves pass: the more dense the rock, the more rapid the motion. The two strong seismic discontinuities show that the earth's interior is made up of three principle layers of differing density: (1) a thin, outer layer of relatively light materials called the **crust;** (2) a deep, central sphere of very heavy materials called the **core;** and (3) a thick, intervening layer of intermediate density called the **mantle,** which makes up the bulk of the earth's interior volume. The core, mantle, and crust, then, are the fundamental structural units of the earth's present-day interior.

FIGURE 1.12

Velocity-depth curves for two types of earthquake waves, *P* and *S* (above), and the inferred internal structure of the earth (below). Major velocity changes occur at the crust-mantle and core-mantle discontinuities.

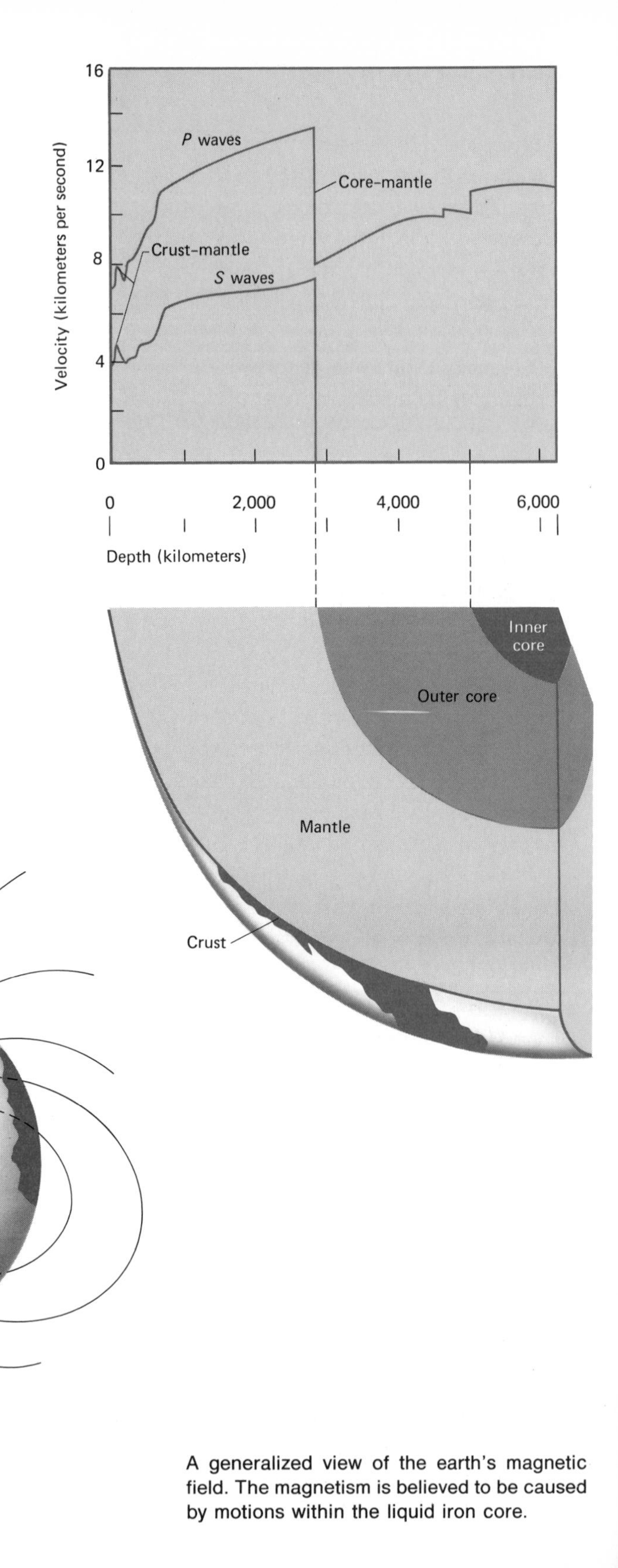

FIGURE 1.13

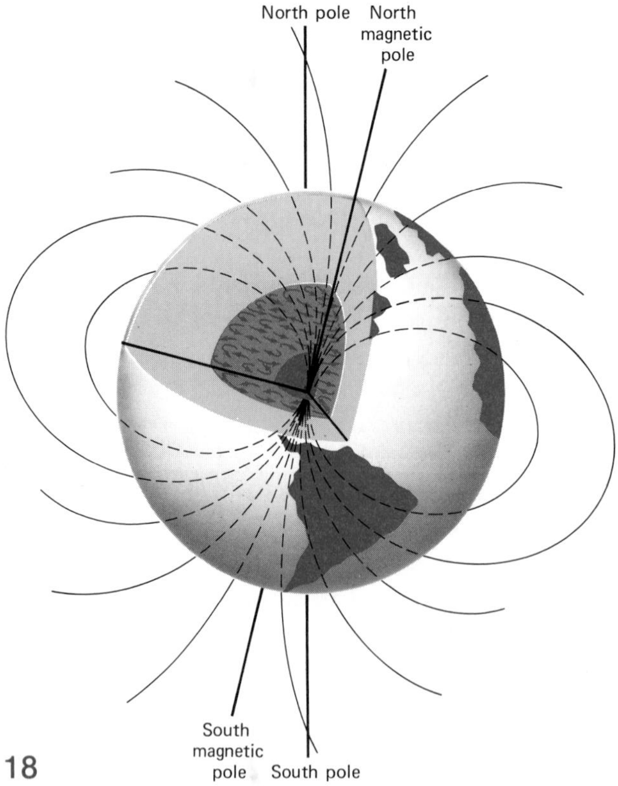

A generalized view of the earth's magnetic field. The magnetism is believed to be caused by motions within the liquid iron core.

The rocks of the crust, as well as some formed in the uppermost mantle, can be studied directly at the earth's surface and are composed largely of relatively light silicate minerals, particularly feldspars, quartz, micas, amphiboles, and pyroxenes. Although there is no way to sample directly the materials making up the deeper mantle and core, several kinds of indirect evidence suggest that the entire mantle, like its uppermost layers, is composed of silicate minerals, whereas the underlying core is largely made up of hot, molten iron.

The molten state of much of the earth's core is indicated by the manner in which it alters passing seismic waves. One of the two principal types of seismic waves, S waves, can be transmitted only through rigid, solid materials; this type of wave disappears as it enters the core, thus providing strong evidence that much of the core is liquid (Figure 1.12). There is no direct evidence about the composition of this liquid, but some indirect clues come from the probable distribution of densities throughout the earth's interior.

Even though seismic wave velocities indicate an increase in density with depth, they do not provide unequivocal density values, because wave velocities are also affected by the elastic characteristics, temperatures, and other properties of the materials through which they pass. These properties are not known with certainty for materials deep within the earth. However, an important limit on internal densities is provided by the earth's average density, which has long been known from astronomical calculations to be about 5.5 grams per cubic centimeter. Because rocks of the crust and upper mantle are much lighter than this, having an average density of only 3.0 grams per cubic centimeter, they must be balanced by more dense materials at depth. Calculations show that a core of liquid iron, overlain by a mantle of less dense silicate minerals, would lead to the observed average density of 5.5 grams per cubic centimeter.

A further suggestion that the core is largely liquid iron comes from the nature of the earth's magnetic field. Measurements of magnetic forces at the earth's surface show that the entire earth acts as if it were a gigantic bar magnet with poles near its poles of rotation. Furthermore, this magnetic field appears to originate deep within the earth's interior, in the region of the core. Electrical currents produced by complex motions in a core of liquid iron are considered to be the most likely source of this magnetism (Figure 1.13).

The final and perhaps most convincing evidence that the core is made of iron and the mantle of lighter silicate minerals comes not from the earth itself, but from observations of materials formed elsewhere, in the nearby reaches of the solar system.

Clues from Beyond the Earth

Until the first rocks from the moon were dramatically returned to the earth by astronauts in 1969, the only direct samples of materials from beyond the earth itself were provided by meteorites, masses of solid matter from space that enter the earth's gravitational field and fall to the surface, where they can be collected and studied. The "shooting stars" seen streaking across the sky on any dark night are composed of such matter, originally left in space during the formation of the solar system and captured by the earth's gravity. Bright streaks are formed as they plunge into the dense atmosphere, where they are heated and usually vaporized by friction with gas particles at heights of 40 to 100 kilometers. Such objects are called meteors if they are completely vaporized and *meteorites* if a part of them survives the fall through the atmosphere and reaches the earth's surface. About 500 meteorites the size of an orange or larger reach the earth's surface each year. Most fall in the ocean; only about 150 fall on land, of which an average of only about 4 per year are recovered for study.

Some meteorites are composed of metallic iron, usually mixed with smaller amounts of nickel and other metals. These are called *iron meteorites* and are much less common than a second type, called *stony meteorites,* which are composed of silicate minerals, principally olivine, pyroxene, and calcium-rich feldspar. The presence of metallic iron in many meteorites formed beyond the earth adds further support to the inference that the earth's dense core is made up largely of iron. Still more indicative about the nature of the earth's hidden interior are studies of stony meteorites.

Chemical analyses show that the proportions of elements in stony meteorites are remarkably similar to the composition of the sun, as revealed by spectrographic analyses of sunlight (Figure 1.14). That is, if the light, volatile elements (principally hydrogen and helium) were removed from the sun, the remaining silicon, aluminum, iron, magnesium, and other nonvolatile elements would be present in the *same relative amounts* found in meteorites. This close similarity suggests that stony meteorites are relatively unchanged fragments of the original solar materials that accumulated to form the earth and other planets. This inference is further strengthened by radiometric measurements of

FIGURE 1.14

The relative compositions of stony meteorites and the sun. If the large volumes of solar hydrogen and helium are omitted, the remaining elements show almost identical abundance, a fact that supports the hypothesis that these meteorites are fragments of solar matter.

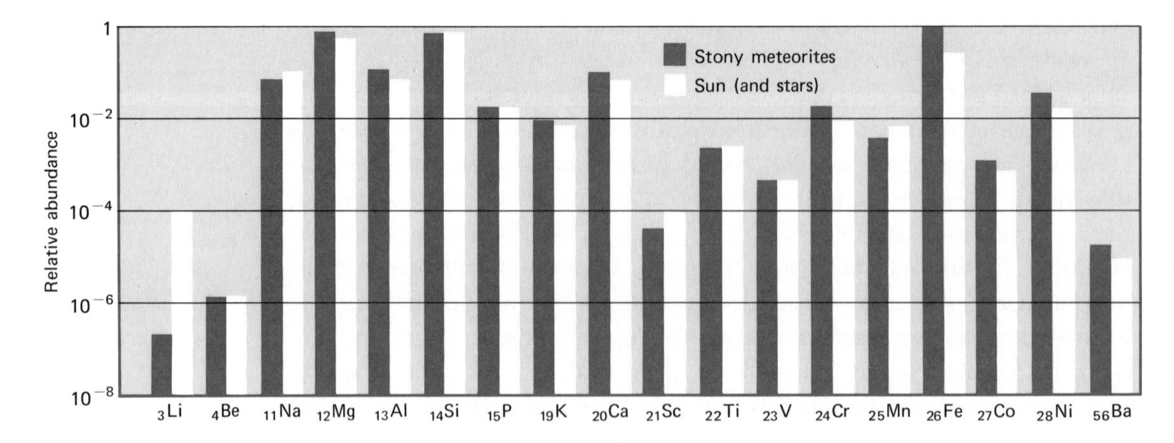

the ages of stony meteorites, which indicate that most of them condensed and solidified at about the same time that the planets are believed to have formed, about 4.6 billion years ago. If stony meteorites represent samples of the materials that originally condensed to make the earth, then the earth's overall composition must be similar. This would be the case if the earth's mantle, which occupies most of its interior volume, were made up largely of silicate minerals having the same general composition as stony meteorites. The evidence from meteorites lends further support to the inferences about mantle composition drawn from seismic discontinuities and the earth's internal density distribution.

Some further clues to the earth's internal differentiation are provided by its larger planetary neighbors. For many years the only evidence of the internal structure of the other planets came from astronomical calculations of their average densities. Such calculations were summarized above in Figure 1.8, which shows that only the earth's nearest planetary neighbors in the inner solar system are composed of relatively heavy materials, as is the earth. The far larger outer planets, in contrast, have much lower average densities, indicating that they are composed predominately of very light elements and are thus unlikely analogues for understanding the earth's density layering.

All of the earth's neighbors in the inner solar system—Mercury, Venus, Mars, and the moon—have average densities in the range of materials making up the solid earth. The moon's density of 3.3 grams per cubic centimeter is the lowest, about the same as that of the rocks in the earth's outer crust. Mercury, the small and poorly understood innermost planet, has the highest density. Mars and Venus, the most earthlike of the planets, have densities of 4.0 and 5.1 grams per cubic centimeter respectively, suggesting that their overall compositions are similar to that of the earth. Recent spacecraft voyages have shown, however, that neither planet has a measurable magnetic field, a sharp contrast to the strong magnetism produced by motions of the earth's core. The reasons for this important difference are still uncertain, but the lack of magnetism could be due either to the absence of a differentiated iron core or, if such a core is present, to the lack of fluid movements within it that produce magnetism. Beyond this, little is known of the deep interiors of the two planets.

Aside from the evidence provided by meteorites, by far the best corroborative external information concerning the earth's interior comes from recent explorations of the moon. Not only are many samples of the moon's surface rocks now available for study, but seismic data on the lunar interior have been provided by small seismometers placed on the surface of the moon during the manned landings of 1969 to 1972. These instruments have shown that the moon is seismically very quiet, compared to the earth. All recorded moonquakes are very small, and their total energy release is about a million times less than that of all earthquakes over a comparable period. Because of their small magnitude, moonquakes reveal relatively little about the moon's internal structure. However, larger shock waves caused by the impact of meteorites or abandoned spacecraft permit a rough determination of changes in seismic-wave velocity with depth in the moon's interior. These changes suggest that the moon has internal layering similar to that of the earth.

A rather sharp discontinuity about 60 kilometers deep appears to separate an overlying lunar crust from an underlying mantle. Wave velocities in the crust are within the range of velocities of crustal rocks on the earth; higher velocities in the moon's mantle suggest somewhat more dense underlying materials. Unlike earthquakes, moonquakes appear to originate very deep within the planetary interior, in a zone between 800 and 1,000 kilometers below the surface. (Most earthquakes occur at depths of less than 100 kilometers.) This zone appears to separate a rigid mantle from a deep, hot, partially melted core, which, like the earth's core, does not transmit certain types of seismic waves. The moon's core, however, must be composed of relatively light silicate minerals, for the planet's low overall density precludes a heavy iron core. In summary then, the moon, like the earth, appears to show a fundamental density layering of its internal materials.

Models of Earth Differentiation

The earth's present internal density layering is most probably a very early feature of earth structure. The oldest known earth rocks, formed about 4 billion years ago, are made up of relatively light silicate minerals, just as are younger crustal rocks; this indicates that a crust and, very probably, a heavier silicate mantle and an iron core were already present at that time. Internal layering must therefore have originated still earlier in earth history, either along with, or soon after, its original accretion about 4.6 billion years ago.

Two general theories have been proposed to explain the origin of the core-mantle-crust layering: **homogeneous accretion** and **inhomogeneous accretion.** Both are closely interrelated with the initial accumulation of the earth from nebular matter (Figure 1.15). If the earth accumulated from a homogeneous mixture of nebular particles, then it must have initially been rather uniform throughout; separation of the core, mantle, and crust must have occurred at some time after initial accretion. On the other hand, if the initial

FIGURE 1.15

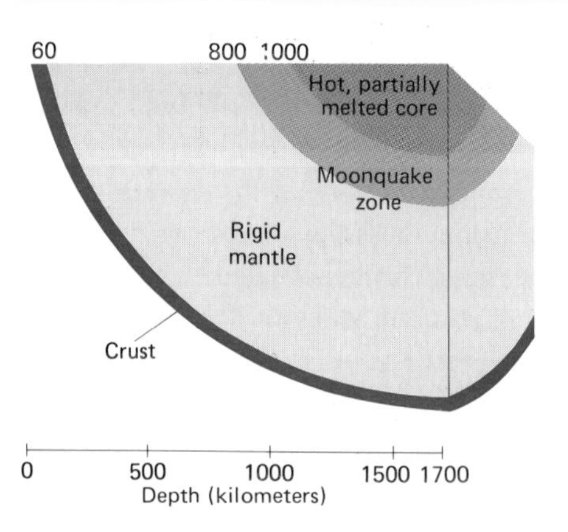

Inferred internal structure of the moon, based on limited seismic observations. Most moonquakes occur deep within the interior in a zone that apparently separates a rigid mantle from a hot, partially melted core. A sharp discontinuity at a depth of about 60 kilometers separates a thin crust from the more dense underlying mantle.

consolidation of the earth took place from inhomogeneous nebular materials, then the iron core might have accreted first, to be subsequently surrounded by a mantle of lighter silicate materials. In this second model, the core, mantle, and crust formed as the earth grew to its present size, rather than afterward. As yet there is no compelling evidence favoring one of these models over the other, and we shall briefly review both.

Homogeneous Accretion According to this hypothesis, the primitive solid earth is considered to have originally had a rather uniform density and composition throughout; that is, the large amounts of iron and nickel now making up the core were dispersed uniformly through a larger mass of lighter silicate minerals (Figure 1.16*a*). This protoearth is visualized as having been rather cool throughout, so that all the elements were present initially as solids, rather than as gases or liquids. In order to separate the heavier iron and nickel and concentrate them in the core, the interior of this initially cool earth must have somehow become hot enough to melt the iron and nickel, allowing them to sink through the lighter, surrounding silicate minerals. These silicates melt at higher temperatures than do nickel and iron but would, nevertheless, soften and slowly flow as the earth's interior became hotter; this flow would permit the much heavier molten metals to sink and displace the lighter silicates at the earth's center.

Such a partial melting of the early earth could have been caused by heat generated through the decay of radioactive materials, which were present in far greater abundance early in earth history than they are today. Calculations based on reasonable estimates of the amount of radioactive material present in the initial earth suggest that, soon after it formed, enough heat might have been generated to begin to melt nickel and iron at depths of about 650 kilometers. The melting would begin first at this relatively shallow depth because much higher melting temperatures are required as pressures become greater toward the earth's center. These molten metals would first accumulate as a layer in the outer part of the earth (Figure 1.16*a*), but being heavier than the soft, underlying silicate material, they would tend to sink slowly as great "drops," to accumulate at the earth's center. This sinking, in turn, would create additional frictional heat that might melt the metals still deeper in the earth, thus continuing the process until all of the free iron and nickel became concentrated in the central core.

As the molten metals sank to form the core, the lighter silicates would have risen into the mantle, which, as a consequence, would have been turbulent and unstable. Under such conditions the lightest and most easily melted elements, such as sodium, calcium, potassium, and aluminum, would have tended to move upward through the turbulent mantle. As a result, they would be concentrated near the earth's surface, which is where they are found today in the rocks of the crust and upper mantle.

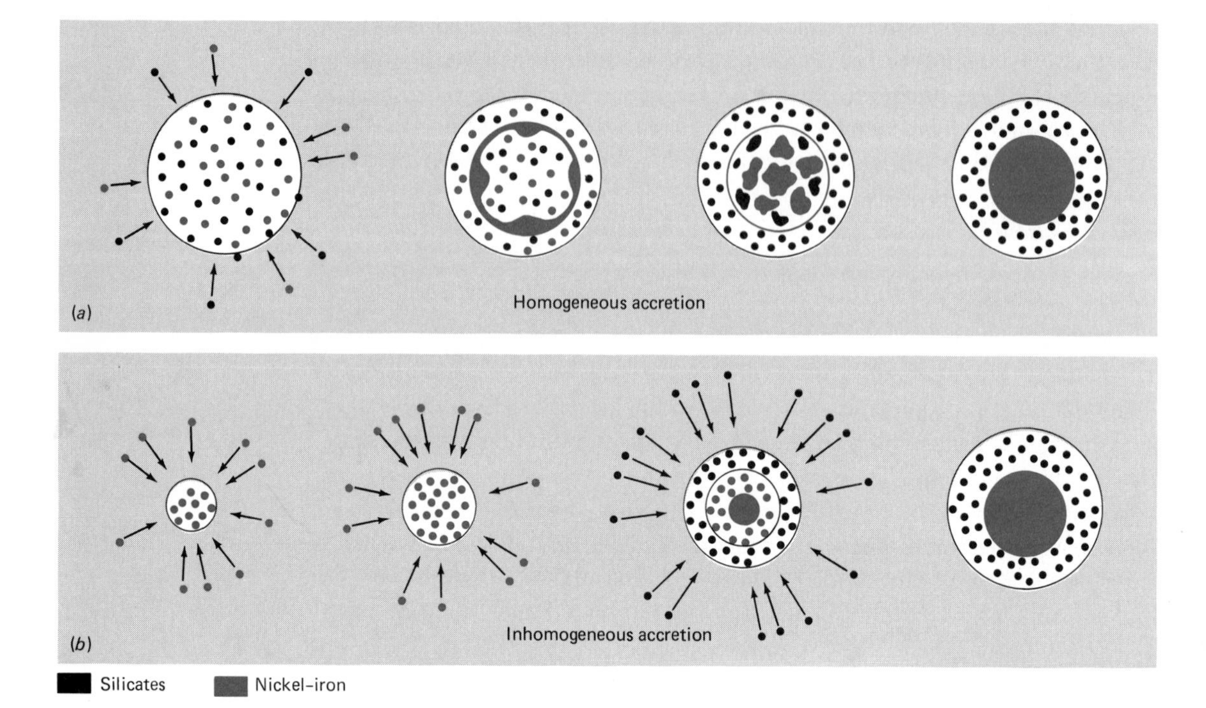

FIGURE 1.16 Two principal hypotheses concerning the initial accretion and internal differentiation of the earth: (*a*) Differentiation of the solid earth by homogeneous accretion. In this hypothesis, the earth was initially a uniform mixture of silicates and nickel-iron. Subsequent radioactive heating then led to a concentration of the nickel-iron in the core, and the silicate minerals in the mantle and crust. (*b*) Differentiation of the solid earth by inhomogeneous accretion. In this hypothesis, the nickel-iron of the core accumulated first and was surrounded by a slightly later accumulation of the silicate minerals of the mantle and crust.

Inhomogeneous Accretion According to this second hypothesis, materials of the core, mantle, and crust are visualized as having condensed sequentially as the hot nebular gases condensed into particles, rather than forming from a uniform mass of solid particles (Figure 1.16*b*). Calculations show that even though iron and nickel melt at lower temperatures than do most silicate minerals, they condense from the gaseous state at slightly higher temperatures. Therefore, in a cooling cloud of hot nebular gases, iron and nickel might condense first and accumulate to form the earth's core, to be surrounded subsequently by later-condensing silicate minerals to form the mantle. The last material to condense and accumulate as the temperature dropped would be the light, volatile elements that are concentrated in the crust and upper mantle. The rate of accumulation of the core, mantle, and crust would depend greatly on the sizes to which the minerals or rocks grew before they began to accumulate. If the particles accumulated while they were still very small, then the entire process of condensation and accumulation may have taken place very rapidly as the original nebular gases cooled—perhaps in only a few hundred or a few thousand years. This hypothesis, therefore, does not require the relatively long interval of radioactive heating and internal reorganization demanded by the homogeneous accretion model.

As yet, no unequivocal evidence favors one or the other of these ideas. In either case, however, it is clear that the internal differentiation of the solid earth took place very early and set the stage for the long chronology of earth history recorded in crustal rocks.

Chapter Summary

The Earth's Setting in Space

The solar system includes as principal components the sun and the planets plus their satellites; the orbital patterns of the planets indicate a common origin with the sun.

Stars and galaxies

Distribution of stars and galaxies: The sun is a star; like other stars, it occurs in an enormous disc-shaped concentration of stars called a galaxy; the sun's galaxy, called the Milky Way, is one of the billions that can be observed with sensitive astronomical instruments.

Composition of stars and galaxies: Analyses of starlight show that all are dominated by the lightest element, hydrogen, which accounts for about 76 percent of the mass of the universe. Helium, the second lightest element, makes up 23 percent, and all remaining elements only 1 percent.

Motions of stars and galaxies: Shifts in the wavelengths of light received from distant stars indicate the velocity and direction of movement; such shifts show that all distant galaxies are receding from each other as the universe expands.

Origin of the universe: Motions of galaxies indicate that all matter in the universe was once concentrated in a single mass; expansion of this mass, called the "big bang," began no less than 18 billion years ago.

Origin of the Solar System

Nebular theory versus encounter theory: The encounter theory, now largely discredited, suggested that the planets arose when a near-collision with another star pulled matter away from the sun; the currently accepted nebular theory suggests that the planets condensed from the same mass of gas and dust that gave rise to the sun.

Accumulation of the planets probably occurred as eddies concentrated material in a gaseous disc around the early sun. The huge outer planets retain much light hydrogen and helium, as does the sun; the smaller and heavier earthlike planets have lost most of their original share of the sun's lighter and more volatile elements.

How did the earth get its moon? This occurred either by original condensation in orbit around the early earth, or by later capture of the moon after it originated elsewhere in the solar system.

Differentiation of the Early Earth

The earth's interior today: Analysis of earthquake waves passing through the earth show it to be divided into three fundamental units: a massive molten core, overlain by a thick solid mantle and a thin solid crust.

Clues from beyond the earth: Evidence from meteorites suggests that the core of the earth is made up mostly of liquid iron, and the mantle of iron-rich silicate minerals.

Models of earth differentiation: The core, mantle, and crust may have separated through early radioactive heating of an initially homogeneous earth; or they may have accumulated as separate units from cooling solar gases.

Important Terms

angular momentum
big bang theory
core
crust
Doppler shift
encounter theory
homogeneous accretion
inhomogeneous accretion
mantle
nebular theory
quasar
seismic discontinuity
seismic wave
seismograph
spectrograph

Review Questions

1. Outline the evidence indicating that the sun and planets had a common origin.
2. What chemical elements dominate the universe? How is this known?
3. What evidence indicates that the universe is expanding? What does this suggest concerning the origin of the universe?
4. How might the planets have originated from the solar nebula?
5. Summarize the evidence for the origin of the moon.
6. Describe the structure of the earth's interior today. How is this known?
7. How might the density layering of the earth have originated?

Additional Readings

Bolt, B. A. "The Fine Structure of the Earth's Interior," *Scientific American,* vol. 228, no. 3, pp. 24–33, 1973. *How earthquake-produced seismic waves reveal the deep structure of the earth.*

Clark, S. P., Jr. *Structure of the Earth,* Prentice-Hall, Inc., Englewood Cliffs, New Jersey, 1971. *Good discussion of the earth's interior.*

Grossman, L. "The Most Primitive Objects in the Solar System," *Scientific American,* vol. 232, no. 2, pp. 30–38, 1975. *Minerals found in the meteorites known as carbonaceous chondrites represent samples of the solid grains that condensed directly out of the gaseous nebula that gave birth to the sun and the planets.*

Head, J. W., J. A. Wood, and T. A. Mutch "Geologic Evolution of the Terrestrial Planets," *American Scientist,* vol. 65, pp. 21–29, 1977. *Comparative surface features of Mercury, Venus, Earth, and Mars and an overview of their evolution as planets.*

Kaufman, W. J. *Planets and Moons,* W. H. Freeman & Company Publishers, San Francisco, 1979. *Up-to-date introductory survey.*

King, E. A. *Space Geology,* John Wiley, New York, 1976. *A well-written and well-illustrated reference on meteorites and planetary geology.*

Schramm, D. N. and R. N. Clayton "Did a Supernova Trigger the Formation of the Solar System?" *Scientific American,* vol. 239, no. 4, pp. 124–39, 1978. *Isotopes found in a few primitive meteorites are probably debris from a large star that exploded about a million years before the meteorites formed.*

Short, N. M. *Planetary Geology,* Prentice-Hall, Inc., Englewood Cliffs, New Jersey, 1975. *An excellent introduction to the topic, with detailed descriptions on an elementary level.*

Wood, J. A. *The Solar System,* Prentice-Hall, Inc., Englewood Cliffs, New Jersey, 1979. *An authoritive introduction stressing recent planetary studies.*

EARTH CHRONOLOGY

2

After the initial formation of the earth and the concentration of its lighter elements near the surface, some of these materials consolidated to form the silicate minerals that made up the earth's original crust of solid rock. We are not certain that any of this earliest rock crust still survives, after billions of years of mountain-building movements, weathering, and erosion. However, rocks have survived that clearly formed not long after the earth's initial consolidation or that may even have been a part of it. With these rocks begins our *direct* knowledge of early events in the earth's past.

Our understanding of such ancient events rests almost entirely on studies of crustal rocks of varying ages that are exposed today at or near the earth's surface. Because certain rocks have survived with little change for millions or even billions of years, they can provide direct records of the earth's past. For example, belts of plutonic igneous and metamorphic rocks now exposed at the earth's surface indicate former intervals of mountain-building deformation deep within the crust, whereas volcanic and sedimentary rocks reflect changing landscapes on the earth's surface at the time they were formed. Sedimentary rocks are particularly useful because they may suggest past climates, thus providing *indirect* clues about earlier states of the ocean and atmosphere.

In order to decipher the long sequence of events recorded in the many kinds of crustal rock, our first task is to organize and interpret the record in local areas where the rock record is accessible. Then we must determine the relative ages of the rock records in numerous localities, so that a worldwide chronology of events can be established.

There are three complementary, but basically independent, methods for establishing age relations of ancient rocks. The first applies primarily to sedimentary rocks, because it relates physical events to the sequential nature of sediment deposition. The second method applies only to the relatively young sedimentary rocks containing **fossils,** the remains of ancient life. Since such animal and plant remnants show progressive changes through time, they provide a direct means of establishing the age relations of fossil-bearing sedimentary rocks. The third method applies primarily to igneous and metamorphic rocks and relies upon the steady rates of decay of certain **radioactive elements** incorporated into the rocks as they were formed.

In this chapter we shall first discuss some overall principles used in organizing and interpreting ancient rocks, and will then elaborate upon the third dating method listed above. Such **radiometric dating** not only provides the quantitative basis for geologic history, but it is by far the most valuable method for deciphering rocks that formed during the first 85 percent of earth history. In Chapter 5 we shall discuss the methods of dating by physical means and by fossils—techniques that have led to a remarkably detailed understanding of the most recent 15 percent of earth history.

Organizing the Rock Record

Field Study of Rocks

An understanding of the geologic history of an area begins with field study of the actual rocks that can be seen at the earth's surface. Not all rocks are exposed, of course; in many regions they are hidden beneath a thin overburden of soil and vegetation. The underlying bedrock is normally exposed only in scattered outcrops where this overburden has either been removed naturally (as in stream valleys) or artificially (as in road cuts and quarries) (Figure 2.1). Even in a region where such exposures are numerous, they commonly reveal only a small portion of the bedrock that underlies the area. It is generally helpful to supplement observations of surface rocks with studies of cores or fragments of rocks that have been brought to the surface from deep drill holes. In general, the best exposures occur in arid or semiarid regions, and the most complete exposures occur in regions of high relief. Both of these conditions prevail on the Colorado Plateau, a portion of which is shown in Figure 2.2. In this region deep canyons expose panoramic outcrops that are among the most spectacular in the world.

FIGURE 2.1

This quarry in the Paris Basin provides a fresh exposure of an Eocene limestone of marine origin that extends widely but is otherwise exposed poorly in this lush agricultural region.

FIGURE 2.2 The Colorado Plateau is an area of high relief and low rainfall. Its extensive natural exposures, largely unobscured by soil and vegetation, greatly enhance the understanding of the region's geologic history.

In field study, all rocks in a region—whether igneous, sedimentary, or metamorphic—are divided by the geologist into discrete, recognizable units that are thick enough and extensive enough to be shown on a geologic map. These mappable rock units—distinguishable from one another on the basis of color, composition, texture, or other properties—are called **formations.** Geologists begin the study of an area by describing in detail all the formations that can be recognized consistently throughout the area, and by measuring the thickness of those units that are layered, or stratified (Figure 2.3). Stratified rocks are chiefly sedimentary in origin, but in some regions they include igneous rocks, such as ash falls and lava flows, that formed in widespread, sheetlike accumulations. They may also include metamorphic rocks in which original stratification is preserved. The thickness of formations generally amounts to a few tens or hundreds of meters.

The method of measuring the thickness of strata depends upon the topography of the area and the complexity of the geologic structure. One method, that of taping the thickness of strata directly, is shown in Figure 2.4. The initial goal is to arrange the rock units chronologically; this chronology is usually illustrated as a vertical column, or **section**, like that shown in Figure 2.5.

Permian Casper Formation, Sand Creek, Wyoming, shows large-scale cross beds in a sandstone deposited as wind-blown dunes.

FIGURE 2.3

Using a measuring tape to determine the thickness of strata.

FIGURE 2.4

FIGURE 2.5

Stratigraphic column for a hypothetical area showing symbols for common rock types.

Dutton Creek Shale
Wheeler Coal
Walcott Sandstone
Gilbert Salt
Meek River Gypsum
Powell Dolomite
King Creek Limestone
Hayden Conglomerate
Hall Peak Granite

Meters
50
25
0

FIGURE 2.6

Three ways of showing the geology of an area: (*a*) geologic map, (*b*) geologic cross section showing vertical slice along A-A′ on map, and (*c*) block diagram of the same area.

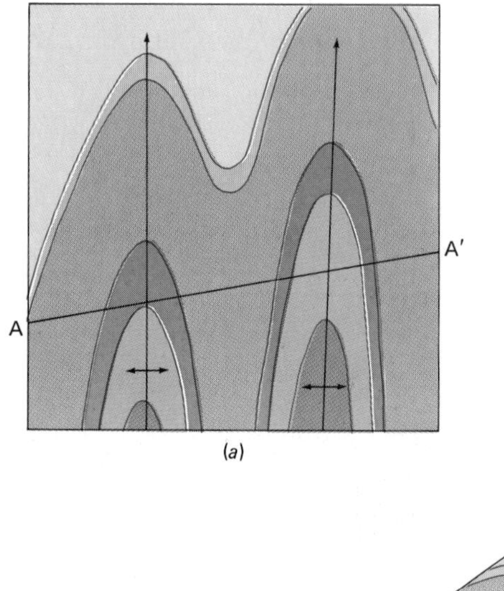

(*a*)

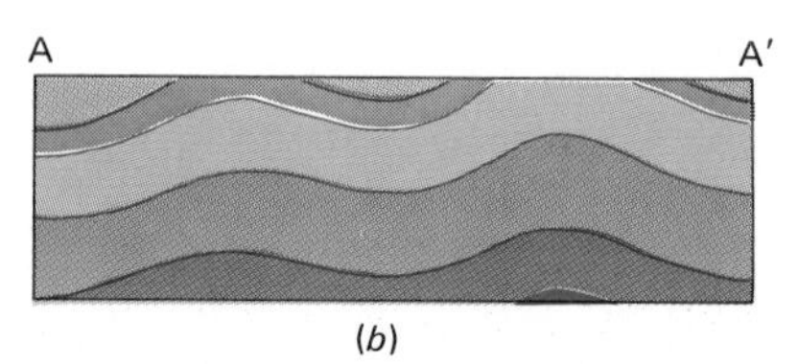

(*b*)

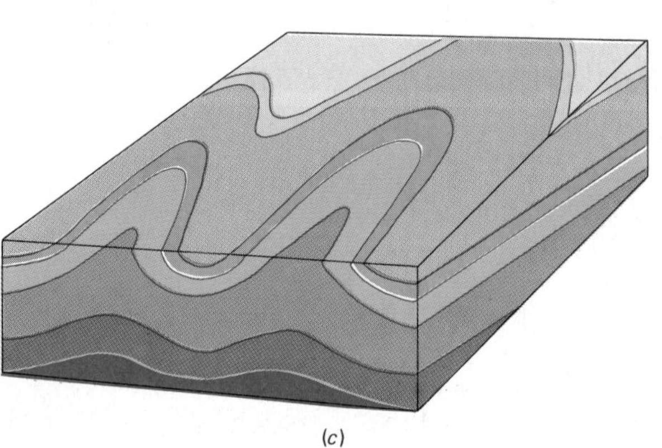

(*c*)

0 1 2
Kilometers

Key
Dutton Creek Shale
Wheeler Coal
Walcott Sandstone
Gilbert Salt
Meek River Gypsum
Powell Dolomite
King Creek Limestone
Hayden Conglomerate

The next major task is the construction of a geologic map that shows where different formations occur. In some cases, maps are supplemented with geologic cross sections or block diagrams to show the inferred structure beneath the surface (Figure 2.6). Together, these techniques illustrate the arrangement and configuration of the rocks in an area.

As the geologist measures the section and prepares to map the area, he or she must work out the **stratigraphy,** that is, the vertical and lateral relationships of the stratified rocks. In areas where the geologic structure is simple and the exposures are good, this is an easy task; but in areas where the rocks are highly deformed and the exposures are poor, the task can be formidable indeed.

Several long-established stratigraphic principles permit researchers to work out the order of events in an area, including the sequence in which the various formations were produced. Three of these principles—**original horizontality, superposition,** and **original lateral continuity**—were perceived in 1669 by Nicolas Steno as he worked on the geology of western Italy. A fourth—the principle of **cross-cutting relationships**—became clear about 1795, following the work of the Scottish geologist James Hutton, who was the first to emphasize the significance of the various kinds of contacts between rock units.

Stratigraphic Principles

Original Horizontality *Sedimentary formations are originally deposited flat. Any dip they now have has resulted from subsequent folding or tilting.*

FIGURE 2.7

Tidal flat, Gulf of California, (*Courtesy of T. R. Walker*)

This principle of original horizontality forms the basis for all structural interpretations in regions of sedimentary rocks. It is valid for a simple reason: Most areas where sediment is accumulating have very low relief, so that deposition occurs on a nearly flat surface (Figure 2.7). **Cross beds,** which are produced by migrating sand waves, bars, dunes, or ripples, may have considerable initial dip, but they are generally small-scale features, and they are bounded by major bedding planes, which are themselves essentially horizontal (Figures 2.8 and 2.9). However, sets of cross strata that are several meters thick may be deposited as submarine debris on the sloping flanks of large reefs or on the slopes of large sand dunes (see Figure 2.3 for example). Dips must be measured with caution in strata deposited in these environments.

FIGURE 2.8

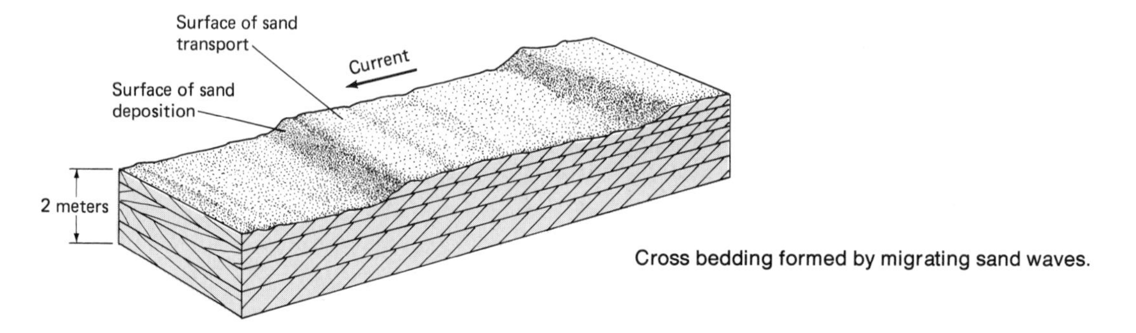

Cross bedding formed by migrating sand waves.

FIGURE 2.9

Lakota Sandstone, Black Hills, South Dakota; tabular cross beds in a Cretaceous sandstone inferred to be a deposit of sand waves like those shown in Figure 2.8.

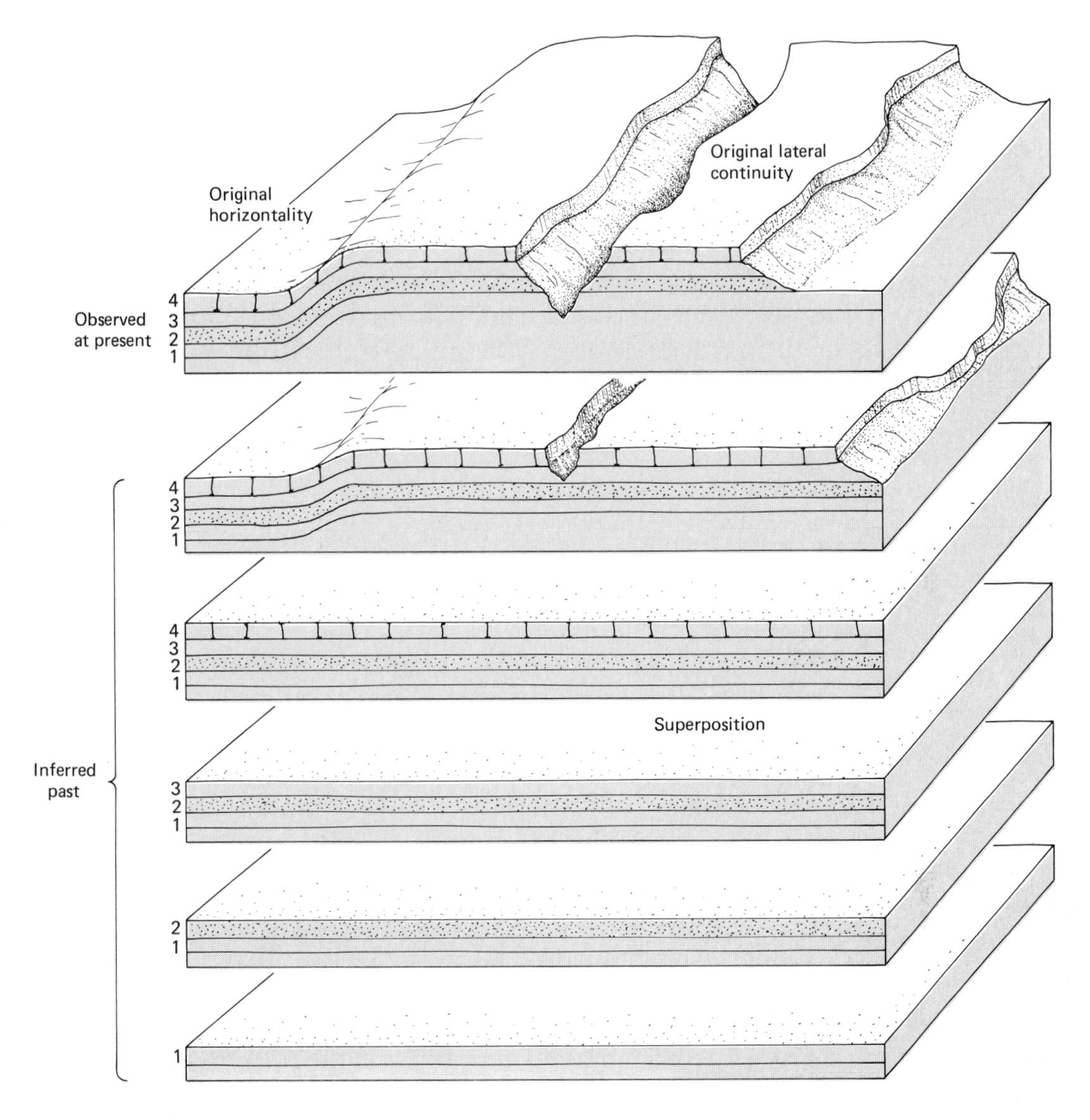

Application of the principles of original horizontality, original lateral continuity, and superposition in reconstructing geologic history.

FIGURE 2.10

Superposition The *principle of superposition states that in a stratified sequence any given bed is younger than those below and older than those above.* As in a layer cake or a brick wall, the bottom layer in a sequence is constructed first, and the top last (Figure 2.10). This is a simple concept, yet it is the most important of all because geologic history is interpreted according to this order of deposition.

The bottom of a rock sequence is easily distinguished from the top in an area of simple structure, but in a highly disturbed area, where beds may be vertical or overturned, "Which way is up?" may be a real problem. If you find a brick wall tilted 90 degrees on its end, how do you determine the sequence in which the bricks were laid? You might look closely and find a telltale place where the mortar had dripped while it was still soft. Similarly, in a sedimentary rock sequence, you might find one of a host of diagnostic features called *sedimentary structures* that enable you to distinguish top from bottom. For example, tracks or trails made by walking or crawling animals will certainly mark the *tops* of beds, as will raindrop imprints, ripple marks, or mud cracks (Figure 2.11*a*). Trough cross bedding is marked by a cut-and-fill structure in which erosional scour of preexisting strata always occurred at the top (Figure 2.11*b*). Graded bedding, in which the grain size of beds becomes finer upward, also provides a good clue (Figure 2.11*c*). Other evidence is provided by organisms preserved in their growth position, such as the algal deposits shown in Figure 2.11*d*.

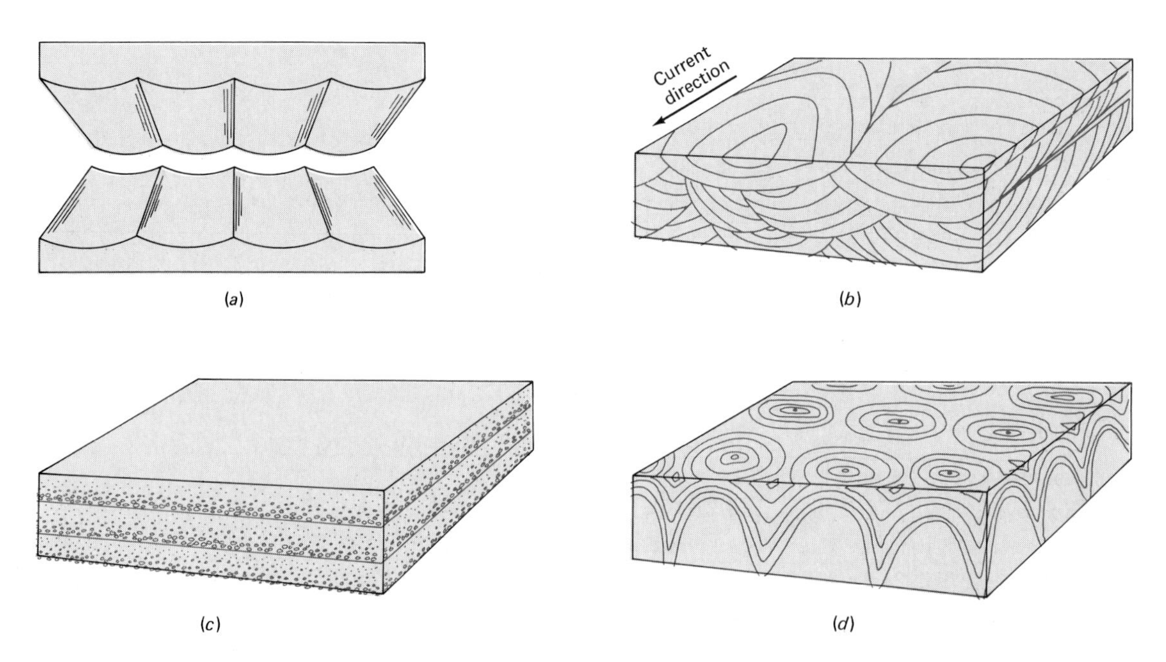

FIGURE 2.11 Representative sedimentary structures useful in distinguishing top and bottom in stratigraphic sequences: (*a*) Oscillation ripple marks and overlying molds are concave upward. (*b*) Trough cross bedding is concave upward with sets truncated at the top. (*c*) Graded beds become finer upward. (*d*) Algal stromatolite heads grow upward.

Original Lateral Continuity *Identical stratigraphic sequences exposed on opposite sides of a valley should be interpreted as remnants of layers that were once continuous across the area through which the valley was cut* (Figure 2.10). With some exceptions, individual formations extend in all directions fairly widely, at least several tens and usually several hundreds of kilometers, before pinching out between other strata or changing laterally into a totally different

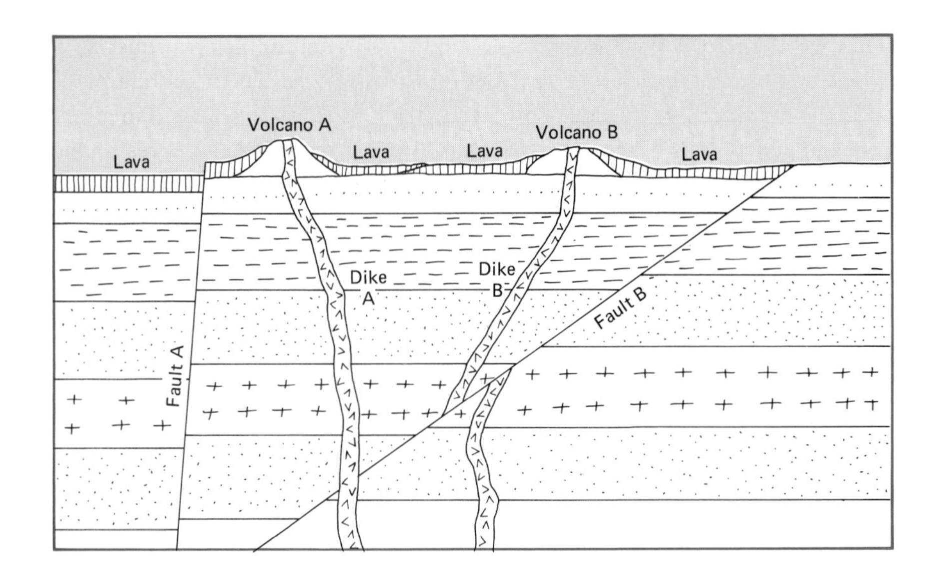

FIGURE 2.12

Law of cross-cutting relationships. Sequence of events is: (1) deposition of sedimentary strata; (2) dike B and volcano B; (3) fault B; (4) dike A and volcano A; (5) fault A.

kind of rock. Steno's recognition that formations that are now interrupted by valleys or mountain ranges were once continuous across these features required a lot of insight. This principle recognizes that areas that are now undergoing erosion were once basins of sediment deposition, indicating that the earth's surface must be changing dynamically and thus be capable of alternately sinking and rising.

Cross-Cutting Relationships A fourth principle important in interpreting the geologic history of an area states that *any rock that is cut by an igneous intrusive body or by a fault is older than that igneous body or fault* (Figure 2.12). James Hutton pointed out the full significance of this principle near the end of the eighteenth century on the basis of a detailed study of igneous intrusive bodies and their contacts with the rocks they intrude.

Cross-cutting relationships are applicable in many contexts. On the moon and planets, for example, where meteorite impact has been a very important process, the relative dating of meteorite craters is accomplished by the very same principle. In Figure 2.13, the overlapping craters can be placed in sequence; the complete small craters are clearly younger than the larger craters whose rims they cut.

Once the rock units in an area have been described and placed in their correct order, and the cross-cutting relationships have established the time of folding, faulting, and igneous intrusion, then the stage is set for reconstructing the geologic history. The study then progresses from the descriptive task of organizing the rock record to the interpretive task of deciphering the geologic processes that produced it.

FIGURE 2.13 Meteorite craters in the Rasena region of Mars can be dated relative to one another by the principle of cross-cutting relationships. (*NASA*)

Deciphering the Rock Record

Sedimentary Environments

Individual sedimentary formations in a typical stratigraphic section like that shown above in Figure 2.5 are distinctive because each is the product of a particular depositional environment (or association of environments) that differed from those that existed before and after. Each formation thus has a story to tell about events at that place on the earth's surface during a brief interval of geologic history. The characteristics of a given sedimentary rock unit depend on: (1) the parent material exposed in the source area; (2) the processes that transported and deposited it; and (3) the chemical processes of cementation and alteration that acted on it after it was deposited. In addition, both **detrital sediments** and **chemical sediments** may have their stratification disrupted or even totally destroyed by burrowing animals after they have been deposited (Figure 2.14). Every sequence of sedimentary formations on earth documents a changing succession of environments.

Ancient surface environments can be reconstructed *only* for those times and places in which there was sedimentation and for which a rock record survives. As documentation of the earth's surface history, the rock record is deficient in that it contains virtually no representatives of the upland and mountain environments that today make up a large part of the earth's surface. Although upland and mountain regions receive temporary deposits that may provide information about recent geologic history, these sediments are not destined to become buried and changed into rock. Over the long term, they will be reeroded and transported to subsiding regions, commonly in or adjacent to the sea, where a preservable historical record *is* accumulating.

In interpreting environments of sediment accumulation we customarily use a classification like that outlined in Figure 2.15. This system of classification first devides sedimentary rocks according to the general environmental setting—whether the sediment formed on land in a *terrestrial environment,* in an oceanic *marine environment,* or in between in a coastal setting, or *transi-*

FIGURE 2.14

X-ray cross section of burrows made within lagoonal muds by burrowing animals. Some of the animals' shells can be clearly seen. (*Courtesy of Donald C. Rhoads*)

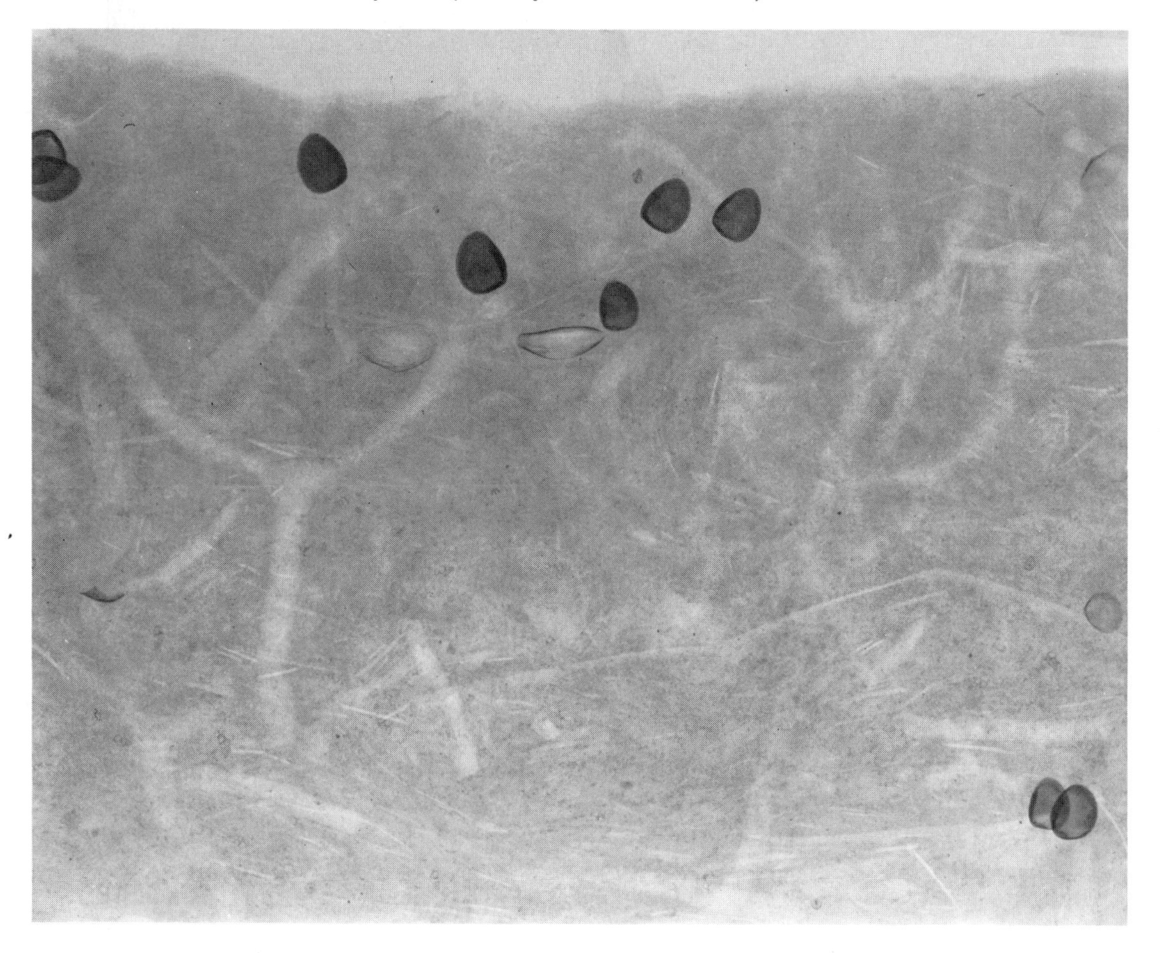

tional environment. We can then stipulate a more specific setting within one of these three broad categories—whether the environment was, for example, a flood plain, a tidal flat, or a shallow sea. Next we ascertain as many details as we can about that environment. What kind of streams flowed on the flood plain? Were they large or small, braided or meandering? What were the temperatures and salinities on the tidal flat? How strong was the current in the shallow sea? Was the water warm or cold? In many cases we can add considerable detail to the general categories shown in Figure 2.15. Geologists are devising many imaginative ways to interpret still more environmental details by careful analyses of sedimentary rocks and associated fossils.

Once an environmental interpretation has been made, the next goal is to draw a map that shows the extent of that particular environment and the environments adjacent to it at that time. Ideally, we would like to be able to do this for the entire earth and for all times in the geologic past, but we are a long way from achieving such a complete synthesis. Although some episodes of geologic history have been reconstructed in detail, the rocks of many regions have received only superficial study.

FIGURE 2.15

CLASSIFICATION OF SEDIMENTARY ENVIRONMENTS	
CONTINENTAL	Alluvial (stream channel and flood plain deposits; alluvial fans) Lacustrine (lakes of humid or arid regions) Eolian (wind-blown deposits, chiefly in deserts) Swamp (in poorly drained areas not associated with the seashore) Glacial (includes deposits of outwash streams as well as the ice itself)
TRANSITIONAL (Mixed or shore-related)	Delta (a composite of alluvial distributaries, marsh, bay, and shallow marine environments) Estuary (drowned valleys) Bay, lagoon Marsh (poorly drained areas at sea's margin) Intertidal and supratidal flat Barrier island and beach Glacial-marine (ice-rafted deposits)
SHALLOW MARINE (Neritic; depth up to 200 meters)	Shelf banks Shelf basin Graded shelf Carbonate shelf and reef Evaporite basin (most shallow, some deep-marine)
DEEP MARINE (Bathyal 200–3,700 meters; abyssal, deeper than 3,700 meters)	Continental slope and submarine canyon Submarine fan Deep-ocean basin Deep enclosed marine basins

The Basis for Interpretation

How do you begin to make environmental interpretations from rock outcrops in the field? The strategy is simple: Compare the physical features and the biological contents of the sedimentary rocks with those of sediments forming today in known environments. If the ancient rocks and the modern sediments are composed of the same materials, and if they have comparable physical features and biological constituents, then they probably formed in similar environmental settings.

In general, marine environments tend to be more widespread than terrestrial environments. Hence, the simple persistence of a given kind of strata over great distances suggests (but does not prove) marine conditions. Some specific kinds of sedimentary rock suggest general environmental settings, although exceptions are common. For example, widespread coal beds generally represent coastal environments, and thick limestone units, marine environments.

Most common sedimentary rock types can originate in a variety of environments and are impossible to interpret on the basis of rock type alone. For example, gray shale forms from gray mud, which may accumulate in flood plains, lakes, lagoons, and shallow or deep ocean environments. How do you distinguish among these possibilities? The most reliable way is by means of the fossils that the shale contains. Land plants and animals have always differed greatly from marine plants and animals. Similarly, transitional environments throughout geologic time have supported distinctive species adapted to the high energies, variable salinities, and rapid sedimentation common to particular coastal regimes. Because animals and plants evolve continually, the older the rock, the more the fauna and flora differ from modern fauna and flora, and, in general, the less certain the environmental interpretations become. Nevertheless, you can infer ancient environmental conditions from the observed adaptations and groupings of fossils, even for long-extinct species.

When you interpret the environment of deposition of a *succession* of rock units, you do so with much greater confidence than if you rely on one individual rock unit alone. A vertical succession of different rock types commonly represents the environments that actually lay adjacent to one another at any given time during deposition. The section on facies that follows elaborates upon this point.

Facies

Just as rock types change vertically due to changing environmental conditions through time, they also change laterally due to differing environmental conditions at any one time. For example, at a given locality you might measure 30 meters of sandstone and silty shale. You might then carefully follow these strata laterally along their outcrop and find that within 2 kilometers the silty shale becomes rich in organic material and the sandstone beds disappear, to be replaced by beds of coal and siltstone. A few kilometers farther, these beds pass laterally into well-sorted sandstone, and this finally passes laterally into gray shale. A shoreline sedimentary model for these lateral changes is shown in Figure 2.16*a*.

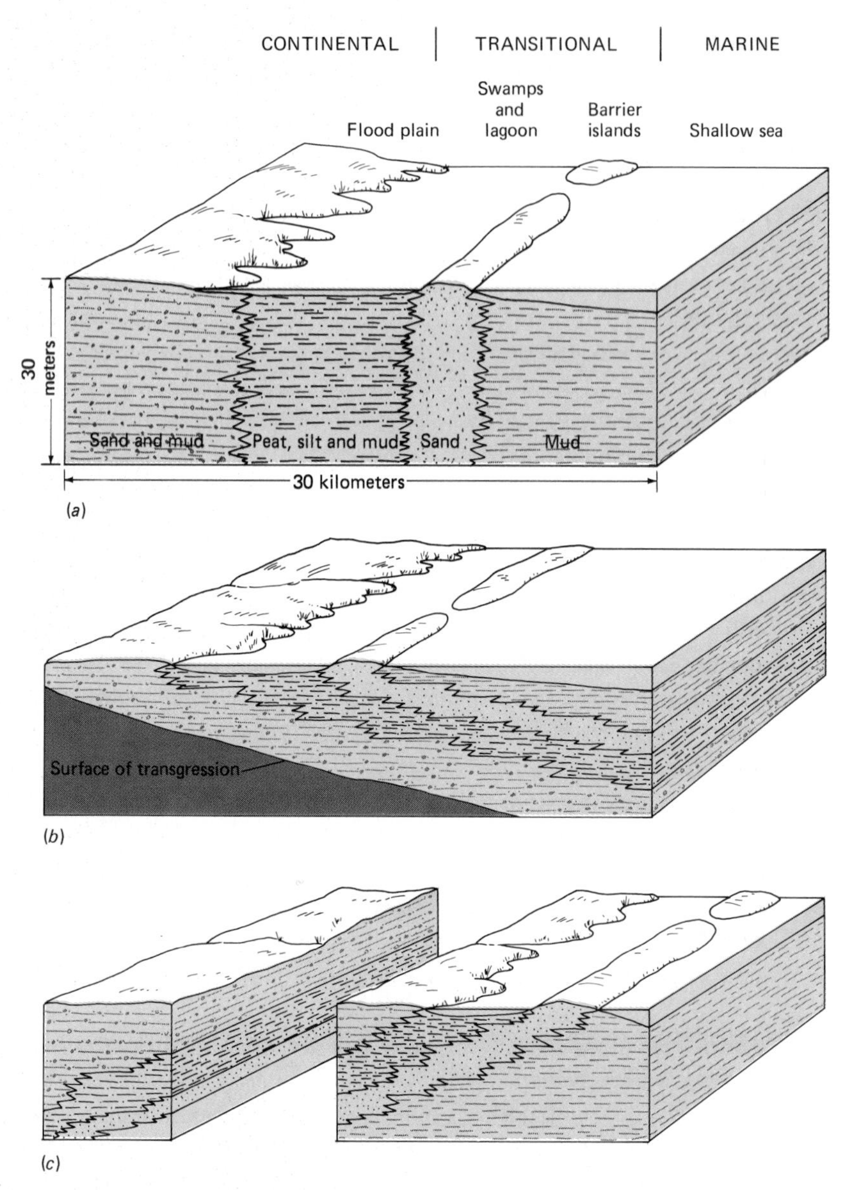

FIGURE 2.16

Block diagrams of facies in near-shore environments showing the stratigraphic successions produced by (*a*) stillstand, (*b*) transgression, (*c*) regression. The vertical scale is exaggerated 300 times.

Different rock types that replace each other laterally within the same stratigraphic interval are called *sedimentary facies.* Many kinds of facies changes occur in rock strata. Some are as striking as those in Figure 2.16*a.* Others, however, are much more subtle and include such things as lateral changes in sedimentary structures, in fossils, in the kinds of heavy minerals in a sandstone, and in the clay mineral content of shales. Lateral changes in physical characteristics may be termed **lithofacies** to distinguish them from changes in the fossil content, or **biofacies.**

To most geologists the term **facies** implies that the different rock types are contiguous laterally; that is, they are the *same age.* In some cases the term may be used loosely, as for example, in discussing the vertical succession at a single locality. But implicit in such usage is the concept that the rock types do indeed replace each other *laterally.* **Microfacies** is a term that has been used for finely detailed aspects (such as grain textures, microfossils, and so on) of sedimentary rocks as they appear in thin section under a microscope. Such usage broadens the term *facies* to refer to the special textural characteristics of a particular rock, without regard to its lateral relationships. In summary, when you read about facies, you must note carefully how the word is used; usually an author's meaning is apparent from the context.

Some of the environments that were listed in Figure 2.15 exist adjacent to one another in nature, but others do not. Hence facies changes do not occur randomly between any two of them. Instead, facies changes reflect lateral environmental relationships that actually prevail today and that evidently prevailed in the geologic past also. For example, marine environments rarely border continental environments directly; a belt of transitional environments similar to those in Figure 2.16*a* nearly everywhere separates the two. Similarly, a deep-sea environment cannot border a beach environment; a shallow-marine environment, however narrow, must lie between them. Just as continental and marine facies or transitional and deep-sea facies do not integrade laterally, their deposits do not normally succeed each other vertically.

If the boundary between adjacent depositional environments remained stationary, the resulting facies would maintain constant geographic positions. The boundaries between facies would be essentially vertical (Figure 2.16*a*). Rarely does this static situation prevail for long, because it requires a precise balance between supply of sediment and rate of subsidence. Far more commonly, the rates of subsidence and supply differ, and environments migrate, their deposits covering earlier deposits of laterally occurring environments. Figures 2.16*b* and 2.16*c* illustrate relationships of typical environments bordering a seacoast. When subsidence exceeds supply of sediment (Figure 2.16*b*), the shoreline moves landward in **marine transgression.** Deposits formed in continental environments are overlain, first by those of transitional environments and then by those of marine environments. When the supply of sediment exceeds subsidence, the shoreline builds seaward, causing **marine regression** (Figure 2.16*c*). Regression forms a vertical sequence that is the reverse of transgression.

Diagrams 2.16 *a-c* illustrate the principle of facies: *Those which succeed each other vertically are those which adjoin laterally.* This law applies as long as sedimentation is relatively continuous and is not interrupted by significant breaks in the sedimentary record.

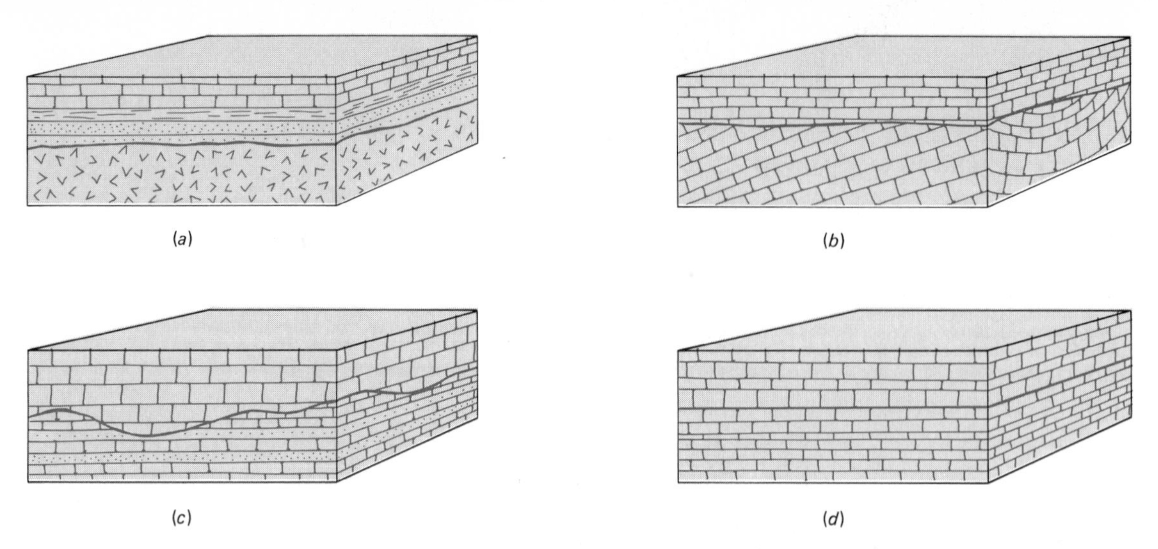

FIGURE 2.17 The four kinds of unconformity: (*a*) nonconformity, (*b*) angular unconformity, (*c*) disconformity, and (*d*) paraconformity.

Areas of Erosion: Unconformities

A major problem in reconstructing ancient environments arises in areas where no sediment accumulates. Much of the surface area of the continents today is not covered by sites of sediment deposition but, instead, is being actively eroded. Obviously, such areas can leave no sedimentary record for the future. Because similar conditions existed throughout much of the earth's past, the continents' veneer of sediments never reveals the complete sequence of surface events, but only those times and places where sediment accumulation was taking place.

As a further complication, even areas that receive sediment at one time may, because of mountain-building deformations, lowered sea level, or other causes, have their sedimentary records destroyed by erosion. For this reason, sedimentary rocks seldom represent a continuous sequence of local sediment accumulation through long intervals of earth history. Instead, the sedimentary veneer in any given area always shows discontinuities representing long time intervals when no sediment was accumulating, and when, in fact, erosion was removing some of the rocks that had previously been deposited. Such discontinuities are called **unconformities.** Even though unconformities reflect intervals of no local sediment accumulation, it is often possible to make inferences about the events they record from the nature of the contact between the rocks above and below the surface of unconformity.

In the field, unconformities are mapped as contacts between formations. The physical relationship of the formations below and above the surface of unconformity fall into one of the types shown in Figure 2.17. **Angular unconformities** are formed by the uplift and tilting of a region, after which the region's sedimentary strata are bevelled by erosion, and a younger stratigraphic sequence is deposited on the tilted and truncated older beds. A **nonconformity** is a surface at which stratified rocks rest on nonstratified intrusive igneous rocks or metamorphic rocks. A **disconformity** is an unconformity in which the strata above and below are parallel. Disconformities commonly show some erosional relief, weathering profiles, or other physical evidence of a substantial break in the sedimentary record. Some unconformities that represent long time breaks may look no different in the field than bedding planes caused by a small pause in sedimentation. Unconformities such as this, which *appear* as if the strata below and above were actually conformable, have been termed **paraconformities.** Paraconformities are recognizable only when fossils from the beds below and above reveal the magnitude of the time break, thereby providing evidence that an unconformity must exist.

FIGURE 2.18

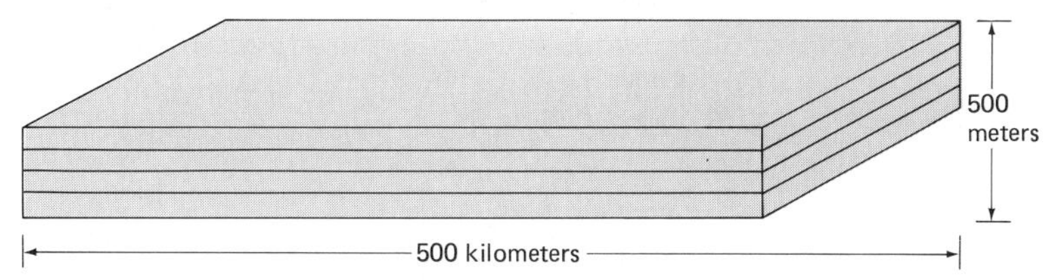

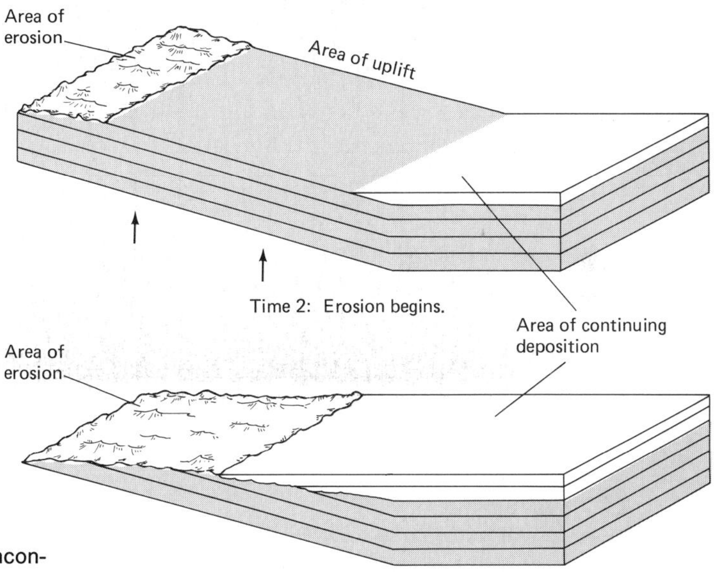

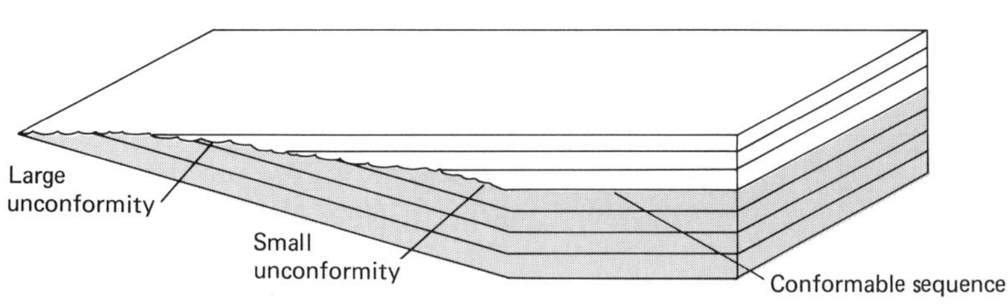

Steps in the production of an unconformity with lateral variation in magnitude. Vertical scale is exaggerated 200 times.

The time period represented at an unconformity varies laterally for two reasons: First, the underlying rock record is eroded more in some areas than in others. Second, the deposition of the overlying stratigraphic sequence resumes earlier in some areas than in others (Figure 2.18). In some cases, an unconformity may be traced into an area of deposition where it disappears into a conformable sequence. A single unconformity may also exhibit different structural relationships laterally in adjacent regions. For example, if an eroded terrain beneath an unconformity surface exposed massive igneous rocks in one area, tilted sedimentary rocks in a second area, flat-lying sedimentary rocks with erosional relief in a third, and flat-lying sedimentary rocks without erosional relief in a fourth area, the end result, after deposition of the overlying sequence, would be a nonconformity, angular unconformity, disconformity, or paraconformity, depending on the locality examined.

The Geologic Time Scale

So far we have been considering how the geologic record is organized, and how the history of an area is interpreted. The next step in this process is to fit local histories into the broader history of the entire region and of the world as a whole. We have noted that the succession of fossil life provides a convenient means of worldwide dating for the sedimentary rocks deposited during the most recent 15 percent of earth history. Early in the nineteenth century, a system was established for subdividing and naming these fossil-bearing rocks that still provides the standard time scale used in all discussions of earth history (Figure 2.19).

The last 15 percent of geologic history is known as the Phanerozoic ("exposed life") Eon. Within the Phanerozoic, three *eras* bounded by relatively profound, sudden, and worldwide changes in the living organisms preserved as fossils are recognized: the Paleozoic ("ancient life") Era, the Mesozoic ("middle life") Era, and the Cenozoic ("recent life") Era. The three eras are further subdivided into *periods,* each bounded by somewhat less profound changes in the living world. Twelve periods are now recognized. The seven periods of the Paleozoic Era, listed in order from oldest to most recent, are: Cambrian, Ordovician, Silurian, Devonian, Mississippian, Pennsylvanian, and Permian. The three periods of the Mesozoic Era are: Triassic (oldest), Jurassic, and Cretaceous. The Cenozoic Era includes only two periods: Paleogene and Neogene. The twelve geologic periods are, in turn, further subdivided according to their fossil contents into still smaller units called *epochs.* The many names of these smaller divisions of geologic time need not concern us; the most important are the seven epochs of the Cenozoic Era, the youngest two of which are the Pleistocene and Holocene Epochs. The Pleistocene includes the recent intervals of ice-sheet expansion and contraction during which modern man evolved, and the Holocene includes only the last 10,000 years, during which the continents were largely ice free.

The oldest sedimentary rocks bearing abundant fossils are those of the Cambrian Period, which begins the Paleozoic Era. Wherever they occur

FIGURE 2.19

THE TIME SCALE OF EARTH HISTORY

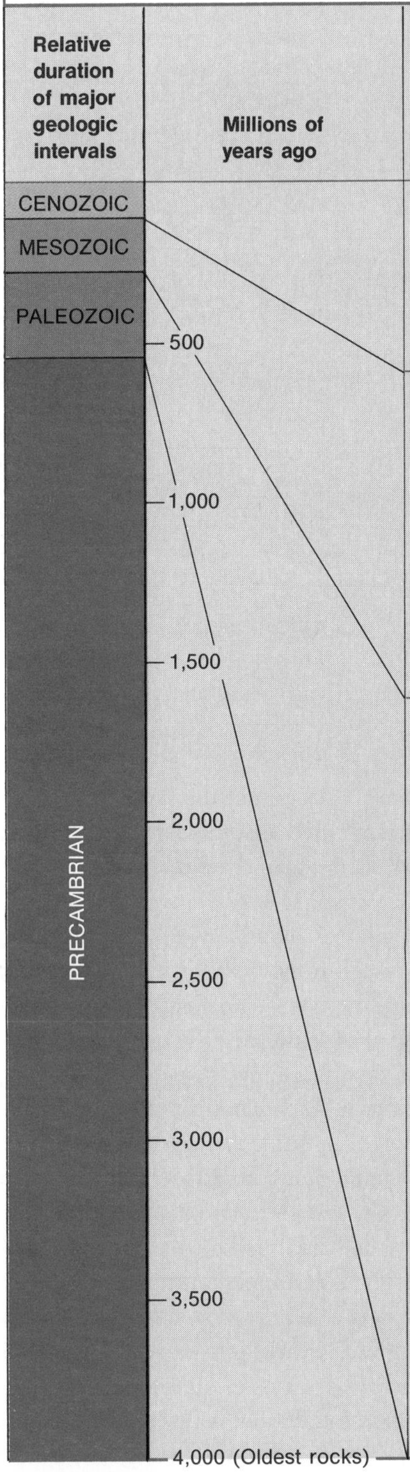

Relative duration of major geologic intervals: CENOZOIC, MESOZOIC, PALEOZOIC, PRECAMBRIAN

Millions of years ago: 500, 1,000, 1,500, 2,000, 2,500, 3,000, 3,500, 4,000 (Oldest rocks)

Era	Period	Epoch	Duration in millions of years (approx.)	Millions of years ago (approx.)
CENOZOIC	Neogene	Holocene	0.01	0.01
		Pleistocene	2.5	2.5
		Pliocene	4.5	7
		Miocene	19	26
	Paleogene	Oligocene	12	38
		Eocene	16	54
		Paleocene	11	65
MESOZOIC	Cretaceous		71	136
	Jurassic		54	190
	Triassic		35	225
PALEOZOIC	Permian		55	280
	Pennsylvanian		45	325
	Mississippian		20	345
	Devonian		50	395
	Silurian		35	430
	Ordovician		70	500
	Cambrian		70	570

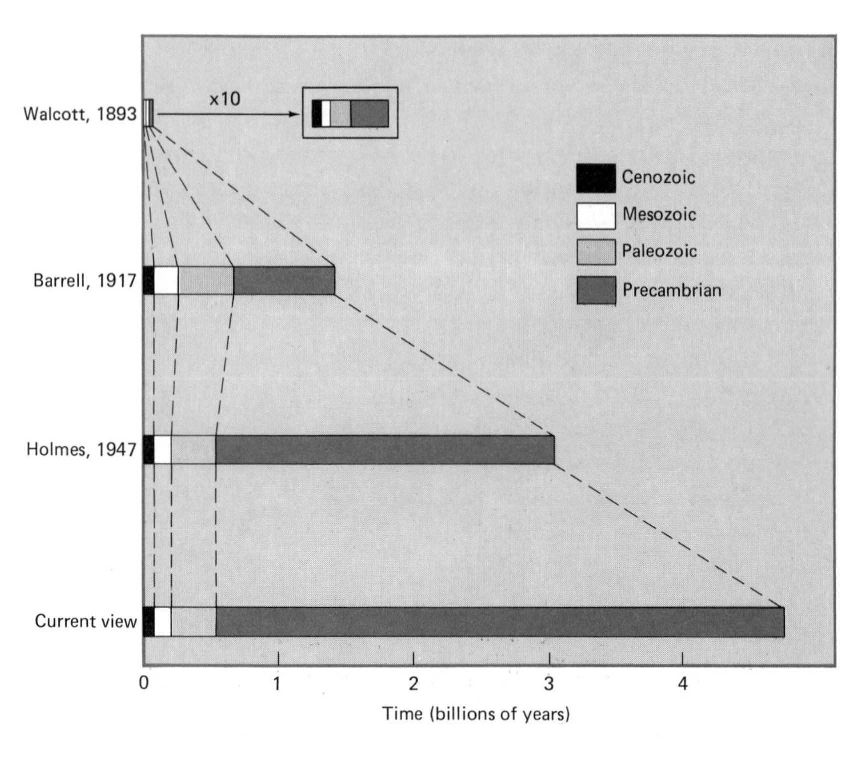

FIGURE 2.20 Changing view of the magnitude of Precambrian time in the twentieth century. *(James, 1960)*

around the world, fossil-bearing Cambrian rocks contain the same distinctive association of shell-bearing animals dominated by the long-extinct trilobites, which were distant relatives of modern crabs and shrimp. These earliest Cambrian sediments always overlie older rocks that lack abundant fossils.

Commonly, Cambrian sediments rest upon a "basement complex" of older igneous and metamorphic rocks that makes up the bulk of the continental crust. In many regions, however, the earliest Cambrian fossils occur above thick sequences of sedimentary rocks that are identical to the overlying Cambrian rocks, except that they contain no fossils or only sparse traces of very primitive life that have virtually no stratigraphic value. The abrupt appearance of relatively advanced kinds of animal life in the sedimentary record of the Cambrian Period is perhaps the most significant milestone in the earth's long history, for it serves to divide the earth's past into two distinct divisions: the Phanerozoic Eon above, and all of Precambrian time below.

In Precambrian exposures, rock units can be mapped and organized into local sections, just as they can in younger rocks, and they can be traced locally so long as continuity of outcrops permits. Without the presence of fossils as a means of comparison, however, they simply cannot be correlated between distant areas. Hence, classification on a worldwide time scale like that established for the Phanerozoic has not been feasible. However, since the 1950s, radiometric-dating techniques have allowed geologists to ascertain the ages of many Precambrian rocks in terms of years. On this basis, we have been able to establish a quantitative time framework that permits Precambrian rocks of widely distant regions to be placed in their proper sequence. These dating tools have resulted largely from the perfection of precise laboratory methods that enable geochemists to analyze vanishingly small quantities of particular elements with remarkable accuracy.

Radiometric dating has contributed greatly to our understanding of the Phanerozoic segment of history as well. Not only did it yield the initially surprising discovery that Phanerozoic rocks represent only 15 percent of geologic time—a fact that we were not aware of until the middle of the twentieth century (Figure 2.20)—but it subsequently permitted a calibration of the Phanerozoic time scale that complements correlation based on fossils. Within the Phanerozoic, radiometric dating has made possible determinations of rates of physical and biological processes that have shed a great deal of light on the past behavior of our planet. The remainder of this chapter outlines the radiometric-dating process and elaborates on some of its most important revelations.

Radiometric Dating

The small nucleus at the center of every atom is composed of two principal kinds of particles: uncharged neutrons and positively charged protons. The number of protons is the same as the number of negative electrons that orbit around the nucleus. A given kind of atom, with a particular number of neutrons and a particular number of protons in the nucleus, is called a **nuclide** (Figure 2.21).

FIGURE 2.21

Schematic representations of four nuclides: hydrogen-1, carbon-12, carbon-14, and silicon-28.

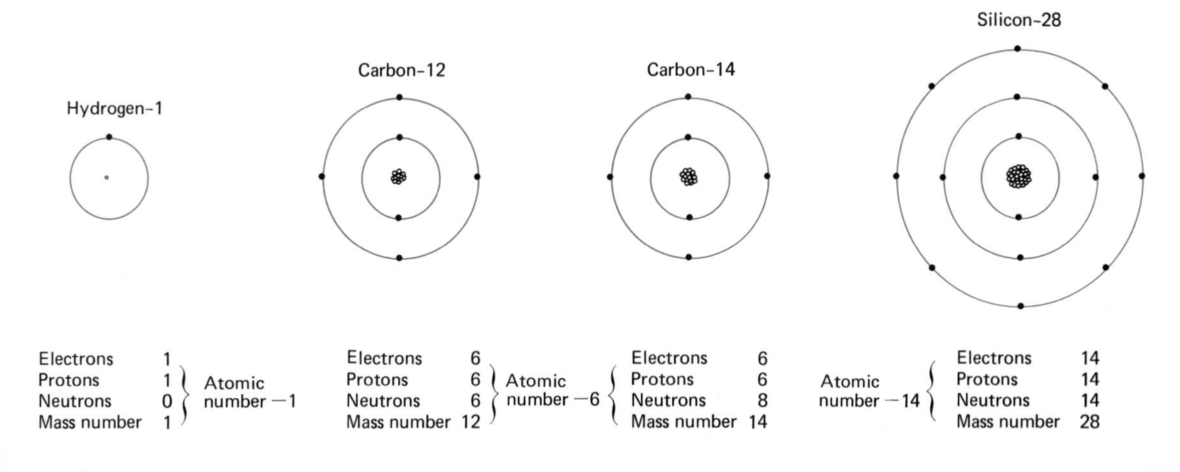

Element						
$_{1}H$	1 / 99.99	2 / 0.01	3 / trace			
$_{6}C$	12 / 98.9	13 / 1.1	14 / trace			
$_{7}N$	14 / 99.6	15 / 0.4				
$_{8}O$	16 / 99.8	17 / 0.04	18 / 0.2			
$_{11}Na$	23 / 100					
$_{12}Mg$	24 / 78.7	25 / 10.1	26 / 11.2			
$_{13}Al$	27 / 100					
$_{14}Si$	28 / 98.2	29 / 4.7	30 / 3.1			
$_{16}S$	32 / 95.0	33 / 0.8	34 / 4.2	36 / 0.01		
$_{17}Cl$	35 / 75.5	37 / 24.5				
$_{18}Ar$	36 / 0.3	38 / 0.06	40 / 99.6			
$_{19}K$	39 / 93.1	40 / 0.01	41 / 6.9			
$_{20}Ca$	40 / 97.0	42 / 0.6	43 / 0.1	44 / 2.1	46 / 0.01	48 / 0.2
$_{26}Fe$	54 / 5.8	56 / 91.7	57 / 2.2	58 / 0.3		
$_{28}Ni$	58 / 67.8	60 / 26.2	61 / 1.2	62 / 3.7	64 / 1.1	

(*a*)

Nuclides are distinguished by two important numbers. The first is the **atomic number,** which is simply the number of protons in the nucleus (or the number of orbital electrons, since the both numbers are always the same); the atomic number determines the element to which the atom belongs. The second is the **mass number,** which is the total of both protons and neutrons in the nucleus. Two different nuclides with the same atomic number but different mass numbers are said to be different **isotopes** of an element. Calcium atoms, for example, always have 20 nuclear protons, whose positive charges are balanced by 20 orbital electrons; thus the atomic number of calcium atoms is always 20. Naturally occurring calcium is, however, a mixture of six different isotopes, which have either 20, 22, 23, 24, 26, or 28 neutrons in the nucleus. These neutrons, added to the constant number of nuclear protons (20), give these six isotopes mass numbers of 40, 42, 43, 44, 46, and 48 respectively (Figure 2.22). For the sake of brevity, nuclides are normally referred to by the name of the element and the mass number; for example, calcium-40 or carbon-14.

Radioactive Decay

Many different nuclides can be prepared artificially in nuclear reactors or particle accelerators by adding protons and neutrons to, or subtracting them from, atomic nuclei. However, most such artificial nuclides are very unstable, or radioactive, which merely means that they decay rapidly to more stable nuclides of other elements by one of the three processes summarized in Figure 2.23. The original radioactive nuclide is called the **parent nuclide,** and the nuclide or nuclides produced by decay are called **daughter nuclides.** Each different radioactive nuclide decays into daughter nuclides at a constant rate that is not affected by physical conditions such as temperature or pressure. The rate is not even altered by chemical changes such as oxidation or reduction of the parent atom, because these involve only the orbital electrons and not the nucleus. Most artificial nuclides decay very rapidly, usually in a few hours or days, to more stable nuclides, but some persist for years.

In contrast to radioactive nuclides produced in reactors and accelerators,

Element	Nuclides (abundance %)			
$_{37}Rb$	85 (72.2)	87 (27.8)		
$_{38}Sr$	84 (0.5)	86 (9.9)	87 (7.0)	88 (82.6)
$_{82}Pb$	204 (1.5)	206 (23.6)	207 (22.6)	208 (52.3)
$_{90}Th$	232 (100)			
$_{92}U$	234 (0.01)	235 (0.72)	238 (99.27)	

(*b*)

FIGURE 2.22

(*a*) Natural nuclides of the 15 most common elements making up the earth, and (*b*) 5 additional elements important in earth chronology. Natural radioactive nuclides are shown in gray. The lower numbers in each box show the percentage abundance of each nuclide in natural mixtures of the element; darker shading indicates dominant nuclides.

most naturally occurring nuclides are stable; that is, they have no tendency to decay to other nuclides (Figure 2.22). Most natural elements that make up the minerals of the earth's crust are mixtures of several (usually from two to eight) stable nuclides of that element. If all naturally occurring nuclides were stable, however, they would be of no value for earth chronology. Fortunately, there are several *natural* radioactive nuclides that make possible the use of radioactive decay as a tool for dating rocks.

The principle of radiometric dating is comparable to that of the hourglass. Turn the glass over and the sand runs from the top chamber to the bottom; as long as some sand remains in the top, the amount in the top relative to the

FIGURE 2.23

The effect on the atomic nucleus of the three types of radioactive decay.

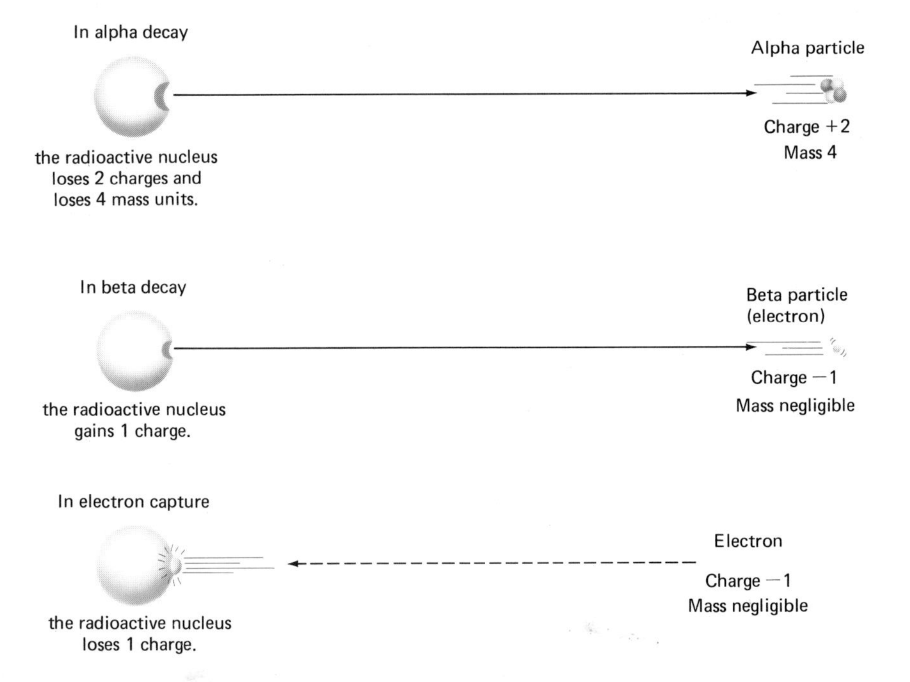

amount in the bottom provides a measure of time elapsed. By analogy, the sand in the top of the hourglass represents decaying radioactive parent atoms, and that in the bottom, accumulating daughter atoms. Just as the hourglass must be sealed so that sand can't escape through the sides, so the mineral grain that contains a radioactive nuclide must be able to hold both parent and daughter atoms without allowing any to escape or, for that matter, to enter from an external source. In other words, the system must be closed.

Unlike the passage of sand through an hourglass, the uniform melting of a candle, or other *linear* rates of depletion, radioactive decay occurs at a *geometric* rate (Figure 2.24). Each individual atom of a given radioactive isotope has the same probability of decaying within a given time period. This probability remains the same, no matter how long the material being dated has been in existence. Probability of decay of a radioactive nuclide is expressed by a number called the **decay constant**, which simply stipulates the *proportion* of atoms of that nuclide that always decay in a given time period, usually expressed in terms of a year.

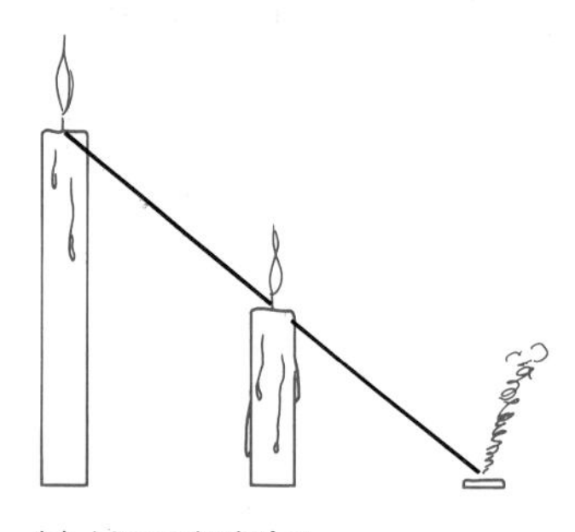

(*a*) Linear depletion

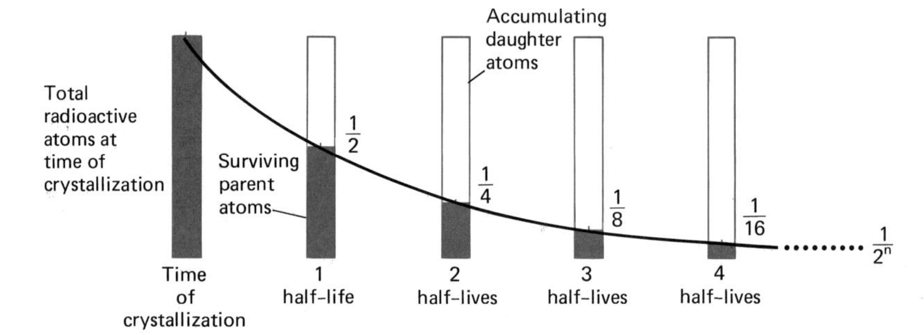

(*b*) Geometric depletion

FIGURE 2.24 Contrast between linear depletion of most everyday processes and geometric depletion of radioactive decay. The end of one half-life interval is the beginning of a new one.

The actual *number* of atoms that decay depends on the number of radioactive parent atoms present in the system at the beginning of the year. At the beginning of each year, the number of radioactive parent atoms is, of course, smaller than it was at the start of the preceding year, for some will have decayed during the course of the year. Hence, the actual number of atoms to decay the second year is smaller, and the number decreases with each successive year.

The total time required for *all* radioactive atoms in a given system to decay cannot be specified. In theory it is infinite, for some atoms will never decay. It is a simple matter, however, to specify the time required for *half* the atoms of a particular radioactive nuclide to decay. This time period is called the **half-life.** If a quantity of radioactive nuclide is segregated, half of the initial number of atoms (N/2) remain after one half-life period; half of those, or one-fourth (N/4), remain at the end of the next half-life period; and half of those, or just one eighth (N/8), remain at the end of the next half-life period, and so on. This simple geometric relationship is the basis of all radioactive dating. Each radioactive nuclide has its own half-life; some have a duration of microseconds; others, trillions of years. A given nuclide is best suited to measure time periods that are of about the same order of magnitude as its half-life.

When mineral grains containing radioactive atoms first crystallize as part of a newly formed rock, they normally contain no atoms of the radiogenic daughter nuclide. The initial daughter:parent ratio is zero, and therefore the indicated age is zero. With time, the decay of radioactive parent atoms produces radiogenic daughter atoms in their place in the mineral grains. Knowing the decay rate of the radioactive parent, we need only to measure the ratio of radiogenic daughter and parent nuclides in the mineral in order to calculate the *radiometric age,* measured in years before the present.

Radioactive Nuclides in Nature

All radioactive nuclides found in rocks, the ocean, and the atmosphere come from one of two sources. The first group, called **primary nuclides,** have such extremely long half-lives that they have persisted since the earth first formed. About 20 such nuclides have been detected, but only four are widespread and abundant enough to be generally useful as chronologic tools. These are potassium-40 (which decays to argon-40), rubidium-87 (which decays to strontium-87), uranium-235 (which decays, through a series of intermediate radioactive nuclides, to lead-207), and uranium-238 (which decays, also through an intermediate series of nuclides, to lead-206) (Figure 2.22).

In addition to these long-lived radioactive nuclides left over from the earth's formation, a second group of much shorter-lived radioactive nuclides is continually being produced in the earth's upper atmosphere by cosmic rays. These are extremely high-energy nuclear particles moving in space from unkown sources. When such particles enter the earth's atmosphere, they collide with atmospheric gas particles and produce nuclear reactions similar to those of laboratory particle accelerators. Some of these reactions produce short-lived radioactive nuclides. At least eight such **cosmic-ray-induced nuclides** have been identified, but only one, carbon-14, has thus far proved to be a widely useful chronologic tool.

Carbon-14 is produced by cosmic-ray neutrons striking atoms of nitrogen, the most abundant atmospheric gas (Figure 2.25). Atmospheric nitrogen is composed of only stable nuclides, principally nitrogen-14, but cosmic-ray neutrons actually displace protons in some of these nitrogen atoms to produce radioactive carbon-14, which then decays back to stable nitrogen-14 with a half-life of about 5,600 years.

Once produced, the carbon-14 atoms quickly combine with atmospheric oxygen to become carbon dioxide. Most atmospheric carbon dioxide is made up of the stable nuclides, carbon-12 and carbon-13; the relatively small quantity of cosmic-ray-induced carbon-14 dioxide quickly mixes with this more abundant and stable atmospheric carbon dioxide and enters into the earth's general carbon cycle. There it may be dissolved in the oceans, precipitated as carbonate minerals, or used by animals and plants. The small amounts of radioactive carbon-14 that ultimately enter animals, plants, and minerals provide an extremely useful tool for dating the last 50,000 years of earth history (Figure 2.25).

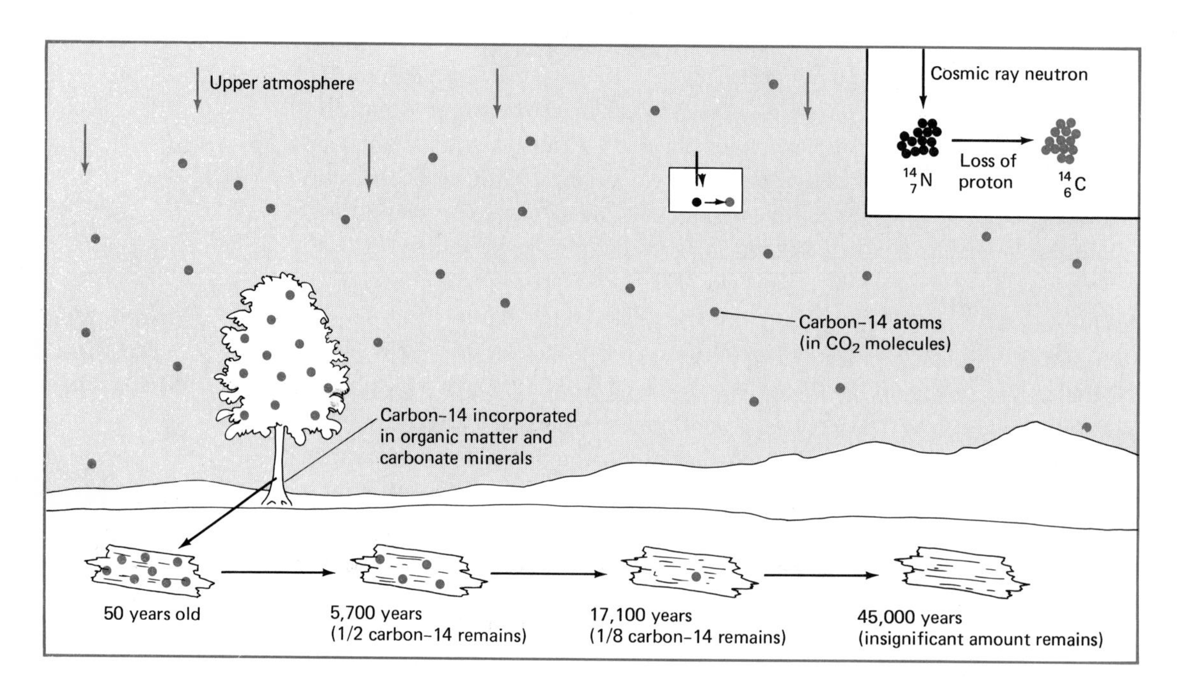

FIGURE 2.25 The carbon-14 cycle. Radioactive carbon-14 is produced from atmospheric nitrogen by cosmic rays (inset). It then enters into CO_2 molecules and becomes incorporated into carbon-bearing sediments and organic remains. The amount of remaining carbon-14 is used to date such materials.

Methods for Dating Ancient Rocks

Four widespread primary nuclides (potassium-40, rubidium-87, uranium-235, and uranium-238) have provided almost all radiometric ages for the earth's ancient rocks (Figure 2.26). Each plays a somewhat different role in deciphering earth chronology. We shall look more closely at these roles before we elaborate upon some of the earth history revealed by these techniques.

FIGURE 2.26

THE CHIEF METHODS OF RADIOMETRIC-AGE DETERMINATION

Parent nuclide	Half-life (years)	Daughter nuclide	Minerals and rocks commonly dated
Uranium-238	4,510 million	Lead-206	Zircon Uraninite Pitchblende
Uranium-235	713 million	Lead-207	Zircon Uraninite Pitchblende
Potassium-40	1,300 million	Argon-40	Muscovite Biotite Hornblende Glauconite Sanidine Whole volcanic rock
Rubidium-87	47,000 million	Strontium-87	Muscovite Biotite Lepidolite Microcline Glauconite Whole metamorphic rock

Potassium-Argon Method Potassium is the seventh most abundant element in the earth's crust; 0.4 percent of this potassium is radioactive potassium-40. Because potassium is a constituent of common rock-forming minerals such as the micas, feldspars, and hornblende, the potassium-argon method can be used to date a great many rocks. Only 11 percent of potassium-40 decays to argon-40 (by electron capture); the rest decays to calcium-40 (by beta decay). The decay of potassium to calcium does not provide a workable dating tool because some original calcium occurs in virtually all common rocks and minerals, and it cannot be discriminated from radiogenic calcium. Argon, however, being an inert gas, is never bound chemically in minerals when they form. Any argon-40 in a mineral or rock is almost certainly a product of potassium-40 decay in that mineral or rock. Consequently, the problem of original daughter contamination is very small, and the potassium-argon method has become an excellent chronological tool.

In general, volcanic rocks and others that have never been buried deeply give reliable potassium-argon ages. However, above 200°C or so (a temperature normally attained at a depth of around 5 kilometers), some of the argon escapes from the crystals in which it is produced. Hot igneous and metamorphic rocks are an open system for argon; that is, they do not retain it. They become a closed system for argon after they have cooled below 200°C or so. If they crystallize at great depth, this temperature may not be reached until several million years after crystallization. For this reason, potassium-argon ages are usually considered to be minimum ages.

Dating based on a specific mineral like biotite or hornblende requires selecting enough individual mineral grains from crushed rock for analysis. But many rocks are so fine grained that individual grains can be separated only with great difficulty. Such rocks are commonly analyzed without separating out individual mineral components. This procedure is called the **whole-rock method.** Because potassium-40 has a half-life of 1.3 billion years, potassium-argon dating is applicable to the oldest Precambrian rocks we know. Tiny amounts of argon can be measured with great precision, and the potassium-argon whole-rock method has been applied successfully to volcanic rocks as young as 100,000 years (Figure 2.27).

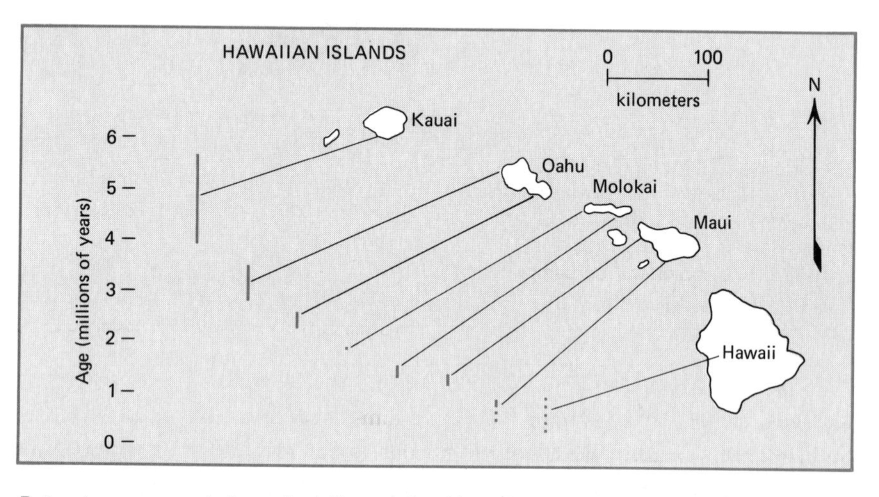

FIGURE 2.27

Potassium-argon whole-rock dating of the Hawaiian Islands shows that they formed progressively from northwest to southeast. (*McDougall, 1964*)

Rubidium-Strontium Method Radioactive rubidium-87 decays in a single step to strontium-87. Rubidium-87 constitutes 28 percent of all rubidium, but rubidium is not very common; the earth's crust contains only about 15 percent as much rubidium-87 as potassium-40. Rubidium occurs in trace amounts in micas, potassium feldspars, pyroxenes, amphiboles, and olivine—all of which may be used for rubidium-strontium age determinations. A trace of original strontium occurs in these minerals also, and it must be measured as part of any rubidium-strontium age determination. The quantity of original strontium is then subtracted in calculating the age.

Original strontium always contains some nonradiogenic strontium-86, and hence it is easily detected. Inasmuch as the quantity of strontium-86 does not change during the life of the mineral, the technique is to analyze a rubidium-poor sample to determine the strontium-87:strontium-86 ratio that prevailed originally in the rock body being dated. Next, a rubidium-rich sample is analyzed, not only for rubidium-87 and strontium-87, but also for strontium-86.

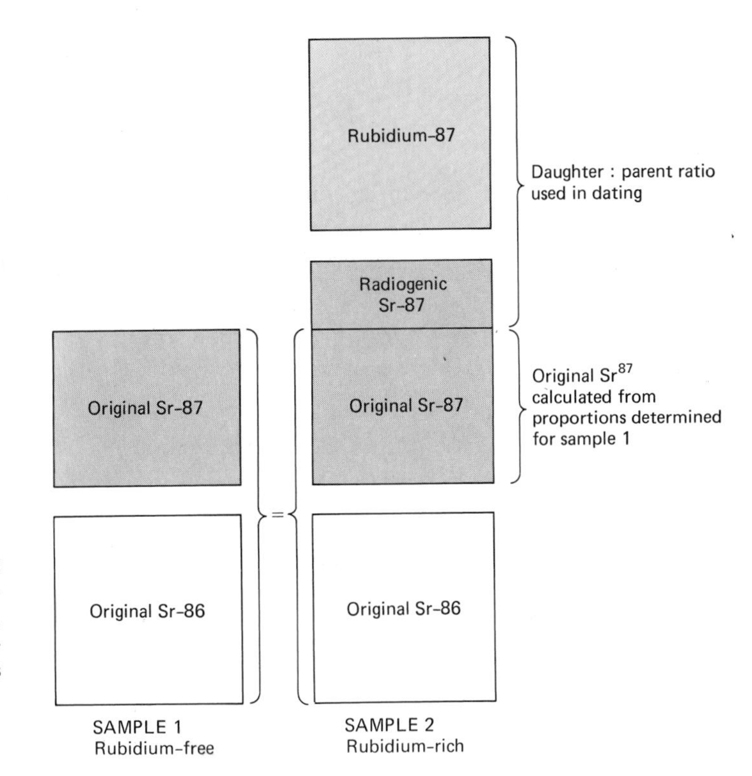

The strontium-87:strontium-86 ratio at time of origin, determined for a rubidium-free sample from the same rock body, is used to determine how much strontium-87 in the rubidium-rich sample is radiogenic.

FIGURE 2.28

Knowing the original strontium-87:strontium-86 ratio for the sample, we can determine the quantity of strontium-87 that is original and the quantity that is radiogenic (Figure 2.28). The age of the sample is then calculated from the daughter:parent ratio. This method works best for rock bodies that contain minerals that are exceptionally rich in rubidium.

A more widely applicable means of determining rubidium-strontium ages is the **isochron method.** Several samples are required, but none need be rubidium free or unusually rubidium rich. The isochron method requires only that the samples vary significantly in relative content of rubidium. As an example, Figure 2.29 shows the rubidium-strontium isochron age of a suite of Paleozoic volcanic rocks from Vinalhaven Island off the Maine coast. At the time these volcanic rocks formed, the strontium-87:strontium-86 ratio was the same in all of them, but the relative rubidium content varied widely. Had a group of samples from these rocks been analyzed immediately after they formed, they would have plotted on a horizontal line on the graph in Figure 2.29, which shows the strontium-87:strontium-86 ratio as the ordinate and the rubidium-87:strontium-86 ratio as the abscissa. With time the rubidium-87 content decreased and the strontium-87 content correspondingly increased in each sample. Shown on the graph, each sample follows its own path upward and slightly to the left, moving at a rate proportional to its relative rubidium content. The series of samples continues to define a straight line whose slope

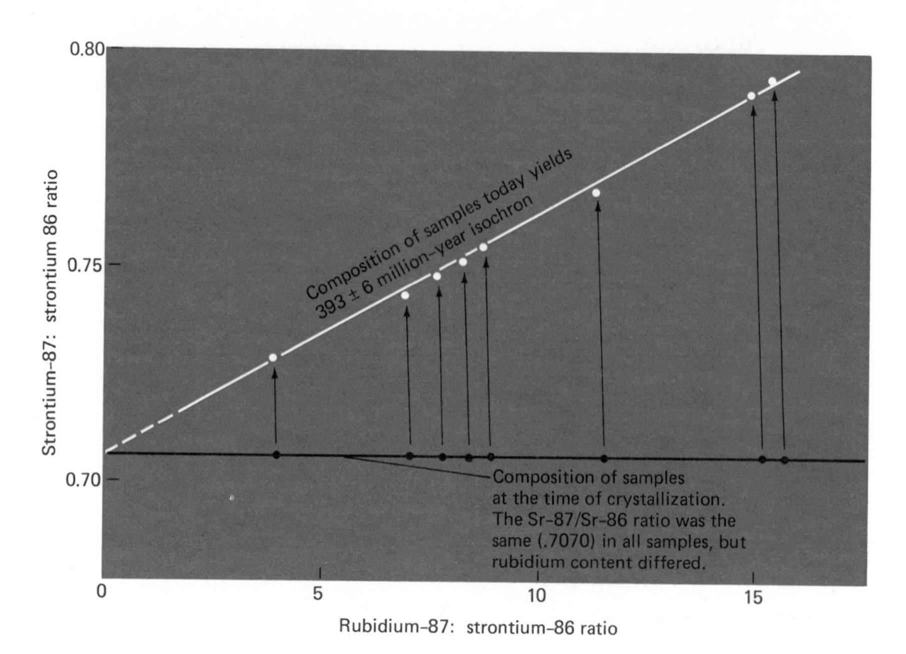

FIGURE 2.29

Rubidium-strontium isochron date based on eight whole-rock samples from Devonian volcanic rocks from Vinalhaven Island off the Maine coast. (*Brookins and others, 1973*)

increases systematically with time. This line is the *isochron* (same time), and its slope actually provides the radiometric age of the rock. In addition, the intercept of the isochron on the ordinate provides the original strontium-87:strontium-86 ratio. The rubidium-strontium isochron age of 393 ± 6 million years* for the volcanic rocks from Vinalhaven Island is similar to ages of other rocks in the Maine coastal volcanic belt. This age-dates the Vinalhaven volcanics as Lower Devonian, indicating that they were erupted during the Acadian Orogeny, which will be discussed in Chapter 7.

The rubidium-strontium method is a versatile tool. It has successfully dated the most ancient rocks, including lunar samples collected by the Apollo missions, and it has been applied to rocks as young as only a few million years. In addition to being widely applicable to both igneous intrusive and extrusive rocks, the rubidium-strontium method provides the most useful technique available for dating metamorphic rocks. Following a metamorphic event, strontium begins to be retained by minerals at far higher temperatures than does argon gas. Hence the rubidium-strontium isochron method establishes the time of metamorphism much more accurately than does the potassium-argon method.

*The "plus or minus 6 million years" appended to the age is the estimated error, and it means that the age of 393 million years is probably accurate to within three percent.

Uranium-Lead Method All naturally occurring uranium contains radioactive uranium-238 and radioactive uranium-235 in a ratio of 138:1. Uranium-238 decays to lead-206, and uranium-235 to lead-207; therefore, these two separate nuclides provide a cross check in determining ages. Most uranium minerals also contain radioactive thorium-232, which decays to lead-208, and this method is occasionally used as still another cross check. Uranium and thorium nuclides decay through a *series* of intermediate nuclides, which are themselves radioactive, before finally decaying to their stable lead daughters. The uranium-238–lead-206 series involves eight alpha-decay steps and six beta steps; the uranium-235–lead-207 series involves seven alpha steps and four beta steps; the thorium-232–lead-208 series involves six alpha steps and four beta steps. Helium, which is produced in the alpha-decay steps of each series, is an additional stable by-product.

Minerals that contain uranium as a chief component are rare, but minerals with uranium in trace quantities are fairly common. The most useful of these is the igneous mineral zircon ($ZrSiO_4$), which typically contains about 0.1 percent uranium. Small quantites of zircon occur in granitic rocks of many ages, and thus uranium-lead dating is widely applicable. Uranium-lead dates have been derived for lunar breccias and fine surface material using whole-rock analyses.

The original lead in a uranium-bearing mineral causes the radiometric age to exceed the true age, unless it is detected. Lead-204, which is never produced radiogenically, provides a convenient way of detecting it. Lead-204 constitutes a tiny proportion of all common lead; if it is present, the other lead isotopes, including lead-206 and lead-207, were present when the mineral formed. The isotopic composition of common lead can be obtained from uranium-poor samples. Then the quantity of lead-204 can be used to calculate the quantities of original lead-206 and lead-207, so that these can be subtracted in calculating the radiometric age.

After allowing for the original lead, the ages of uranium-235–lead-207 and uranium-238–lead-206 should agree, provided the mineral or rock has remained a closed system. If they do agree, the ages are said to be *concordant*, and the probability is high that the radiometric age is the same as the true age. On a graph showing the Pb-207:U-235 ratio as the ordinate and the Pb-206:U-238 ratio as the abscissa, the loci of all concordant ages define a curve called the **concordia.** If the uranium-lead ages do not agree, then they do not fall on the concordia, and they are said to be *discordant*.

An example of concordant ages from the moon is shown in Figure 2.30. Uranium-lead dates for samples collected by Apollo 11 lie between 4.6 and 4.7 billion years. A corroborative age of 4.65 billion years is obtained by thorium-232–lead-208 dating. The 4.6- to 4.7-billion-year age is the same as ages determined from meteorites. This is inferred to be the age of the moon and of the rest of the solar system as well.

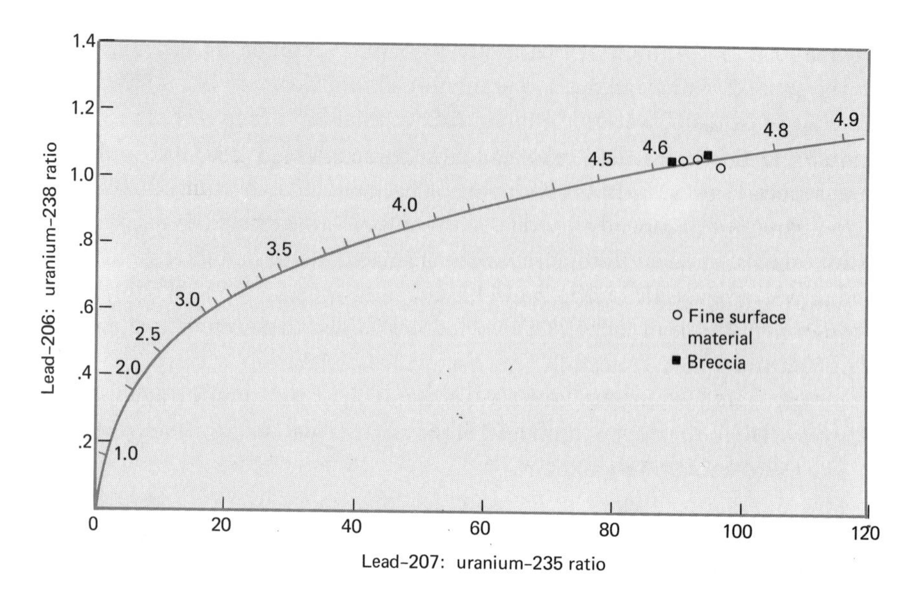

FIGURE 2.30

Uranium-lead concordia showing concordant ages (in billions of years) of lunar soil and breccia collected by Apollo 11. Age between 4.6 and 4.7 billion years accords with the age of the solar system. (*Wetherill, 1971*)

Age of the Earth

The oldest crustal rocks so far dated, granites from southwestern Greenland, have radiometric ages of about 4 billion years. In other regions, metamorphosed sedimentary rocks surround similar ancient granites that were intruded into, and thus are younger than, the deformed sediments that surround them. This would indicate that erosion of the crust, deposition of sediment, and the formation of sedimentary rocks were all taking place very early in earth history. Apparently, these processes have completely recycled and thus obliterated any original rocks formed as the crust first consolidated. Direct dating of crustal rocks can therefore only indicate that the earth is *older* than 4 billion years. To answer the question "How much older?" we must turn to less direct evidence of two sorts.

The first comes from radiometric dating of meteorites and rocks recovered from the surface of the moon. In Chapter 1 we saw that all the solid material of the solar system (including the earth, the moon, and the small particles that fall on the earth as meteorites) is believed to have had a common origin from gases and dust that condensed at the same time as the sun. Presumably, neither the rocks of the moon nor the meteorites have been subjected to the processes of recycling that have affected rocks of the earth's crust. For this reason, direct dating of these rocks could indicate when they first consolidated, and by extrapolation, when the earth and the other planets formed. Most significantly, almost all meteorites have radiometric ages around 4.6 billion years, suggesting that the earth has about the same overall age. In addition, the oldest rocks so far recovered from the surface of the moon have ages of about 4.6 billion years, which suggests that the earth is at least that old.

A second kind of indirect evidence for the overall age of the earth is based on the present-day abundance of the various nuclides of lead that occur in minerals of the earth's crust. Natural lead is a mixture of four stable nuclides: lead-204, lead-206, lead-207, and lead-208. Three of these (206, 207, and 208) are produced by the radioactive decay of uranium and thorium. The fourth (lead-204) is not produced by radioactive decay; all of it present on the earth today originated when the earth was formed. Only a part of the present-day lead-206, -207 and -208 originated at that time; the rest has been slowly added through the course of earth history by radioactive decay of other elements (Figure 2.31).

Researchers have determined the present-day abundance of the four lead isotopes relative to one another and to the uranium and thorium isotopes that produce three of them. If there were some means of determining the original relative abundance of the four lead isotopes at the time the earth was formed, then we could simply calculate the time required for the additional lead-206, -207 and -208 to have been produced by radioactive decay. (The decay rates producing each isotope are constant and are precisely known from laboratory measurements.) Although there is no direct way to estimate the original lead abundances, certain meteorites containing neither uranium nor other radioactive elements that decay into lead are thought to provide a reasonable

FIGURE 2.31

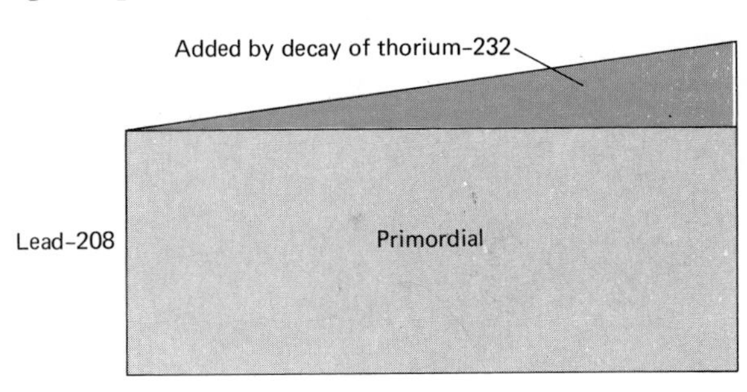

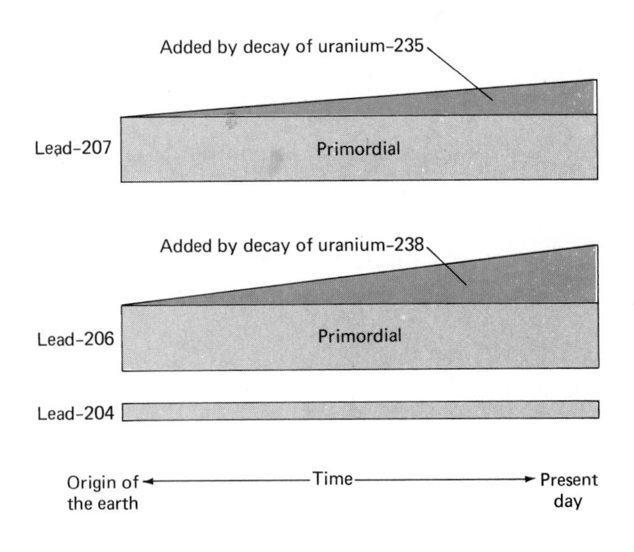

Determination of the age of the earth from its initial and present-day abundances of lead nuclides. The initial abundance is estimated from meteorites, and the present abundance from the average of many rock measurements. The difference represents additions from the decay of radioactive uranium and thorium through earth history.

approximation of the earth's original abundance of lead. Using this information, we can calculate that about 4.6 billion years of radioactive decay would be necessary to produce lead having the average nuclide abundances found on earth today. This estimate further confirms the suggestion that the earth originated at least 4.6 billion years ago.

Age of the Elements

As we saw in the preceding section, the relative amounts of uranium and lead isotopes on the earth today are known accurately. The proportions of lead isotopes that were originally present and that were subsequently produced by the radioactive decay of uranium are also known. Before the solar system formed, the nebular cloud of interstellar dust that gave rise to it contained all the nuclides that were to form the sun and the planets. In this dispersed nebula, uranium isotopes were surely decaying into lead at their unwavering, constant rate, even before they condensed to become part of the sun and planets. From the known quantities of parent uranium-235 and daughter lead-207 that existed when the earth formed, we can determine that, even if all the lead-207 that was around at the time was radiogenic, it would have been produced from the uranium-235 that existed then in about one billion years.

Apparently there was an episode of element production not very long before the formation of the solar system, during which some of the original lead-207 that became part of the earth must have also been formed. Not all of it came from the decay of uranium-235 in the nebula. Hence, the element-producing episode must have predated the formation of the solar system by substantially less than 1 billion years—a comparatively short time on the cosmic scale.

Throughout the universe elements can be produced only at the very high temperatures that are attainable in stars. The same nuclear processes that cause stars to shine can produce the lighter elements up to and including iron (atomic weight 56) by fusion. Heavier elements are produced by adding individual neutrons to the atomic nucleus, thereby building increasingly heavy elements piecemeal from lighter ones. This process of **neutron capture** requires the enormous energy available only when stars explode. Extremely heavy elements such as uranium and thorium can be produced only in the incredibly violent stellar explosions called **supernovas.**

Neutron capture in supernova explosions produces not only the kinds of nuclides that make up the present solar system, but also the short-lived radioactive nuclides that decay quickly to undetectable quantities. If the solar system formed soon after the episode of element production, then some of these short-lived nuclides would have been incorporated into the planets and meteorites. Although these nuclides themselves would no longer exist, their daughter products might be recognizable in meteorites, inasmuch as they are samples of primordial material—stuff that formed at the beginning and has since been essentially undisturbed.

Of several short-lived radioactive nuclides that would have been common for a few tens of millions of years after the element-producing episode, iodine-129 would be the most easily detectable. Iodine-129 decays to xenon-129 by beta decay, with a half-life of 17 million years. If a meteorite contained iodine-129 when it formed, it would be recognizable now as excessive xenon-129. The search for excess xenon-129 began in the mid-1950s; in 1959 excess xenon-129 was detected in the Richardton meteorite, a small stony meteorite that had fallen in North Dakota several years earlier. This meteorite contained proportionately 50 percent more xenon-129 than does xenon from the earth's atmosphere (Figure 2.32). Subsequently, other meteorites have been found with even greater excesses of xenon-129.

FIGURE 2.32

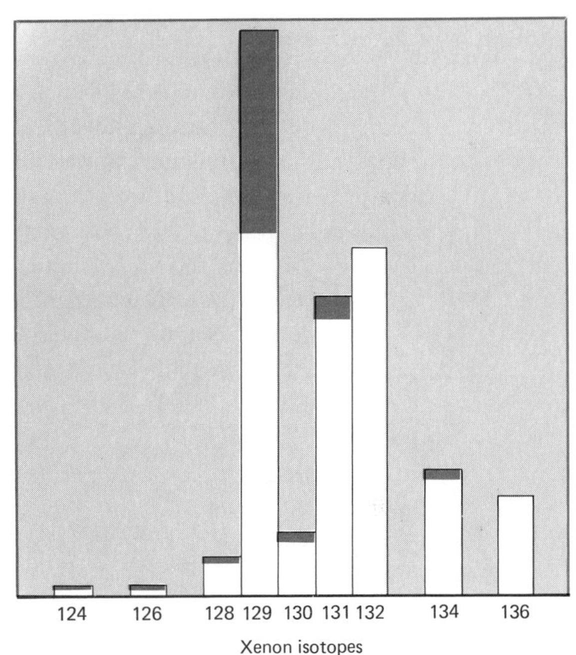

Shaded portion of each isotope is the excess in the Richardton, North Dakota meteorite over that present in the earth. Xenon-129 shows by far the greatest excess; xenon-132 is the standard. (*Reynolds, 1960*)

How do we know it is the meteorite's xenon and not the earth's xenon that is out of proportion? There are nine nuclides of xenon, and it is easy to see how one of them could be supplied in excess to a meteorite from a local radioactive source, but there is no process that could *remove* the xenon-129 from the earth's atmosphere without disturbing the other eight. In addition, many other meteorites, which apparently contained no iodine-129 initially, have a xenon-129 content similar to the earth's. Thus the nuclides of the earth's xenon appear to be representative of the solar system as a whole.

Meteorites with excess xenon-129 also contain stable iodine-127. This would be expected if the excess xenon-129 were initially iodine-129, because the two iodine nuclides would have been formed together and would behave the same chemically. Assuming that the excess xenon-129 began as iodine-129, the iodine-129:iodine-127 ratio when the meteorites formed was around 1:10,000. This figure provides a means of estimating the time that elapsed between the production of the elements and their incorporation into meteorites. Iodine-129 and iodine-127 must have been produced in approximately equal quantities. Following the episode of element production, the 17-million-year life of iodine-129 would reduce the ratio from around 1:1 to 1:10,000 in a little more than 200 million years. Of that time, the collapse of the primordial cloud to form the solar system probably required about 50 million years.

Besides meteorites that contain an excess of xenon-129, there are meteorites that have a simultaneous excess of four other xenon nuclides: xenon-131, -132, -134, and -136. When these were first discovered, geochemists could account for them only by postulating the spontaneous fission of a very heavy nuclide. Indirect evidence favored plutonium-244, which has a half-life of 82 million years and hence no longer exists in nature. Recently, direct evidence that plutonium-244 was the source of the excess xenon-131, -132, -134, and -136 has come from laboratory experiments. Plutonium-244 produced at Oak Ridge National Laboratory was isolated and allowed to decay for two years. At the end of this time, spontaneous fission in the sample had produced xenon-131, -132, -134, and -136 in proportions that agree almost exactly with the excesses of these isotopes in meteorites (Figure 2.33). This proves that plutonium-244 existed when the solar system formed.

FIGURE 2.33

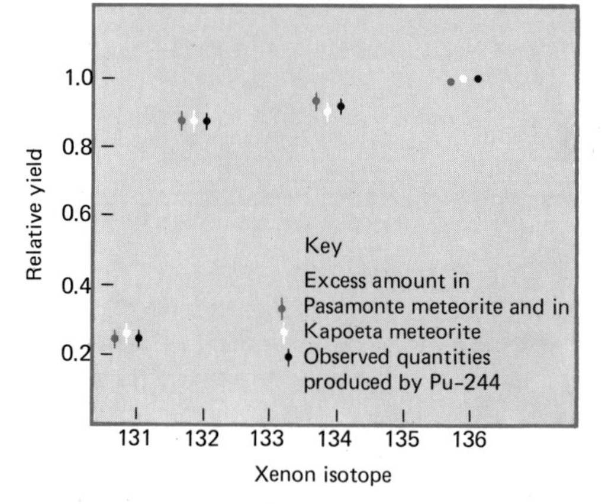

Excess xenon-131, -132, -134, and -136 observed in the representative achondrite meteorites agrees perfectly with the xenon produced by spontaneous fission of plutonium-244. (*Alexander and others, 1971*)

Estimates of the primordial abundance of plutonium-244 and estimates from the abundance of iodine-129 corroborate that the element-producing episode occurred approximately 200 million years before the solar system formed. Plutonium-244 is an extremely heavy nuclide that could only have been produced in a supernova explosion. This would seem to indicate that the cosmic dust cloud that contracted to form the sun and planets had been blasted into interstellar space about 200 million years earlier by a violently exploding star.

Chapter Summary

Organizing the Rock Record

Field study of rocks involves measuring the thicknesses of differing rock units, called formations, and plotting their distribution on maps.

Stratigraphic principles help to establish the age relations of formations.

Original horizontality: sediments are normally deposited in horizontal sheets; any dip must have resulted from postdepositional folding or tilting.

*Superposition:*Younger sedimentary units overlie older ones.

Original lateral continuity: Most formations interrupted by valleys or mountain ranges were originally continuous.

Cross-cutting relationships: Any rock cut by faults or intrusive igneous rocks is older than the fault or intrusion.

Deciphering the Rock Record

Sedimentary environments determine the types and distributions of sedimentary rock. Sediments accumulate on land in terrestrial environments or on the sea floor in marine environments; these and other environmental distinctions can usually be recognized in ancient sediments.

The basis for interpretation: Ancient environments are usually recognized by comparing the rocks and fossils with sediments forming today in similar environments.

Facies are different sediment types that replace one another laterally; they have usually accumulated at the same time but in differing environments. Facies are particularly evident where shallow seas and their associated patterns of environments move landward in transgressions or seaward in regressions.

Areas of erosion: Unconformities are gaps in the sedimentary record caused by erosion, rather than sediment deposition; the erosional surface may represent a small or large time gap, and the length of the gap may vary from place to place.

The geologic time scale is based on the changing fossil life found in sedimentary rocks. Unlike the physical principles considered so far, this scale provides a worldwide rather than a local means of dating discoveries. The time scale is divided into two eons, Precambrian and Phanerozoic; the Phanerozoic Eon is further divided into three eras and twelve periods. Precambrian rocks, which lack abundant fossils, are dated primarily by radiometric techniques

Radiometric Dating

Radioactive decay is the spontaneous transformation of an unstable parent atom into a daughter element through the gain or loss of nuclear particles; each radioactive nuclide decays at a constant rate that can be measured and used to date ancient rocks through measuring their ratios of parent and daughter elements.

Radioactive nuclides in nature: About 20 long-lived natural nuclides are known; the four most useful for dating are potassium-40, rubidium-87, uranium-235, and uranium 238; carbon-14 is a useful short-lived nuclide produced by cosmic rays in the upper atmosphere.

Methods for dating ancient rocks: Potassium-40 occurs in many igneous and metamorphic minerals, as do trace amounts of rubidium-87; the techniques are useful both separately and combined to date all but the very youngest igneous and metamorphic rocks. Uranium is less widely distributed but occurs in small quantities in the mineral zircon, which is found in many granitic rocks.

The age of the earth: Radiometric dating of earth rocks, moon rocks, and meteorites indicates that the earth originated about 4.6 billion years ago.

The age of the elements: Heavy elements are produced in violent stellar explosions called supernovas; parent/daughter nuclide analysis of meteorites suggests that the materials making up the sun and planets formed from a supernova explosion around 4.8 billion years ago.

Important Terms

angular unconformity
atomic number
biofacies
chemical sediment
concordia
cosmic-ray-induced nuclide
cross-cutting relationship
cross beds
daughter nuclide
decay constant
detrital sediment
disconformity
facies
formation
fossil
half-life
isochron
isotope
lithofacies
marine regression
marine transgression
mass number
microfacies
neutron capture
nonconformity
nuclide
original horizontonality
original lateral continuity
paraconformity
parent nuclide
potassium-argon method
primary nuclide
radioactive element
radiometric dating
rubidium-strontium method
section
stratigraphy
supernova
superposition
unconformity
uranium-lead method
whole-rock method

Review Questions

1 How are formations defined and recognized?
2 Summarize the stratigraphic principles that can be used to establish the age relations of ancient rocks.
3 What are the two principal types of sedimentary environments?
4 What significance do unconformities have in interpreting geologic history?
5 What are the principal subdivisions of geologic time?
6 Describe the process of radioactive decay.
7 What are the principal natural radioactive nuclides? How can they be used to date ancient rocks?

Additional Readings

Berry, W. B. N. *Growth of a Prehistoric Time Scale,* W. H. Freeman & Company Publishers, San Francisco, 1968. *A well-written review of the founding of the time scale.*

Dunbar, C. O. and J. Rodgers *Principles of Stratigraphy,* John Wiley, New York, 1957. *An excellent reference book on organizing and deciphering the stratigraphic record.*

Faul, H. *Ages of Rocks, Planets and Stars,* McGraw-Hill, New York, 1966. *A brief but comprehensive introduction to radiometric-dating methods.*

Faure, G. *Principles of Isotope Geology,* John Wiley, New York, 1976. *An intermediate-level text on the principles of radiometric dating and other uses of isotopes in geology.*

Harland, W. B., A. G. Smith, and B. Wilcock (eds.) *The Phanerozoic Time Scale.* Geological Society of London, 1964. *A scholarly review of how the time scale is defined, organized, and dated radiometrically.*

Harper, C. T. (ed.) *Geochronology: Radiometric Dating of Rocks and Minerals,* Dowden, Hutchinson and Ross, Stroudsburg, Pennsylvania, 1973. *A volume of reprints of the major scientific papers marking the development of concepts and techniques in radiometric dating.*

Laporte, L. F. "Paleoenvironments and Paleoecology," *American Scientist,* vol. 65, pp. 720–28, 1977. *Good summary of the history of studies of past environments.*

THE PRECAMBRIAN EARTH

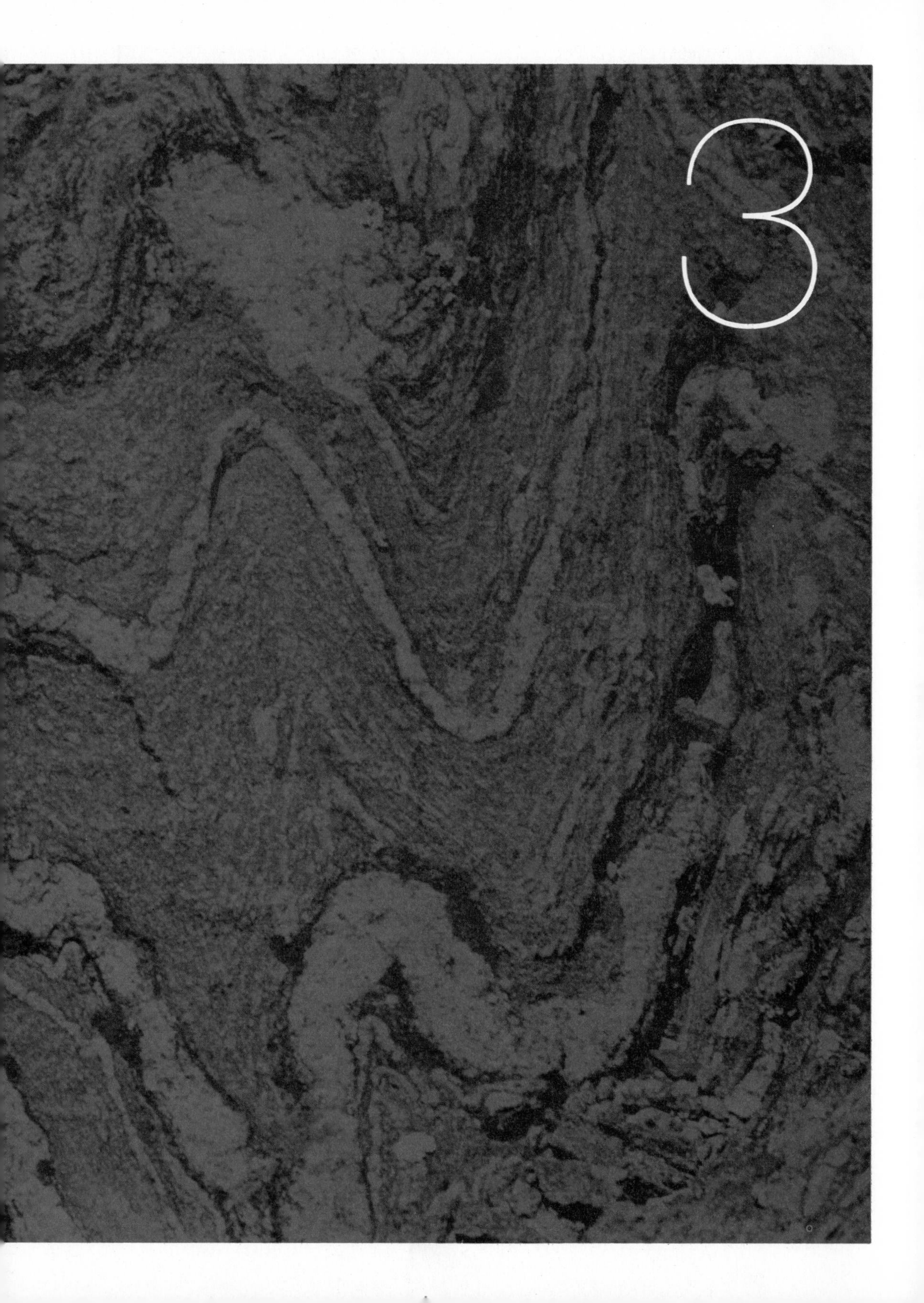
3

The Precambrian phase of earth history begins with the oldest rocks, formed about 4 billion years ago, and ends with the expansion of shell-bearing animal life, about 600 million years ago. Precambrian rocks thus account for 85 percent of the total recorded span of earth history (Figure 3.1).

As the name suggests, *Precambrian* was originally coined as something of an afterthought to refer to the complex igneous and metamorphic rocks usually found below the base of Cambrian strata. In the early days of geology, the span of time represented by these rocks was grossly underrated. They were viewed as nothing more than a foundation or "basement" on which the fossil-bearing records of geologic time—Paleozoic, Mesozoic, and Cenozoic sedimentary rocks—were deposited. This viewpoint first gained acceptance because most Precambrian exposures in western Europe, where the geologic time scale was conceived, consisted of structurally complex crystalline rocks—chiefly granites, schists, and gneisses like those shown in Figure 3.2—that show no clear sequence of events. The viewpoint persisted into the twentieth century because of the low estimates of the overall age of the earth that were fashionable in the years prior to the extensive use of radiometric dating. The better-

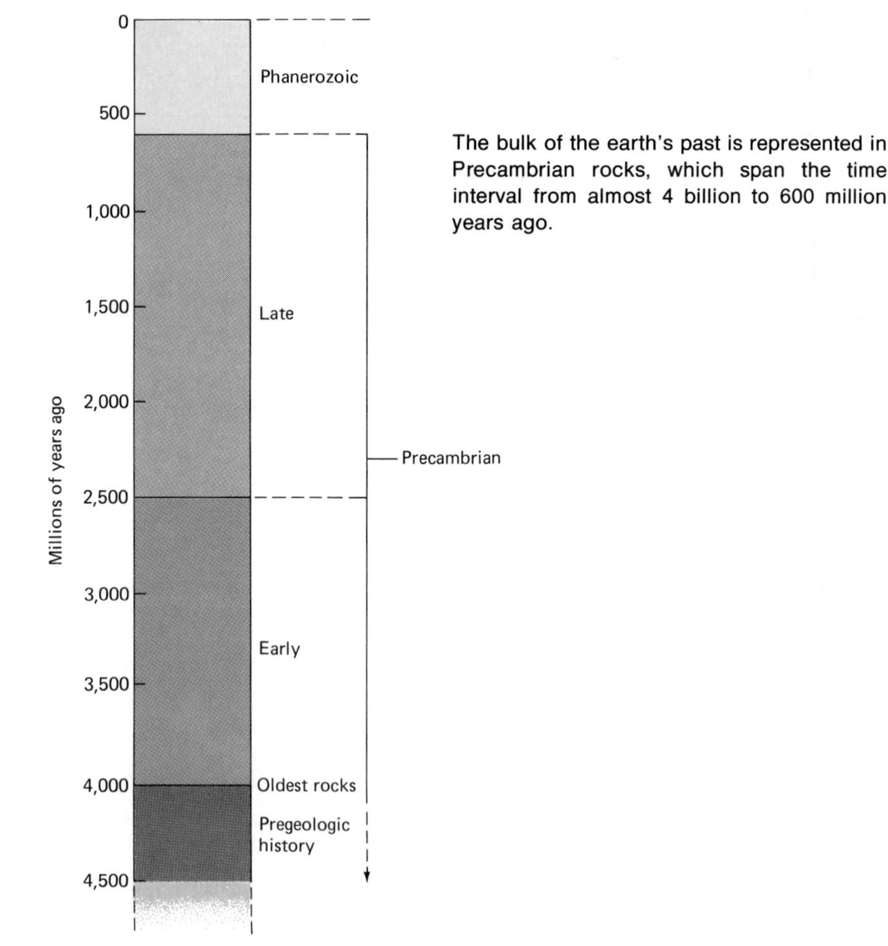

FIGURE 3.1

The bulk of the earth's past is represented in Precambrian rocks, which span the time interval from almost 4 billion to 600 million years ago.

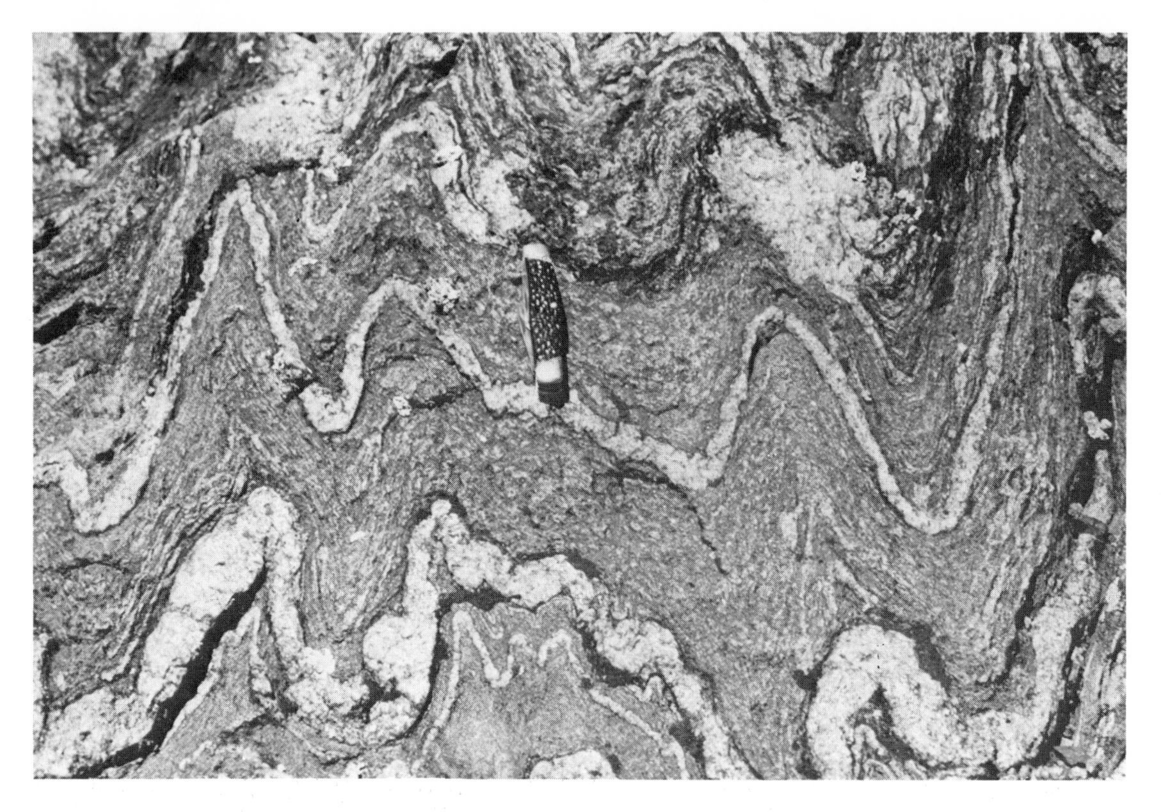

Precambrian metamorphic rocks that crystallized at great depth, now exposed in the Colorado Front Range. (*Courtesy of W. A. Braddock*)

FIGURE 3.2

preserved Phanerozoic record required most of the total time span as then conceived, thus allowing little time for the Precambrian.

In recent years, radiometric-dating techniques have led to an awakening interest in, and knowledge of, the long Precambrian phase of earth history. In addition, a wide variety of remarkably preserved microscopic organisms has recently been discovered in ancient Precambrian cherts. Although these microfossils have very little potential use in time-stratigraphic correlation, they attest to an extremely long period of biological development prior to the Paleozoic Era. Far from being a dormant time during which the organic and inorganic worlds were gathering strength for the forthcoming Phanerozoic "main event," the Precambrian now appears as a time when physical, chemical, and biological systems were involved in a dynamic interplay whose outcome was to shape the much more visible Phanerozoic world that followed.

Precambrian Rocks and Geography

All continents contain large regions that have undergone little mountain-building deformation since the close of Precambrian time, 600 million years ago. These are called stable areas, or **cratons.** Phanerozoic sedimentary rocks that occur on these cratons are typically only a few hundred meters thick and

FIGURE 3.3

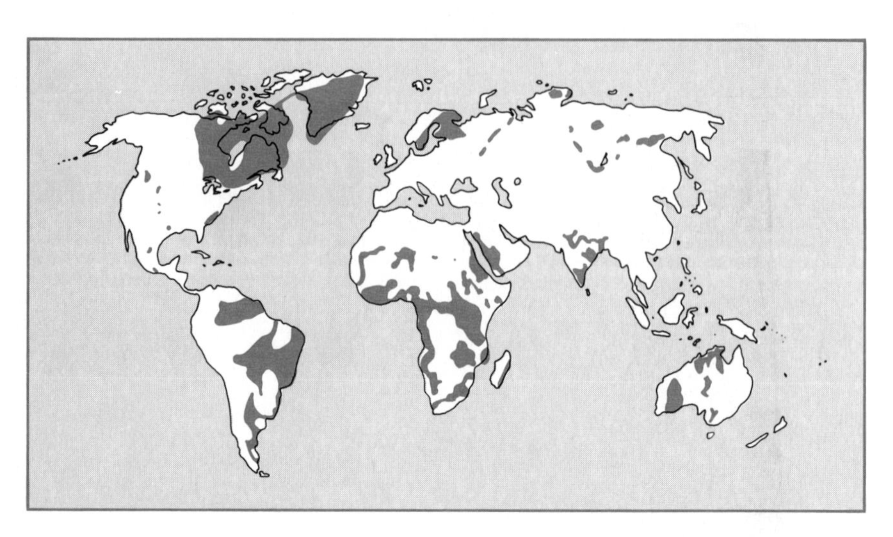

World distribution of Precambrian rocks in modern continental cratons.

are flat-lying or only very gently tilted and folded. Each continental craton includes a large area where Precambrian rocks—chiefly metamorphic and granitic—are exposed without a cover of younger strata. These are termed **Precambrian shields.** Such shields constitute about 20 percent of the earth's land surface (Figure 3.3).

The portions of continents that are not part of the undeformed craton are called **mobile belts.** Mobile belts are elongate regions in which thick sequences of sedimentary and volcanic rocks, chiefly of Phanerozoic age, have been folded, metamorphosed, and intruded by plutonic rocks. The distinction between cratons and mobile belts is sharpened by the general coincidence of cratons with areas of low relief and of mobile belts with mountains. Precambrian shield areas are today part of the continents' stable cratons, yet detailed geologic mapping reveals linear patterns of deformed rocks within them that almost certainly represent the roots of ancient mountains—mountains that have long since been worn away. Thus, present-day shield areas, although stable since Precambrian time, were themselves once the sites of ancient mobile belts.

In addition to shield areas, smaller exposures of Precambrian basement rocks occur in some younger mountain ranges, such as the Appalachians and Rockies, and in deeply dissected cratonic plateaus, such as the Colorado Plateau in Arizona. Most Precambrian exposures consist of igneous and metamorphic basement rocks, but some consist of sedimentary rocks that have been deformed or metamorphosed only slightly or not at all (Figure 3.4). These rocks are of special interest because they constitute a record of Precambrian surficial environments. Many of the sedimentary sequences are tens of thousands of meters thick and represent hundreds of millions of years of Precambrian time.

(*a*)

FIGURE 3.4

(*a*) Precambrian strata (Unkar Group) in Grand Canyon, Arizona extend above the inner gorge to include the prominent scarp in the middle ground. (*b*) Strata of the Precambrian Uinta Mountain Group in northern Utah are about the same age as the Grand Canyon strata shown in (*a*), approximately 1 billion years.

(*b*)

FIGURE 3.5

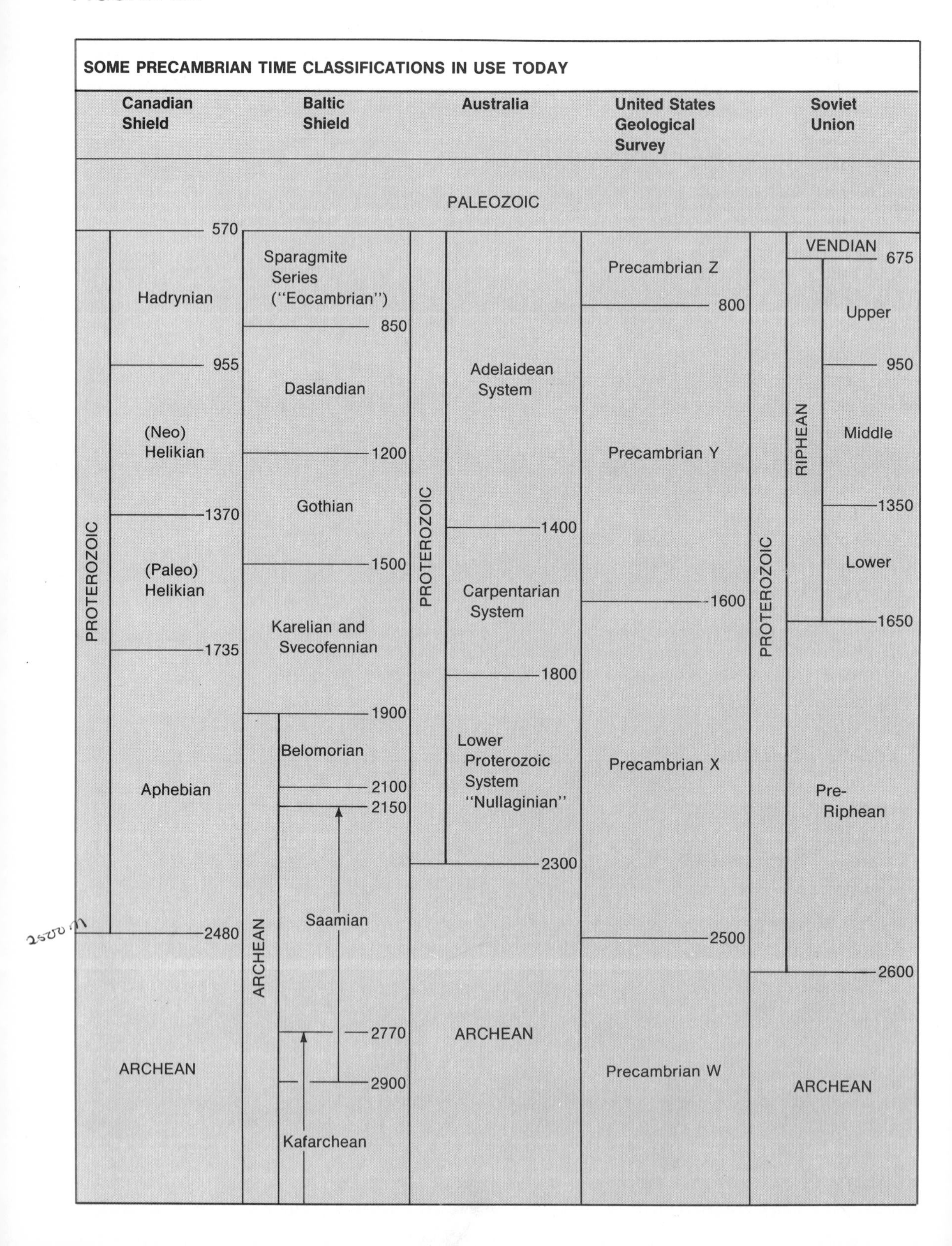
SOME PRECAMBRIAN TIME CLASSIFICATIONS IN USE TODAY
Canadian Shield
Baltic Shield
Australia
United States Geological Survey
Soviet Union
PALEOZOIC
570
Hadrynian
955
(Neo) Helikian
1370
(Paleo) Helikian
1735
Aphebian
2480
PROTEROZOIC
ARCHEAN
Sparagmite Series ("Eocambrian")
850
Daslandian
1200
Gothian
1500
Karelian and Svecofennian
1900
Belomorian
2100
2150
Saamian
2770
2900
Kafarchean
ARCHEAN
Adelaidean System
1400
Carpentarian System
1800
Lower Proterozoic System "Nullaginian"
2300
PROTEROZOIC
ARCHEAN
Precambrian Z
800
Precambrian Y
1600
Precambrian X
2500
Precambrian W
VENDIAN
675
Upper
950
Middle
1350
Lower
1650
RIPHEAN
Pre-Riphean
2600
PROTEROZOIC
ARCHEAN

Precambrian Chronology

Because Precambrian sedimentary sequences lack fossils that are useful for correlation, it has not yet been possible to fit them into a common time classification scheme that would be applicable worldwide; instead, Precambrian rocks of each region are still classified according to local criteria. Representative regional classifications are shown in Figure 3.5. The only feature they have in common is that, with the exception of the "alphabet soup" scheme of the United States Geological Survey, they are divided into an **Archean Eon** below and a younger eon above. In Canada, Australia, and the Soviet Union, the more recent eon is termed the Proterozoic. But even the top of the Archean is not dated identically from one shield region to the next, and the subdivisions of post-Archean time vary. This makes communication difficult among Precambrian geologists working in different parts of the world. Moreover, it exemplifies our lack of knowledge about this enormous body of rocks that represents 85 percent of recorded earth history, makes up 50 percent of the earth's sedimentary rock volume, and is exposed on 20 percent of the land surface. The knowledge we have recently acquired of Precambrian chronology has come largely from radiometric-age determinations of igneous and metamorphic rocks.

One of the most significant discoveries through use of radiometric dating is that each Precambrian shield region can be divided into subregions based on rock ages. In the Canadian Shield, for example, seven provinces of differing age can be recognized (Figure 3.6). Within each, the style of deformation and tectonic-fold direction is distinct from adjacent areas, and hence they are usually called **structural provinces.** Where they join, these structural provinces are separated by abrupt metamorphic or fault boundaries, termed **orogenic fronts.** Although structural provinces can be recognized from their mode of deformation, their identity is most pronounced in radiometric ages. The igneous and metamorphic rock ages within each province indicate the date of the most recent episode of deformation, at which time widespread igneous and metamorphic activity melted or recrystallized the preexisting minerals and thus "reset" their radioactive "clocks" (Figure 3.6). In some provinces, older radiometric dates survive locally and illustrate that younger provinces have simply been superimposed on older ones.

On the Canadian Shield, the radiometric dates cluster around 2,480, 1,735, 1,370, and 955 million years ago, indicating four great orogenic episodes. The oldest **orogeny,** the Kenoran, was the last period of folding, metamorphism, and plutonic intrusion that affected the Superior and Slave Provinces. The Hudsonian was the last to affect the Churchill, Southern, and Bear Provinces. The Elsonian was the last to affect the Nain Province, and the youngest orogeny, the Grenville, was the last to affect the Grenville Province. The ages of these orogenies help determine the boundaries for the subdivisions of the Precambrian in the modern Canadian Shield classification shown in Figure 3.5.

The provinces have also been recognized well beyond the Canadian Shield in areas where most Precambrian rocks are hidden by a cover of younger strata.

FIGURE 3.6

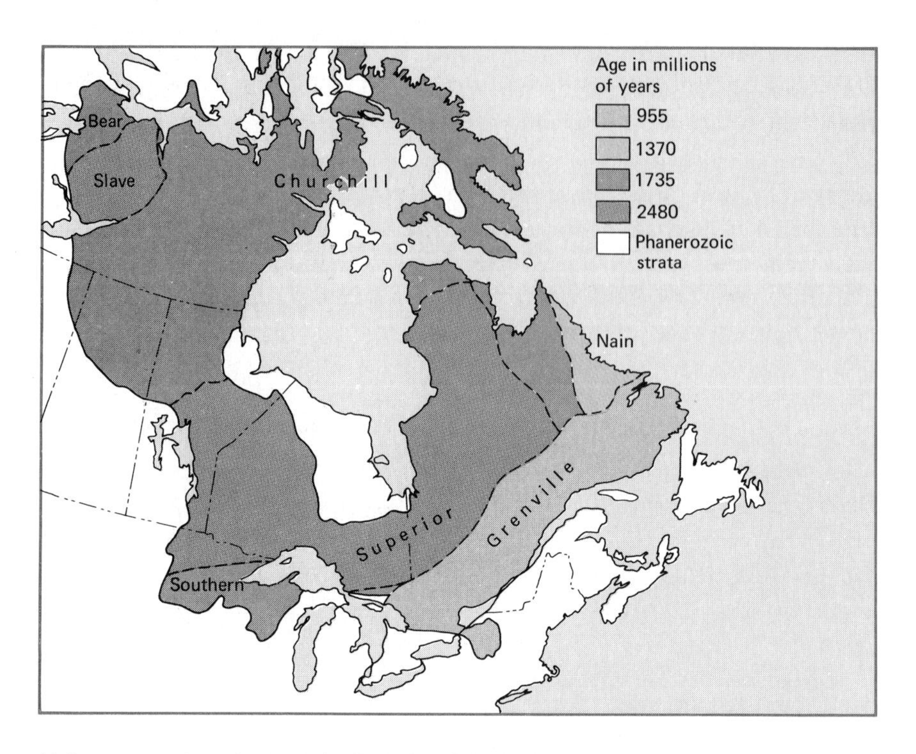

Major structural provinces of the Canadian Shield and their radiometric ages. (*Stockwell, 1970*)

Radiometric dates of samples taken from scattered exposures and from deep wells have enabled geologists to trace the shield provinces and to map them throughout much of western Canada, the United States, and Greenland.

Radiometric dating has also contributed to our understanding of the thick sequences of sedimentary rocks of Precambrian age. Although these are difficult to date directly, their ages can often be ascertained from interbedded volcanics, from igneous rocks that they overlie, and from intrusives that cut them. Late Precambrian sedimentary rocks, like Phanerozoic sedimentary rocks, exhibit a wide variety of rock types and sedimentary structures. This variety reflects a host of sedimentary environments—marine, terrestrial, and transitional.

As we explore farther and farther back into the Precambrian sedimentary record, we find that, with increasing age, limestone and dolomite become less common, and that marine evaporites, which are rare even in the late Precambrian, disappear altogether. Bedded cherts, on the other hand, become increasingly abundant. In rock sequences older than about 2 billion years, they commonly contain thin red or black layers that are rich in iron. These iron-rich cherty rocks that occur only in the older Precambrian sedimentary record are called **banded iron formations.**

Looking still further back in time, we find that sedimentary rocks older than about 2.5 billion years are monotonously similar wherever they are found.

These Archean sedimentary rocks document considerable mountain-building activity and volcanism, and they appear to have formed in oceanic environments at a time before stable cratons first developed on the earth. The record of Archean time is dim, but it appears to reflect the truly formative stages of the earth's crust. In that respect, it differs fundamentally from later records of geologic history.

The discussion that follows will consider Precambrian geography and environments, beginning with Archean rocks and concluding with younger Precambrian strata.

Greenstone Belts

The best preserved Archean sedimentary rocks occur in thick sequences that also contain mildly metamorphosed volcanic rocks called greenstones. In each of the Precambrian shield areas, these volcanic and sedimentary assemblages occur in great elongated downwarps which are referred to as **greenstone belts.** Greenstone belts are commonly several tens of kilometers long; they lie between and surround much more extensive, domelike masses of Archean granite. On the Canadian Shield, greenstone belts are concentrated in large tracts several hundred kilometers long and several tens of kilometers across (Figure 3.7). Between the broad greenstone tracts lie vast regions of granite and high-grade metamorphic rocks.

FIGURE 3.7

Archean greenstone belts (in black) of the Canadian shield. (*Baragar and McGlynn, 1976*)

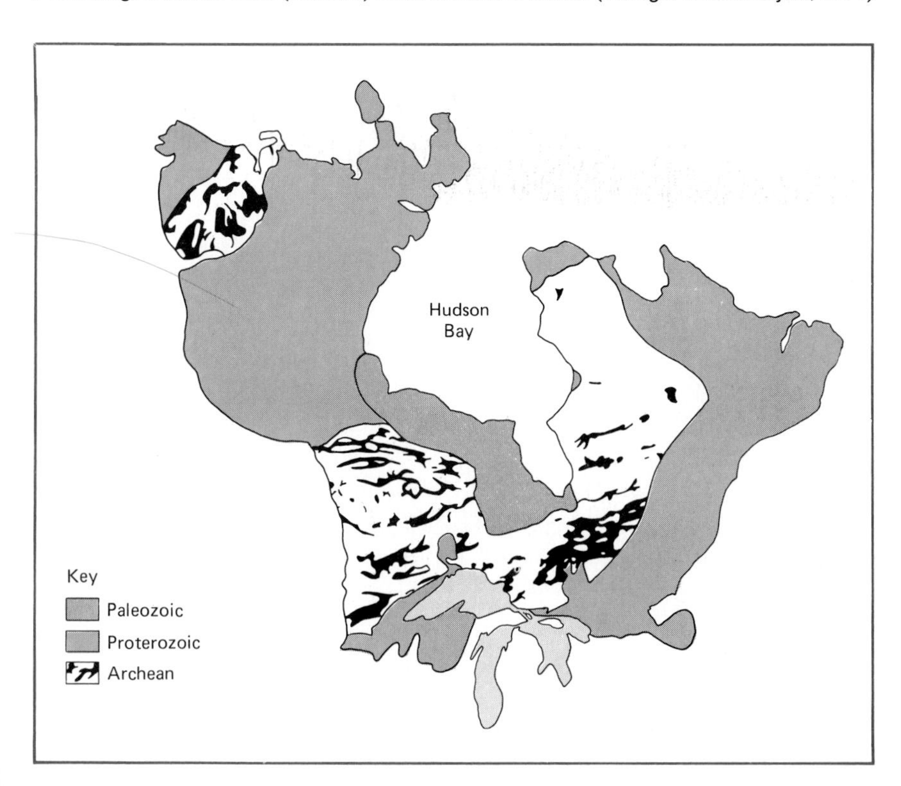

The volcanic rocks that make up the bulk of most greenstone belt sequences include abundant **pillow lavas,** which indicate underwater extrusion (Figure 3.8). They also include **volcaniclastic rocks**—breccias, sandstones, and tuffs—made up of volcanic debris. Sedimentary rocks do not occur in the lowermost portions of greenstone-belt sequences, but they become increasingly common farther upward, and they dominate in the uppermost portion.

By far the most common sedimentary rocks in the greenstone sequences are **greywackes** (poorly sorted sandstones), which are typically interbedded with slates. Individual greywacke beds are thin, rarely more than one meter in thickness. They are commonly graded; that is, grain size in each bed becomes finer from bottom to top. **Graded bedding** probably results from the sudden deposition of individual beds by a **turbidity current,** a mass of water made dense by suspended sediment that moves downslope under gravity. Turbidity currents deposit their coarsest sediment first, and then progressively finer sediment as they slow down and finally stop (Figure 3.9). Beds deposited by this mechanism are called **turbidites.** Turbidites occur throughout the geologic record, but in the Archean they are dominant in all sedimentary sequences. The graded, poorly sorted texture of turbidite beds can be preserved only if they are not reworked on the sea floor by currents (or by organisms, for those

FIGURE 3.8 Basaltic pillow lavas of Cenozoic age overlain by white limestone near Oamaru, New Zealand. Similar pillow lavas are common in Precambrian greenstone belts. (*Courtesy of Wesley LeMasurier*)

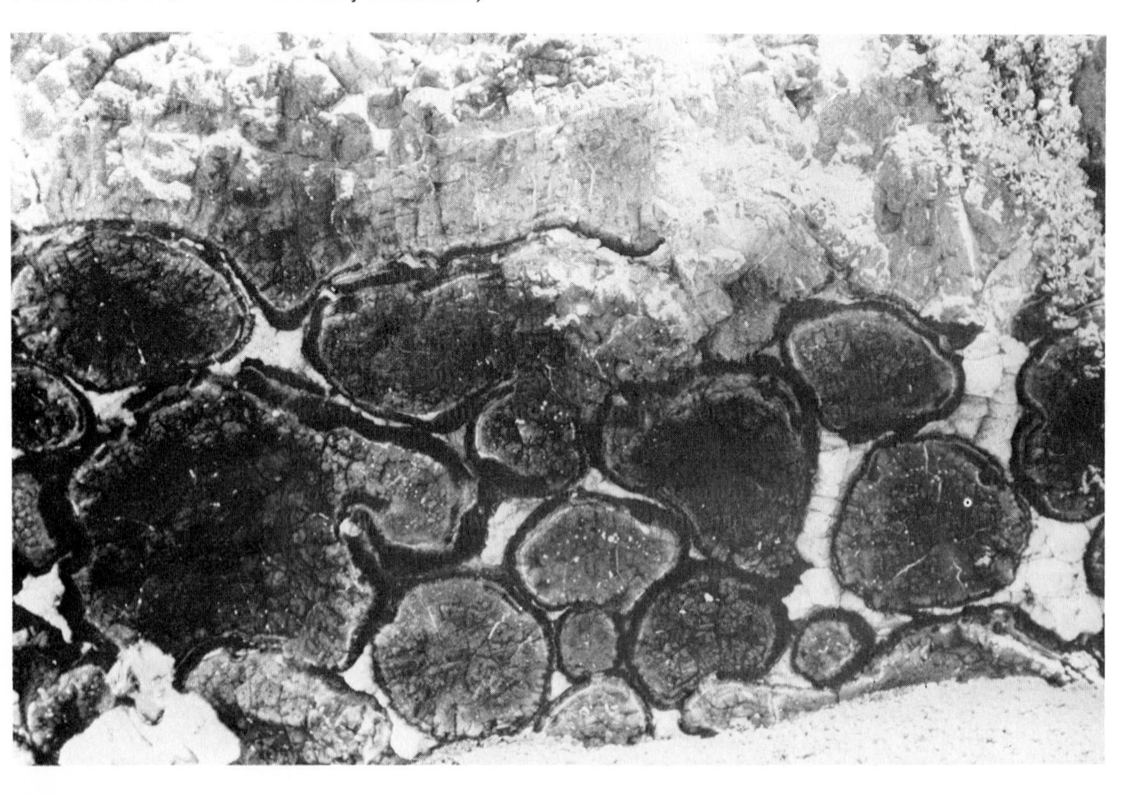

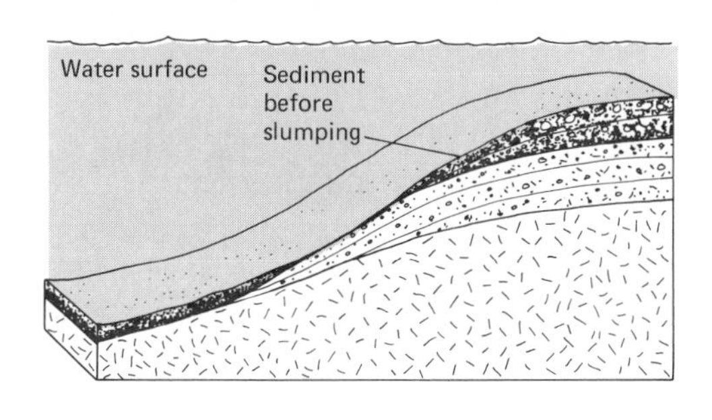

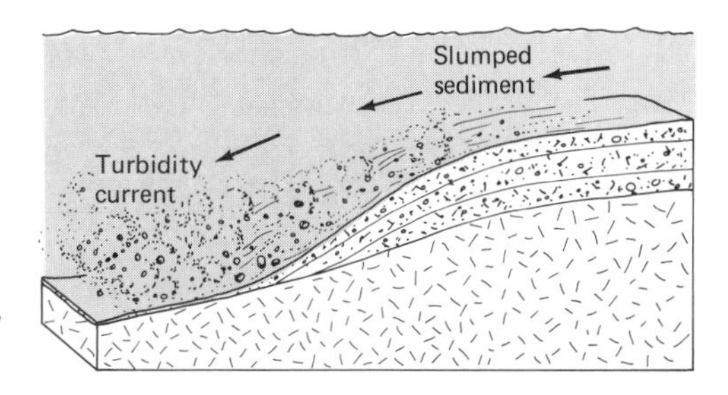

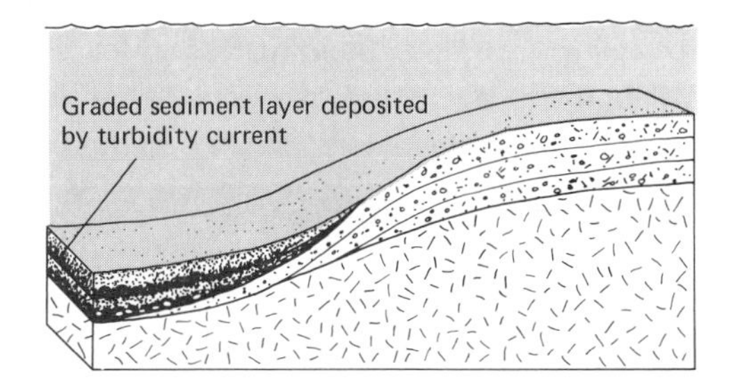

FIGURE 3.9

Origin of graded bedding from sediment slumping and turbidity current flow.

deposited in the Phanerozoic). Most turbidites are believed to form in relatively deep water. They are common today on continental slopes and are primarily responsible for the construction of continental rises.

In addition to greywacke and slate, Archean greenstone belts also contain units of coarse conglomerate up to several hundred meters thick that may extend over several kilometers. Because they are typically interbedded with turbidites, they too are believed to be the product of turbidity-current transport. The only other sedimentary rock types common in the Archean are the chert- and iron-rich banded iron formations. Banded iron formations are not well understood, but they apparently reflect conditions that prevailed only during the Archean and early Proterozoic, for they are virtually unknown in younger rocks. They will be discussed further in the next section.

Because greenstone belts are strongly faulted and folded, the exact sequence in which the rocks formed is difficult to work out. Careful study of greenstone belts in many areas of the world, however, has shown them to be remarkably similar. First, all greenstone-belt sequences are thick, ranging typically from 6,000 to 20,000 meters, and in some cases reaching 30,000 meters. Second, their sedimentary rocks are similar, everywhere consisting dominantly of turbidites. Finally, the sequence in which the various kinds of igneous rocks occur is also similar.

FIGURE 3.10

MAJOR TYPES OF IGNEOUS ROCKS

Mafic mineral content (Percent of rock)	
0	
	Felsic rocks (example: granite)
20	
	Intermediate rocks (example: andesite)
45	
	Mafic rocks (example: basalt)
75	
	Ultramafic rocks (example: peridotite)
100	

Figure 3.10 shows a general scheme for classifying igneous rocks based on the relative proportions of mafic and felsic minerals. Mafic minerals are dark minerals that are rich in iron and magnesium, such as olivine, pyroxene, and hornblende; felsic minerals are light-colored minerals that contain little or no iron and magnesium, such as feldspar and quartz. The order in which the major igneous rock types are listed in Figure 3.10, from ultramafic at the bottom to felsic at the top, is the same order in which they occur in greenstone-belt sequences. The basal part of greenstone belts consists of **ultramafic igneous rocks,** such as peridotite, which occur as lava flows and sills. These are succeeded by **mafic rocks,** chiefly basalts, which also occur as flows and sills. These are overlain by andesite and other intermediate igneous rocks, which occur both as flows and in volcaniclastics. These, in turn, are overlain by **felsic igneous rocks,** which occur chiefly in volcaniclastics. This succession of rock types is a gradational one, but the general sequence characterizes greenstone belts everywhere, strongly suggesting progressive evolution of molten rock at depth.

We do not know the composition of the primitive crust on which the greenstone belts were deposited. Some geologists believe that the ultramafic and mafic rocks at the base of greenstone belts represent surviving fragments of a thin, formerly widespread crustal layer that had the composition of sima (mafic igneous rocks), much like modern oceanic crust. Others believe that, before the oldest greenstone belts formed, the earth had already developed a thin crust of sial (felsic igneous rocks), much like modern continental crust, and that the greenstone belts were produced on it by localized volcanism and orogeny (Figure 3.11). Indeed, the oldest rock we know is a 4-billion-year-old granitic gneiss, which indicates that at least local differentiation of sialic material must have occurred very early. The nature of the rocks of greenstone belts suggests that, whatever the composition of the underlying crust, it was relatively thin and unstable.

The ultramafic igneous flows that occur in the basal portion of greenstone belts are extremely unusual rocks. Their origin is difficult to explain because they could have formed only by the nearly total melting of mantle rock. Ultramafic intrusive rocks are believed to be a common constituent of the lower part of modern oceanic crust; they also occur commonly in narrow belts in many Phanerozoic fold mountain chains. Only in the Archean greenstone belts, however, were ultramafics ever extruded as molten material onto the earth's surface.

Ultramafic lavas cannot be generated under ordinary circumstances, because when ultramafic mantle rocks begin to melt, they immediately form a magma of mafic composition. Crystals of very heavy magnesium- and iron-rich minerals such as olivine, which are the last to liquefy, quickly settle out when about 30 percent of the rock has melted. Ultramafic magma could only have formed at the point at which 60 percent of the rock had melted, which

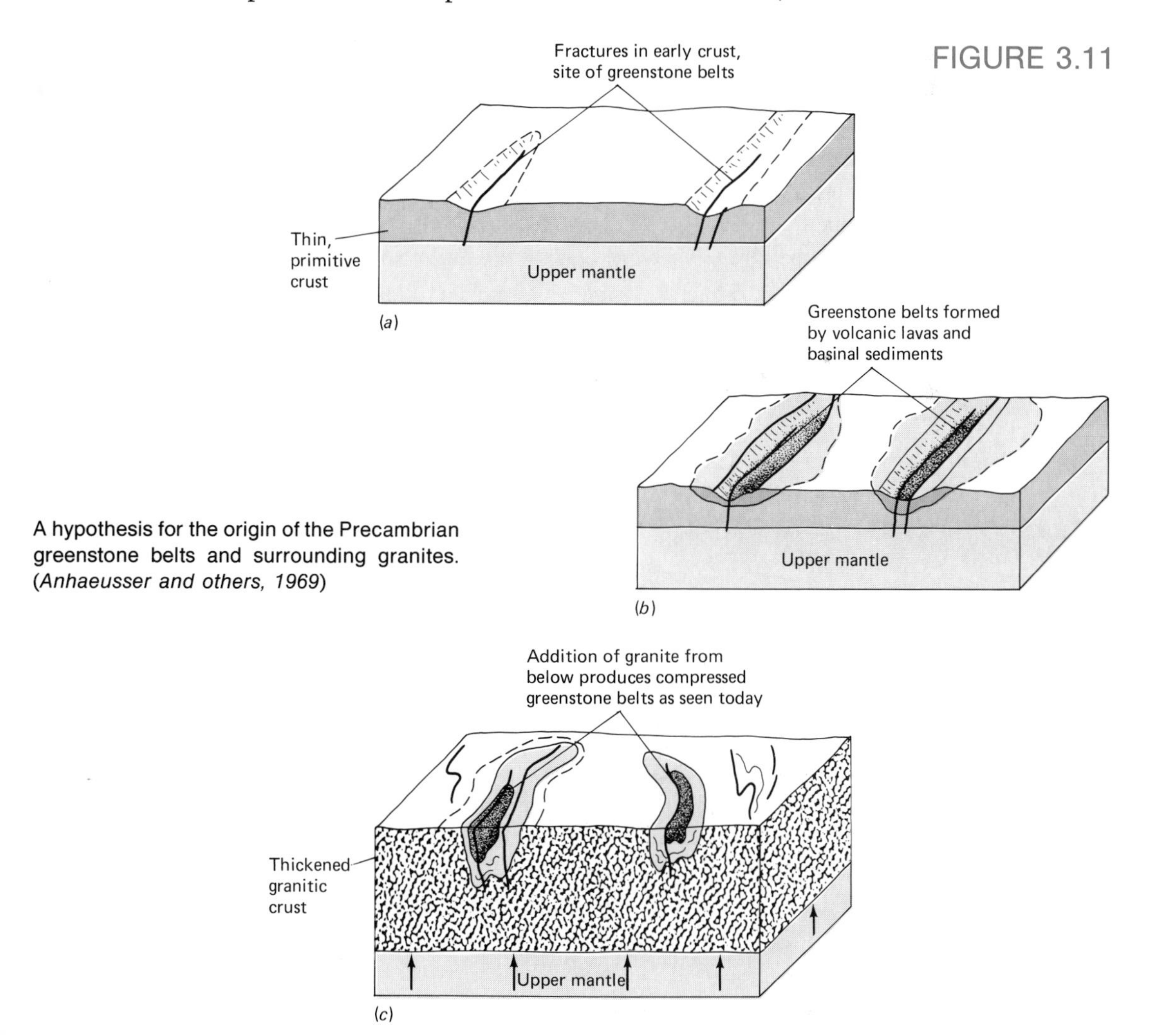

FIGURE 3.11

A hypothesis for the origin of the Precambrian greenstone belts and surrounding granites. (*Anhaeusser and others, 1969*)

would occur at a temperature of about 1600°C. (Basalt from modern volcanoes, by comparison, comes out at about 1100°C.) The only way magma of ultramafic composition could be produced and transported to the earth's surface would be to achieve 60 percent melting of the mantle rocks very suddenly, and to transport the melt upward quickly as a sort of slush before gravitational settling could occur. The problem, therefore, is to account for this sudden melting.

One possibility is that the sudden melting was triggered by large meteorite impacts. Most large meteorite craters on the moon date from around 4 billion to 3.8 billion years ago, at which time the earth was probably bombarded at least as intensively. Those impacts surely generated a great deal of heat at the earth's surface, but the sudden melting of the underlying mantle probably resulted, not from heating, but from unloading. Whether the ultramafic mantle material is solid or liquid depends on confining pressure as well as temperature. In some cases, melting can be achieved by simply reducing the pressure, instead of increasing the temperature. A powerful meteorite impact that blasted away crustal material and made a deep crater would cause a sudden decrease in pressure in the underlying mantle. If the mantle rocks were hot enough, this could have caused immediate partial melting.

Figure 3.12*a* shows the liquid-solid boundary in ultramafic rock as a function of temperature and pressure. It also shows the postulated temperature and pressure environment of the early Precambrian mantle as a gentle curve

FIGURE 3.12

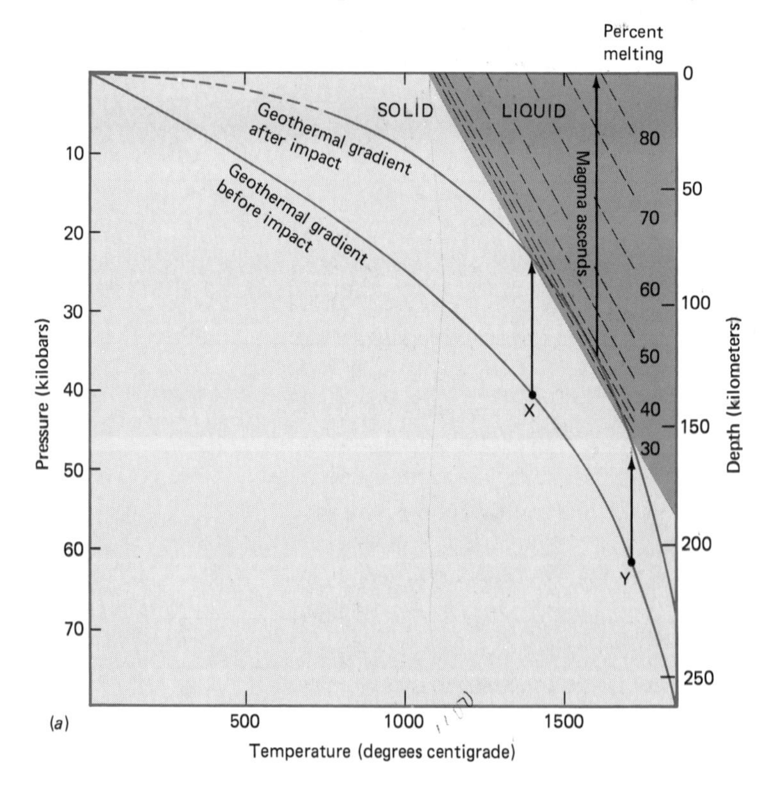

(*a*) Liquid and solid phases of ultramafic mantle. Rocks between points X and Y begin to melt solely by pressure decrease, continue melting during ascent, and extrude at the surface. (*b, c*) A meteorite explosion causes unloading of the crust. (*Green, 1972*)

FIGURE 3.12 *(cont.)*

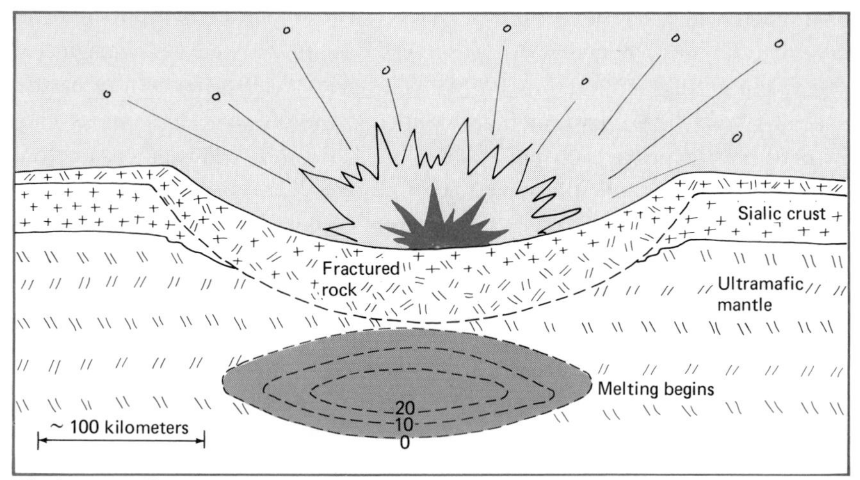

(*b*) Impact and impact-triggered melting

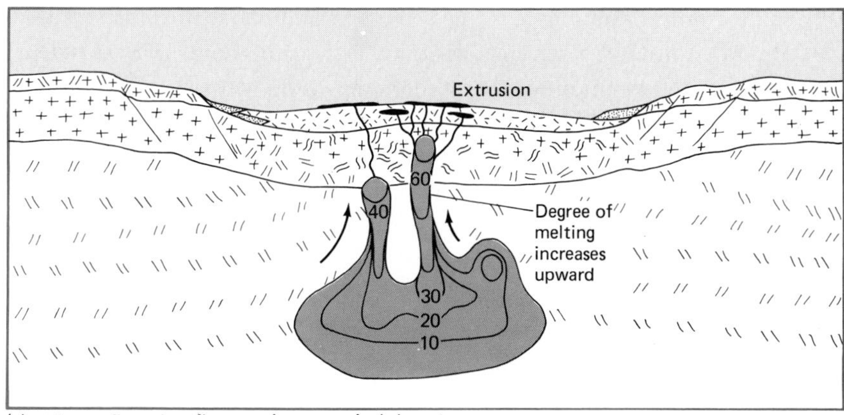

(*c*) Ultramafic and mafic extrusions, marginal slumping

(the geotherm). The mantle rocks between points X and Y on the curve could be melted by an increase in temperature (which would displace the curve to the right on the diagram) or by a decrease in the confining pressure (which would displace the curve upward). The newly melted material would ascend quickly, and as it ascended, it would encounter still lower pressure, melt even more, and become more mobile. This process would continue until perhaps 80 percent melting was achieved, by the time the ultramafic material was extruded at the surface (Figures 3.12*b* and *c*). The meteorite crater and its ultramafic floor would then persist as a basin of accumulation of basaltic volcanics and sedimentary rocks, the site of an evolving greenstone belt.

The chief problem with this idea is in the timing. After 3.8 billion years ago, meteorite bombardment of the moon (and presumably of the earth) declined greatly. However, some greenstone-belt ultramafic flows appear to be as young as 3 billion years. However, dates of the ultramafic flows are still only tentative, and if they eventually prove to be older than they now appear, meteorite cratering may be the most plausible mechanism for their production.

Archean Granitic Rocks

In contrast with the thin crust on which the greenstone-belt rocks were deposited, seismic evidence indicates that the granites that now surround them make up the thickest and most stable parts of the continental crust. During deposition of the greenstone-belt sequences and shortly thereafter, these granites were emplaced as huge, domelike blobs that welled upward from depth. This granitic magma surrounded the greenstone belts, which, being more dense, subsided among the granitic bodies and acquired synclinal configurations (see Figure 3.11). In this way, the earliest stable cratonic crust gradually accreted from below to form a granitic layer more than 40 kilometers in thickness that has been little disturbed since. What was the origin of this enormous volume of granite?

Today granitic magmas are believed to form chiefly beneath volcanic arcs, such as those which rim the Pacific Ocean basin. (We shall see in Chapter 6 that island arcs and their bordering oceanic trenches are considered to be zones where two of the earth's major lithospheric plates converge, so that one passes below the other and descends into the underlying mantle.) Modern island arc-oceanic trench systems are built chiefly on oceanic crust, and in this respect the general environment of granitic intrusion is similar to that in the basal part of greenstone belts.

In other respects, however, those geologic processes which produced modern island arc-trench systems differ fundamentally from those which produced Archean granitic-greenstone belt assemblages. In island-arc settings, granites are intruded at depths of several kilometers, and the accompanying metamorphism in the surrounding rocks typically extends outward several kilometers. The Archean granites, however, metamorphosed the greenstone belt volcanics and sedimentary rocks outward for only a kilometer or so—not because the Archean granitic magmas were cooler than younger magmas (thermal metamorphism is equally intense in immediately adjacent rocks), but because the Archean magmas intruded to extremely shallow depths, so that the heat dissipated rapidly. Finally, the granites beneath volcanic arcs today probably come largely from the melting of preexisting crustal material that was carried to great depth by descending plates. The Archean granites, on the other hand, appear to represent a formative period during which much of the earth's sialic crust was differentiated for the first time from the underlying mantle.

Archean granitic material was probably derived from the upper mantle by **partial melting.** Although the primitive mantle consisted largely of ultramafic minerals, it must have contained a proportionately minor, but volumetrically huge, quantity of felsic minerals (chiefly quartz and feldspar). Felsic minerals not only have lower densities than mafic minerals, but they also melt at lower temperatures. Consequently, partial melting provided a ready mechanism for differentiation. As mantle rock melted, felsic minerals liquefied first and mafic minerals later; consequently, the composition of the first melt was granitic. Once produced, the low-density granitic liquid could have readily ascended into the thin primitive crust overhead to form the huge masses of Archean granite, thereby producing the crustal thickening depicted in Figure 3.11.

Lunar History

By 3.1 billion years ago, while Archean greenstone belts were still being produced on earth, virtually all the rocks on the surface of the moon had already been formed. Since that time, the moon's surface has been subjected only to very gradual pulverization due to impacts of small meteorites. Inasmuch as few surface rocks on the earth are as old as 3.1 billion years, the moon's older record provides a valuable supplement to the earth's—and a revealing insight into the early history of the solar system.

Major features of moon's surface are its smooth, dark **maria** and rugged, light highlands. The maria were produced by the outpouring of huge lakes of basaltic lava. The highlands are composed largely of anorthosite, an igneous rock made up chiefly of calcic feldspar, which appears to have formed by early differentiation of the moon's interior. Both the maria and the highlands contain meteorite-impact craters, but the craters are much larger and more abundant on the highlands. The maria basalts have flowed over and around highland craters and are, therefore, younger. From this and similar observations, a fourfold sequence of major events in lunar history has emerged: First, the highland rocks differentiated and solidified. Then, they were intensively bombarded by meteorites, some of which were very large. Next, the maria basalts erupted and filled some of the largest craters. Subsequently, both the maria and highlands were bombarded still further by meteorites, producing the small craters and fine material and breccias that now cover much of the lunar surface. These events are summarized in Figure 3.13.

Radiometric dates of lunar rocks collected by the Apollo missions indicate that the highlands formed around 4.1 billion years ago, and that large meteorite impacts were frequent until about 3.8 billion years ago. Most maria basalts extruded between 3.7 and 3.3 billion years ago, and the last significant eruption occurred about 3.1 billion years ago. Occasional meteorites have continued to fall up to the present day, as they have on earth. Figure 3.14 relates the major events in lunar history to the earth's geologic time scale.

FIGURE 3.13

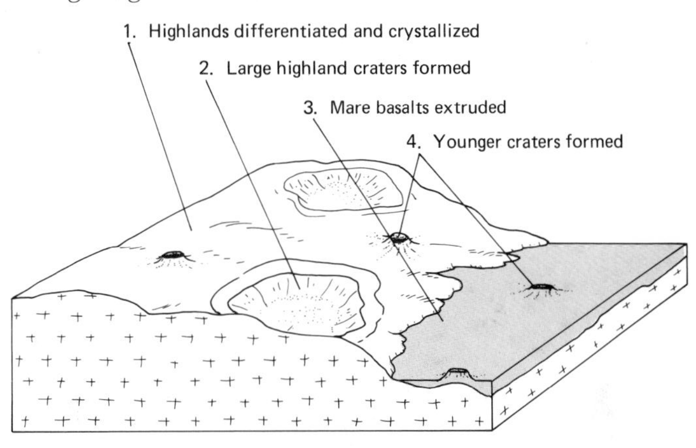

The sequence of major geologic events on the moon.

FIGURE 3.14 **PRINCIPAL EVENTS IN LUNAR HISTORY, RELATIVE TO THOSE ON EARTH**

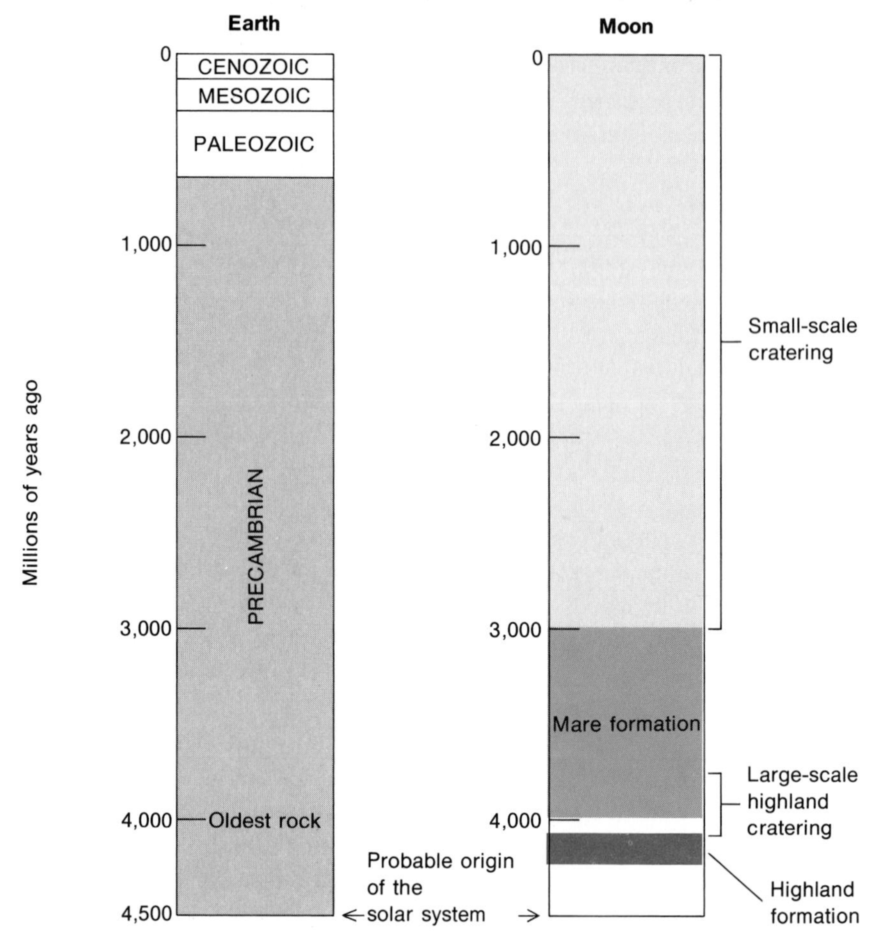

Anorthosites, which apparently dominate the lunar highlands, are fairly rare rocks on earth, but several anorthosite occurrences on earth have been carefully studied, and their origin is well understood. They form when basaltic magmas cool very slowly beneath the earth's surface. During this slow cooling, the first crystals to form are heavy magnesium- and iron-rich silicates, which tend to sink to the bottom of the magma body. The remaining calcium- and silicon-rich fluid crystallizes later as an upper layer of feldspar-rich anorthosite. This process is called **fractional crystallization.** It occurs on earth in certain deep-seated magma chambers, some of which have been exposed at the surface by later erosion.

If the lunar highlands consist largely of anorthosite, as now appears probable, then wholesale melting and differentiation of a thick outer layer of lunar mafic material must have occurred prior to 4.1 billion years ago. This large-scale melting and the later melting and extrusion of the maria basalts indicate that the outer part of the moon was very hot early in its history. Some of the heat may have been produced by intense tidal friction, and some may have been generated by the impacts of large meteorites. Most, however, was probably produced by radioactivity.

The First Continents

The same kinds of energy sources that heated the early moon must also have heated the earth. Segregation of the vast amounts of Archean granite from the underlying mantle required enormous heat energy. Most of this probably came from radioactive nuclides in the rocks of the upper mantle. At the time of this major crustal differentiation, radioactive heat production was two to three times as great as it is now (Figure 3.15).

The earth's continental cratons probably began as small granitic nuclei, or protocontinents, that eventually became welded together into one or more large continental masses. Some geologists believe that the first protocontinents enlarged systematically by marginal additions of successive belts of newly formed granitic crust. Others believe that numerous small protocontinents formed piecemeal and joined together more or less randomly in time. Radiometric dating may eventually show which idea is more nearly correct. However, radiometric dating of the Archean is especially difficult because intense, worldwide thermal activity at the end of the eon, around 2.5 billion years ago, obliterated earlier dates. The few earlier Archean dates that have survived suggest that thick Archean granites were produced over a period of about one billion years, from about 3.4 billion to 2.4 billion years ago (Figure 3.16).

FIGURE 3.15

Radioactive heat production within the earth has decreased throughout geologic time.

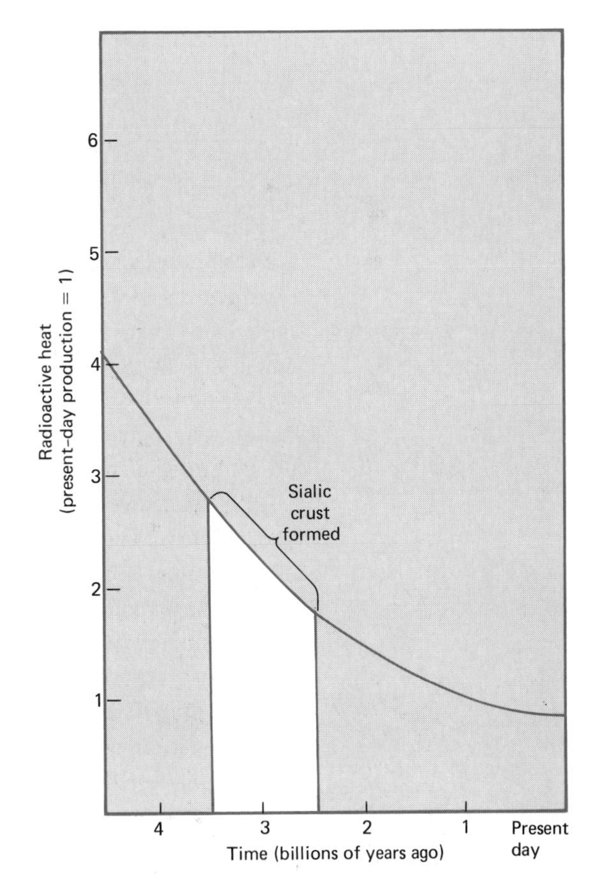

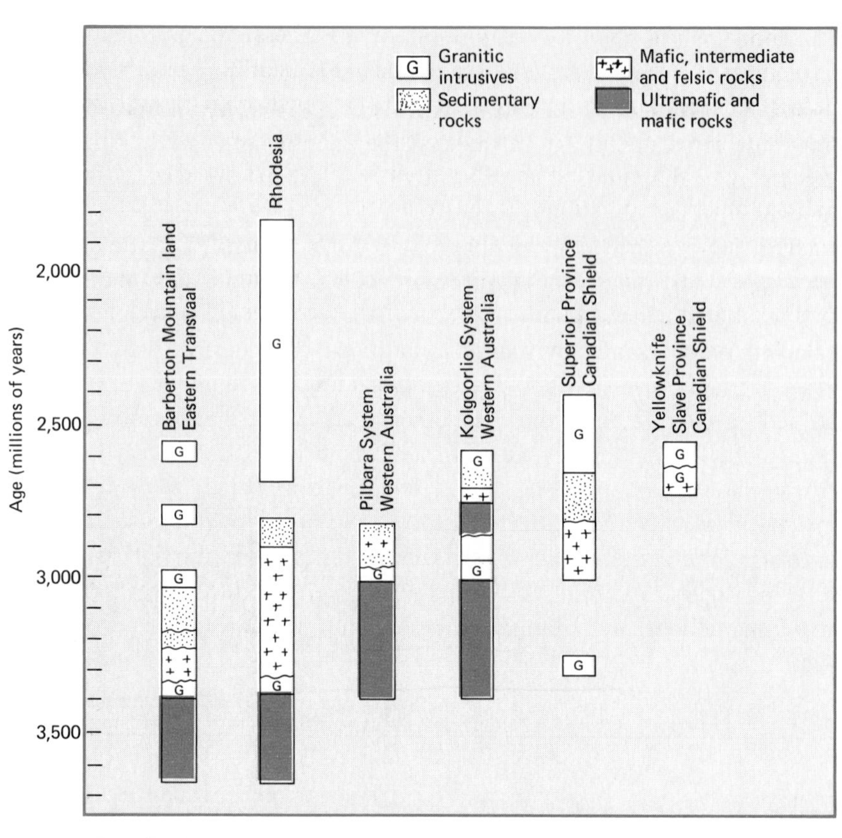

FIGURE 3.16 Age distribution of six well-dated Archean greenstone belts. (*Glickson, 1972*)

Following the worldwide episode of extraordinary thermal activity that closed the Archean, the continental cratons became stabilized and thereafter stood high, platform-style, with respect to the surrounding deep ocean basins. The effect on the sedimentary rock record was dramatic. The pervasive turbidites that dominated the Archean record gave way in the Proterozoic to a wide variety of shallow-marine, marginal-marine, and continental deposits. This substantially more varied rock record has continued to characterize all subsequent geologic history.

Post-Archean sedimentary rock assemblages can be placed in one of two fundamental categories: **platform** or **geosynclinal.** Platform sequences form on stable cratons. Since the early Proterozoic, the cratons have apparently remained relatively near sea level. At times the cratonic platforms subsided gently and were covered by extensive shallow seas called **epeiric seas;** at other times they emerged and underwent subaerial erosion. In these stable platform settings, unconformities (gaps in the sedimentary record) are numerous, and total sedimentary accumulations are generally thin, on the order of one or two thousand meters. Locally, however, strata thicken in broad cratonic basins where thicknesses reach several thousand meters. Much of the Proterozoic sedimentary record now preserved was probably produced in similar basins.

Mobile belts form at the margins of cratons. These are unstable areas that have alternately been submerged to considerable depths beneath the sea and uplifted to mountainous heights (Figure 3.17). During episodes of subsidence, mobile belts commonly receive sedimentary thicknesses in excess of 15,000 meters. These thick accumulations of sedimentary rocks, which usually form in belts several hundred kilometers long, are called **geosynclines.** Following the accumulation of their thick sedimentary sequences, many geosynclines have been uplifted and subjected to large-scale folding and faulting. The mountains produced during these episodes have simultaneously been intruded at depth by granitic magma.

FIGURE 3.17

Major Phanerozoic mobile belts surrounding North America.

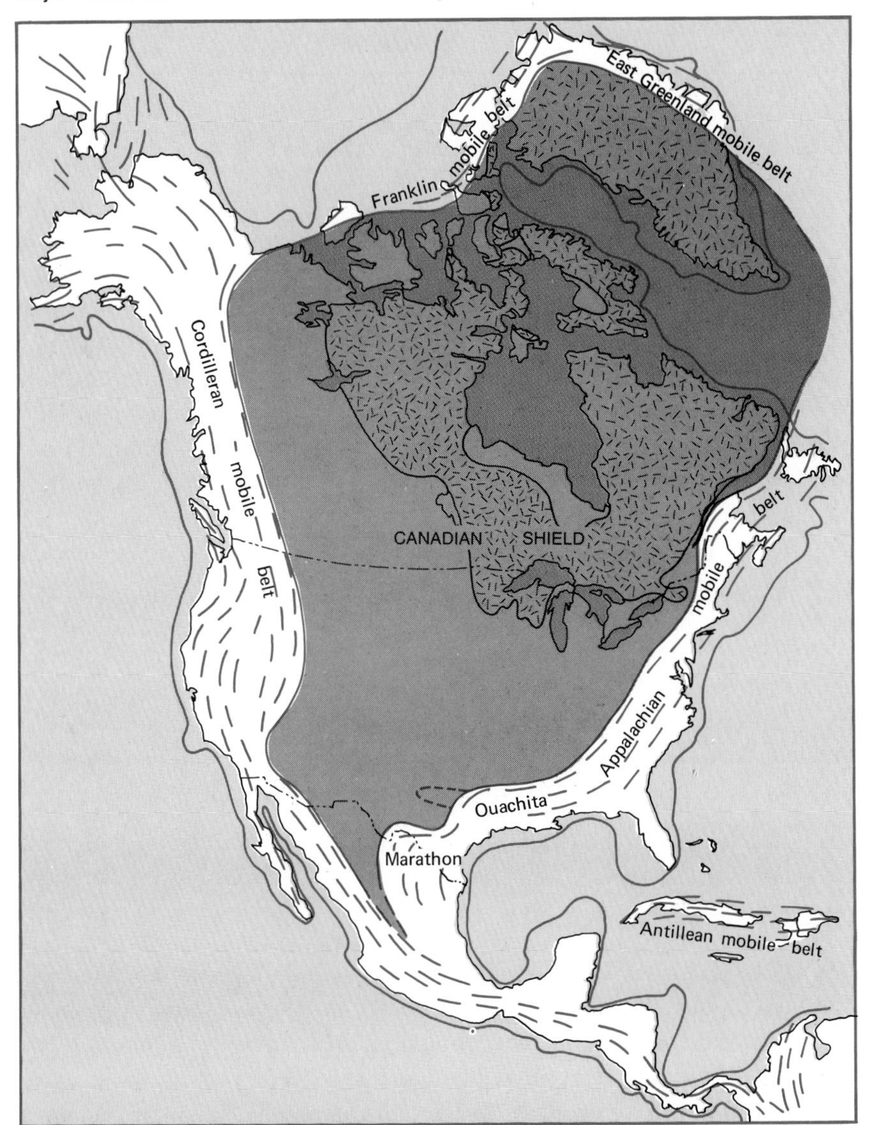

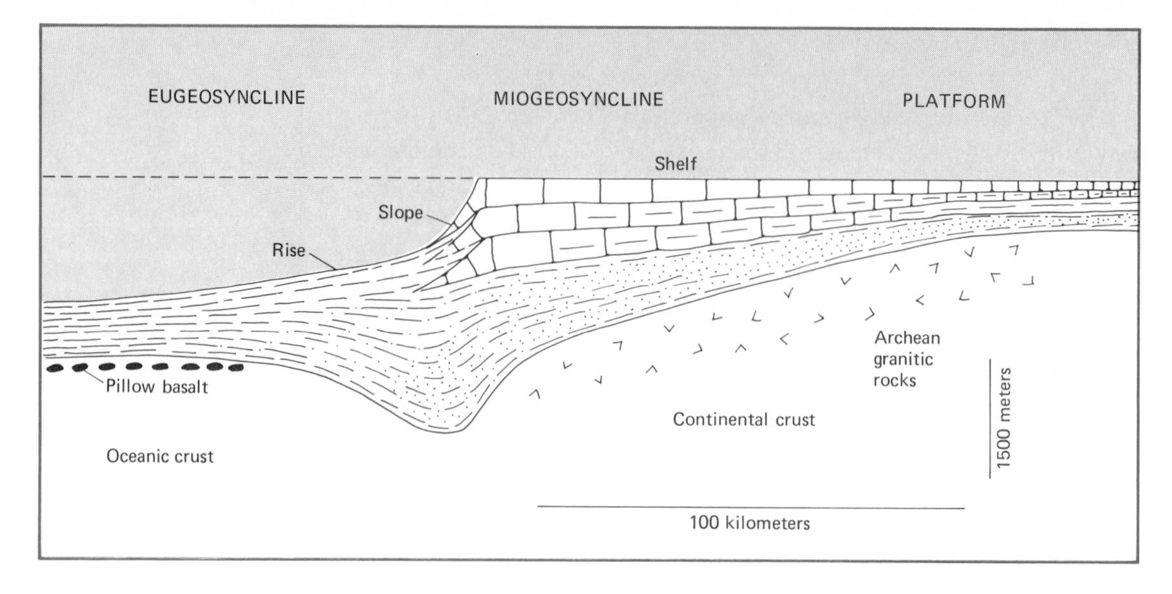

FIGURE 3.18 Restored cross section of the Coronation Geosyncline of northwestern Canada. These 2-billion-year-old rocks were strongly folded and faulted in the Hudsonian Orogeny, about 1.7 billion years ago. (*Fraser and others, 1972*)

Fold mountains, which are made of deformed geosynclinal deposits, commonly consist of two distinctive rock suites: **miogeosynclinal** deposits and **eugeosynclinal** deposits. The miogeosynclinal suite occupies the inner margin of the mobile belt nearest the craton, and the eugeosynclinal suite occupies the outer portion of the mobile belt. Miogeosynclines contain mainly shallow-water limestones, dolomites, shales, and clean, well-sorted sandstones—all of which formed in relatively quiet settings, free from igneous activity. Eugeosynclines, by contrast, consist mainly of shales and poorly sorted sandstones that are inferred to be deposits of relatively deep water. Carbonate rocks are rare in eugeosynclines, but bedded cherts are common. Eugeosynclines also contain a significant component of volcanic rocks, chiefly submarine lava flows.

In Chapter 6 we shall examine in more detail the dynamic processes that led to this sequence of sediment accumulation and subsequent deformation. Such processes have dominated the growth of the continents since the close of Archean time. Figure 3.18 shows in cross section an early Proterozoic geosyncline on the Canadian Shield as it may have appeared during its accumulation on the edge of the Archean continent.

Evolving Precambrian Environments

The stable continental platforms that first appeared in earliest Proterozoic time were sites of both shallow-marine and, for the first time, widespread terrestrial sediment accumulation. Samples of such Proterozoic rocks are preserved today in thick depositional basins (Figure 3.19). Proterozoic sedimentary rocks were at one time almost certainly deposited much more widely on

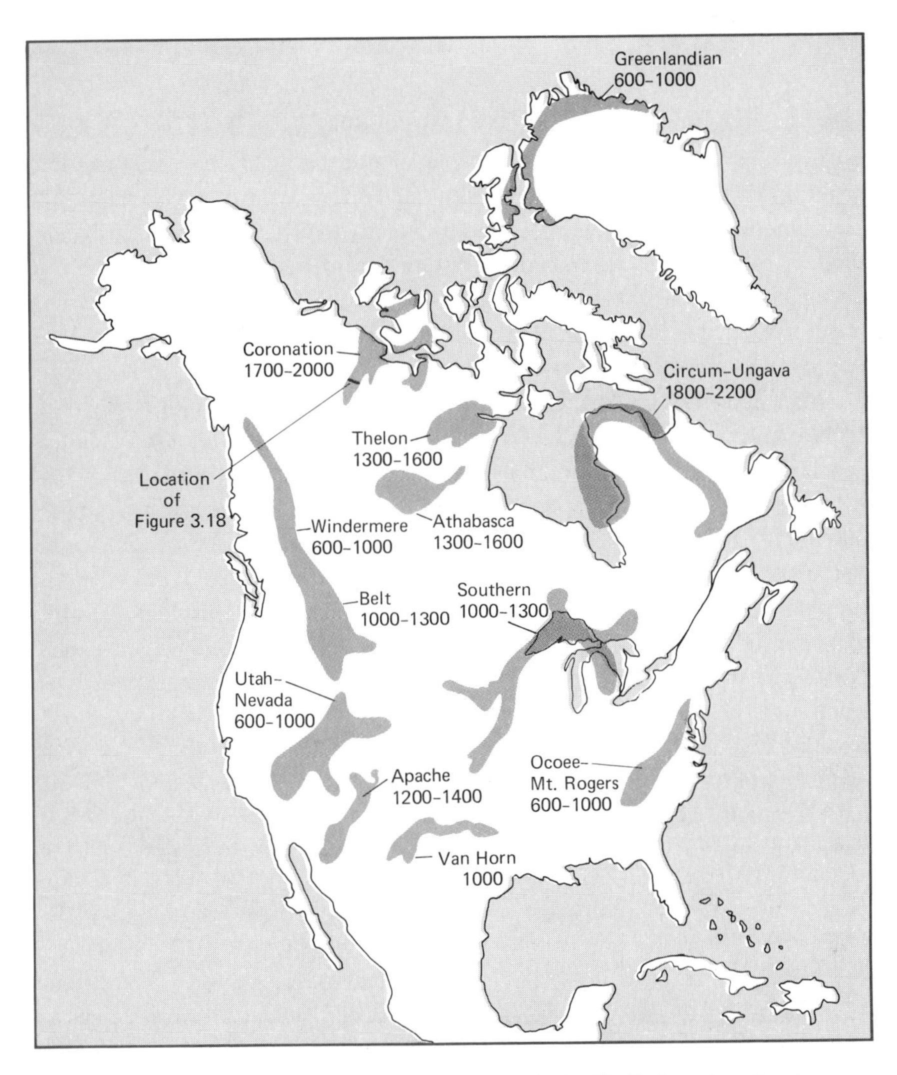

FIGURE 3.19

Major occurrences of Proterozoic sedimentary rocks in North America. Numbers are approximate ages in millions of years; names refer to sedimentary basins. (*Condie, 1976*)

the cratons, but have since been removed. Large unconformities, which separate underlying Precambrian rocks from overlying Phanerozoic strata in most continental regions, indicate that the late Precambrian continents were intensively eroded. The Proterozoic sedimentary rocks that have survived include marine sandstones and argillites, abundant banded iron formations, and widespread dolomites and limestones made up of distinctive moundlike structures. Continental rocks include conglomerates, sandstones, and, surprisingly, tillites (deposits of glacial debris), which testify to widespread continental ice sheets. Abundant banded iron formations and the wide variety of moundlike carbonates are unique to the Proterozoic. Tillites, though not unique to this time, are the best indication we have of Proterozoic climates. In this section we shall explore the important environmental clues provided by these and other Proterozoic sedimentary rocks.

Banded Iron Formations

Banded iron formations (Figure 3.20) are extensive sequences of bedded cherts, usually hundreds of meters thick, made up of alternating black or red, iron-rich layers and light gray, iron-poor layers, or bands. The layers range from less than 1 millimeter to several centimeters in thickness, and they are remarkably continuous; some have been traced for distances up to 300 kilometers. The iron occurs chiefly as iron oxide; in the iron-rich layers, this typically constitutes about one-third of the rock. Like the chert itself, the iron oxide appears to have precipitated directly from ocean water.

Most banded iron formations are more than 2 billion years old. Today these rocks constitute the greatest reserves of iron ore in the world. After 2 billion years ago, banded iron formations became scarce; ultimately they ceased to form entirely. This important change in the sedimentary record was apparently caused by the appearance of *free oxygen* in the atmosphere and waters of the earth. In the presence of free oxygen, very little dissolved iron would be present in the oceans, because oxygen causes iron to form insoluble, rustlike (ferric iron) oxides, which remain in rocks and in soils, and which cannot dissolve to accumulate in the oceans. In the absence of free oxygen, however, iron (ferrous iron) is readily soluble, and it could easily have been weathered from iron-rich rocks and transported in solution by streams to the oceans. Thus, quantities of iron sufficient to produce the banded iron formations could only have accumulated in the Precambrian oceans *before* the atmosphere and waters of the earth attained significant concentrations of free oxygen.

FIGURE 3.20 A typical banded iron formation exposed in western Australia. The iron is concentrated in the darker layers. (*Courtesy of B. J. Skinner*)

Actually, the precipitation of iron in the open ocean was accomplished by oxidizing it. In the absence of any available free oxygen in the oceans, the oxygen necessary for this precipitation was probably supplied by primitive, single-celled phytoplankton (planktonic plants), which had evolved the process of **photosynthesis**, but which had not yet developed a means of coping with the oxygen waste produced in that process.

Photosynthesizing plants convert water and carbon dioxide into carbohydrates using solar energy. The general process can be illustrated as follows: $CO_2 + H_2O + \text{light} = (CH_2O)_n + O_2$. The $(CH_2O)_n$ represents organic compounds that form the actual plant cells and the O_2 is a waste product. Modern photosynthesizing plants exhale waste oxygen through their cell walls; however, they do this without oxidizing their own tissues, because they manufacture special oxygen-mediating enzymes that counteract the poisonous effects of the oxygen. Prior to the development of such enzymes, early Precambrian phytoplankton may have used the abundant ferrous iron dissolved in the water all about them as a protective oxygen sink. Oxygen produced by the organism during photosynthesis was instantly combined with the dissolved ferrous iron, thereby oxidizing it to ferric iron and causing it to precipitate to the sea floor. Petrologic thin-section studies reveal tiny spherical bodies around 30 microns in diameter in many banded iron formations that may be the remains of the very phytoplankton that caused precipitation of the iron.

When at last photosynthesizing plants began to manufacture oxygen-mediating enzymes, they no longer needed to use dissolved ferrous iron to dispose of their waste oxygen. Instead, they could simply expel oxygen directly into the water in which they lived. Free oxygen began to accumulate thereafter in the earth's ocean and atmosphere; this must have severely affected many organisms that had not yet learned to manufacture oxygen-mediating enzymes. Some of these organisms probably became extinct, and others withdrew to local anaerobic environments, where their descendants still exist today. As significant quantities of free oxygen began to accumulate, the last of the oceanic iron was precipitated. After this, banded iron formations ceased to form. Iron could then be oxidized in place on land, forming soils of a modern type and producing terrestrial **redbeds**, which are common in subsequent geologic history.

Stromatolites

Although fossil animal and plant remains are extremely rare in Precambrian rocks, certain widespread Precambrian carbonate rocks are now known to have been deposited by living organisms. These are **stromatolites**, which originate as distinctive, typically moundlike accumulations of calcium carbonate layers formed by algae.

Stromatolites are being deposited today in shallow-marine and intertidal environments where the sediment surface is coated with a layer of cells of blue-green algae. The algal cells are organized into threadlike filaments that

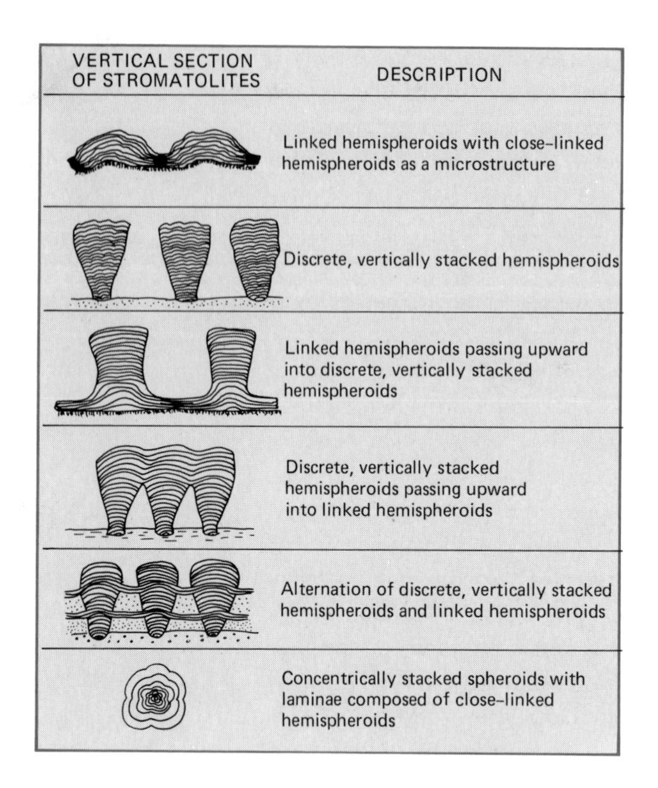

FIGURE 3.21

Representative shapes of stromatolites. Their heights may range from a few centimeters to 2 meters or more. (*Logan, Rezak, and Ginsburg, 1964*)

FIGURE 3.22

Ranges of some stratigraphically significant columnar stromatolites in the Precambrian of the Soviet Union. Width of the vertical area indicates abundance. (*Semikhatov, 1976*)

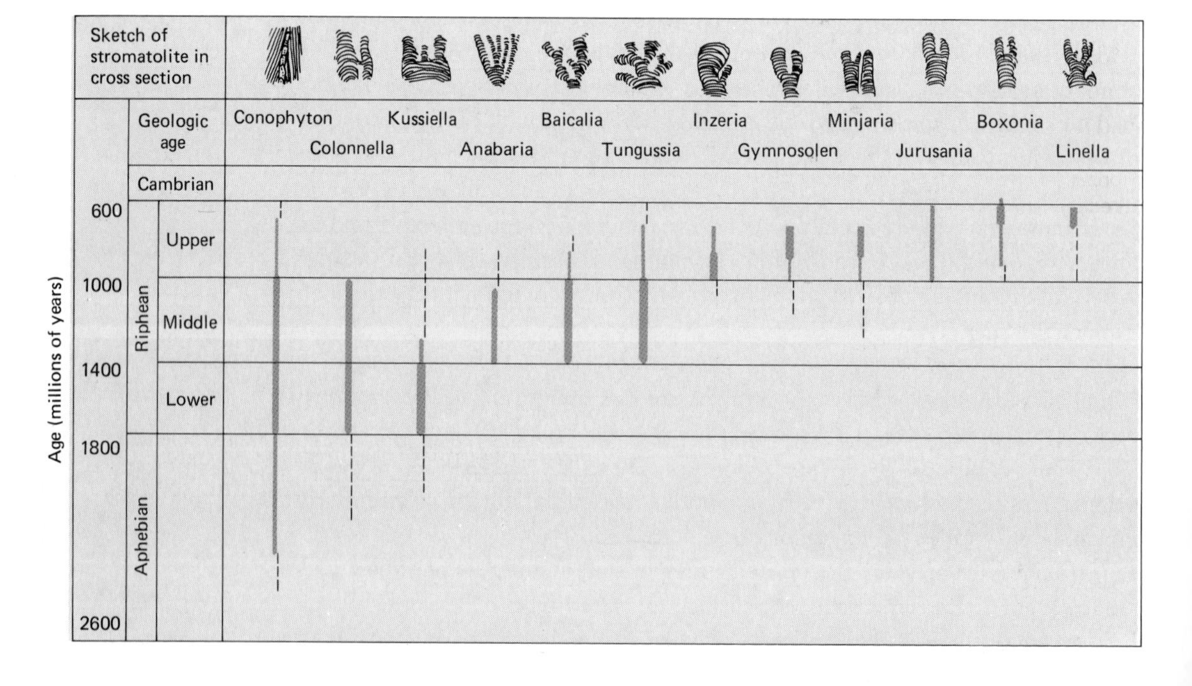

bind the surrounding particles of carbonate material into a coherent layer, called an **algal mat.** When a high tide or storm brings in a load of fine-grained calcium carbonate particles, a thin layer of them is trapped by the filaments on top of the algal mat. Quickly, within hours, the algae extend their filaments through this fresh layer, thereby incorporating it into the mat as they repopulate the new top surface. With each new influx of carbonate-laden water the process repeats, producing the characteristic laminated structure. The algal layers themselves decay quickly, but their former presence is indicated by the pronounced layering of the carbonate particles.

Algal stromatolites occur in a variety of forms: flat, wavy, moundlike, and even columnar forms called heads (Figure 3.21). Today living columnar stromatolites are very rare, but in the late Precambrian many kinds occurred, some of which appear to have existed for comparatively short ranges in time. Russian geologists in particular have found these variations useful in correlation (Figure 3.22). Others feel that many of the morphologic differences among stromatolites may result merely from differing local environments. For example, columnar stromatolites today form elongate heads whose asymmetry reflects the direction and strength of local water currents (Figure 3.23). Many small-scale differences among stromatolites, such as thickness and extent of individual laminae, appear to be biologically (genetically) controlled. This is best documented in Precambrian occurrences, where stromatolites of similar morphology are found in several different kinds of rocks that must represent significantly different local environments. Thus, although gross shapes of stromatolites may be environmentally controlled, different small-scale structures within them probably reflect genetic differences.

Stromatolite diversity reached a peak in the late Precambrian, when widespread stromatolite colonies occupied huge areas of shallow seas. At that time the algae apparently had no competitors, for more advanced forms of life had not yet appeared. In the very latest Precambrian, however, the number of

FIGURE 3.23

Horizontal and vertical sections of a stromatolite showing the relation of shape to current direction. (*Hoffman, 1976*)

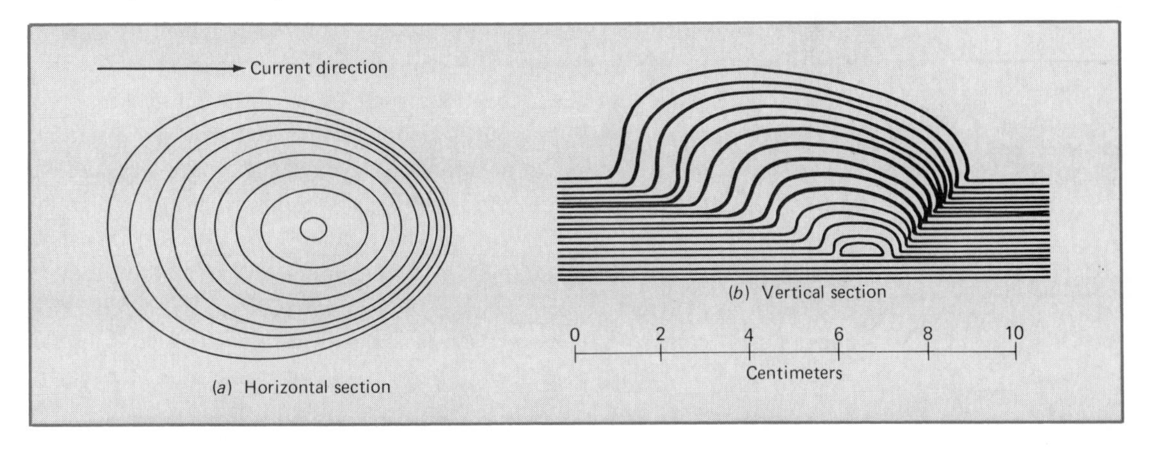

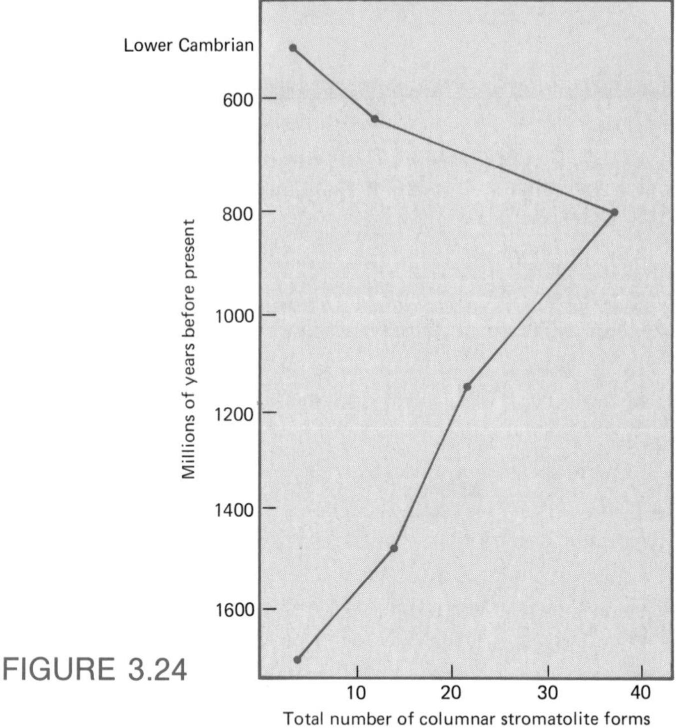

FIGURE 3.24

Changes in diversity of late Precambrian columnar stromatolites. (*Awramik, 1971*)

species of stromatolites declined abruptly (Figure 3.24). This probably reflects the evolution of small grazing animals that found the stromatolites a convenient source of food and caused the extinction of many of them.

Tillites

About 30 percent of the earth's land surface was covered by continental glaciers in recent geologic history, and much of this ice remains today in Greenland and Antarctica. Were it not for this, we would probably be hard put to interpret the thick, unstratified, poorly sorted conglomerates that are widespread in Precambrian rocks, but which closely resemble modern glacial sediments called tills. Like modern tills, these ancient sedimentary rocks commonly contain boulders that are not in contact with one another but are totally surrounded by a matrix of silt and clay (Figure 3.25). Precambrian boulder conglomerates, which are inferred to be lithified glacial till and are known as **tillites**, have been found on almost every continent.

Besides ice, another mechanism that produces unstratified and completely unsorted boulder-laden sediments is debris flow. Debris flows (sometimes called mud flows) consist of water-saturated mud and rock that moves quickly downslope under the influence of gravity. Because these are short-lived, rather rare events, few geologists have observed them. Nevertheless, debris flows transport huge quantities of material, and in populated areas they may be tremendously destructive. Today most debris flows occur following heavy rains in arid regions where vegetation is slight or in regions of easily eroded volcanic rocks. In the Precambrian, before vegetation mantled the earth's surface, they were probably much more frequent and widespread.

FIGURE 3.25

Precambrian tillite from the Headquarters Schist of southern Wyoming.

It is nearly impossible to distinguish deposits of debris flows from deposits of tillites. Some of the tillitelike rocks in the Precambrian are probably debris-flow deposits. Nevertheless, numerous Precambrian boulder conglomerates are universally accepted as tillites, because the rock surfaces on which they rest are polished and marked with parallel scratches and grooves exactly like those found in glaciated regions today (Figure 3.26). Associated with many tillites are thinly bedded shales and siltstones that contain widely scattered cobbles and boulders that appear to have dropped onto the accumulating finer sediment from above. The shales and siltstones are interpreted as marine (or in some cases, lake) deposits. The dropstones they contain are believed to have come from melting icebergs as they passed above (Figure 3.27). Similar deposits are forming today in iceberg-infested waters along the Alaskan coast and in other regions. Hence, these distinctive dropstone-bearing rocks provide additional evidence for glaciation in the Precambrian.

Widespread Precambrian glaciation occurred at least twice—the first time in the early Proterozoic, around 2.2 to 2.3 billion years ago, and the second in the late Proterozoic, around 700 million years ago, based on radiometric ages. Although the actual dates of the Precambrian glaciations have been supplied only in recent years, the existence of such glaciation has been known since the middle of the last century. Tillites provide an explicit record of paleoclimates, inasmuch as they record extended periods of freezing temperatures. Even though these deposits were probably produced in high latitudes, their existence would indicate that the earth's surface was approximately the same temperature in the early Proterozoic as it is today. In the midnineteenth

FIGURE 3.26 Grooves and striations on the rock floor beneath this Precambrian boulder conglomerate in southern Australia testify to its glacial origin.

FIGURE 3.27 Dropstone in marine shales of Permian age in Australia.

century, at the time the evidence for Precambrian glaciation was first discovered, the earth was believed to be gradually and progressively cooling, following its molten origin. The evidence for ancient glaciation showed that, far from indicating hotter temperatures with increasing age, the old rocks testify that the earth's climate has been much the same for a very long time.

Evolution of the Ocean and Atmosphere

So far we have made only passing reference to the history of the earth's fluid cover of water and air. Many Precambrian rocks survive to give direct testimony of their origins, but the Precambrian ocean and atmosphere are gone forever, leaving only indirect traces on the surviving rocks. We have seen, for example, that banded iron formations provide clues to the increase of free oxygen in the atmosphere. From similar inferences has grown a body of knowledge about the origin and changing nature of the earth's fluid cover of water and air.

Fluids from the Earth's Interior

At first glance it seems reasonable to suppose that the ocean and atmosphere are mere residues of light, volatile materials left over from the original accretion of the solid earth. There is, however, strong evidence that the earth's surface fluids did *not* originate at the time of its initial accumulation but, instead, were added somewhat later, by volcanic **outgassing** from the materials of the solid crust and mantle.

This evidence comes from the earth's comparative scarcity of the inert gases, particularly neon, argon, krypton, and xenon (Figure 3.28). These extremely stable gases never form natural compounds with other elements. Moreover, since they are too heavy to escape the earth's gravitational attraction and be lost in space, any that were present in the earth's earliest atmosphere should *still* be present. There is no known way that these gases could either have escaped into space or been taken up by the solid crust. However, these elements are about a million times *less* abundant today in the earth's atmosphere than in the sun and other stars. For this reason, scientists believe that the earth's gaseous atmosphere is not merely a residue of the initial solar gases but is, instead, an accumulation of volatile materials released by the volcanic heating and outgassing of elements originally combined into the minerals of the solid earth (Figure 3.29).

There is abundant evidence that local heating and melting of the earth's solid crust and upper mantle do take place, for these processes are responsible for present-day volcanoes—all of which release large quantities of volatile elements at the earth's surface. Such volcanic outgassing, over the long course of earth history, is the most probable source of our oceans and atmosphere. Water vapor and carbon dioxide are the principal constituents of modern volcanic gases; hydrogen, nitrogen, ammonia (NH_3), methane (CH_4), chlorine, and many other gases occur in smaller quantities. If melting of the earth's

1 H Hydrogen																	2 He Helium
3 Li Lithium	4 Be Beryllium											5 B Boron	6 C Carbon	7 N Nitrogen	8 O Oxygen	9 F Fluorine	10 Ne Neon
11 Na Sodium	12 Mg Magnesium											13 Al Aluminum	14 Si Silicon	15 P Phosphorus	16 S Sulfur	17 Cl Chlorine	18 Ar Argon
19 K Potassium	20 Ca Calcium	21 Sc Scandium	22 Ti Titanium	23 V Vanadium	24 Cr Chromium	25 Mn Manganese	26 Fe Iron	27 Co Cobalt	28 Ni Nickel	29 Cu Copper	30 Zn Zinc	31 Ga Gallium	32 Ge Germanium	33 As Arsenic	34 Se Selenium	35 Br Bromine	36 Kr Krypton
37 Rb Rubidium	38 Sr Strontium	39 Y Yttrium	40 Zr Zirconium	41 Nb Niobium	42 Mo Molybdenum	43 Tc Technetium	44 Ru Ruthenium	45 Rh Rhodium	46 Pd Paladium	47 Ag Silver	48 Cd Cadmium	49 In Indium	50 Sn Tin	51 Sb Antimony	52 Te Tellurium	53 I Iodine	54 Xe Xenon
55 Cs Cesium	56 Ba Barium	57 La Lanthanum	72 Hf Hafnium	73 Ta Tantalum	74 W Tungsten	75 Re Rhenium	76 Os Osmium	77 Ir Iridium	78 Pt Platinum	79 Au Gold	80 Hg Mercury	81 Tl Thallium	82 Pb Lead	83 Bi Bismuth	84 Po Polonium	85 At Astatine	86 Rn Radon
87 Fr Francium	88 Ra Radium	89 Ac Actinium															
				58 Ce Cerium	59 Pr Praseodymium	60 Nd Neodymium	61 Pm Promethium	62 Sm Samarium	63 Eu Europium	64 Gd Gadolinium	65 Tb Terbium	66 Dy Dysprosium	67 Ho Holmium	68 Er Erbium	69 Tm Thulium	70 Yb Ytterbium	71 Lu Lutetium
				90 Th Thorium	91 Pa Protactinium	92 U Uranium	93 Np Neptunium	94 Pu Plutonium	95 Am Americium	96 Cm Curium	97 Bk Berkelium	98 Cf Californium	99 Es Einsteinium	100 Fm Fermium	101 Md Mendelevium	102 No Nobelium	103 Lw Lawrencium

FIGURE 3.28 Table of the elements showing the rare gaseous elements (neon, argon, krypton, and xenon) that are far less abundant on the earth today than in the sun and other stars. This depletion indicates that the earth's atmosphere is not a residue of gases left over from its formation out of solar materials.

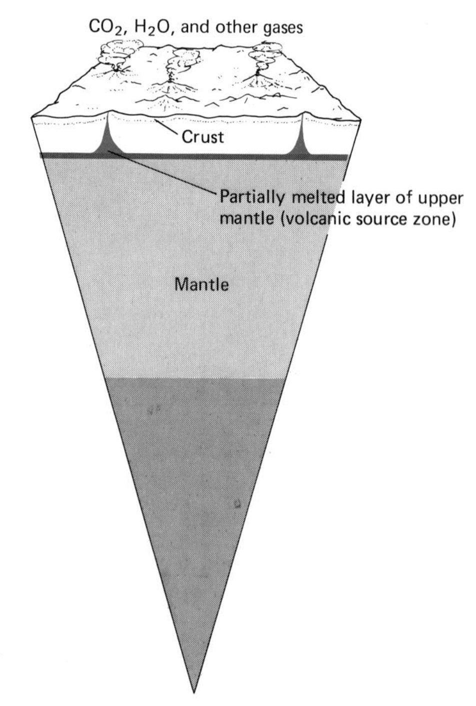

FIGURE 3.29 Origin of the ocean and atmosphere by volcanic outgassing of the upper mantle.

early interior produced similar gases, then the abundant water vapor would have condensed to form the initial ocean as the underlying crustal rocks cooled below 100°C, leaving an initial gaseous atmosphere composed largely of carbon dioxide.

The Ocean Through Earth History

Recall for a moment that the oldest Archean sedimentary rocks consist of particles that were weathered from preexisting rocks and were deposited in fairly deep water. This means that, even before the earliest rocks came into existence, there was already an atmosphere and an ocean. The early ocean probably differed little from its modern counterpart, inasmuch as water comprised a portion comparable to today's 96.5 percent. However, the large quantity of bedded cherts (massive, noncrystalline quartz) in the early Precambrian suggests that the early ocean contained more silica than today's ocean does (Figure 3.30). Also, the earliest oceans must have contained vastly more iron than they do today, judging from the abundance of banded iron formations in Archean and lower Proterozoic rocks. Later, in the middle and late Precambrian, when carbonate rocks became common in the sedimentary record, they consisted almost entirely of dolomite, $CaMg(CO_3)_2$, instead of

FIGURE 3.30

Volume percent of sedimentary rocks as a function of age. (*Ronov, 1964*)

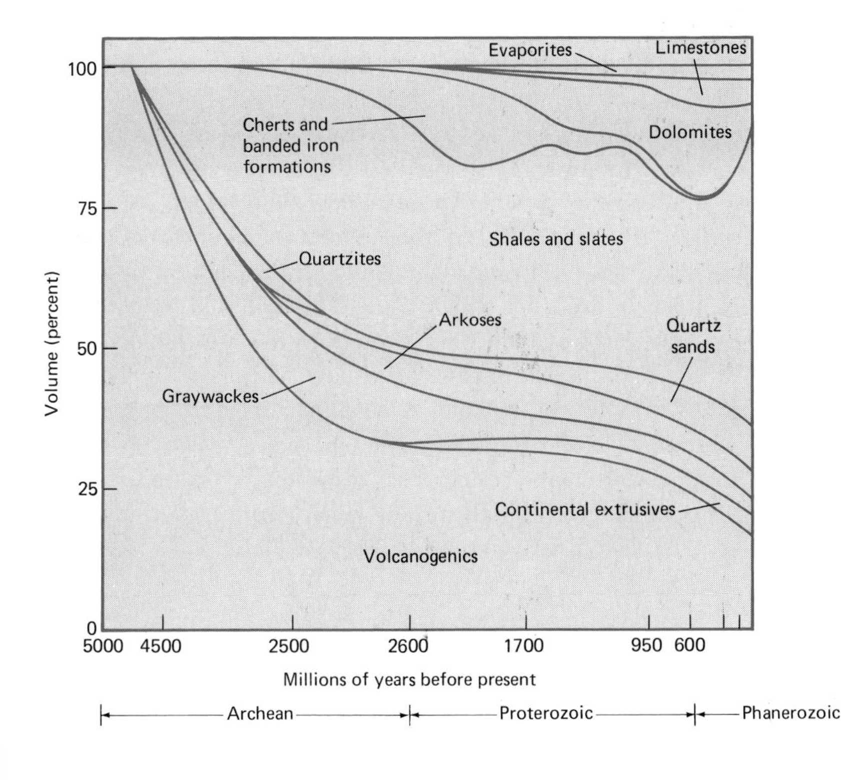

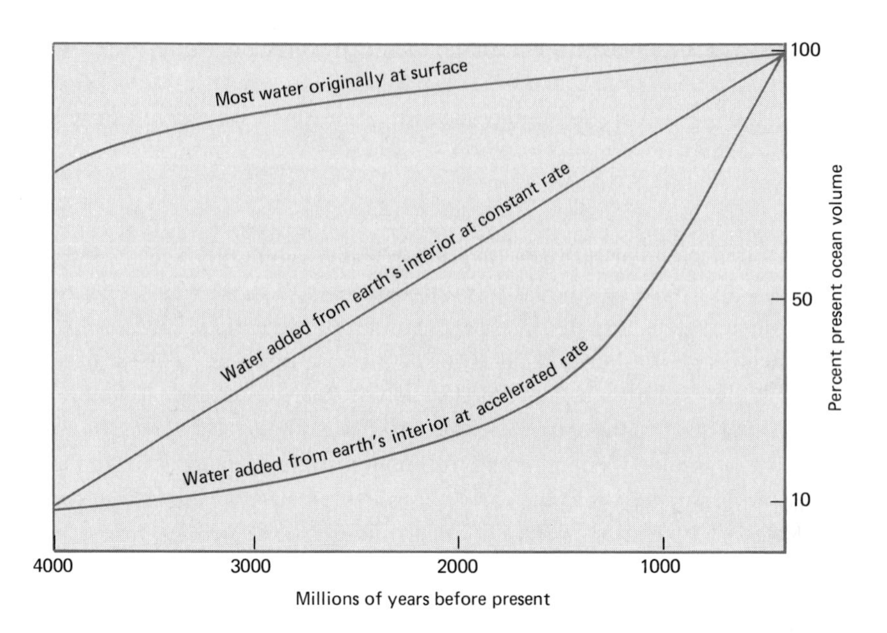

FIGURE 3.31 Possible volumes of the ocean during earth history according to different hypotheses.

calcite ($CaCO_3$), which is dominant in recent geologic history. At first glance this suggests that Precambrian oceans may have contained more magnesium than modern oceans. However, most geologists believe that the dolomite is a secondary replacement of limestone, and that the ocean need not have contained more magnesium than it does today. Hence, they feel that overall changes in the total composition of the world's oceans have been slight.

A still more fundamental question concerns possible changes, not in the composition, but in the overall *volume* of the ocean throughout earth history. Here, unfortunately, we have still less evidence. Extensive volcanism soon after the earth's accretion—an event for which there is no evidence—might have released large quantities of water vapor to produce a large initial ocean. According to this hypothesis, the ocean would have had a nearly constant volume throughout earth history, with only relatively minor additions of water from volcanic eruptions. Or the initial ocean may have been relatively small, with steady volcanic additions causing it to grow to its present volume gradually (Figure 3.31). As yet there is no unequivocal evidence favoring one of these hypotheses over the other.

The Changing Atmosphere

The rocks of the solid earth and the waters of the ocean have probably undergone only relatively minor changes in bulk composition since very early in earth history. In contrast, there is strong evidence that the gases of the atmosphere have shown profound historical changes. Other than water vapor,

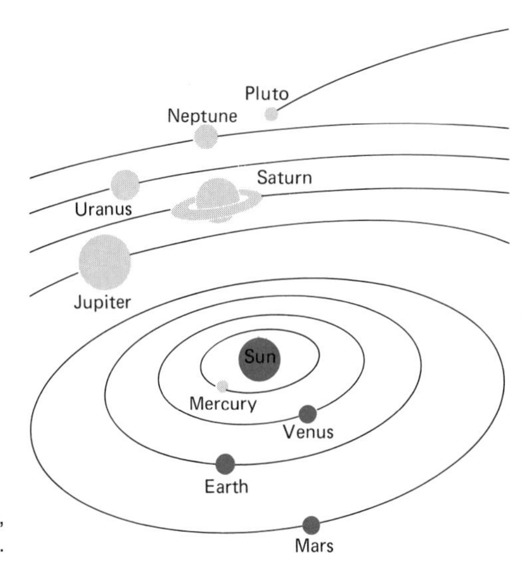

FIGURE 3.32 Relative positions of Earth, Venus, and Mars within the solar system.

the dominant gas emitted into the atmosphere by present-day volcanoes is carbon dioxide, and this has probably been true throughout much of the earth's past. Assuming that volcanic outgassing of the earth's interior was the original source of the atmosphere, there must have been mechanisms that acted throughout earth history to change these carbon-dioxide-dominated volcanic gases into our present atmosphere of nitrogen and oxygen. Some of the strongest clues to the nature of these mechanisms have come from recent spacecraft observations of the earth's closest and most similar planetary neighbors, Venus and Mars (Figure 3.32).

The Earth, Venus, and Mars Unlike the moon and Mercury, both of which lack an appreciable atmosphere, Venus and Mars, the two remaining earthlike planets, have long been known to be surrounded by atmospheric gases. Early telescopes showed that both have atmospheric clouds just as does the earth. Venus is perpetually hidden beneath dense and continuous clouds, but Mars shows occasional clouds that only rarely obscure the planetary surface. These and other observations indicate that Venus, lying closer to the sun than does the earth, has a more dense atmosphere than does the earth, whereas Mars, farther from the sun, has an atmosphere that is less dense.

It was long believed, without real evidence, that the atmospheres of both Venus and Mars were dominated by nitrogen, just as the earth's atmosphere is. Beginning in the 1960s, however, observations made by spacecraft showed conclusively that the atmospheres of both planets consist largely of carbon dioxide, rather than nitrogen. As on the earth, these planetary atmospheres most probably originated from volcanic outgassing of the solid materials that make up the principal mass of the planets. However, on Venus and Mars the volcanic carbon dioxide has accumulated as an atmospheric gas, whereas on

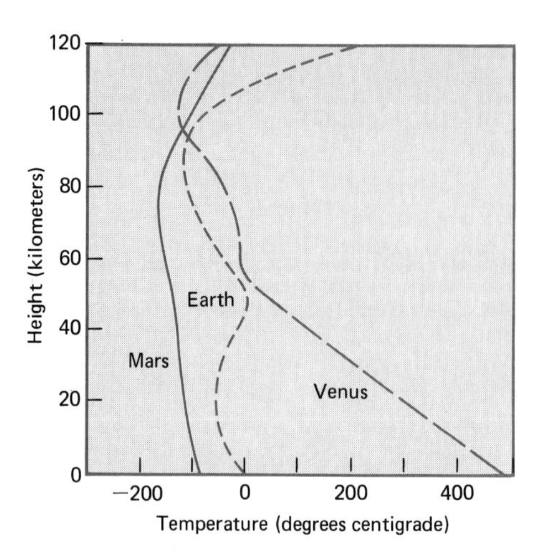

FIGURE 3.33 Variation of atmospheric temperature with height in the atmospheres of Venus, Mars, and the earth. The atmospheres of Venus and Mars are composed largely of carbon dioxide, in contrast to the nitrogen-oxygen earth atmosphere.

the earth it has been steadily removed from the atmosphere, leaving nitrogen as the dominant gas.

This difference is most probably related to the earth's unique ability to retain liquid water upon its surface. Before turning to the earth, however, we need to consider briefly the question of why liquid water has not accumulated to form oceans on either Venus or Mars. On both planets we may assume that, as on the earth, much water vapor has been added to the atmosphere, along with carbon dioxide, by volcanic outgassing. The ultimate fate of this water is, however, closely dependent on planetary temperatures, which, in turn, are largely determined by relative distance from the sun.

Venus, lying closer to the sun than does the earth, has probably always been too hot for the condensation of water as a liquid (Figure 3.33). Thus, volcanic water would be retained in the atmosphere as water vapor. Measurements by spacecraft show, however, that even though the surface and atmosphere of Venus are today extremely hot, they lack water vapor. The most probable explanation is that the water added to the atmosphere by volcanic outgassing during Venus's history has been steadily removed as water molecules were broken into separate hydrogen and oxygen atoms by the planet's intense solar radiation. Following such chemical dissociation, the light hydrogen atoms would tend to escape the planet's gravitational field and be lost into space. The oxygen might be lost rather quickly as well, but by chemical reaction with the hot planetary surface and with other atmospheric gases.

On Mars the situation is less complex. The outgassing that produced its relatively thin atmosphere (about 1/100 the mass of the earth's atmosphere), would have produced relatively small amounts of atmospheric water vapor. Because of the planet's low surface temperature (Figure 3.33), it would quickly condense to form frost, rather than accumulating as an ocean. Thus, because of their relative distances from the sun, neither Venus nor Mars has the earth's cover of *liquid* water.

Carbon Dioxide and the Ocean The earth's ocean is the unique product of its average temperature and its distance from the sun, for our planet is neither so hot that its water is all gaseous vapor, nor so cold that it is solid ice. The presence of large quantities of liquid water, in turn, have profoundly influenced the history of the earth's atmosphere, principally through the steady removal of carbon dioxide.

The ultimate source of all the earth's carbon is carbon dioxide gas released into the atmosphere by volcanoes (Figure 3.34). Today, however, only a very small fraction—far less than 1 percent—of the earth's surficial carbon occurs in the atmosphere. A larger amount, about 60 times as much, occurs in the ocean as dissolved carbon dioxide. But even taken together, these two carbon reservoirs still account for much less than 1 percent of the total carbon at the earth's surface. The bulk occurs as solid materials in sedimentary rocks. About 77 percent is found in carbonate minerals that make up the extensive limestone and dolomite deposits of the earth's sedimentary veneer. The remainder, about 23 percent, occurs as finely divided, carbon-rich particles from ancient animals and plants that have been buried in sedimentary rocks, particularly dark-colored shales.

In the presence of large quantities of liquid water, most carbon dioxide is dissolved, rather than retained as an atmospheric gas. Calcium carbonate forms by precipitation from the dissolved carbon dioxide in the ocean and, on a

FIGURE 3.34

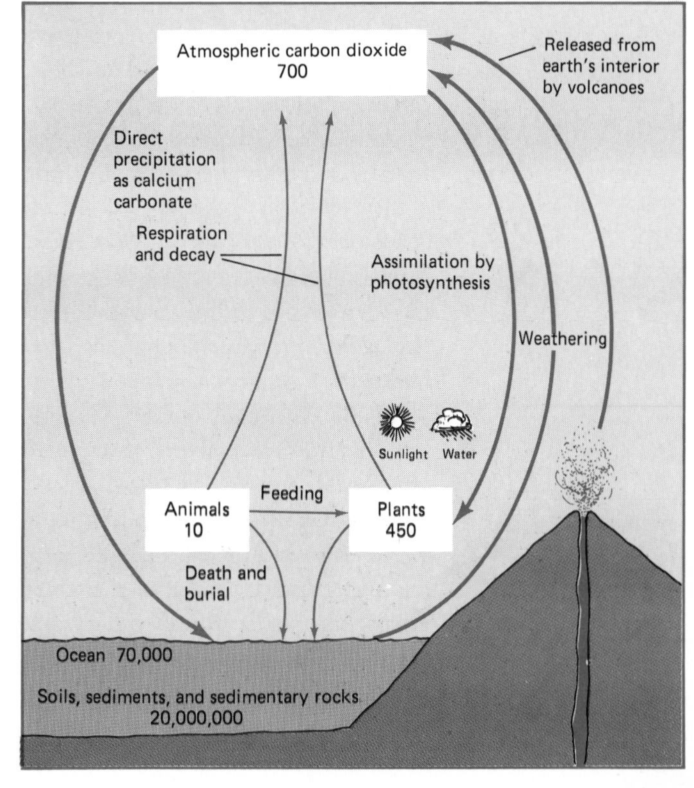

Removal of carbon dioxide from the atmosphere. The numbers show the present-day distribution of carbon in billions of metric tons. Much of the CO_2 released through geologic time by volcanoes has been dissolved in the ocean and then been precipitated as calcium carbonate sediments. Still more CO_2 is removed from the atmosphere by the metabolic activities of animals and plants, whose carbon-rich remains also accumulate in soils and sediments. Through such physical and biological processes, most of the earth's surficial carbon has come to be concentrated in sedimentary rocks, rather than occurring as atmospheric carbon dioxide.

much smaller scale, from that dissolved in lakes and streams on land. In water, carbon dioxide (CO_2) first reacts to become the bicarbonate ion (HCO_3^-), and finally the carbonate ion (CO_3^{--}), which then joins a calcium (Ca^{++}) ion to become $CaCO_3$. Today nearly all calcium carbonate being deposited is precipitated organically as skeletons of invertebrate animals and plants, most of them microscopic marine plankton, whose skeletons settle to the ocean floor.

In addition, plants remove dissolved carbon dioxide directly from the water during the process of photosynthesis, where CO_2 + H_2O + light react to form the various organic compounds that make up living cells. On the death of the organism, most of these compounds quickly oxidize and again become carbon dioxide. Some of them, however, are actually buried and are stored in sediments, usually in the form of fine particulate matter. Both of these processes—limestone deposition and burial of organic carbon—thus remove carbon dioxide from the oceans (and on a much smaller scale, from lakes and streams as well). Since the original source of the carbon dioxide was atmospheric gases dissolved in water, these processes, in turn, continuously remove carbon dioxide from the atmosphere (Figure 3.34).

Because of the abundance of both limestones and carbon-rich dark shales in rocks of almost all ages, it is clear that large-scale oceanic removal of atmospheric carbon dioxide has been occurring throughout most of the earth's long history. Most significantly, calculations show that, if the large quantities of carbon now stored in sedimentary rocks were returned to the atmosphere, then the earth's gaseous cover, like that of Venus and Mars, would be made up largely of carbon dioxide. If the carbon dioxide were removed from the atmospheres of Venus and Mars, it is probable that nitrogen, a chemically rather inert and unreactive gas, would be the dominant gas, just as on the earth.

The Atmosphere and Life Although the presence of liquid oceans most probably accounts for the relative lack of carbon dioxide and the dominance of nitrogen in the earth's atmosphere, other processes are still required to explain the second most abundant atmospheric component, oxygen. Unlike nitrogen, a chemically sluggish gas, free oxygen is unusually reactive. For this reason, it would be lost rather quickly through combination with oxygen-deficient surface minerals and volcanic gases, were it not continuously replenished by some large-scale opposing process.

Only two sources of free atmospheric oxygen are known; in both, the oxygen originates from the splitting of water molecules into their component hydrogen and oxygen atoms. Such splitting is known to occur in the upper atmosphere, as scattered water molecules are subjected to high-energy ultraviolet radiation from the sun. Although some free oxygen has undoubtedly been added to the atmosphere by this process, it contributes only an insignificant fraction of the atmosphere's enormous volume of oxygen, according to current theory.

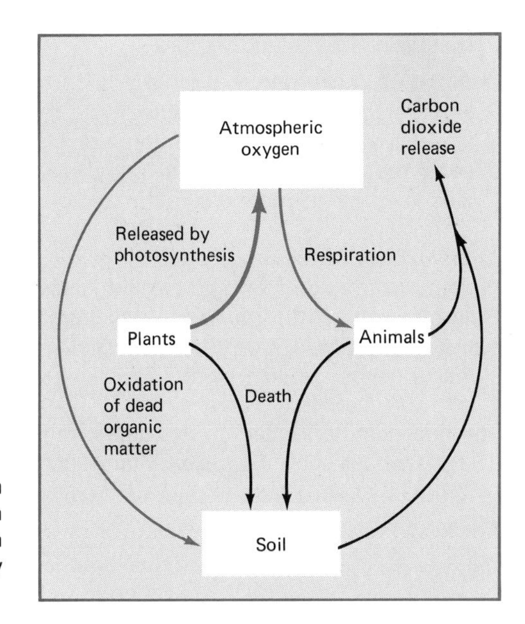

FIGURE 3.35

Interactions of atmospheric oxygen and the living world. All the oxygen of the atmosphere cycles through plants and animals about once every 3,000 years.

In the second process, oxygen is released from water molecules, not directly by solar energy, but indirectly through the life processes of green plants (Figure 3.35). During photosynthesis, plants use solar energy to separate water molecules into hydrogen and oxygen; the hydrogen is combined with carbon dioxide in a complex series of reactions to form carbohydrates, and the oxygen is released as a waste product. The carbohydrates (principally starches, sugar, and cellulose), in turn, provide a fundamental energy source, not only to the synthesizing green plants themselves, but also to animals, who are all ultimately dependent on such plant materials for their nutrition.

In discussing Precambrian banded iron formations (p. 78), we saw that the initial free oxygen produced by green plants was probably consumed in oxidizing iron dissolved in the surrounding waters. As this material was depleted, photosynthesis began to release oxygen into the atmosphere. Most of this free oxygen, then and now, does not long remain in the atmosphere. Instead, as the plants that produce the oxygen and the animals that eat the plants die and decay, their carbon-rich organic materials are recombined with atmospheric oxygen to produce carbon dioxide, and the cycle begins anew (Figure 3.35). However, a small fraction of the carbon-rich organic material does *not* recombine with atmospheric oxygen, but becomes buried in sedimentary rocks. This burial of organic carbon not only removes carbon dioxide from the atmosphere (along with the other processes of carbon removal mentioned earlier), but it also produces a long-term excess of atmospheric oxygen. Thus, the slow accumulation of plant-produced organic materials in sedimentary rock has indirectly permitted free oxygen to accumulate in the atmosphere over the long course of earth history.

Chapter Summary

Precambrian Rocks and Geography

Precambrian chronology: Each Precambrian shield region can be divided into subregions based on radiometric dates that indicate the most recent episode of deformation of the subregion.

Greenstone belts are Archean (early Precambrian) deep-water sedimentary and volcanic sequences that formed on the thin, unstable crust of the early earth.

Archean granitic rocks surround the greenstone belts and make up the thickest parts of the continental crust; they may represent the initial separation of lighter continent-forming minerals from heavier materials of the mantle.

Lunar history: Most of the moon's rocks and features originated before 3.1 billion years ago in the early Precambrian phase of earth history; thus they provide clues to the early earth.

The first continents had consolidated and stood high above the surrounding ocean basins by 2.4 billion years ago; they received shallow-water sediments that contrast sharply with those of the earlier greenstone belts.

Evolving Precambrian Environments

Banded iron formations are thick sequences of iron-rich sedimentary cherts that are abundant only in early Precambrian rocks; they may reflect conditions in the ocean before free oxygen was present in the atmosphere.

Stromatolites are moundlike accumulations of calcium carbonate produced by algae; they are especially abundant in late Precambrian deposits.

Tillites were deposited by ancient glaciers and ice sheets at least twice during Precambrian time, at about 2200 and 700 million years ago.

Evolution of the Ocean and Atmosphere

Fluids from the earth's interior: The paucity of heavy, stable gaseous elements in the present atmosphere indicates that the ocean and atmosphere are not original features left over from the earth's initial accumulation but have, instead, been released from its interior through volcanic activity.

The ocean through earth history: Oceans have been present since early Precambrian time, but the volume and precise composition of these early oceans are uncertain.

The changing atmosphere

The earth, Venus, and Mars: The present atmospheres of Venus and Mars are dominated by carbon dioxide, as most probably was the early atmosphere of the earth.

Carbon dioxide and the ocean interact to remove CO_2 from the atmosphere and deposit it as calcium carbonate sediments; this removal has left nitrogen as the principal atmospheric gas.

The atmosphere and life: The oxygen of the present atmosphere has been added through geologic time by the life processes of green plants.

Important Terms

algal mat
Archean Eon
banded iron formation
craton
epeiric sea
mobile belt
orogenic front
orogeny
outgassing
partial melting

eugeosyncline
felsic igneous rocks
fractional crystallization
geosyncline
geotherm
graded bedding
greenstone belt
greywacke
mafic igneous rocks
maria
miogeosyncline
photosynthesis
pillow lava
platform
Precambrian shield
redbed
stromatolite
structural province
tillite
turbidite
turbidity current
ultramafic igneous rock
volcaniclastic rock

Review Questions

1. What are greenstone belts? What do they suggest about the earliest phases of earth history?
2. Summarize the nature and possible origin of the earliest continental rocks.
3. What do banded iron formations indicate about the early atmosphere?
4. How did stromatolites form? What do they suggest about the Precambrian atmosphere?
5. What is the most probable source for the fluid of the ocean and atmosphere? Why?
6. How do the present atmospheres of Venus and Mars differ from that of the earth?
7. What processes have acted to remove most carbon dioxide from the earth's atmosphere?

Additional Readings

Anhaeusser, C. R. "Precambrian Tectonic Environments," *Annual Review of Earth and Planetary Sciences,* vol. 3, pp. 31–53, 1975. *Comprehensive advanced review.*

Moorbath, S. "The Oldest Rocks and the Growth of Continents," *Scientific American,* vol. 236, no. 3, pp. 92–104, 1977. *A summary stressing the similarities of ancient and more recent rock-forming processes.*

Ponnamperuma, C. (ed.) *Chemical Evolution of the Early Precambrian,* Academic Press, New York, 1977. *An authoritative advanced review.*

Society of Economic Geologists "Precambrian Iron Formations of the World (A Symposium)," *Economic Geology,* vol. 68, pp. 913–1179, 1973. *Technical review of these enigmatic sediments.*

Tarling, D. H. *Evolution of the Earth's Crust,* Academic Press, New York, 1978. *An advanced survey stressing Precambrian history.*

Walker, J. C. G. *Evolution of the Atmosphere,* MacMillan, New York, 1977. *An advanced review with extensive bibliography.*

Walter, M. R. "Interpreting Stromatolites," *American Scientist,* vol. 65, pp. 563–71, 1977. *Excellent survey of these Precambrian fossils and their present-day relatives.*

Windley, B. F. (ed.) *Early History of the Earth,* John Wiley, New York, 1976. *A review with papers at varying levels of detail.*

THE EXPANSION OF LIFE

4

In Chapter 3 we saw that the life processes of plants and animals are directly related to the cycles and quantities of the earth's atmospheric gases. In addition, the chemistry of the oceans and the deposition of sedimentary rocks are also profoundly affected by living systems. Thus, the history of each principal component of the earth—rock, water, and atmosphere—is closely linked to the history of life, a major topic in the chapters that follow. This chapter begins our survey of ancient life by considering how organisms may have originated and expanded during the earliest phases of the earth's physical history.

Earliest Life

Like the earliest differentiation of the solid earth, ocean, and atmosphere, life on earth probably originated in the obscure beginnings of earth history. Since the remains of ancient organisms are found in sedimentary rocks of almost all ages, including some formed over 3 billion years ago, the beginnings of life must have taken place still earlier. Because we have no direct record of the transition between nonliving and living systems, we are forced to rely on the indirect clues provided by the probable chemical environment of the early earth.

Components of Life

In outward appearance, the living world is extraordinarily complex. Pine trees bear no resemblance to mosquitoes, nor zebras to amoebas. Thus, it is hard to imagine that all life is made up of only a few essential chemical compounds. Chemically, nevertheless, all organisms are endlessly varied combinations of five principal constituents: *water, carbohydrates, fats, proteins,* and *nucleic acids* (Figure 4.1). These major constituents, and most minor ones as well, are compounds dominated by the elements hydrogen, oxygen, carbon, and nitrogen. Carbon, in particular, is fundamental because of its unique chemical properties that enable it to link with other elements into the large and complex molecules characteristic of living organisms. The first requirement for life, then, was a supply of these chemical elements in forms that might combine

FIGURE 4.1

PRINCIPAL CHEMICAL CONSTITUENTS OF LIFE

Compounds	Functions	Composition
Water	Universal solvent	Hydrogen, oxygen
Carbohydrates	Energy source	Hydrogen, oxygen, carbon
Fats	Energy storage	Hydrogen, oxygen, carbon
Proteins	Structural; facilitation of chemical reactions	Hydrogen, oxygen, carbon, nitrogen, phosphorus, sulfur (organized into 20 amino acid "building blocks")
Nucleic acids (DNA, RNA)	Patterns for protein construction	Hydrogen, oxygen, carbon, nitrogen, phosphorus (organized into 5 nucleotide "building blocks")

into the components of living systems. The most likely form for these light elements would have been in atmospheric gases.

In Chapter 3 we saw that the earth's atmosphere was derived from volcanic outgassing of rock material. Present-day volcanic gases are made up largely of carbon dioxide, plus smaller amounts of water vapor, nitrogen, hydrogen, methane (CH_4), and other gases. The atmospheres of Venus and Mars, the earth's neighboring terrestrial planets, are similarly dominated by carbon dioxide. These facts have led most geologists to conclude that the earth's earliest atmosphere probably contained an abundance of carbon dioxide, along with smaller quantities of nitrogen, water vapor, and other gases.

"Primitive Earth" Experiments

It is, of course, a long step from an atmosphere containing hydrogen, carbon, oxygen, and nitrogen to even the simplest living systems. Atmospheric evidence suggests that all the principal elements now found in organisms were abundant, in some form, on the early earth. This supposition was about as far as one could reasonably go until 1953, when S. L. Miller performed a now-classic experiment on the origin of life. Miller constructed an apparatus for circulating steam through a mixture of ammonia, methane, and hydrogen (Figure 4.2). The steam-gas mixture was subjected to a high-energy electrical spark and condensed to a liquid; the cycle was then begun again by heating the water to steam. After a week of repeatedly being cycled through the gases and electrical

FIGURE 4.2

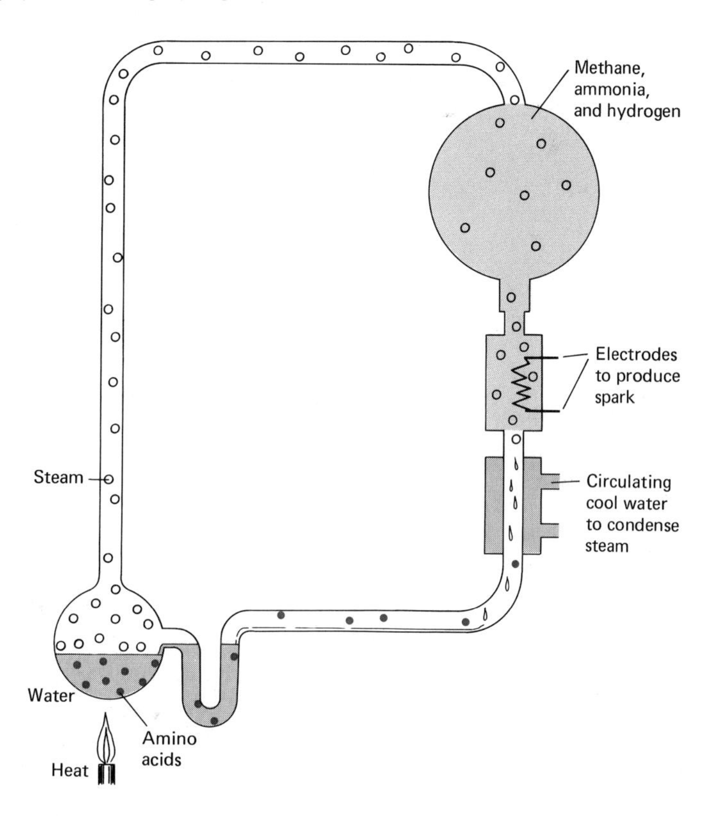

Miller's apparatus for producing the organic building blocks of life from a simulated "primitive atmosphere" of methane (CH_4), ammonia (NH_3), and hydrogen (H_2). When exposed to a high-energy electrical spark simulating natural lightning, amino acids are formed.

discharges, the condensed water in the apparatus had become deep red and turbid. On analysis, the water was discovered to contain a complex mixture of **amino acids**, the basic structural units of proteins. Miller's simple experiment showed that an electrical discharge such as lightning in a primitive atmosphere of ammonia, methane, and hydrogen could have led to the production of some of the complex molecules of living systems. This discovery opened a new field of investigation, the experimental synthesis of the chemical constituents of life under primitive earth conditions.

Since 1953, many such experiments have been performed employing various mixtures of gases that probably were present in the early atmosphere (such as carbon dioxide, water vapor, methane, nitrogen, ammonia, and hydrogen) and various possible energy sources (such as strong solar radiation, lightning, thunderlike sound waves, or meteorite shock waves), any of which might have caused the relatively simple atmospheric compounds to combine into larger and more complex molecules. Although none of these experiments has yet produced anything approaching the complexity of even the simplest organism, they have succeeded in showing that a variety of the complex chemical building blocks (amino acids) that make up life could have been present in the early ocean and atmosphere.

These studies have led scientists to visualize a time early in earth history when the surface was covered with oceans or lakes rich in the molecules that are fundamental to life. The waters of these oceans or lakes have often been described as a "dilute organic soup." This concept was first developed in the 1920s and 1930s by the English biologist J. B. S. Haldane and the Russian biochemist A. I. Oparin, pioneer workers on the origin of life. In the burst of interest that followed Miller's experiment, the ideas of Haldane and Oparin were greatly expanded, and a variety of hypotheses were formulated to explain how the first self-duplicating organisms developed from the nonliving building blocks of the early organic soup.

However, most of these hypotheses have difficulty in explaining how relatively small and uniform amino acid molecules became linked into much larger and more complex proteins. Such linkage seldom takes place spontaneously in the presence of water. but it does occur when amino acids are partially dried. For this reason, some researchers stress the importance of periodically dried environments, such as beach or tide pools, as sites for the synthesis of larger molecules.

A still more fundamental difficulty concerns how such molecules, once formed, were capable of self-duplication. It is known that, under certain rather restricted conditions, organic compounds suspended in water tend to aggregate spontaneously into small spheres bounded by a wall or membrane that separates this material from the more dilute surrounding liquid. How such spherical concentrations of organic compounds could develop the extraordinarily complex reproductive mechanisms of even the simplest living cell remains a puzzle.

Simple Present-Day Life

Because there is no direct experimental evidence of the transition from complex carbon molecules to actual living cells, perhaps that best glimpses of early life are provided by simple organisms that still exist today. All organisms must have a continuous supply of the basic components of life—carbon, hydrogen, oxygen, and nitrogen—for essential structural proteins and enzymes. However, living systems obtain these essential materials in two fundamentally different ways (Figure 4.3).

FIGURE 4.3

PRINCIPAL SOURCES OF STRUCTURAL MATERIALS IN PRESENT-DAY LIFE

Inorganic CO_2 (Autotrophs)	**Organic compounds (Heterotrophs)**
Green plants Some bacteria	Animals Nongreen plants (fungi) Most bacteria

Some organisms, called **autotrophs**, can synthesize structural materials *directly* from atmospheric carbon dioxide and water. The most familiar present-day autotrophs are green plants, which utilize solar energy to convert water and atmospheric carbon dioxide into carbohydrates through a complex series of reactions known as photosynthesis. The carbohydrates so produced then become a secondary energy source for the manufacture of structural proteins; the necessary nitrogen is supplied as dissolved nitrogen compounds in the water utilized by the plant. (This is the reason why nitrogen is a principal component in fertilizers used to stimulate plant growth.) Although most autotrophs obtain nitrogen from nitrogen compounds dissolved in water, a few, called nitrogen fixers, are able to utilize gaseous nitrogen from the atmosphere for the production of proteins. These autotrophs have the simplest possible nutritional requirements, for they can subsist on atmospheric carbon dioxide and nitrogen, plus water. Included in this group are certain microscopic bacteria, as well as some algae.

In contrast to the diverse array of autotrophs, an equally diverse spectrum of organisms, called **heterotrophs**, derive their structural materials in a different way. They cannot utilize atmospheric carbon dioxide as a carbon source; instead, they are dependent on organic molecules produced by autotrophs, principally proteins, for their structural materials. All animal life, as well as most bacteria and some nongreen plants such as fungi, fall into this category.

At first glance it would seem that the earliest life might have been simple autotrophs, perhaps similar to some present-day photosynthetic bacteria that manufacture all their components from atmospheric carbon dioxide and nitrogen. Biochemists are convinced, however, that the chemistry of photosynthesis is far too complex to have arisen in the earliest living systems. It appears more probable that the first organisms were heterotrophs that depended on nonbiological carbon compounds for structural materials. In other words, they must have "eaten" the organic soup from which they arose.

From a structural standpoint, the simplest modern organisms are microscopic bacteria and blue-green algae. These two groups, collectively known as **procaryotes**, have patterns of cell structure and function that sharply separate them from all other life. So fundamental are these differences that they serve to divide all present-day life into two great subcategories whose distinctions are far more basic than are the familiar differences between the plant and animal kingdoms.

The cells of bacteria and blue-green algae lack a chromosome-bearing nucleus to transmit genetic messages during cell reproduction. All other organisms, from single-celled plants and animals to oak trees and elephants, are made up of cells having such a nucleus. Nucleus-bearing organisms are collectively known as **eucaryotes**. In addition to the presence or absence of a cell nucleus, procaryotes and eucaryotes differ in many other basic structural features, some of which are summarized in Figure 4.4. These differences indicate that procaryotes generally have less complex genetic systems and metabolic processes than do eucaryotes. This distinction further suggests that the earliest life may have had procaryotic structure, and that eucaryotes arose from procaryotic ancestors. We shall see that there is independent geological and biological evidence for this conclusion.

FIGURE 4.4

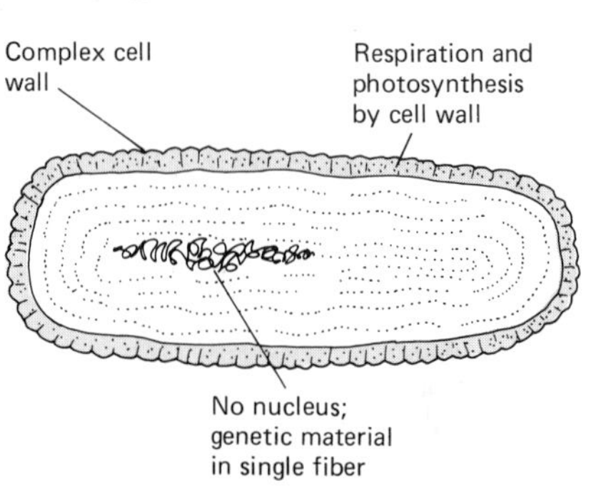

Procaryotic and eucaryotic cells

Limits of Life

As comfort-loving animals, we human beings tend to define the optimum conditions for life in terms of those that *we* find most satisfying: temperature around 20°C, humidity relatively low, and plenty of pure, fresh air. In spite of these rather narrow preferences, the human species has adapted to an unusually wide range of environmental extremes, from frozen tundras to steamy rain forests. Far greater adaptive versatility, however, is found among simpler forms of life, particularly procaryotic bacteria and blue-green algae, some of which survive in extraordinarily rigorous settings.

Temperature and humidity, critical factors in human comfort, affect all living systems, but with differing limits. Thus, certain bacteria thrive in hot spring waters at temperatures above the boiling point, but only in those deeply buried waters where the pressures of the overlying rock cause the water to remain liquid, even though it is heated above 100°C. The presence of water *in the liquid state,* rather than temperature, appears to set the upper thermal limit for bacterial life. (Only bacteria, it should be noted, can survive in such extremes; blue-green algae are not found at temperatures above 75°C, eucaryotic algae do not survive above 60°C, and animal life ceases at 50°C.)

A somewhat similar pattern is found at extremely low temperatures. A great variety of animals and plants can survive at average temperatures only slightly above freezing; in addition, certain bacteria and eucaryotic fungi can survive indefinitely at constant temperatures well below freezing. Here, again, the *presence of liquid water* is critical. Such forms have high concentrations of internal salts that maintain liquid water at temperatures as low as −20°C.

Even in less rigorous temperatures, water remains the critical limiting factor. The only completely sterile regions on earth are the few desert and polar environments that totally lack water. However, certain bacteria, as well as a few higher plants and animals, have developed dormant stages in which their tough walls resist drying, so that they can survive indefinitely without water. When water is again present, they quickly revive and renew their life processes.

In addition to extremes of temperature and humidity, some procaryotes survive in a host of other seemingly hostile environments. Most blue-green algae, for example, are extraordinarily tolerant of strong ultraviolet radiation; they readily survive doses that would be dangerous or lethal to human beings and most other organisms. Certain bacteria can grow in concentrated sulfuric acid so strong that it would cause serious burns on human skin. Still others thrive in high concentrations of carbon monoxide or other noxious gases that would be fatal to most higher organisms.

Such examples emphasize that we must not judge the ultimate limits of life by our own narrow human requirements. Clearly, some forms of early procaryotes could have thrived under environmental rigors that would never be tolerated by the familiar eucaryotes that arose much later in earth history.

Fossil Evidence of Early Life

The only direct evidence of early life is that provided by the study of fossils, remains of past organisms that were preserved in ancient sedimentary rocks. We have already seen that fossils are not equally abundant throughout the earth's long sedimentary record. They are rare in Precambrian rocks, being mostly concentrated in rocks of the past 600 million years. This comparatively abundant fossil record provides a universal means of dating and subdividing

FIGURE 4.5

Stromatolites (layered masses of limestone deposited by the metabolic activities of blue-green algae). These are 2-billion-year-old examples from Great Slave Lake, Canada. (*Courtesy of Paul Hoffman*)

the Phanerozoic interval of earth history; such dating will be the subject of Chapter 5.

Until rather recently it was believed that Precambrian sedimentary rocks contained *no* unequivocal indications of ancient life. Since the 1950s, however, it has become clear that simple procaryotes, both bacteria and blue-green algae, were abundant in certain favorable environments throughout most of the long Precambrian phase of earth history.

The first clues to Precambrian life came from stromatolites, the layered masses of Precambrian limestone discussed in Chapter 3. Similar layered deposits occur also in Phanerozoic sedimentary rocks and are being formed today in shallow, warm oceans by the metabolic activities of blue-green algae. Because of their resemblance to these younger deposits, Precambrian stromatolites have long been suspected to be the metabolic remains of Precambrian blue-green algae (Figure 4.5). Somewhat similar materials, however, can also originate by direct precipitation of silica or calcium carbonate without the intervention of life. For this reason, the algal origin of Precambrian stromatolites was long a subject of debate. This uncertainty has been resolved by the discovery of actual preserved remains of ancient algae and bacteria in many Precambrian stromatolites during the past two decades. This discovery has revolutionized our understanding of the living world in the long Precambrian interval of earth history.

Fossil Procaryotes

For at least the past 50 years, scattered findings of microscopic fossils from Precambrian rocks were reported, but it was not until 1954 that the first convincing discovery of ancient fossil procaryotes was made. These fossils were found in the Gunflint Chert, a formation of Precambrian stromatolites almost 2 billion years old exposed along the shores of Lake Superior. Since 1954, several dozen additional localities, ranging in age from about 600 million to over 3 billion years, have also yielded Precambrian procaryotes, yet the original Gunflint discovery remains one of the most diverse and well-preserved records of these early fossils (Figure 4.6).

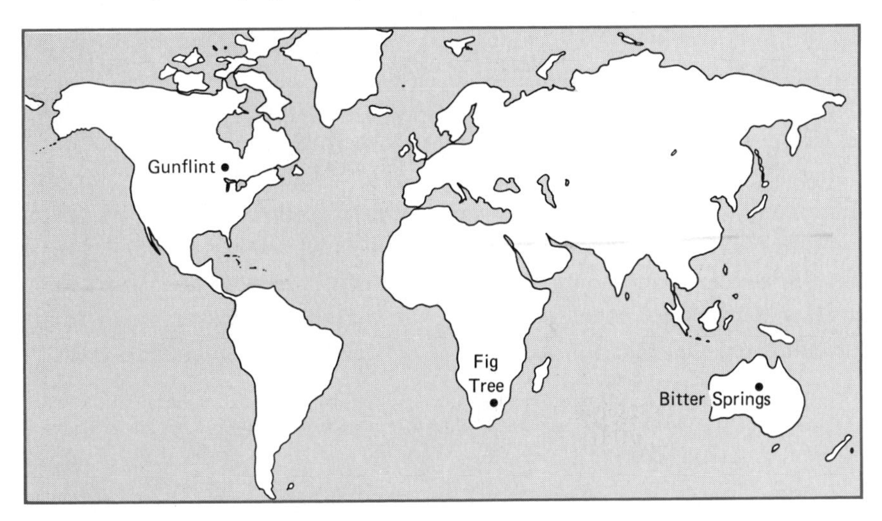

FIGURE 4.6

Important fossil-bearing exposures of Precambrian rocks.

Gunflint Chert Fossils The Gunflint Chert is exposed in northern Minnesota and adjacent Ontario as part of a series of Precambrian sedimentary rocks about 15 meters thick that overlies an older volcanic and granitic terrain. The fossil-bearing unit occurs in the middle of the sedimentary sequence as a three-meter layer made up of thin, alternating layers of white, red, and black cherts and iron-bearing carbonates. The cherts occur in large, dome-shaped stromatolitic masses, which, like similar Precambrian stromatolites, were long presumed to be algal deposits. In 1954 S. A. Tyler, a field geologist, and E. S. Barghoorn, a paleobotanist, dissolved some of the chert in hydrofluoric acid and discovered a residue of tiny filaments and spheres that were obviously fragments of primitive organisms. The fragments had apparently become embedded soon after death in a gelatinlike silica, which must have hardened rapidly into a dense chert that protected the delicate structures from extensive deformation and decay. Most Precambrian procaryotes discovered since have been found in similar cherts. They are best studied by examining very thin slices of the chert under high magnification with optical and electron microscopes. Such studies have shown that Gunflint organisms fall into three broad categories:

1 Thin threads, some with wall-like partitions, that closely resemble present-day filamentous bacteria and blue-green algae (Figure 4.7*a*).
2 Spherical bodies, probably of diverse origin. Some resemble present-day bacteria and unicellular blue-green algae; others may be reproductive spores of blue-green algae; still others are unlike any present-day organism (Figure 4.7*b*).
3 Star-, umbrella-, and parachute-shaped bodies of unknown affinities (Figure 4.7*c*).

The similarities to present-day procaryotic bacteria and blue-green algae of the first two kinds of structure are extremely important, for these are precisely the kinds of organisms that you would expect to find early in the history of life. Although some modern bacteria are autotrophs, most are not, deriving their nutrients instead as parasites or from the decay of dead organisms. Modern blue-green algae are all photosynthetic autotrophs. It is probable that many of the Gunflint fossils similarly derived their nutrients by photosynthesis from atmospheric carbon dioxide and solar energy. Radiometric age determinations show the rocks of the Gunflint Chert to be about 1.9 billion years old; the process of photosynthesis was, therefore, almost certainly developed by that time.

FIGURE 4.7

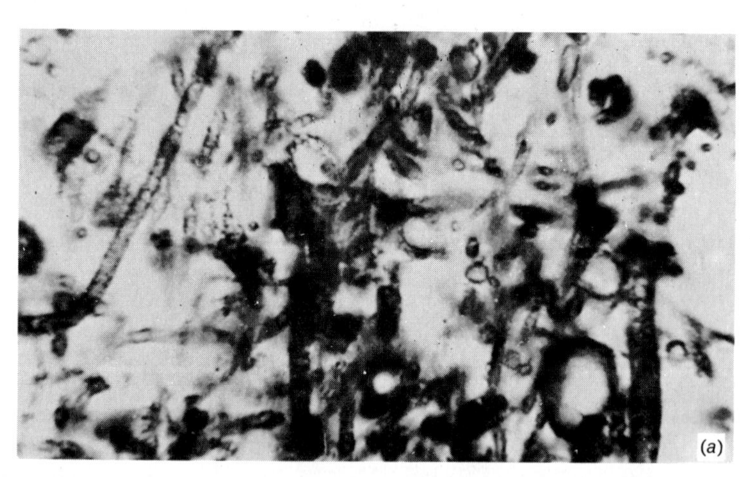

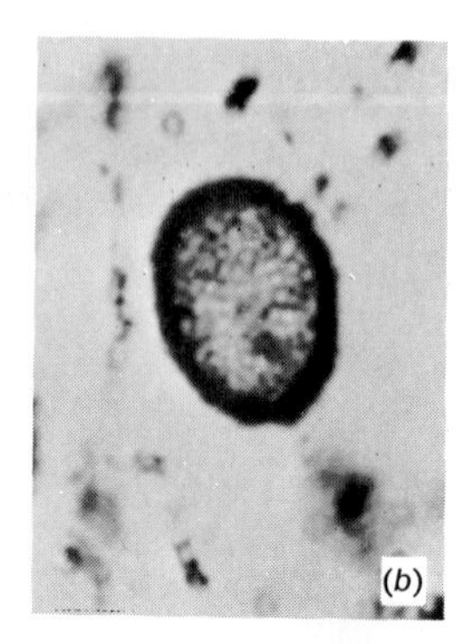

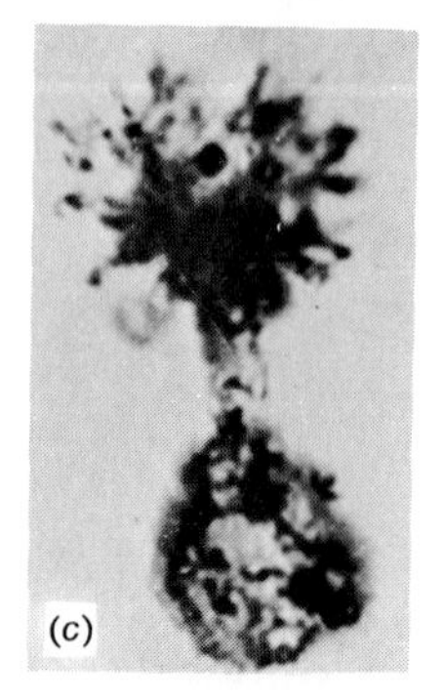

Procaryote remains from the 1.9-billion-year-old Gunflint Chert of Ontario, Canada magnified from 1,000 to 2,000 times: (*a*) thread-shaped forms that closely resemble modern filamentous bacteria and blue-green algae; (*b*) spherical form resembling a modern bacterium; (*c*) parachute-shaped form of unknown affinities. (*Courtesy of Elso Barghoorn*)

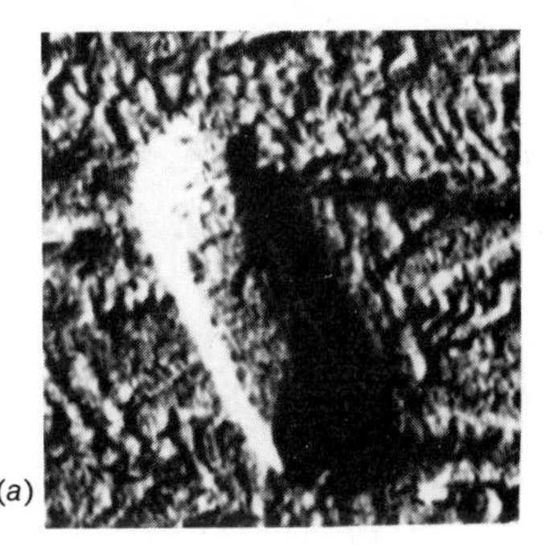

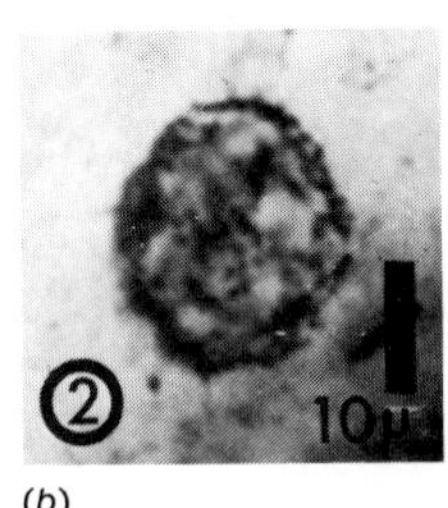

FIGURE 4.8

Probable procaryote remains from the Fig Tree Group of South Africa, magnified from 1,000 to 2,000 times. These fossils, over 3.2 billion years old, are among the oldest preserved remains of life. (*a*) Rod-shaped forms resembling bacteria. (*b*) Spherical forms with a granular surface, possibly a simple blue-green alga.

Fig Tree Group Fossils Because the Gunflint Chert fossils include rather *advanced* photosynthetic procaryotes, there is great interest in finding older Precambrian fossils that might reflect still earlier stages of procaryotic development. Unfortunately, only a few fossil-bearing localities have been found in rocks significantly *older* than the Gunflint. Of these, perhaps the most important occur in the Fig Tree Group of South Africa, a thick series of sedimentary shales, sandstones, and cherts exposed in the Barberton Mountains of the Transvaal Region (Figure 4.6). Radiometric dating shows that the Fig Tree sediments were deposited more than 3.2 billion years ago, which makes them among the oldest sedimentary rocks yet discovered.

As with the Gunflint fossils, the microscopic Fig Tree organisms are preserved in cherts; however, they are far less diverse and well-preserved than are the Gunflint fossils. Two principal forms occur:

1 Rod-shaped structures whose size, shape, and cell-wall structure suggest that they are bacterial remains (Figure 4.8*a*).
2 Spherical structures with a granular surface that resemble certain simple blue-green algae in size and shape (Figure 4.8*b*).

The presence of bacterialike procaryotes so early in earth history is not surprising. But if the spherical Fig Tree structures are indeed the remains of blue-green algae (a conclusion that is still uncertain), then photosynthetic autotrophs must have been present from close to the beginning of the Precambrian sedimentary record. If that is so, the origin of photosynthesis, like the origin of life itself, must have occurred in the dim pregeological interval of earth history. It is also significant that the oldest known stromatolites, which were almost certainly formed by blue-green algae, were deposited about 2.8 billion years ago; thus they are only slightly younger than the Fig Tree microfossils. This too indicates that algal photosynthesis developed very early in the preserved record of life.

Bitter Springs Formation Fossils In addition to the Gunflint Chert, preserved microfossils have been found in several dozen younger Precambrian localities. Of these, perhaps the most significant occur in the Bitter Springs

Formation of central Australia (Figure 4.6). This formation consists of limestones, sandstones, and cherts deposited only about 900 million years ago, not long before the sudden expansion of animal life that began the Phanerozoic Eon about 600 million years ago.

Like the Gunflint Chert, the Bitter Springs Formation contains procaryotic bacteria and blue-green algae preserved in cherts. As might be expected, these bacteria and algae are more modern forms than the much older Gunflint fossils. The blue-green algae in particular are more diverse and show closer similarities to still-living representatives of the group. Several forms, in fact, are practically indistinguishable from certain present-day blue-green algae, which makes these the longest-lived of any modern organism.

The Bitter Springs Formation also contains small spheres that enclose what appear, at first glance, to be degraded remains of cell nuclei. Such structures would be an important discovery, because only eucaryotic cells have nuclei. These fossils were first accepted as the earliest eucaryotes, since they had the same general size and shape as certain eucaryotic green algae living today. However, restudy has suggested to some paleontologists that the supposed cell nuclei are actually the shrunken remains of the *entire* original cell. Thus it appears likely that the Bitter Springs deposits, as all older Precambrian rocks yet examined, do not contain fossil eucaryotes.

Early Eucaryotes

The first unmistakable fossil eucaryotes are relatively complex animals and plants found in late Precambrian rocks. Their earlier eucaryote ancestors were most probably single-celled forms similar to certain simple green algae* that still exist today. Such simple forms probably gave rise to a host of microscopic, single-celled eucaryotic plants and animals in late Precambrian time, which, in turn, developed into larger, multicelled plants and animals (Figure 4.9). In order to distinguish them from their smaller, single-celled relatives, such multicelled eucaryotes are known as **metaphytes** ("changed plants") or **metazoans** ("changed animals").

Scattered and rather poorly preserved remains of both metaphytes and metazoans are found in late Precambrian rocks from almost every continent. The first really good look at eucaryotic life is provided by a remarkable association of metazoan fossils from Australia. These were discovered in 1947, when an Australian geologist found some rounded impressions resembling fossil jellyfish in sandstones of the Ediacara Hills in south Australia. These sandstones were presumed to be of Cambrian age; because such impressions are relatively common in Cambrian rocks, the discovery passed without notice. Several years later, two private collectors discovered not only jellyfish, but also impressions of segmented worms and other, more puzzling organisms in the same rocks. These additional fossils, which did not resemble any known Cam-

*See the classification of organisms in Figure 4.12. Note that only one algal group, the blue-green algae, are procaryotes. All other algae, as well as all higher plants and animals, are eucaryotes.

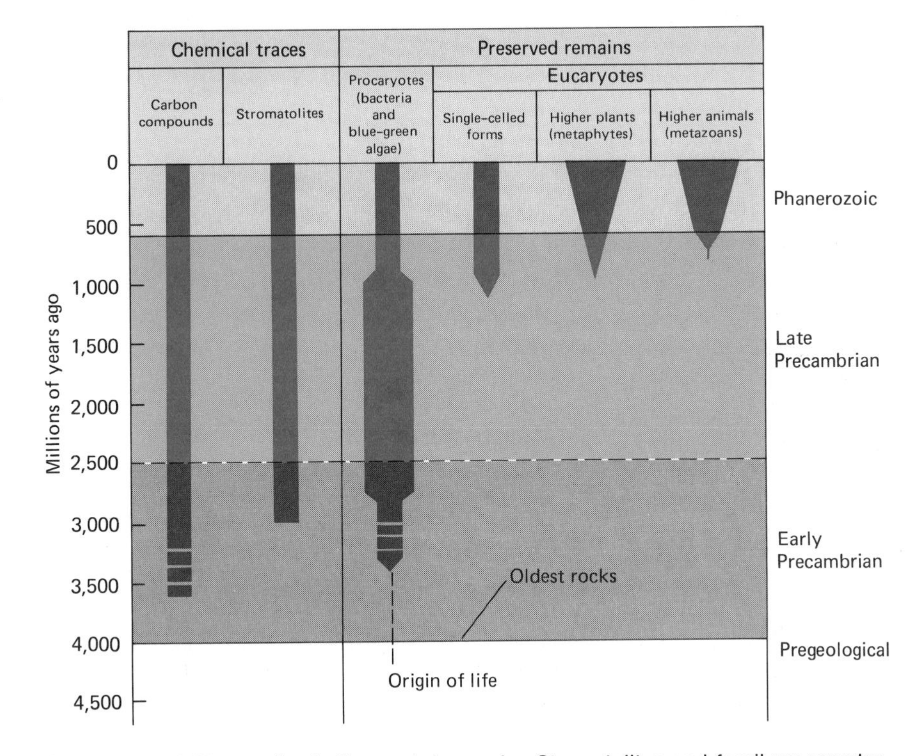

FIGURE 4.9

Distribution of life remains in Precambrian rocks. Stromatolites and fossil procaryotes (bacteria and blue-green algae) are found through much of the Precambrian record, but eucaryotic remains occur only in late Precambrian deposits.

brian organisms, prompted an intensive restudy of the Ediacara material by M. F. Glaessner, a paleontologist at the University of Adelaide. Glaessner's work soon showed that the strange fossils occurred in a continuous sequence of flatlying sedimentary rocks in which the only other fossils were typical early Cambrian forms. These other fossils were found about 150 meters stratigraphically above the problematic fauna. Because of the distinctiveness of the Ediacara fossils and their occurrence *below* typical Cambrian forms, it seems highly probable that they represent late Precambrian organisms; their estimated age is 650 to 700 million years.

Several thousand specimens have thus far been collected from the Ediacara Hills, mostly rather large organisms measuring several centimeters in length. All are preserved as impressions of the original animal along bedding planes of sandstone. The principal types discovered include:

1 Rounded impressions, some with radiating grooves, that resemble modern jellyfish (primitive animals of the group known as coelenterates). These are the only fossils in the Ediacara fauna that are also common in younger rocks (Figure 4.10*a*).
2 Impressions of stalklike fronds with grooved branches that suggest present-day sea pens (primitive colonial marine animals that are also coelen-

terates) (Figure 4.10*a*). Similar, but less well-preserved, impressions have been reported from late Precambrian rocks of southwest Africa, England, and elsewhere.

3 Elongate, wormlike impressions consisting of a horseshoe-shaped head followed by about 40 identical segments. These impressions probably represent animals similar to present-day segmented worms of the advanced group called annelids (Figure 4.10*b*).

4 Rounded, flattened, wormlike impressions with a central groove and strong segmentation. Traces of an intestinal tract have been reported, suggesting a link to somewhat similar-looking present-day annelid worms (Figure 4.10*c*).

The Ediacara fossils thus include two recognizable present-day animal groups: (1) the *coelenterates*, a group of primitive, multicellular organisms that includes modern corals, jellyfish, and sea anemones; and (2) the *annelids*, a more advanced group of segmented worms that includes the modern earthworms.

FIGURE 4.10

The oldest known fossil animals from the late Precambrian Ediacara Hills of south Australia. The fossils are preserved as impressions in sandstone. (*a*) Rounded, jellyfish-like forms and a branched sea pen-like form (upper right) (one-third natural size). (*b*) Elongate, worm-like form (magnified two times). (*c*) Rounded, worm-like form (actual size). (*Courtesy of M. F. Glaessner*)

(*b*)

(*a*)

(*c*)

These earliest fossil metazoans, although relatively simple and unspecialized animals, were nevertheless enormously more complex than were their presumed single-celled eucaryotic ancestors. Not only were the Ediacara animals many-celled creatures with highly differentiated systems of muscles, nerve cells, and food gathering organs, but they also were probably capable of sexual reproduction, as are all present-day metazoans. Regrettably, these dramatic evolutionary advances are not yet documented by still earlier Precambrian fossils, which might show the transitional stages between such highly developed animals and the simpler eucaryotes.

The End of Precambrian Time

The Ediacara fauna is made up entirely of soft-bodied, wormlike and jellyfishlike animals preserved as impressions in sandstone. Preservation of this type of animal requires rather special conditions of burial and original sediment chemistry. Consequently, fossil remains of soft-bodied life are rare, compared to the more easily preserved animal hard parts, such as shells or skeletons. Apparently shell or skeleton-bearing animals had not yet evolved when the Ediacara sediments were deposited (about 650 to 700 million years ago), because such remains are everywhere absent from rocks of that age or older. The Ediacara animals, however, clearly foreshadow what is probably the most dramatic single event in the entire history of life—the rapid expansion of shell-bearing animals that marks the end of the long Precambrian interval. Such animals appear abundantly and almost simultaneously in rocks about 550 to 600 million years old that are found today on every continent. Their entry upon the scene marks the beginning of the Cambrian Period and the start of the Phanerozoic Eon—our subject throughout the remainder of this book.

The Cambrian expansion of animal life consists of not just one or two new adaptations. Remarkably, most shell-bearing marine animals still found in the oceans today are represented in primitive forms in Cambrian sedimentary rocks. In Chapter 7 we shall look more closely at these ancestral marine animals and trace their history through the Cambrian and succeeding periods. For the present we need only emphasize that the Cambrian fossil faunas are dominated by many-legged trilobites, which, although now extinct, were distant relatives of modern crabs, and by brachiopods, two-shelled, superficially clamlike animals that are common today in some regions (Figure 4.11). In addition, early relatives of such familiar modern groups as sponges, clams, snails, and starfish are also found.

The transition from Ediacara soft-bodied metazoans to the shell-bearing metazoans of Cambrian time raises the question of why advanced metazoans suddenly appeared so late in earth history. Most speculations offered to explain this puzzle fall into two categories: One emphasizes important evolutionary changes in the organization or interaction of the animals and plants themselves, rather than external, physical changes. The second relates the rapid metazoan expansion either to some facilitating change in the physical world, such as a more benign climate, or a change in the chemistry of the oceans or the

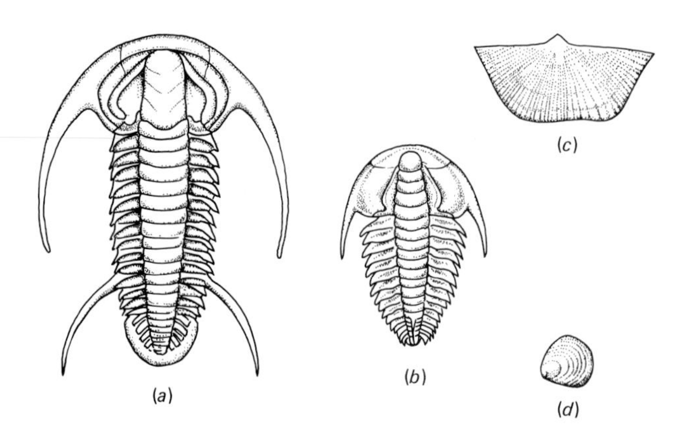

FIGURE 4.11 Typical trilobites (*a*,*b*) and brachiopods (*c*,*d*), the most abundant animal fossils in Cambrian rocks.

atmosphere. To date, no hypothesis from either category has been widely accepted. We shall here briefly review one suggestion of each type.

One recent idea suggests that the rise of multicellular animals and plants is related to the development of **sexual reproduction.** Simple one-celled organisms, whether procaryotes or eucaryotes, reproduce asexually by duplication of critical cell components, followed by a splitting of the cell into two separate individuals. More complex multicellular eucaryotes, particularly plants, sometimes reproduce similarly by budding, in which a part of the parent expands and breaks off to form a new individual. More typically, however, multicellular eucaryotes multiply by sexual reproduction, in which specialized germ cells (eggs and sperm) from different individuals unite to form a new individual. Sexual reproduction offers enormously greater potential for modifying the genetic material, structure, and adaptability of the offspring than does simple **asexual reproduction.** For this reason, some scientists have suggested that the rapid expansion of multicellular animals and plants in late Precambrian time occurred when eucaryotic life first developed the complicated mechanisms of sexual reproduction.

The second hypothesis relates the metazoan expansion to changes in the concentration of atmospheric oxygen. In Chapter 3 we noted that the earth's early atmosphere was probably made up of carbon dioxide, water vapor, and perhaps nitrogen released from the melting and volcanic outgassing of the solid earth. The present-day atmosphere contains carbon dioxide, water vapor, and nitrogen, but it differs profoundly from the early atmosphere in that it also contains large quantities of gaseous oxygen. It is assumed that most of this oxygen has been released into the atmosphere through photosynthesis by green plants (see pp. 107–109).

Some geologists now believe that the great expansion of metazoan life that marks the end of Precambrian time occurred because oxygen produced by plants first accumulated in the atmosphere at about that time. Most modern metazoans require free oxygen for their life processes. In contrast, many present-day bacteria and blue-green algae (modern descendants of the procaryotes present through much of Precambrian time) do *not* require free oxygen, either in the atmosphere or dissolved in the water surrounding them. Blue-green algae release oxygen by photosynthesis as a waste product; bacteria have different biochemical processes that do not release oxygen.

Initially, oxygen released by the blue-green algae would have combined quickly with dissolved iron and other oxygen-deficient elements and compounds in the surrounding waters and not been released to the atmosphere. (We saw in Chapter 3 that this process may account for the unique banded iron formations of early Proterozoic time.) Eventually most of these elements would have become oxidized by the algae-produced oxygen; free oxygen would then have begun to accumulate slowly in the oceans and would have bubbled to the ocean surface, to be released into the atmosphere. The development and expansion of oxygen-dependent metazoan life late in Precambrian time may thus mark the time when the atmosphere and oceans first accumulated enough free oxygen to support such forms.

The Diversification of Life

Attempts to explain the rapid diversification of late Precambrian life immediately raise the still broader question: *How can one form of life change into another?* For the remainder of this chapter we shall consider this extremely important question by reviewing current knowledge of the mechanisms that enable organisms to increase in diversity and complexity—in short, the topic of **organic evolution.**

For thousands of years, careful observers have noticed that individual animals and plants, although differing from one another in detail, tend to fall into natural groupings of similar individuals. These groupings are called **species.** Thus, all individual peach trees represent one species, all domestic dogs another, and all edible American oysters still a third. Following a scheme first proposed by the eighteenth-century Swedish botanist Carolus Linnaeus (1707–1778), scientists have customarily given each living or fossil species a distinctive two-part Latin name. For example, the peach tree is *Prunus persica;* the dog, *Canis familiaris;* and the oyster, *Crassostrea virginica.* The second part of each name *(persica, familiaris, virginica)* identifies the species; the first part *(Prunus, Canis, Crassostrea)* identifies groupings of similar species called **genera**(singular, **genus**). According to this system, the genus for peach trees *(Prunus)* also applies to plum trees *(Prunus domestica)* and to other closely related species. Similarly, the genus *Canis* includes other doglike animals such as the coyote *(Canis latrans). Crassostrea* includes, among other species, the Portuguese oyster *(Crassostrea angulata)* and the Japanese oyster *(Crassostrea gigas).* Just as related species are combined into genera, so are similar genera combined into **families,** families into **orders,** orders into **classes,** and classes into **phyla** (singular, **phylum**), the highest category (Figure 4.12). However, only two of these categories, genus and species, comprise the formal Latin name of each species.

FIGURE 4.12

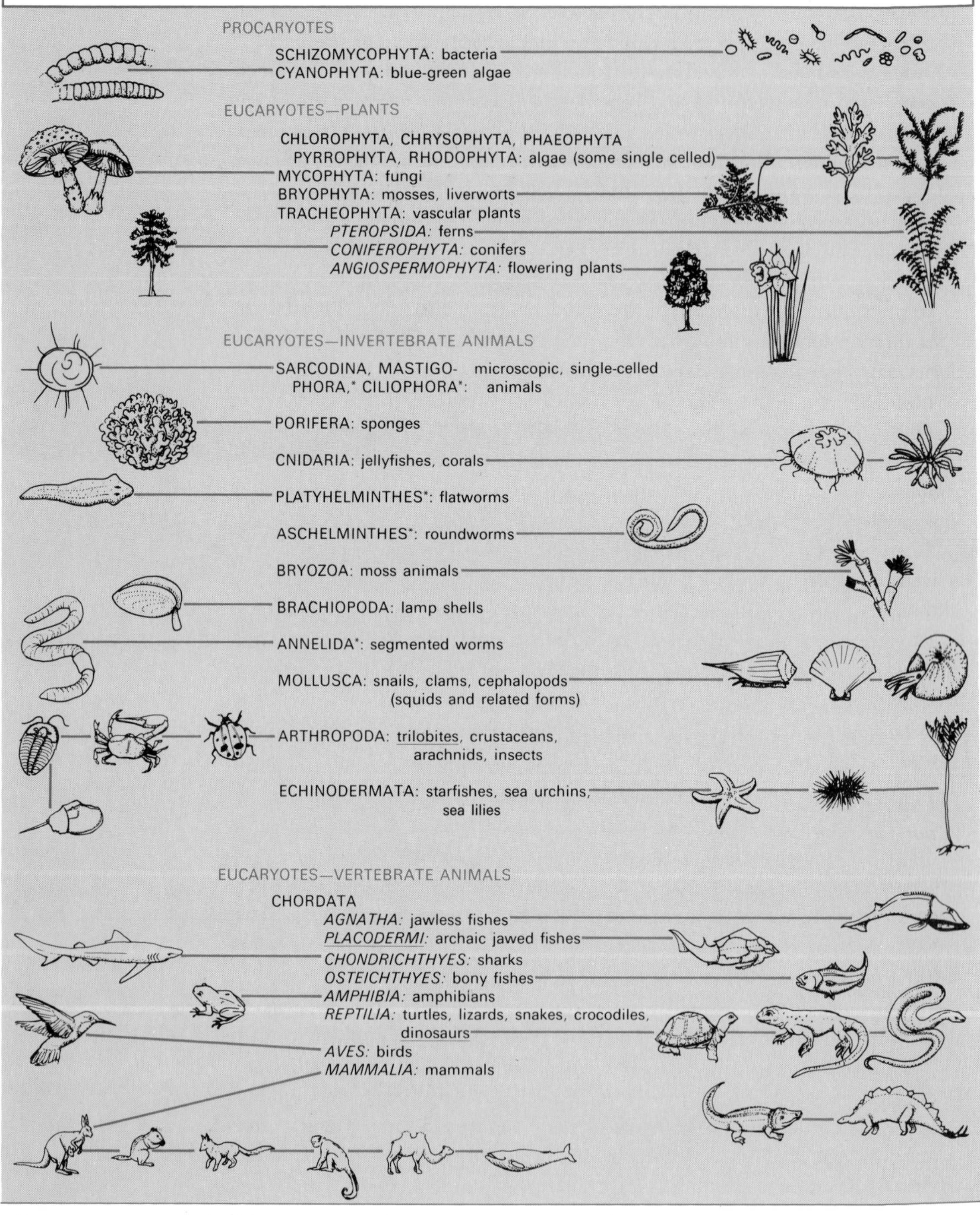

Extinct groups are underlined. Some important classes are listed for vascular plants and animals.
*Predominantly soft-bodied phyla with little or no fossil record.

Many species were carefully named and described in the early eighteenth century; but until the middle of the nineteenth century, naturalists generally agreed that each species had been created separately. They attributed such creation either to a Supreme Being or to "spontaneous generation," a vague concept according to which fully developed organisms were believed to have sprung from water, soil, or other nonliving matter. The idea that complex animals and plants developed by gradual change from simpler forms had been suggested in classical times by Greek philosophers, but it was not widely accepted until the midnineteenth century, when it was popularized by the writings of the English naturalist Charles Darwin.

Darwin and Natural Selection

Like most scientists of his day, Darwin had often debated the revolutionary idea that one form of life might change into another, but, like his contemporaries, he began by disbelieving it. The turning point came in 1837, when he realized that certain puzzling facts about South American fossil mammals and the distribution of living species on isolated islands could best be explained by the heretical idea of *species change.* The more Darwin considered the idea, the more reasonable it seemed. Yet there was one key still missing—an adequate mechanism to explain just how one species changes into another.

Two years of intensive reading and thinking were to pass before he conceived of a suitable mechanism. Recognizing the wide variation in size, shape, and strength among individuals of the same species, and observing that most organisms produce many offspring that die before maturity, he reasoned that the individuals that survive to reproduce the species must be those with the most successful combinations of variable traits. Thus, *only the "fittest" individuals pass their desirable variations on to the next generation.* To use Darwin's own example:

Let us take the case of a wolf, which preys on various animals. . . and let us suppose that the fleetest prey, a deer for instance, had . . . increased its numbers, or that other prey had decreased in numbers, during that season of the year when the wolf was hardest pressed for food. I can under such circumstances see no reason to doubt that the swiftest and slimmest wolves would have the best chance of surviving and so be preserved or selected . . . I can see no more reason to doubt this, than that man can improve the fleetness of his greyhounds by careful and methodical selection.[*]

Over many generations, this selective reproduction by successful survivors would lead to adaptive changes in the species, which would ultimately lead to new species. Darwin called this process **natural selection,** in order to distinguish it from the artificial selection practiced by breeders of domesticated animals and plants.

*Charles Darwin, *The Origin of Species,* reprint of 6th ed. (New York: New American Library, 1958), p. 95.

The Science of Genetics

It is a tribute to Darwin's genius that natural selection is still the cornerstone of modern thinking about evolution. However, early in this century, Darwin's ideas were overshadowed by equally dramatic discoveries in the new science of genetics.

The principal contribution of genetics to an understanding of evolution has been its explanation of how variability in individual organisms of the same species is inherited. It is this variability that provides the raw materials for natural selection and the rise of new species.

Geneticists have discovered that the entire process of development of the individual organism is controlled by hereditary regulators known as **genes.** Genes are constructed of the nucleic acid called **DNA (deoxyribonucleic acid).** Genes are normally found in the cell nucleus, where they are organized into larger, paired, threadlike units called **chromosomes,** each of which may contain thousands of genes. The number of chromosomes is usually constant for each species. However, it may vary between species, ranging from as few as one pair to as many as several hundred pairs. The usual number is between 5 and 30 pairs. For example human beings have 23 pairs of chromosomes. When a cell divides during normal growth, the chromosomes reproduce themselves exactly, giving the two new cells the same number and kinds of chromosomes as the original parent cell. This process of exact chromosome reduplication is called **mitosis.** In most organisms that multiply by sexual reproduction, a more specialized kind of cell division called **meiosis** takes place in the organs where

FIGURE 4.13 The two kinds of cell division, mitosis and meiosis.

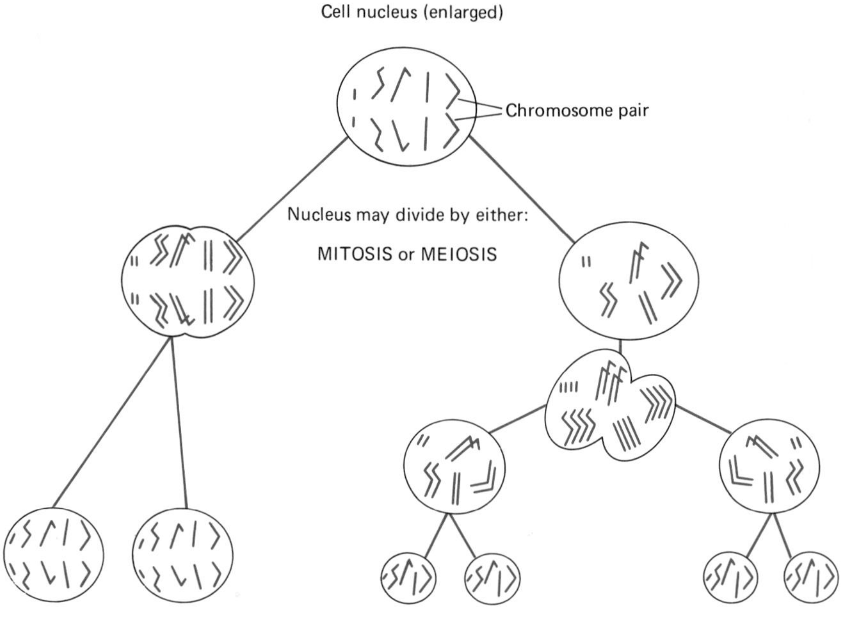

gametes (specialized reproductive cells such as eggs and sperm) are produced.

Two cell divisions take place in meiosis. The first step is similar to mitosis in producing two new cells, each with a complete complement of paired chromosomes. The two new cells then further divide in a manner that leaves only one set of each of the original pairs of chromosomes in each of the four offspring cells (Figure 4.13). These become the gametes, carrying only *half* the number of chromosomes necessary for the final organism. When two gametes are joined in the process of fertilization, the new organism that is produced has the normal complement of chromosomes, half of them from each of the two parents. Thus, meiosis and subsequent fertilization provide a means of *interchanging genetic material* between organisms, whereas mitosis provides a means of *exactly duplicating* cells within an individual organism. In humans, all cell divisions are achieved by mitosis, except those in the testes of the male and ovaries of the female, where sperm and eggs are produced by meiosis.

The Geologic Record of Evolution

During the past 40 years much progress in understanding the process of evolution has resulted from combining Darwin's mechanism of natural selection with the most recent discoveries of geneticists concerning the inheritance of individual variations. A fundamental theme has been the study of inheritance not only in individual organisms but in populations, which are interbreeding groups of individuals of the same species. Yet, the fossil record of animals and plants reveals some surprising evolutionary patterns that would never be suspected from examining evolution only in populations of present-day species. Such studies show that the changes leading to new species normally take place very slowly over many generations. We might, therefore, expect that the more complex sequence of changes leading to entirely new kinds of life would be very gradual indeed. Instead, the fossil record reveals that these radically new adaptations tend to arise rather rapidly. Furthermore, the times of their origin are not distributed randomly throughout geologic time. Instead, they tend to be clustered, so that many new groups appear simultaneously over a short period of time, and then persist with relatively little change for much longer periods. In other words, the rate of evolutionary change in most organisms is not constant; rather, it is extremely variable. Most of the phyla of metazoan animals, for example, originated in late Precambrian and Cambrian time. Although there have been many evolutionary changes within the phyla since Cambrian time, only one or two new phyla have evolved. Likewise many classes of shell-bearing marine invertebrates that are common today evolved in Late Cambrian and Early Ordovician time; similarly, most present-day families of flowering plants originated during the Cretaceous Period, and most modern orders of mammals evolved in the Eocene Epoch. These rapid diversifications are called **evolutionary radiations.**

Just as the origins of major groups are clustered in time, so too are their ultimate fates. The fossil record reveals that organisms tend to die out simultaneously in relatively sudden, worldwide extinctions. Furthermore, these periods of extinction are often followed by periods of rapid evolutionary radiation, suggesting that the vacant environments left behind by extinct animals and plants provide an ideal setting for new evolutionary experimentation. The most dramatic extinctions—near the end of the Permian and Cretaceous Periods—separate the Paleozoic, Mesozoic, and Cenozoic Eras. Smaller extinctions mark the boundaries of other units of geologic time. As we shall see in the next chapter, the sharp changes in the fossil record that accompany widespread extinctions have been used as criteria for recognizing geologic time units. As with evolutionary radiations, the causes of these periodic extinctions are unknown.

Chapter Summary

The Earliest Life

The components of life are water, carbohydrates, fats, proteins, and nucleic acids; all are compounds dominated by the elements hydrogen, oxygen, carbon, and nitrogen.

"Primitive earth" experiments: When energy is supplied to various mixtures of gases containing the elements required for life, complex molecules—the building blocks of life—are formed.

Simple present-day life. Bacteria and blue-green algae, together known as procaryotes, are the simplest surviving organisms; they provide clues to early life.

The limits of life. Modern bacteria survive environmental extremes that would quickly kill more advanced animals and plants; such extremes may have been present early in earth history.

Fossil Evidence of Early Life

Fossil procaryotes are found throughout most of the sedimentary record, the earliest being over 3 billion years old.

Gunflint Chert Fossils were the first well-preserved Precambrian procaryotes to be discovered; they occur in 1.9-billion-year-old rocks of northern Minnesota and adjacent Ontario.

Fig Tree Group fossils are less well preserved, older procaryotes (over 3.2 billion years) from South Africa.

Bitter Springs Formation fossils are typical of many younger Precambrian localities (under 1.5 billion years) that contain diverse associations of fossil procaryotes.

Early eucaryotes: The first undoubted eucaryotic fossils are jellyfishes, segmented worms, and related forms from latest Precambrian rocks of Australia.

The end of Precambiran time is marked by a worldwide expansion of shell-bearing eucaryotic animals that took place about 600 million years ago.

The Diversification of Life

Darwin and natural selection provide the basis for modern understanding of the diversity of the living world.

The science of genetics added to Darwin's insights by providing an understanding of the mechanisms of inheritance in eucaryotic life.

The geologic record of evolution shows that changes in the living world have not been continuous but have, instead, been concentrated in relatively brief intervals of extinction and subsequent evolutionary expansion.

Important Terms

amino acid
asexual reproduction
autotroph
chromosome
class
DNA (deoxyribonucleic acid)
eucaryote
evolutionary radiation
family
gamete
gene
genus, genera
heterotroph
meiosis
metaphyte
metazoan
mitosis
natural selection
order
organic evolution
phylum, phyla
procaryote
sexual reproduction
species

Review Questions

1. What are the essential chemical elements of life? How might they have become organized into living systems?
2. What clues to early life are provided by the structure and habits of the simplest surviving organisms?
3. Summarize the Precambrian fossil record of procaryotic life.
4. How and when did eucaryotic life arise? What might have caused the late expansion of eucaryotes?
5. Outline Darwin's contributions to our understanding of organic evolution.
6. Summarize the most significant advances in evolutionary theory since Darwin's time.

Additional Readings

Cloud, P. *Cosmos, Earth and Man,* Yale University Press, New Haven, Connecticut, 1978. *Chapters 11–14 provide good summaries of the origin and early evolution of life.*

Dobzhansky, T., F. J. Ayala, G. L. Stebbins, and J. W. Valentine *Evolution,* W. H. Freeman & Company Publishers, San Francisco, 1977. *An up-to-date introductory text; several chapters deal with the fossil record.*

Folsome, C. E. *The Origin of Life,* W. H. Freeman & Company Publishers, San Francisco, 1979. *A nontechnical introduction.*

Folsome, C. E. (ed.) *Life, Origin and Evolution,* W. H. Freeman & Company Publishers, San Francisco, 1979. *A collection of articles originally published in* Scientific American; *some of the earlier ones are now a bit dated.*

Mayr, E. (ed.) *Evolution, A Scientific American Book,* W. H. Freeman & Company Publishers, San Francisco, 1978. *Nine articles provide a nontechnical review of the latest thinking in the field.*

Schopf, J. W. "Precambrian Paleobiology: Problems and Perspectives," *Annual Review of Earth and Planetary Sciences,* vol. 3, pp. 213–49, 1975. *An authoritative review of Precambrian life.*

THE PHANEROZOIC RECORD

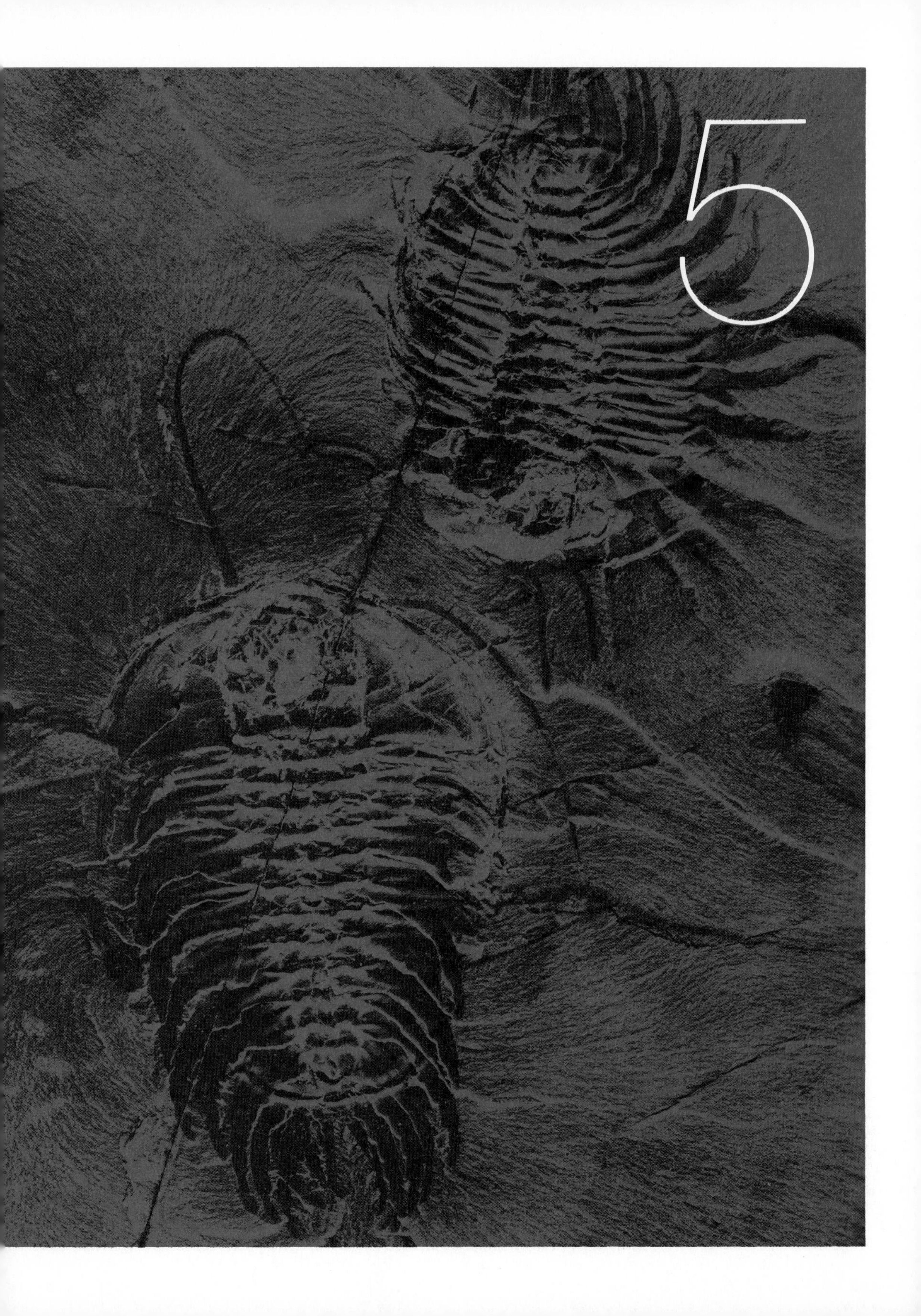
5

The late Precambrian rise of metazoan life set the stage for the Phanerozoic, or "exposed life," phase of earth history—a phase that began with the expansion of shell-bearing animals about 600 million years ago. The presence of abundant fossils makes Phanerozoic rocks much easier to date and to place in a chronological setting than Precambrian rocks. But even if they lacked fossils, Phanerozoic rocks would still be better known than Precambrian because they are so much more widely exposed. Although they represent only the most recent 15 percent of earth history, *they make up fully one-half of all sedimentary rocks on earth.* They are spread out as a relatively thin sheet over vast areas covering about 80 percent of the land areas of the earth. Phanerozoic strata normally range in thickness from a few hundred to a few thousand meters, but in some areas they thicken to more than 15,000 meters. In general, Phanerozoic rocks have never been buried as deeply in the crust as have Precambrian rocks; hence, they are less metamorphosed and better preserved.

How can one-half of the earth's sedimentary rocks represent only the last one-sixth of its geologic history? Could it be that more sedimentary rocks formed in this comparatively recent time? Probably not. The answer instead lies in the workings of the **rock cycle,** which is a way of viewing the turnover of materials at the earth's surface. Before proceeding further, let us briefly examine this concept.

The Rock Cycle

At any given time in the geological past, some portions of the earth's surface were undergoing erosion and some portions were undergoing sedimentation, thereby continually producing new sedimentary records. Of course, no material on earth is truly new. In this context, the "new sedimentary record" simply represents a *new arrangement of old material* that has been around since the earth began. Old rocks are destroyed not only by being uplifted and eroded into sediments, but also by being buried deeply enough to encounter high temperatures and pressures sufficient to recrystallize them in the process of metamorphism. Still greater heat may melt the rocks, and the resulting magma may either rise to the surface to become a volcanic rock, or it may crystallize at depth into plutonic rock. Because the earth's crust is continually being deformed and uplifted, even plutonic rock, given enough time, will eventually become exposed to erosion at the surface. There it will erode to become sediment again, thus completing the cycle of rock change.

This simplified rock cycle (shown in Figure 5.1) is not a methodical machine that processes all the earth's crust in prescribed periods of time. It contains shortcuts; for example, from sedimentary rock to sediment and back again. In addition, the ages of the materials processed in a given time period vary randomly. Some crustal materials have been recycled many times, but other very old rocks have survived with little change. Because surficial erosion and deep metamorphism have always been going on, old materials have had

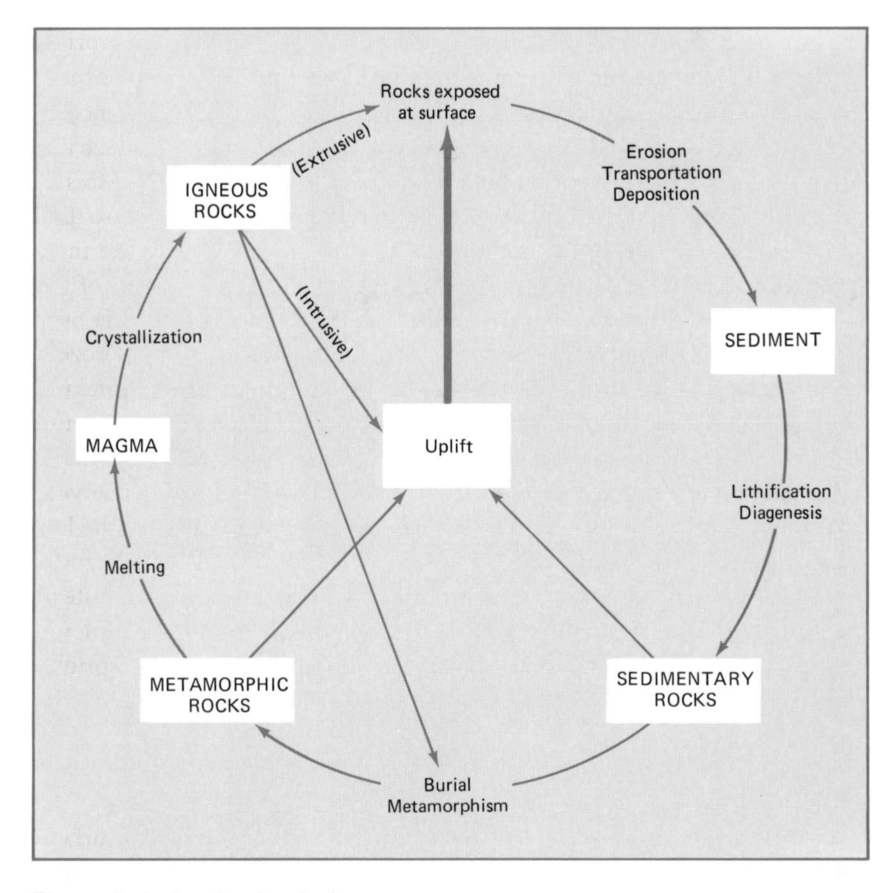

FIGURE 5.1

The geological cycle of rock change.

much greater opportunity to be recycled than have young materials. Therefore, the older a body of rock, the less likelihood of its being preserved. Consequently, as one explores farther back in time, the preserved record becomes more and more sparse.

Roughly half of the total sedimentary rock mass that is actually preserved today is Precambrian in age, and half is Phanerozoic in age. The mass half-age of all crustal rocks is thus about 600 million years. We extend this concept to mean that three-fourths of all crustal rocks are younger than 1200 million years, seven-eighths younger than 1800 million years, and so on. At the end of any one of the geologic time intervals shown in Figure 5.2, half of the crustal rocks were less than 600 million years old. In the ensuing time interval, those rocks were themselves recycled, and at the end of another 600-million-year-period, only about half of them survived.

If these rates are approximately correct, then in the time since the oldest rocks formed, the total sedimentary rock mass that has been deposited is about five times that of the existing mass. Crustal materials include very resistant rock types, such as shales and sandstones, as well as extremely easily eroded

FIGURE 5.2

Worldwide distribution by age of all sedimentary rocks today (far right) and for four times in the geologic past, showing how age distribution was achieved, assuming a rock half-life of 600 million years.

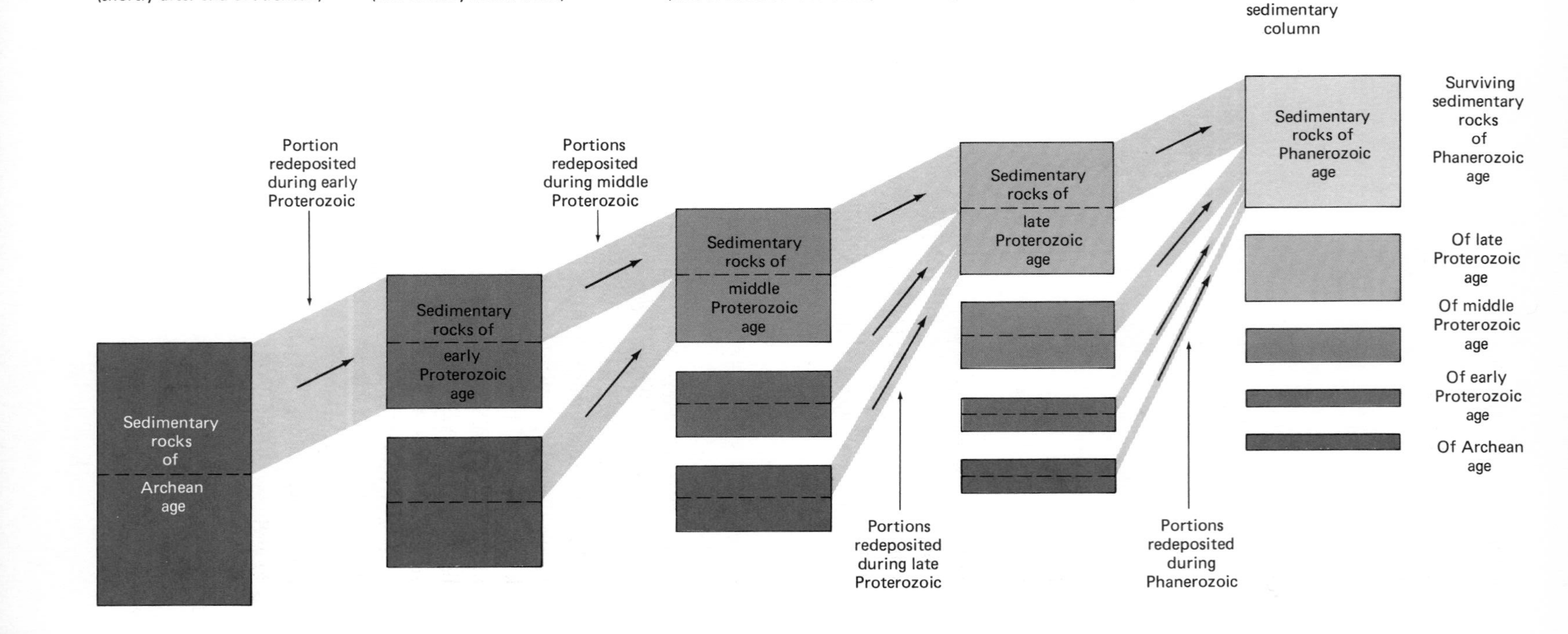

rocks. Carbonate rocks (dolomite and limestone) are soft and soluble, and probably have a mass half-age of about 300 million years. Evaporites (chiefly salt and gypsum), which are still softer and much more soluble, have a mass half-age of perhaps 200 million years. Such rapid turnover of evaporites and carbonates may well explain their comparative abundance only in the Phanerozoic record (see Figure 3.30). The dearth of evaporites in Precambrian strata prompted some people to suggest that very little gypsum and salt formed in the Precambrian; in fact, much may have formed, but it simply could not survive destruction over the long time that followed.

Let us now turn to that half of the sedimentary rock record that constitutes the Phanerozoic Eon. Phanerozoic strata are organized according to the stratigraphic principles enumerated in Chapter 2: original horizontality, superposition, original lateral continuity, and cross-cutting relationships. The application of these principles permits us to place rock units in their proper chronological sequence and to decipher their historical record. In Chapter 2 we noted that *formations* are the basic rock units, both for geologic mapping and for the study of geologic history. In this chapter we shall examine how rock units relate to geologic time units, and how time units are identified and interpreted on the basis of the sequence of fossil life.

Stratigraphic Units

Rock Units

Formations are defined solely on the basis of a distinctive rock identity. A formation is considered worthy of recognition and naming if *its rocks are distinguishable from units above and below, and are thick enough to be plotted on a large-scale topographic map.* Formation names consist of two parts: the first identifies the geographic locality where the rock is well exposed; the second describes the general rock type. The Burlington Limestone and the Prospect Mountain Quartzite are typical formation names. Where no single rock type predominates, the second part of the name becomes simply "Formation," as in Green River Formation. Because formations are composed of different rock types, they commonly have different colors and topographic expressions, and they weather to different soils that support distinctive kinds of vegetation (Figure 5.3). These characteristics all facilitate mapping.

An accurate geologic map shows exactly the extent of the various formations and how they are arranged. The map is a model of how the rocks would look if all the soil were removed from the surface. An indication of the distinctiveness of most formations is that they can be mapped not only on the surface, but from photographs taken several kilometers above the ground.

Rock units that are distinctive enough to warrant recognition, but are generally too thin to be mapped are called **members** of formations. For convenience in small-scale mapping, two or more formations may be lumped into a single larger rock unit called a **group.**

FIGURE 5.3 Two thick distinctive rock units in southern Utah: the Straight Cliffs Sandstone (forming vertical cliffs), underlain by the Tropic Shale (forming barren slopes), both of late Cretaceous age.

FIGURE 5.4 The Pine Limestone thickens from Area 1, where it is not mappable, to Area 2, where it is mappable. Elevating it to formation status necessitates the addition of two new formation names, Oak Creek Shale and Cornville Shale.

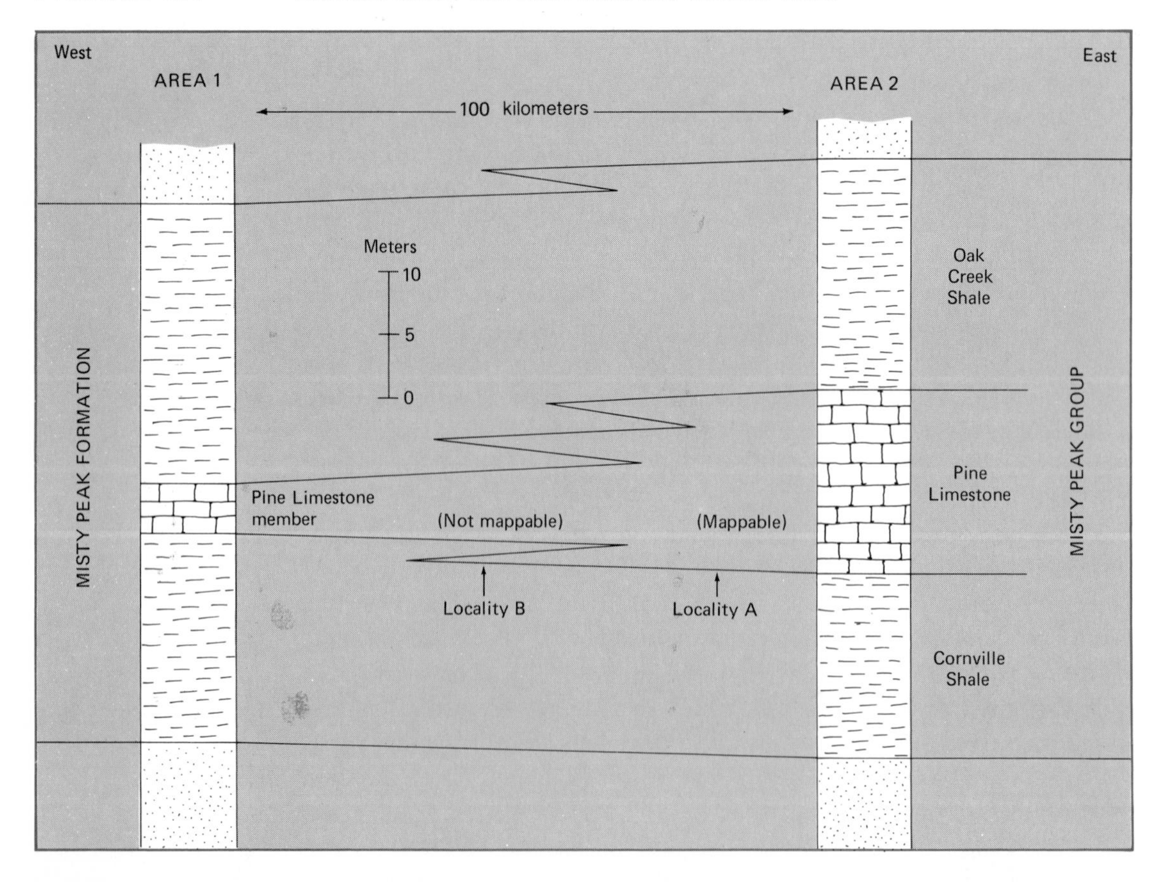

A member of a formation in one area may thicken sufficiently to be mappable in another area, and there, under its old member name, it may be considered a formation. Figure 5.4 shows such a case. The Pine Limestone Member of the Misty Peak Formation in Area 1 is traced eastward to Area 2, where it merits formational rank. The overlying and underlying portions of the Misty Peak require formational rank as well, because the Misty Peak Formation cannot be retained with another formation (the Pine Limestone) inside of it. If the Pine Limestone is considered mappable in Area 2, so are the units above and below; hence, these shale units must be designated as new formations and named. In this case, the Misty Peak Formation is elevated to the status of a Misty Peak Group, a common but not mandatory practice. All formations undergo thickness and facies changes laterally, and one commonly encounters name changes when tracing rock units over wide regions.

The boundary between two successive formations represents an environmental change; for example, from deposition on tidal flats to deposition in shallow-marine water, as at localities A and B in Figure 5.4. The boundary might be an abrupt contact between alternating beds of silty shale and ripple-marked sandstone below and limestone above, as at locality A. Or the boundary might be gradational, if tidal-flat deposition alternated with shallow-marine deposition for a time before the shallow-marine environment alone became established, as at locality B. In either case, the depositional environment did not change everywhere at one time; deposition in the marine environment began earlier in the east than in the west, and it ended later. As we saw in Chapter 2, the boundaries of rock units do not commonly coincide with time boundaries, but usually transgress time boundaries as they are traced laterally.

Time-Stratigraphic Units

If formations extended worldwide, and if their deposition began and ended everywhere at the same time, then rock units and time units would be equivalent. In tracing rock unit boundaries, one would be following exact time units. In nature, however, two factors combine to complicate this simplistic picture, creating a need for one kind of stratigraphic unit for rocks and another kind for time. The first factor is that sediment deposition is seldom continuous at any one place over long periods of time. As a result, unconformities occur in virtually every stratigraphic section, although the age and quantity of missing record differs greatly from place to place. The second factor is facies. Most rock units represent deposits of a single environment or a suite of closely related environments; as such, they commonly change to different kinds of rocks that represent different environments as they are traced laterally along time planes.

Early in the nineteenth century, after geologists had come to realize the time-transgressive quality and limited distribution of rock units, they also discovered that *fossils provide an independent key to age comparisons* of Phanerozoic sedimentary rocks. The pioneers in this endeavor were William

Smith in Britain and Georges Cuvier in France. Building on their work, geologists were able to define major units of sedimentary rocks. Using distinctive fossil species (commonly called **guide fossils** or **index fossils**) that occur widely but are characteristic of a short span of geologic time, they were able to distinguish rocks of comparable time periods in distant regions, even though the rock types might be entirely different.

The major time-stratigraphic units defined by this method became known as geologic **systems**; these include, for example, the Cambrian, Ordovician, and Silurian Systems. Most of the geologic systems were defined in Europe in the middle of the nineteenth century. The actual rock outcrops that define a system are called the **type section**, and the area in which they occur is called the **type area**. The systems have been traced from their type areas to virtually all parts of the world on the basis of fossils.

During the period in which they were naming the systems, geologists were simultaneously developing a concept of geologic time. This time framework is based upon the great system of rocks, each represented by a considerable thickness of strata. Each system overlies older strata below and underlies younger strata above, indicating that each system was deposited during a discrete portion of geologic time. Any rock unit, anywhere, that was formed within that time range belongs to that particular system.

A time-stratigraphic unit is independent of rock type or thickness; that is, those factors are not part of the definition. A system can be recognized by the kinds of rocks it contains only at the type area where it was defined; everywhere else it must be recognized by time correlation. The type section of the Ordovician System in Wales, for example, has been defined as a particular body of strata. Its sedimentary rocks and the fossils they contain provide a detailed local history. Strata immediately below and above the Ordovician reveal the Cambrian events that preceded and the Silurian events that followed. Without any doubt, the rocks in question belong to the Ordovician System, because they *define* the Ordovician.

To demonstrate this technique, let us apply it hypothetically to a stratigraphic section in the central United States. We can distinguish successive rock types, recognize local rock units, study their faunas, interpret the sequence of environments they represent, and map them. In short, we can work out the complete local history. If we can then determine, through time correlation, that a portion of this section was deposited contemporaneously with all or a part of the type section of the Ordovician in Wales, we can call these Ordovician rocks. In using the name "Ordovician," we are saying, "These rocks were deposited within the same time interval as the body of strata in Wales that constitutes the type."

To recognize time-stratigraphic units, therefore, we must establish correlations, either directly with the type section or with intermediate sections, which have in turn been correlated with the type. We have several ways of correlating rocks, all of which are subject to some measurement error. By far the most reliable of these methods, especially over long distances, is the correlation and dating of strata through the use of **zones** based on fossils.

Biostratigraphic Units

Zones are defined solely on the basis of fossils. They are considered separate kinds of stratigraphic units that are called **biostratigraphic units.** The kind of strata that contain the fossils does not enter into the definition. For example, a zone may encompass several units of limestone and shale, or it may fall entirely within a part of a single unit of shale. Thus defined, some zones have time-stratigraphic value but others do not, and instead represent long-lived depositional environments. Neither does a zone's potential time significance, or lack of it, enter into the definition, but is an interpreted characteristic.

Zones that are, in fact, interpreted to have time significance are called time-stratigraphic zones. In actual practice, these represent the smallest time-stratigraphic units that can be recognized. Zones may be defined on the total stratigraphic range of a single species or by various combinations of overlapping ranges of several species. (The use of zones will be discussed later in this chapter under the heading *Biostratigraphy.*)

Geologic Time Units

The same intervals of geologic time represented by each of the systems occurred everywhere in the world, even if the rocks representative of that system are missing locally as a result of nondeposition or later erosion. For example, the Triassic System is missing from the Canadian Shield, which was probably an emergent source area during Triassic time. But, of course, the Triassic time interval occurred there. (In referring to this geologic time unit, note that we call it the *Triassic Period,* because by convention, the abstract geologic time units that correspond to the systems are called **periods.**

Time terminology is a valuable tool in referring to historical events and circumstances. For example, we might wish to state that fish were abundant during the Devonian Period (because their remains are abundant in rocks of the Devonian System). Or, if the Permian System is absent in a given area, we might seek evidence of whether the Permian Period was a time of nondeposition in that area, or whether rocks were deposited but later eroded away. Of course, we are aware that a geologic period occurred only because we have a rock sequence formed during that time somewhere else. In summary, we find it convenient in many cases to employ abstract time terminology, as well as time-stratigraphic terminology.

Geologic systems are divided into **series,** commonly referred to as Lower, Middle, and Upper applied to the system name. The parallel time divisions are **epochs,** usually Early, Middle, and Late. For example, the Devonian System, which was deposited during the Devonian Period, is commonly divided into the Lower Devonian, Middle Devonian, and Upper Devonian Series, whose rocks were deposited during the Early Devonian, Middle Devonian, and Late Devonian Epochs (Figure 5.5). Epochs and series are further subdivided into **ages** and **stages.** Like the systems, many of the series apply worldwide. Because they are smaller units, stages generally can be recognized only within single continents or regions, but some stages can be traced intercontinentally. Stages in most regions comprise several time-stratigraphic zones, bodies of strata that contain short-ranging species of fossil animals or plants.

FIGURE 5.5

COMMONLY USED CATEGORIES OF STRATIGRAPHIC UNITS WITH EXAMPLES

Time-stratigraphic units	Corresponding geologic time units	Biostratigraphic units	Rock units
	Era (Mesozoic Era)	Biostratigraphic zones (*Baculites reesidei* Zone)	Group (Dakota Group)
System (Cretaceous System)	Period (Cretaceous Period)		Formation (South Platte Formation)
Series (Upper Cretaceous Series)	Epoch (Late Cretaceous Epoch)		Member (Plainview Sandstone Member of the South Platte Formation)
Stage (Campanian Stage)	Age (Campanian Age)		Bed (Mostly informal terminology such as "the third coal bed")
Time-stratigraphic zone (*Baculites reesidei* Zone)			

Unconformity-Bounded Units

The Phanerozoic time-stratigraphic record has been assembled from thousands of local sections, so that virtually all parts of Phanerozoic time are represented someplace by rock strata. As the time-stratigraphic record has become more complete and more refined, it has become a tool for investigating the extent and magnitude of unconformities in various regions. The largest unconformities document continent-wide cycles of uplift and erosion that divide the Phanerozoic into large units. In North America, six such unconformity-bounded units have been recognized (Figure 5.6). These units are known as **sequences.** Sequences have been named for convenience in referring to them, but the names are only used occasionally. (See Figure 5.6 for sample sequence names.) Some sequences consist of several geologic systems, and others of only part of one system. Figure 5.6 shows that the rocks that make up the North American sequences represent far more time at the margins of the continent than they do in the interior, where unconformities are largest.

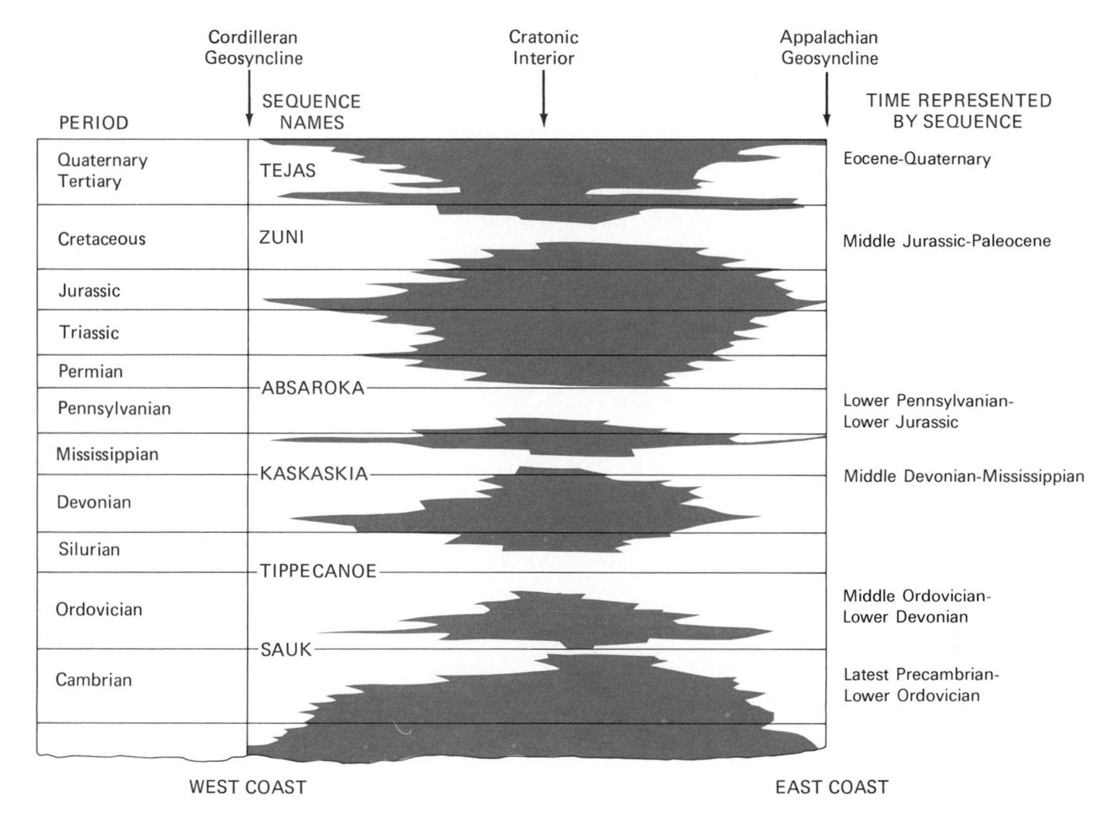

FIGURE 5.6

Time-stratigraphic relationships of unconformity-bounded sequences in North America. Dark areas represent large gaps in the stratigraphic record, which become smaller toward the continental margins; white areas represent strata. (*Sloss, 1963*)

There is no reason to assume that continent-wide unconformities like those shown in Figure 5.6 were limited to the Phanerozoic. They almost surely occurred in the Precambrian as well. However, due to the lack of fossils, we cannot delineate them. The greatest unconformity of all in Figure 5.6 lies at the base of the Cambrian, above the underlying Precambrian rocks (which are not shown). Uplift at this time was so prolonged and erosion so intense that most of the late Precambrian record was removed from the craton.

Biostratigraphy

Biostratigraphy is the science of correlating and dating sedimentary rocks by means of fossils. Too often in the past, paleontologists have considered fossils as isolated treasures to be salvaged from the field and brought to the safety of the laboratory, where they could be studied independently, quite apart from the rocks. In some cases, neither the kinds of rocks nor the stratigraphic levels in which the fossils occurred were noted, and a great deal of information has been lost as a result. In fact, fossils are as much a part of the rock as are sand grains or clay particles, and they need to be interpreted in terms of rock type and exact position in the stratigraphic section. Today the rich fossil record of the Phanerozoic has largely been arranged in proper chronological order; this record provides an elaborate document of organic evolution and makes accurate correlations possible. Biostratigraphy thus constitutes the very foundation of the time-stratigraphic framework for the Phanerozoic.

Not all changes in the fossils that occur in a stratigraphic succession are caused by organic evolution. Like species of animals and plants living today, ancient organisms were distributed in complex local patterns governed by environments. Fossils, therefore, provide much information about the ever-changing conditions on the earth's surface. Through studies of paleoecology and biogeography, fossils add significantly more to the environmental record than do rock types and sedimentary structures alone.

Paleoecology

In existing environments that occur throughout the world today, certain species of plants or animals repeatedly occur together. Each of these assemblages of species favors a particular habitat having specific environmental conditions. Adjacent areas with differing environmental conditions support different assemblages. Figure 5.7, for example, shows the preference of some present-day marine invertebrates for various shallow-water environments that have different salinities and substrates. Each of the species shown represents an

FIGURE 5.7

Distribution of some genera of invertebrate animals in various nearshore environments in the Gulf of Mexico. (*Parker, 1959*)

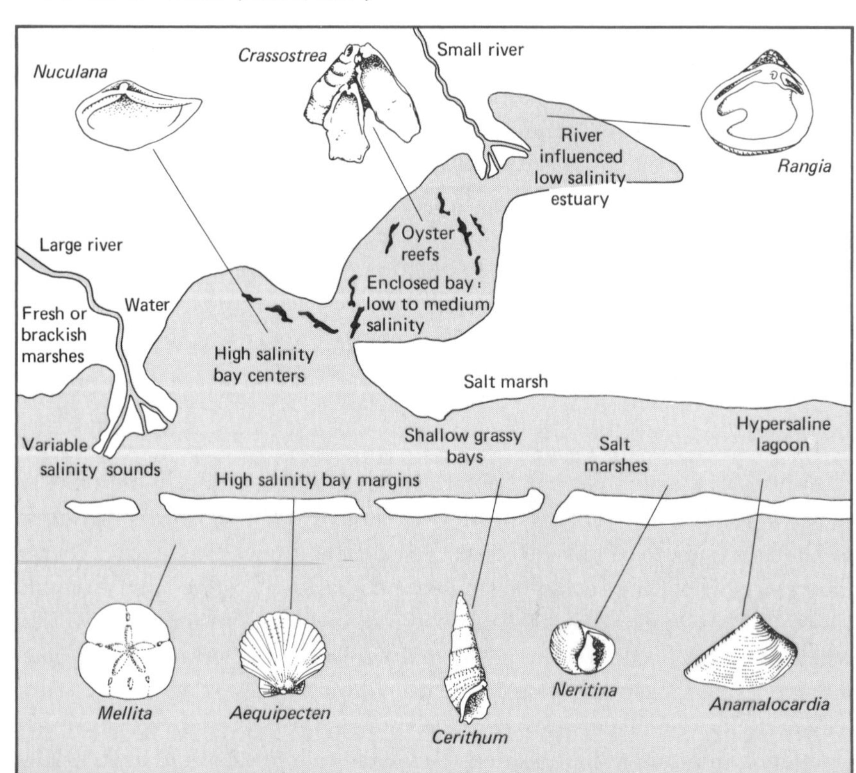

FIGURE 5.8

(*a*) *Radiolites,* A typical Cretaceous rudistid. (*b*) Localities where rudistids are common elements of Cretaceous faunas. (*Coates, 1973*) (*c*) Two reefs in the Lower Cretaceous Mural Limestone of northern Chihuahua, Mexico that contain abundant rudistids.

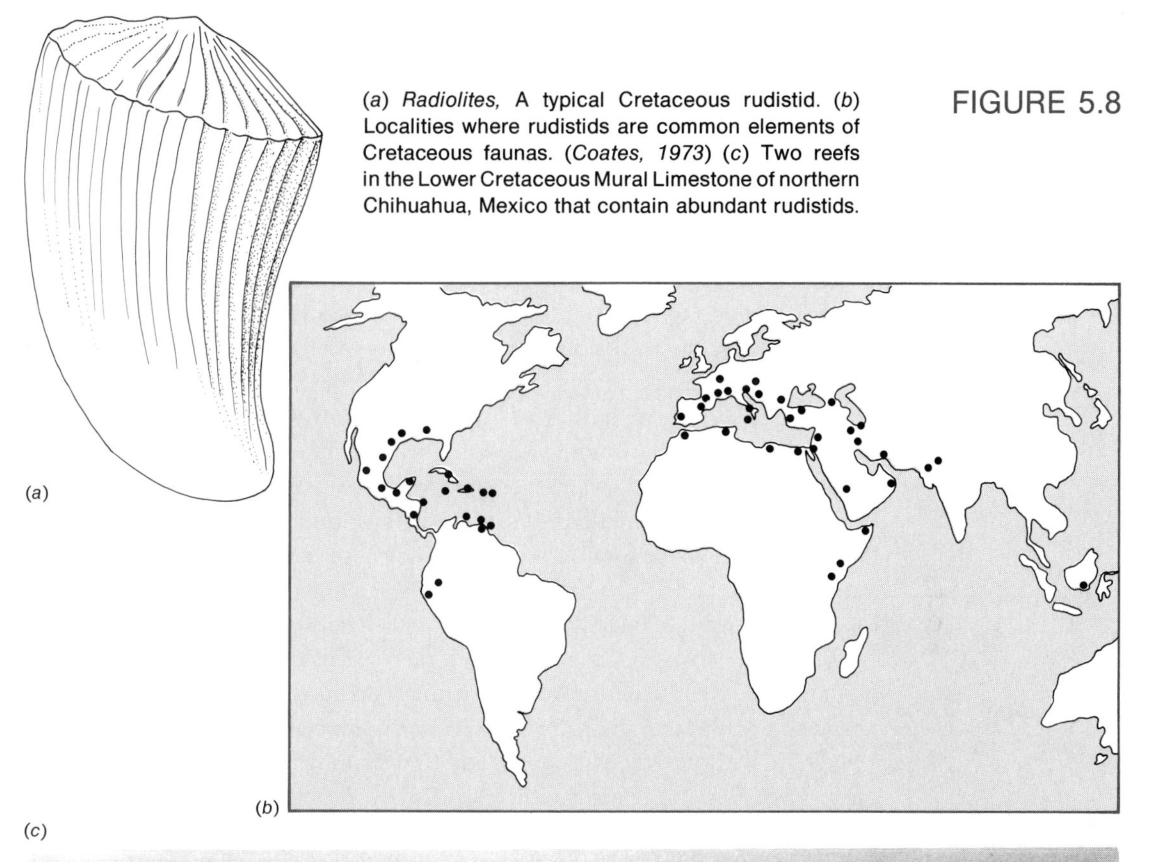

(*c*)

assemblage of several species adapted to that particular environment. Sediments accumulating in each of these environments thus contain the remains of characteristic fauna, and these will appear in the future rock record. Distribution of species of animals and plants throughout the geologic past was similarly controlled by environments. Interpreting the relationship of fossils to their ancient environments is the concern of **paleoecology.**

We cannot observe directly the environmental preferences of extinct species. In general, the older the strata, the greater the proportion of extinct species it contains. For example, it is easier to reconstruct the environments for late Cenozoic faunas and floras, many of whose species can be observed directly today, than it is for older strata, which contain only remote ancestors of modern species. However, even in very old fossiliferous rocks, environmental preferences can be inferred from features of the organisms themselves and from their associations with other species whose environmental preferences are known.

For example, Cretaceous rudistids are unusual bivalves with one deep cone-shaped valve and one simple cap that covers the open end of the cone (Figure 5.8*a*). Although rudistids have no modern descendants, their overall coral-like shape leads to the inference that, like many modern corals, they lived chiefly in shallow, tropical waters, and they commonly occurred in reef habitats. Their distribution in Cretaceous tropical seas (Figure 5.8*b*) and the occurrence of some in huge mounds believed to be reefs (Figure 5.8*c*) support this interpretation.

Biogeographic Provinces

Every species of animal or plant produces more young than can survive in its living space, and thus each tends to expand into all areas where environments permit. Highly mobile animals like birds have obvious means of rapid dispersal. But even immobile organisms like land plants produce seeds adapted for wide dispersal by animals and by wind; many attached, bottom-dwelling marine animals have free-swimming larvae that can be carried great distances by ocean currents. Yet almost no species of plants or animals occurs in all parts of the world. Most are confined to large regions called **biogeographic provinces.** Biogeographic provinces are surrounded by barriers to dispersal, which may be physical barriers or climatic barriers. Land areas are barriers to the dispersal of marine organisms, and open water is a barrier for most land-dwelling organisms. For a marine species adapted only to shallow water, a deep ocean basin poses as great a barrier as a land mass. Inhospitable climates constitute a barrier to both terrestrial and marine animals and plants.

An example of an existing biogeographic barrier is the Isthmus of Panama, which separates the marine faunas that live in the Pacific Ocean from those in the Caribbean Sea, although both areas provide similar environmental conditions (Figure 5.9). Of the 805 Pacific species and 517 Caribbean species of molluscs that live in this region, only 24 species are common to both provinces.

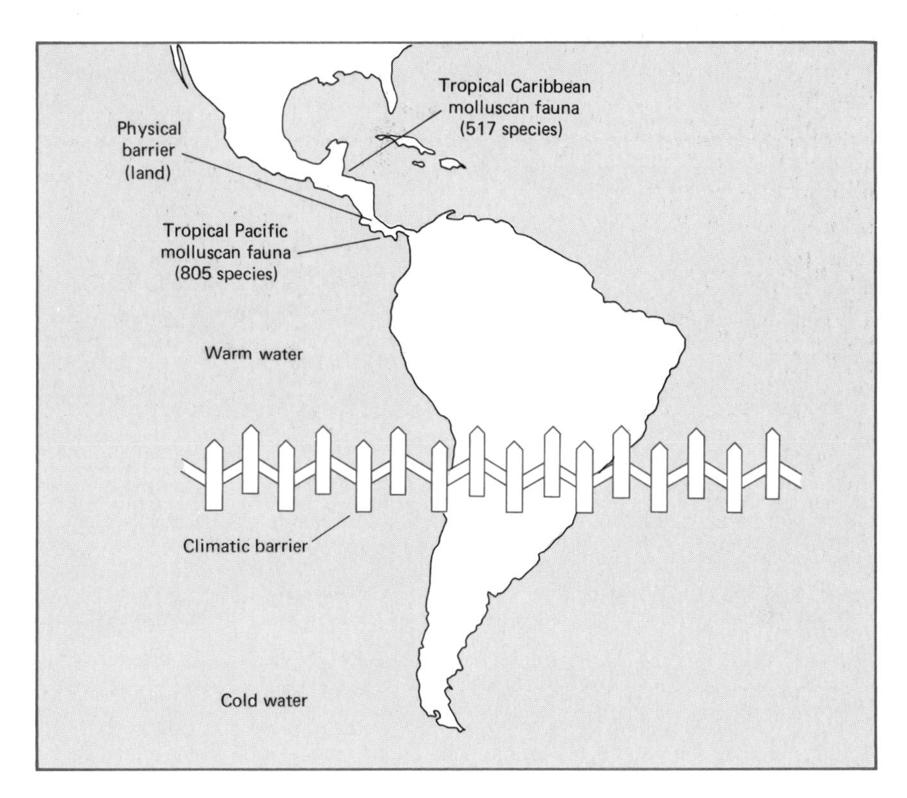

FIGURE 5.9

Tropical Pacific and Caribbean molluscan faunas are prevented from mixing by a physical barrier between them and a climatic barrier to the south.

So long as the Isthmus of Panama remains above sea level, the faunas on either side will remain isolated from one another. Should the Isthmus subside below sea level sometime in the future, the Pacific species would quickly invade the Caribbean, and vice versa. The records of such suddenly introduced species in accumulating marine sediments would mark a time horizon in the sedimentary record (Figure 5.10). Similarly, the breakdown of a climatic barrier, such as that which prevents migration of the warm Pacific and Caribbean faunas around the southern tip of South America, could occur either by general warming of the climate or by shifting of warm ocean currents into higher latitudes.

Once a species is introduced into a province, whether by evolution or by immigration, its dispersal throughout the province is instantaneous, in terms of geologic time. For example, a common snail of the New England shore, *Littorina littorea,* was first brought from Europe into the Halifax, Nova Scotia area in the middle part of the nineteenth century. Its subsequent spread is well documented. By 1868 it had reached the coast of Maine, and by 1880 it had migrated southward around Cape Cod and westward into the Long Island Sound, a distance of 360 miles in 12 years. Even at this snail's pace, *Littorina littorea* could, if environments permitted, migrate halfway around the world in just 400 years—a geologic eyeblink.

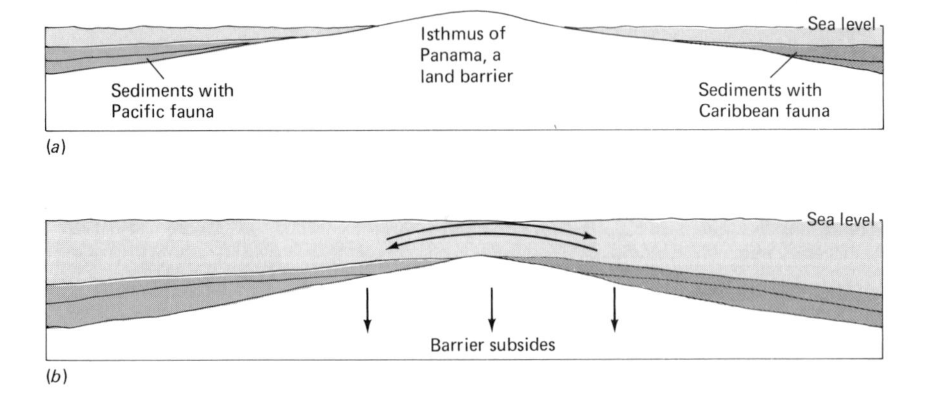

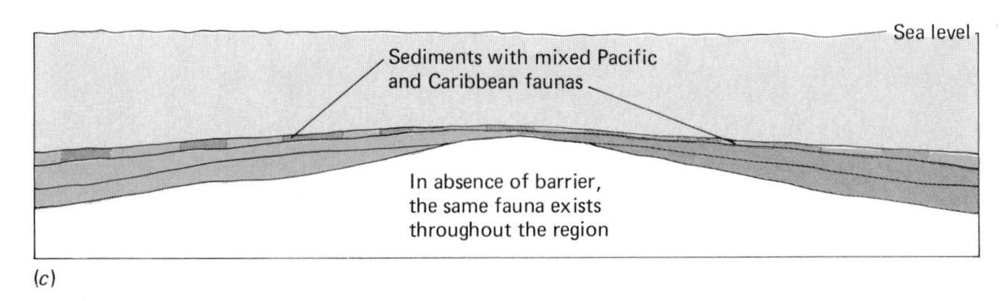

FIGURE 5.10 The destruction of a geographic barrier such as the Isthmus of Panama would be recorded by the geologically instantaneous mixing of formerly isolated faunas. (*a*) Fauna on either side of the Isthmus of Panama are separated by a land barrier. (*b*) Tectonic sinking of barrier below sea level permits faunas on either side to migrate across former barrier. (*c*) The destruction of a geographic barrier such as the Isthmus of Panama would be recorded by the geologically instantaneous mixing of formerly isolated faunas.

The same biogeographic restrictions on the global distribution of species that are in operation today also operated throughout geologic history. The presence of ancient barriers where none exist today and the absence of ancient barriers where today there are vast oceans or high mountains provide evidence of different climates and different arrangements of the continents on the earth's surface in the distant past, a subject we shall examine in Chapter 6.

Fossils and Time

Because of the great importance of fossil remains as tools for dating Phanerozoic rocks, we need to examine further exactly how they are used as indicators of geologic age.

No plant or animal has existed for all geologic time. Each species evolved from some ancestor, and thus had a discrete beginning; if it is extinct, it also had an ending. Each extinct species, therefore, establishes a three-part division of geologic time: the time before it evolved, the time during which it existed, and the time since it became extinct. Any rocks that contain the species must have been deposited within the time during which the species existed. If the total time during which that species existed was short, then the rocks in which it occurs can be located precisely in the geologic time scale. But if the total time during which the species existed was long, then that species can be found in sedimentary rocks that differ widely in age.

Rates of evolution vary greatly. Some groups of organisms have evolved very little over long stretches of time; others have evolved at comparatively rapid rates and have left a fossil record of striking change. Rapidly evolving organisms, if they are widely dispersed, serve as valuable short-ranged species for time-stratigraphic correlations. Paleontologists very early became aware that every geologic system has its characteristic short-lived groups of organisms that are of great stratigraphic value. Figure 5.11 illustrates the time-stratigraphic significance of major groups of marine invertebrates for various times in the Phanerozoic.

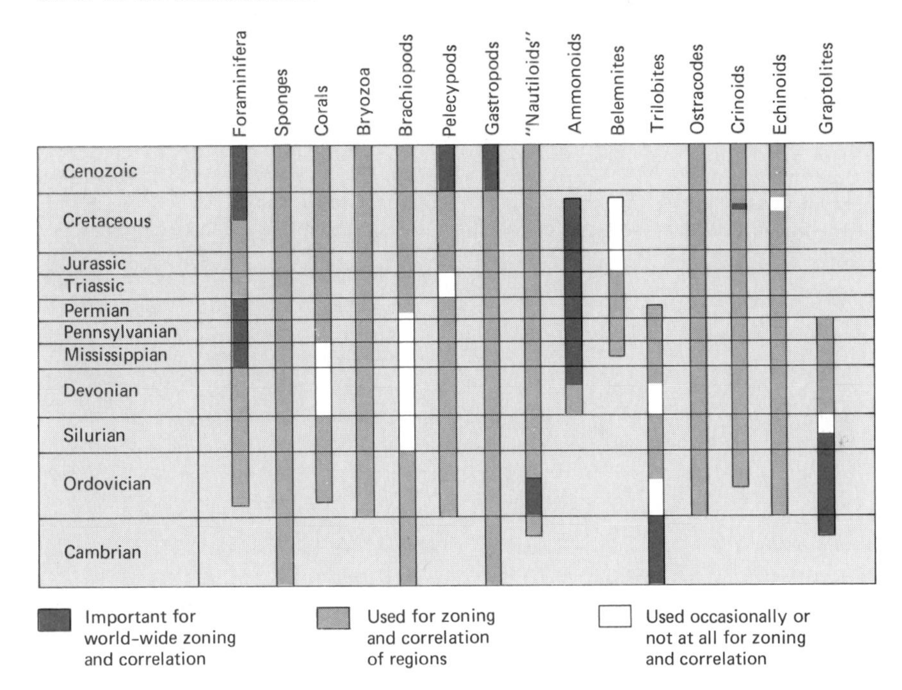

FIGURE 5.11

Relative correlative value of major groups of marine invertebrates during the Phanerozoic. (*Teichert, 1958*)

Range Zones

All strata that contain the fossil remains of a particular species constitute the **range zone** of the species. The range zone at most localities represents only *part* of the total time during which the species existed, because the species did not appear everywhere or die out everywhere at exactly the same time. The total stratigraphic range, which represents the entire life span of the species, can be ascertained only by extensive collecting over wide regions in order to find the very oldest and youngest strata that contain it.

Ranges of most fossil species are fairly well known, but virtually all are subject to small adjustments based on new finds. In general, the upper bound-

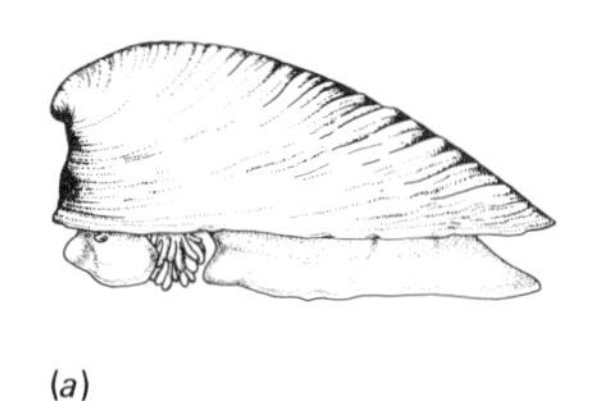

(*a*)

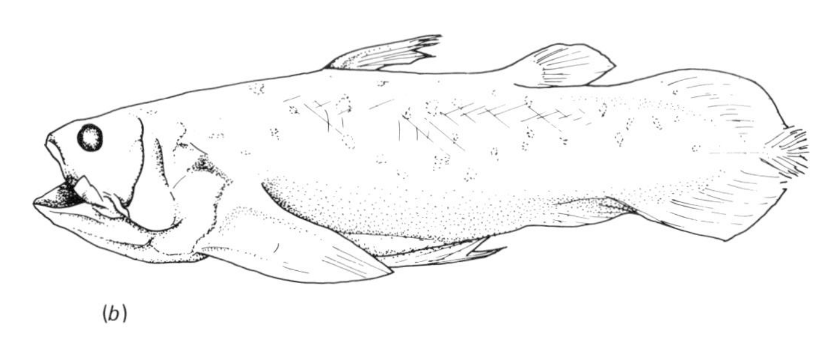

(*b*)

FIGURE 5.12 Living fossils: (*a*) *Neopilina,* a deep-sea mollusc, and (*b*) *Latimeria,* a coelocanth fish, represent ancient stocks that were long believed to be extinct.

FIGURE 5.13 Relationship of a hypothetical graptolite's range zone to its total range in time in two biogeographic provinces. (*Hedberg, 1964*)

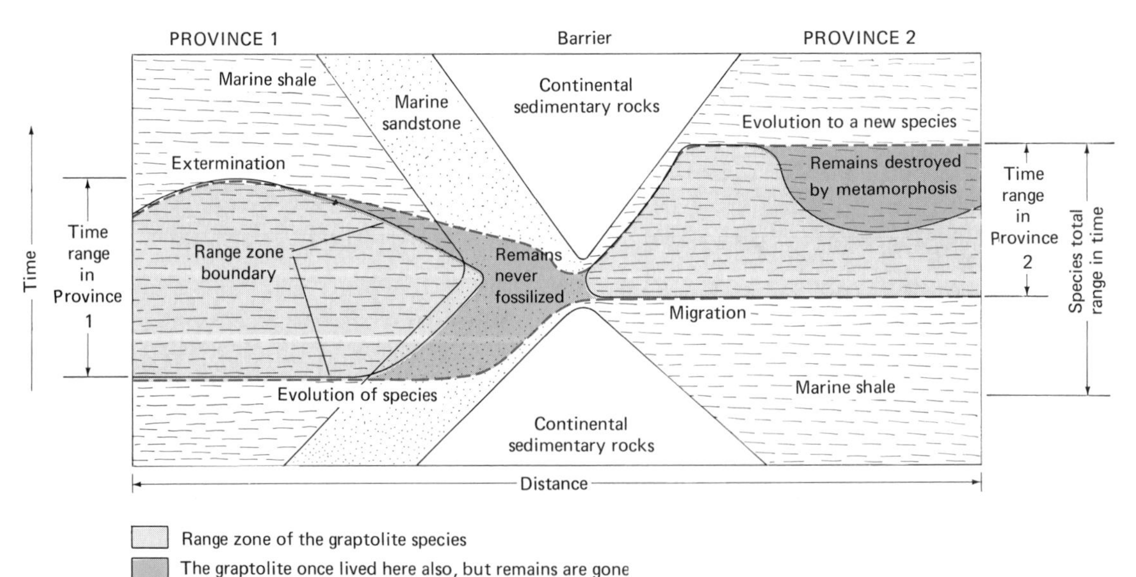

ary of a species' range is most subject to change. When a successful species first appears, it tends to spread quickly to the limits of its environmental tolerance. Extinction is more likely to be a piecemeal process, in which isolated populations may survive for a time after the species has disappeared from most places. It is remotely possible, for example, that trilobites, which we have never found in rocks younger than Paleozoic, still live in some unexplored oceanic deep. This would indeed be a surprise, but we have been surprised before. Not long ago *Neopilina,* a mollusc whose family was believed to be extinct since the Paleozoic, was found living at great depths in modern oceans. And *Latimeria,* a coelocanth fish whose family tree was never supposed to have extended into the Cenozoic, was found swimming contentedly in the Indian Ocean (Figure 5.12). These finds were dramatic and exceptional. Paleobiologists have, by now, done a lot of looking, both at the fossil record and at the modern world, and are fairly confident that the stratigraphic ranges of the relatively short-ranging organisms that are used in chronology will not change significantly with future work.

At a locality where a species' range zone is overlain by strata containing its direct descendants and underlain by strata containing its immediate ancestors, the range zone very likely represents the species' total range in time. In most cases, however, a species appears in the fossil record fully developed, without a trace of any ancestors. The range zone of such an abruptly appearing species must therefore represent only a part of its total range in time. The top of a species' range zone may likewise be abrupt. If its last appearance accompanies a lithologic change, that might signify nothing more than an environmental shift; but if regional evidence indicates that the disappearance is universal, then it probably represents an extinction event, which has chronologic value.

In summary, time-significant events based on fossils may be grouped into three categories: evolutionary events, extinction events, and migration events. Figure 5.13 diagrams these events for a hypothetical species; it shows the relation of the rocks that contain the species (range zone) to the total time during which the species existed and to its time range in two biogeographic provinces.

Concurrent Range Zones

Maximum accuracy in fossil correlation is achieved by utilizing the simultaneous occurrence of two or more species, rather than the occurrence of one species alone. Such co-occurrences represent only those portions of range zones that overlap. Generally, this is a smaller stratigraphic interval than the range zone of any one of the constituent species (Figure 5.14). Furthermore, range zones based on several species are less subject to error, because the probability of extending the known range of several species upward or downward at a new locality is lower than that of extending the known range of just one of them. Zones based on overlapping ranges are called **concurrent range zones.**

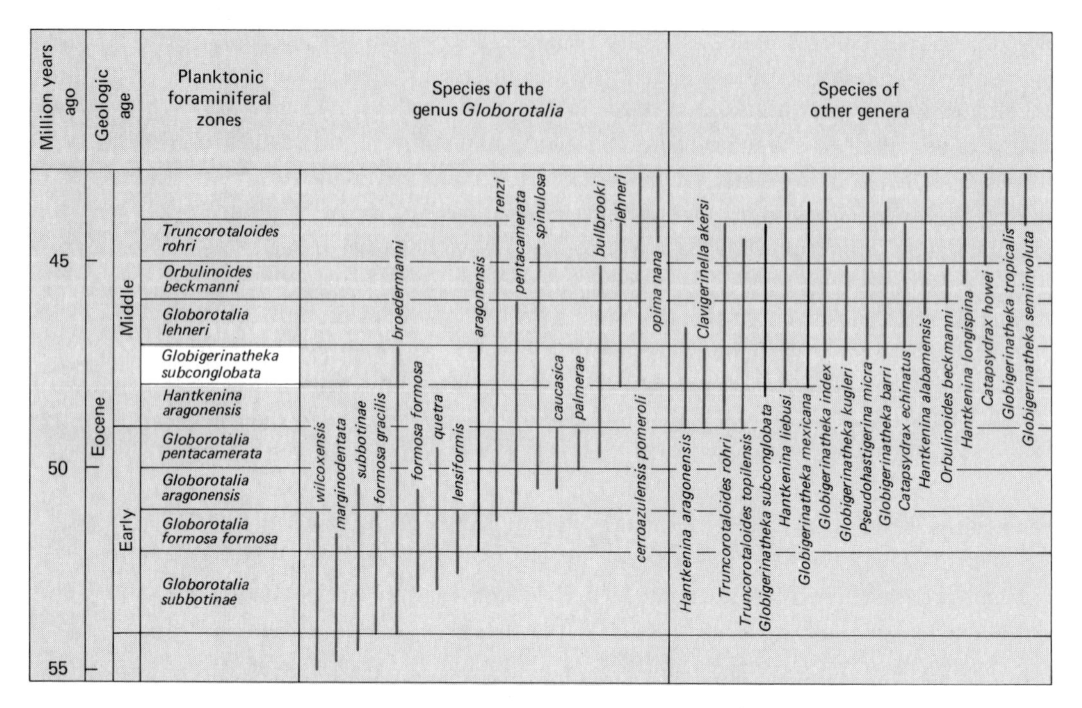

FIGURE 5.14 Planktonic foraminiferal zones for the Early and Middle Eocene are based on overlapping ranges of numerous species, only some of which are shown. By way of example, marker species for the *Globigerinatheka subconglobata* concurrent range zone are noted. (*Stainforth and others, 1975*)

Concurrent range zones are generally based on species that belong to a single group of organisms, such as graptolites or foraminifera. Some, however, are based on species belonging to several groups; the strata that comprise such zones may represent two or three different environments. Although some of the characteristic species may never occur together in the same bed, boundaries of such zones are relatively immune to inaccuracies caused by minor differences in depositional environments. The name of a concurrent range zone is taken from one of its constituent species, but that does not mean that the name giver has more significance than other members of the diagnostic association in recognizing the zone. The name giver need not be confined within the zone. For example, the Middle Eocene *Globigerinatheka subconglobata* Zone is defined as the interval between the first occurrence of *Globigerinatheka mexicana* and the last occurrence of *Globorotalia aragonensis* (Figure 5.14).

Methods of determining the boundaries of concurrent range zones vary, depending on the particular task at hand and the kind of information available. Figure 5.15 illustrates four different ways in which fossil ranges in a single section could be assembled into concurrent range zones:

1 The zones represented by Arabic figures on the left side of the chart are based on first appearances. Each zone is bounded below by the lowest

appearance of a significant species, and above by the lowest appearance of another significant species.

2 The capital letters in the second column on the chart illustrate a method based on the tops of ranges. This method is commonly used by paleontologists who deal routinely with subsurface samples, because they encounter the geologic record in reverse sequence in drilling, and because cave-ins make identification of the lower limit of range zones difficult.

3 Some paleontologists, observing great numbers of particular species in some beds and relatively few in others, have incorporated the idea of *peak abundance* into the concept of concurrent ranges. A method that takes abundances of species into account is shown by the Roman numerals on the right side of Figure 5.15. This method might be particularly accurate in very small areas, but it clearly depends more on environmental parameters than on actual stratigraphic ranges, and hence is not reliable over even moderate distances.

4 A method designed to include as many species as practicable within each zone is shown by the lower-case letters at the far right of Figure 5.15. This

FIGURE 5.15

Four methods of recognizing concurrent range zones in a single section. Width of vertical areas indicates the relative abundance of specimens. (*Schenck and Graham, 1960*)

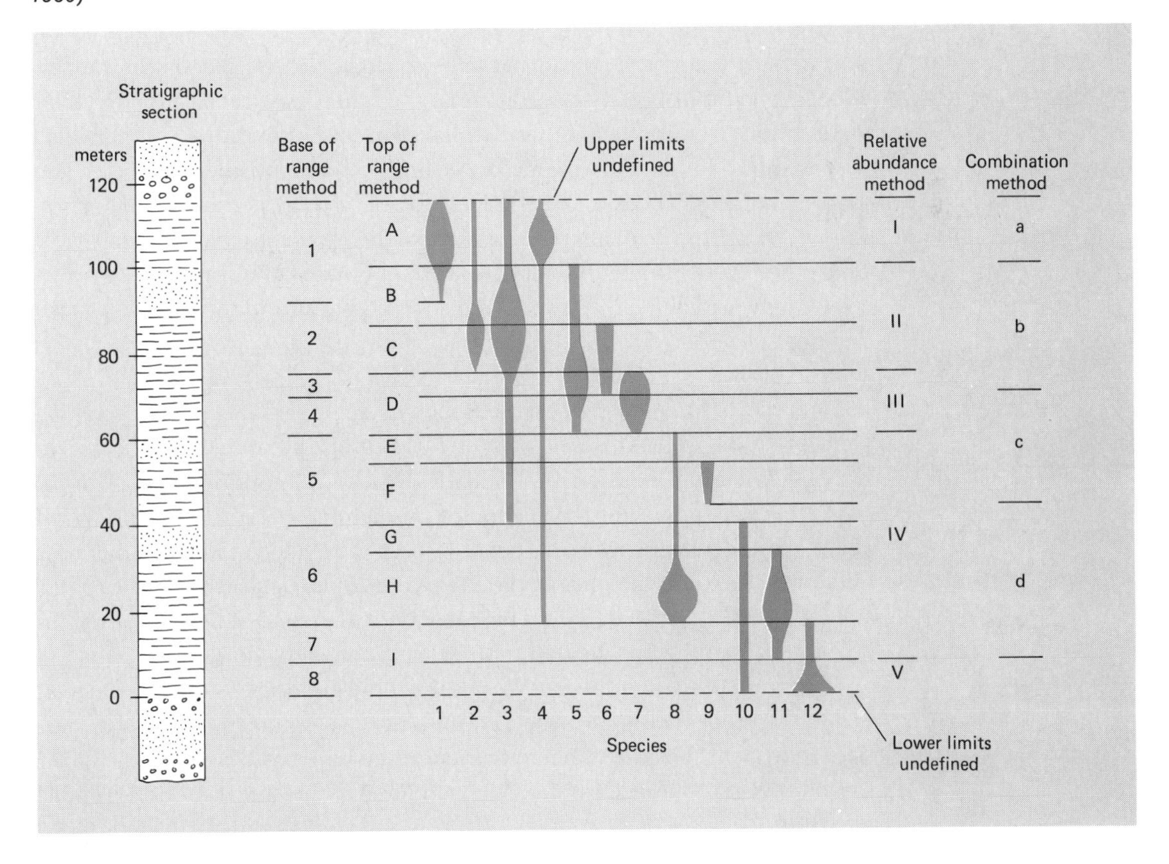

method includes more upper and lower limits of ranges within each zone than do the others. Although this method does not yield very many zones, the zonal boundaries might prove to be particularly stable in the time-stratigraphic framework, as they are traced very widely.

Most concurrent range zones can be recognized only within a single biogeographic province, because the migration and evolution events on which they are based typically have no counterparts in adjacent provinces. When characteristic species that have evolved within a province migrate into a second province, they provide valuable tie points between the zonal sequences of each, but exact zonal boundaries can rarely be carried from one province to another.

Physical Evidence for Time Equivalency

Time correlation between two stratigraphic sections in different parts of the world must be based on an event that affected both areas simultaneously. In addition to the biological events discussed above, certain physical events record themselves in strata in widely separated areas. We shall briefly discuss two different kinds of physical events that make correlation possible: *sedimentary* events and *magnetic* events.

Correlation by sedimentary events begins with the mapping and actual tracing of formational contacts in the field. However, the correlation process does not end there, because as sedimentary rock units are traced laterally, their upper and lower contacts typically change in age. Nevertheless, individual beds within a formation were produced by single depositional events. Commonly, individual beds can be traced for only short distances in a given outcrop. Beds that do extend over several kilometers are rarely distinguishable from other beds of similar lithology in areas of isolated outcrops. Occasionally, beds extend widely and also possess distinguishing characteristics that make them useful in correlation; these are called **key beds.**

Key Beds

In Figure 5.16, the Alpha Sandstone is shown "climbing in section" (becoming younger) to the west. This is obvious on the diagram, because the vertical scale is exaggerated and because the individual beds that represent depositional events are drawn to show that deposition began first at East Mesa and ended last at West Mesa. However, it is not obvious to the geologist studying outcrops in the field, because the beds at the top of the Alpha Sandstone at both West and East Mesas, although of different ages, look the same at both localities, as do those beds at the bottom of the Alpha Sandstone.

To say that the sandstone at West Mesa is the same age as the sandstone at East Mesa is not totally incorrect, because the two outcrops are in large part time equivalent. But the statement obscures the important fact that the lower part of the sandstone at East Mesa was deposited at the same time as the upper

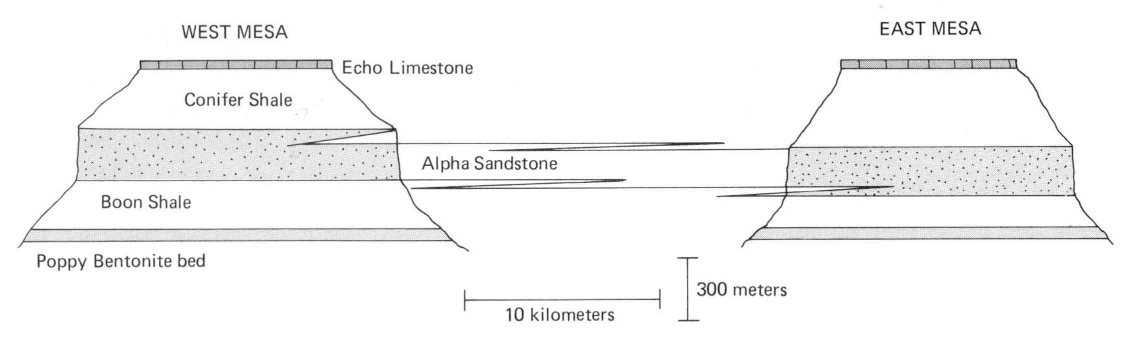

FIGURE 5.16

The Poppy Bentonite Bed and the Echo Limestone are inferred to be key beds that are nearly synchronous throughout, thus paralleling time planes. The Alpha Sandstone crosses time planes. Vertical exaggeration × 50.

part of the Boon Shale at West Mesa. The reason this detail is important is that it indicates the *direction* of movement of the shoreline, a significant element in reconstructing the geologic history of the region.

Figure 5.16 also shows the use of key beds as one method of detailed correlation. A key bed is an individual bed that can be recognized over a wide area; ideally, it results from a single depositional event and has correlative value. For example, a bed deposited by a single large submarine slump may extend over tens of thousands of square kilometers on the deep ocean floor. Or a bed of volcanic ash may cover a similarly large area as a result of a single eruption. In ancient marine rocks, beds of volcanic ash have been altered to distinctive white clays called **bentonite.** Bentonite beds are distinctive and represent essentially instantaneous deposits over a broad area. In Figure 5.16 the Poppy Bentonite Bed establishes precise correlation between East and West Mesas. The greater distance stratigraphically below the Alpha Sandstone at West Mesa suggests that the base of the Alpha there is younger. Similarly, the convergence of the Echo Limestone and the Alpha Sandstone to the west suggest that the top of the Alpha might be younger westward.

Position in a Transgressive-Regressive Cycle

Successful physical correlation among localities requires that the same event be recorded in all of them. Beds that record any kind of common event may be used. The event may not have produced precisely the same kind of deposit in all localities but may still be reflected in more subtle ways. A sedimentary cycle of marine transgression and regression is a good example.

If the level of the sea changes, such a change necessarily occurs worldwide. This is called **eustatic change,** a simultaneous event at every shoreline on earth. Hence it is a potential tool for correlation. Other sea-level changes may occur only regionally as a result of upward or downward warping of the earth's crust. In these *tectonic* changes, it is the land, not the sea, that moves up or down, and the effect is regional, not worldwide. In the affected region, however, the resulting sedimentary record shows a relative rise or fall of sea level exactly as if the change were eustatic. A relative rise in sea level produces *transgression* of the sea over the land, and a relative lowering in sea level produces *regression* of the sea from the land.

The best documented eustatic changes in the geologic record were produced by the alternate growth and melting of huge continental ice caps during the last few million years. Sea level fell one hundred meters or more when ice accumulated on the land, and it rose again when the ice melted. In the more distant geologic past, changes of greater magnitude occurred, probably as a result of changes in the capacity of the ocean basins themselves.

FIGURE 5.17

(*a-c*) Deposition of a transgressive-regressive cycle. (*d*) Correlation on the deepest water environment at each locality, based on environmental analysis of depth changes in each section.

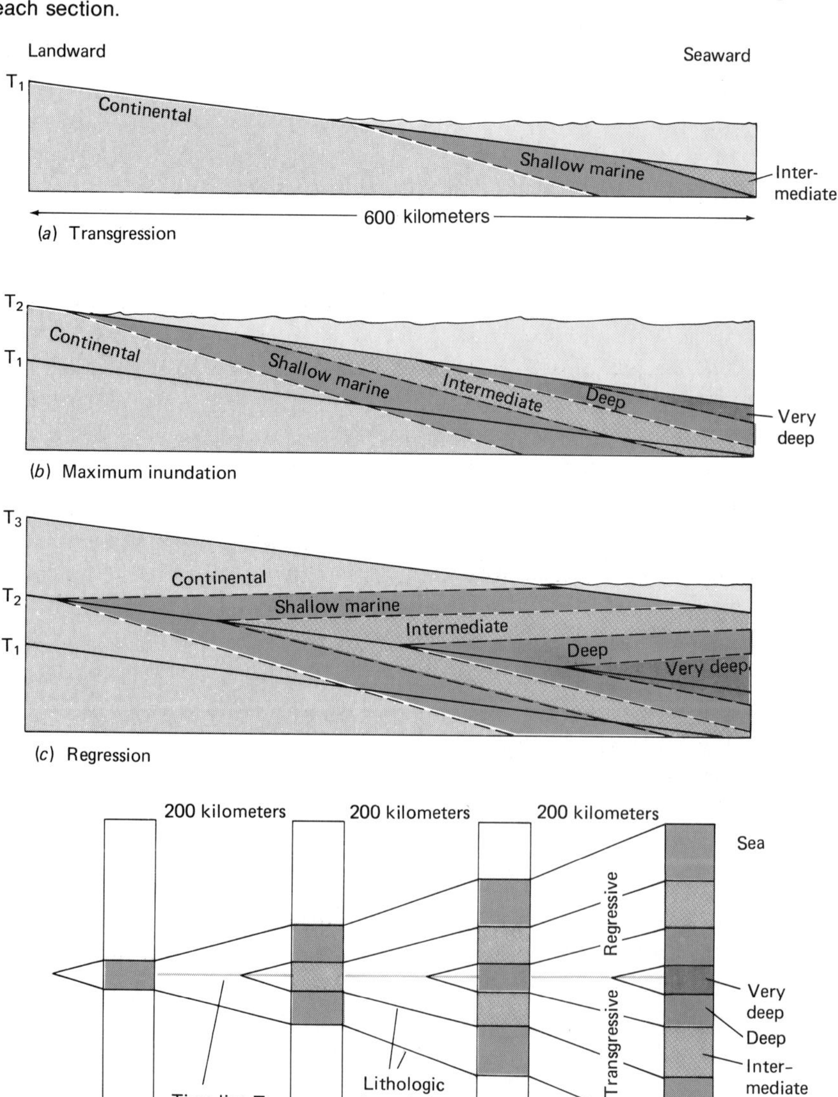

Eustatic changes caused by glaciation probably take place over several tens of thousands of years, whereas significant changes in the volume of the ocean basins have perhaps required several millions of years. Eustatic sea-level changes, therefore, cannot provide as precise a basis for correlation as can a key bed, but their effect may be much more widespread.

For correlation within a single depositional basin, such as the Gulf Coast of the United States or the Anglo-Paris basin of northern Europe, a transgressive-regressive cycle caused by eustatic sea-level changes contains an inherent time marker: the time of maximum inundation of the sea. During transgression, the water becomes deeper and deeper until maximum inundation is reached; then, on regression, it becomes shallower. The maximum depth achieved is not the same everywhere, of course; near the edges of the basin, the maximum water depth may be only one meter, and near the center, several hundred meters. Yet, regardless of how great or how small (unless there is local tectonic warping of the depositional basin during the cycle), the greatest depth near the edge of the basin will be attained at the same time as the greatest depth in the center, and at all localities in between (Figure 5.17). Correlation in this case depends on the ability to identify the deepest part of the resultant sedimentary record, based on environmental analysis of fossils and rock types.

Paleomagnetic Correlation

Magnetic iron oxide minerals occur in very small quantities in most sedimentary and igneous rocks. When tiny particles of these minerals settle to the ocean floor, or when they cool within a newly formed igneous body, their magnetic polarity aligns itself with that of the earth's magnetic field at that place. This produces a preferential direction of magnetization in rocks that contain the minerals. This property is called **remanent magnetism.** Dark igneous rocks show the strongest remanent magnetism, but that of many other kinds of rocks can be measured easily. In Holocene lava flows and sediments, the alignment of the remanent magnetism consistently parallels that of the earth's present magnetic field (Figure 5.18). But in older rocks, the alignment of the remanent magnetism departs from that of the present field. In general, the older the rock, the greater the departure.

Thousands of paleomagnetic measurements from rocks of different ages throughout the world indicate that the earth's magnetic poles have gradually changed position with respect to the continents through geologic time. This phenomenon is commonly called **polar wandering.** However, geologists believe that, in fact, the poles have remained relatively stable, and it is the continents that have moved. Computed polar positions for rocks of various ages from a particular continent delineate a curving or twisting path across the earth's surface. Polar-wandering paths have been reconstructed for most continents. Many of these still contain large uncertainties, chiefly because some of the rocks for which paleomagnetic poles have been determined were recrystalized a long time after they formed. In these cases, the magnetization is not original, but dates from some later, undeterminable time. In spite of this

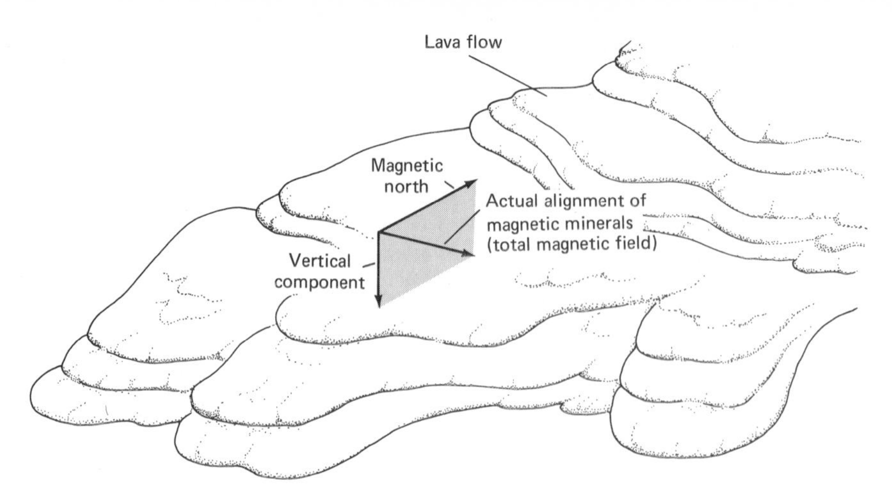

FIGURE 5.18

A sample taken from any part of this young lava flow will be magnetized in the direction of the earth's present magnetic field.

problem, polar-wandering paths are being refined continually. They show promise as an aid in correlation because the paths are time dependent. Once a polar-wandering path has been determined for a given continent, rocks of unknown age from that continent may be dated by ascertaining the position of their paleopoles on the established path.

Figure 5.19*a* shows the Phanerozoic polar-wandering path for North America and for Europe, and Figure 5.19*b* shows one use to which polar-wandering paths can be put. When combined as a single polar-wandering path, the individual paths from both continents help to establish the fit of the North American and European continents in their predrift reconstruction; they also aid in timing the continents' breakup (see Chapter 6).

FIGURE 5.19

(*a*) North pole views showing Phanerozoic polar-wandering paths for North America and for Europe. (*b*) Arrangement of the continents when the two paths are superimposed. (*McElhinny, 1973*)

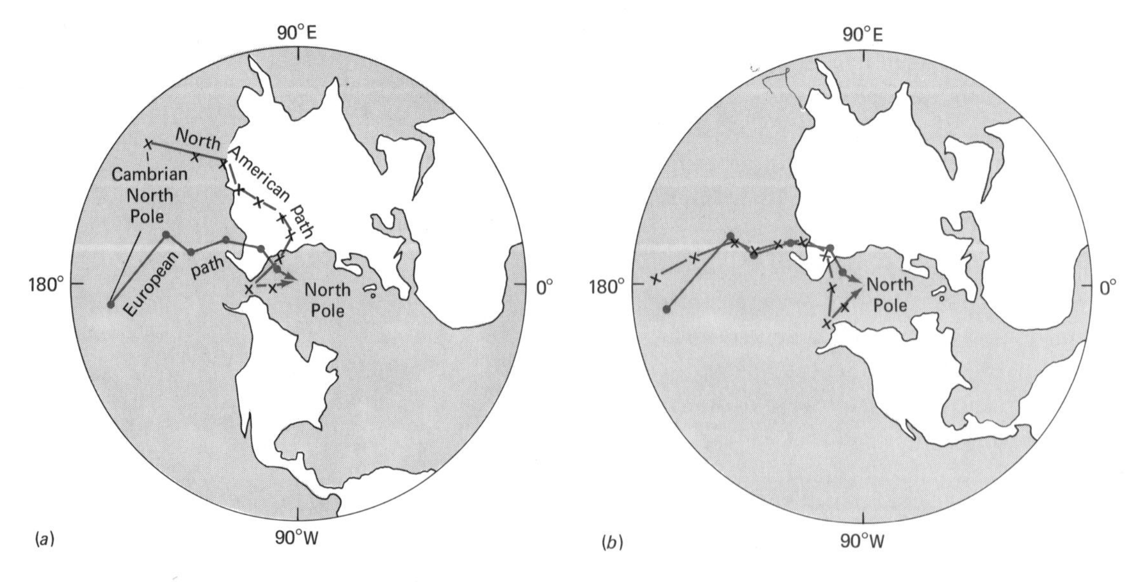

In addition to changing positions relative to the continents, the earth's north and south magnetic poles have abruptly switched polarity many times in the geologic past. During these times, the north needle on a compass would have actually pointed to the south pole! Each of these **magnetic reversals** simultaneously affected the field for the entire earth. Evidence of numerous reversals of the earth's magnetic polarity during late Cretaceous and Cenozoic time has been obtained independently from the remanent magnetization in such diverse rocks as terrestrial lava flows and deep-sea sediments. The sequence of reversals for the Cretaceous and Cenozoic is now well estab-

FIGURE 5.20

Correlation of seven deep-sea cores from the Antarctic region based on polarity changes in the earth's magnetic field. (*Opdyke and others, 1966*)

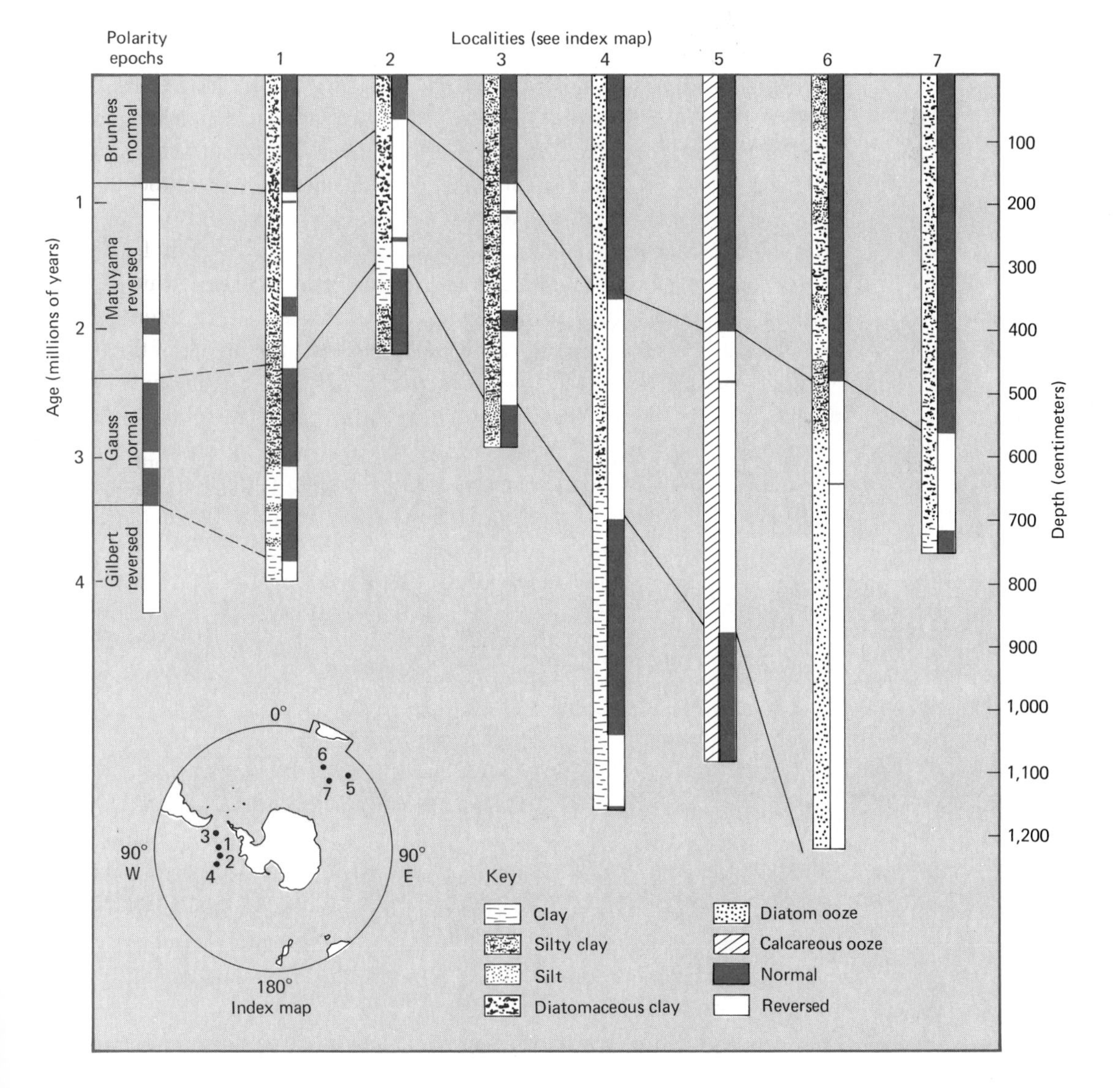

lished. For considerable lengths of time, the polarity was either normal, as today, or reversed. These extended intervals during which the polarity was either dominantly normal or dominantly reversed are called **magnetic polarity epochs.** Within these epochs, there were short-lived polarity changes, which have been termed magnetic polarity events. The mechanism for the reversals of the earth's magnetic field is unknown, but it is assumed to relate somehow to changes in circulation patterns in the earth's liquid core, where the main magnetic field originates.

Figure 5.20 shows the correlation of seven widely spaced cores of deep-sea sediment from the Antarctic region, based on reversals in their remanent magnetism. The cores represent approximately the last 3.5 million years. The chronological accuracy of the polarity epochs is verified by their constant relationships to time-stratigraphic zones based on fossils. The polarity changes are independent of the markedly different kinds of sediment found in the deep-sea cores. Paleomagnetic reversal thus appears to be an extremely promising tool for correlating and dating deep-sea sediments. Because polarity of the earth's magnetic field is a global phenomenon, the method has worldwide potential. Cores from ocean floor sediments from many oceanic regions and rocks from continental regions have been correlated with the paleomagnetic events shown in Figure 5.20.

Throughout most of the Cenozoic Era, the frequency of paleomagnetic reversals was comparable to that of the last 4 million years (Figure 5.21). Because this frequency is relatively high, the correlation potential of the reversals prior to a few million years ago quickly falls to zero, because there are only two kinds of polarity epochs. After establishing that a rock's magnetism is either normal or reversed, one cannot tell which normal or reversed epoch it represents without an independent means of correlation. In these older rocks, the time span between paleomagnetic reversals is generally less than the time span that can be resolved by fossil zones or by radiometric dating, and hence the identity of the paleomagnetic event cannot be determined.

FIGURE 5.21

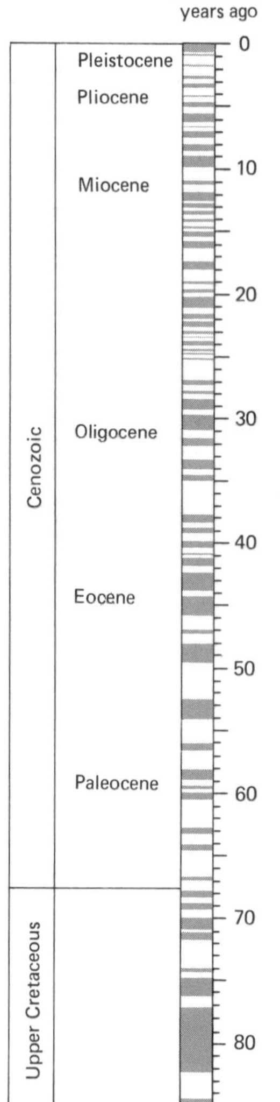

A time scale for magnetic reversals during the last 85 million years. Shading represents time of normal polarity. (*Larson and Pitman, 1972*)

Radiometric Calibration of the Phanerozoic Time Scale

If the Phanerozoic part of the geologic time scale were calibrated in detail with accurate radiometric dates, then any rock for which a new radiometric date became available—whether it was igneous, metamorphic, or sedimentary—could immediately be placed in the proper stage or zone as confidently as if it contained diagnostic fossils. We do not yet have a quantitative time scale of such accuracy, chiefly because most radiometric dates cannot be tied to sedimentary rocks whose exact position in the geologic time scale is known. The Phanerozoic time scale is defined on sedimentary rocks and implemented

FIGURE 5.22

A bentonite bed in the Cretaceous Greenhorn Formation of southern Colorado. In fossiliferous strata, similar bentonites have yielded key radiometric dates for calibrating the Phanerozoic time scale.

worldwide by correlations based on fossils. Most radiometric dates, on the other hand, are from igneous or metamorphic rocks, whose correlation with fossiliferous sedimentary sequences is problematical. How can fossiliferous sedimentary rocks be dated by radiometric calibration so that we can quantify the Phanerozoic time scale?

In some places, lavas or volcanic ash have been introduced suddenly into sedimentary environments without interruption in sedimentation (Figure 5.22). Igneous material datable by the potassium-argon or rubidium-strontium methods thus becomes interbedded with sedimentary rock datable by fossils. Stratigraphically well-dated sedimentary sequences containing interbedded lavas or ash deposits, although not very common, have provided some of the most valuable reference points in the radiometric time scale.

Igneous intrusives provide minimum ages for strata they intrude and maximum ages for strata that overlie them. In circumstances in which the time spread between deposition of older strata, intrusion of an igneous body, and deposition of younger strata is small, bracketed igneous intrusives yield valuable dates.

Finally, radiometric ages can be obtained in some cases from **authigenic minerals** (minerals that crystallize within sediments at the time the strata are deposited). The most commonly used of these is glauconite, a silicate of potassium, aluminum, and iron, which is datable by the potassium-argon method. Unfortunately, glauconite loses argon easily when it is buried to moderate depths, and hence glauconite ages are normally given credence only as minimum ages.

Dates of the Phanerozoic system boundaries, as they are presently known, appear adjacent to the geologic time scale on page 49.

Presenting Stratigraphic Data

Once rocks have been accurately described, a framework exists for subsequent interpretations, including the vitally important one of time correlation. Once the correlations are made, the lateral facies equivalents become apparent and can be illustrated on a series of stratigraphic sections that are linked together by correlation lines into a stratigraphic cross section. Figure 5.23 shows an example of a limestone that has been interpreted as an ancient organic reef and its associated environments that were actively building seaward.

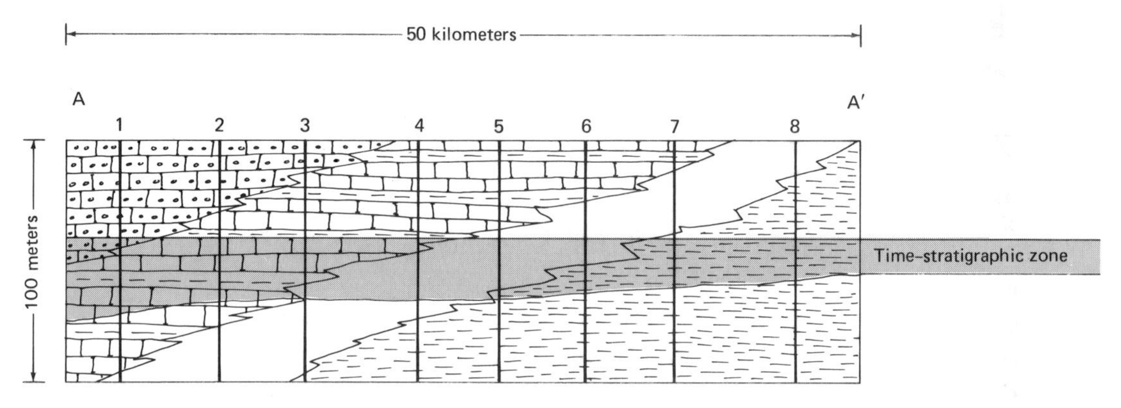

FIGURE 5.23 A stratigraphic cross section showing the correlation of a single time-stratigraphic zone. Location of sections is shown in Figure 5.24.

If the data are sufficient, lithofacies can actually be plotted on a map like that shown in Figure 5.24. Lithofacies maps simply show the distribution of rock types of a given age in an area. Their accuracy depends on: (1) the accuracy of the correlations used to recognize the time interval, and (2) the number of stratigraphic sections available. The areal distribution patterns on facies maps greatly aid environmental interpretations, because both the sedimentary associations and the shapes of the deposits at a given time can be compared directly to those accumulating in similar environments today. Commonly, the thickness of the mapped interval is also shown on lithofacies maps by means of **isopach** (equal-thickness) lines like those in Figure 5.24. Finally, the location of all stratigraphic sections used in the making of the map are shown, to verify the adequacy of the control.

From inferences based on the facies, it is possible to make an environmental map showing marine, transitional, and terrestrial environments. Maps that show the gross environmental geography inferred from the facies relationships

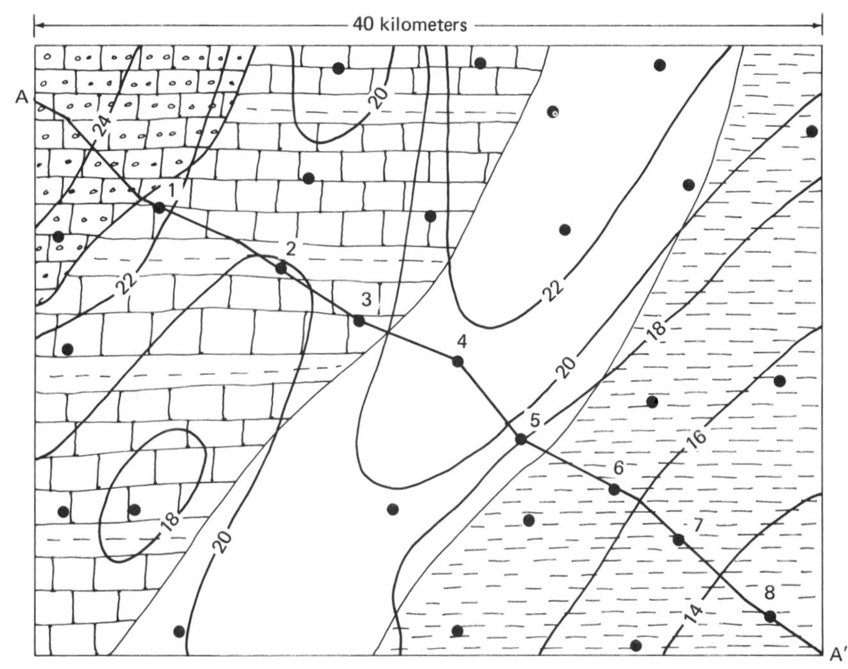

FIGURE 5.24

A lithofacies map of the time-stratigraphic zone shown in Figure 5.23. Dots represent locations of sections. Isopach interval equals 2 meters.

of ancient rocks are called **paleogeographic** maps. Figure 5.25 shows the paleogeographic interpretations based on the lithofacies map in Figure 5.24. Here a reef tract separates the shallow sea on the east from lagoonal and mud-flat environments on the west. Oolitic shoals are slowly advancing over the lagoon and mud flats from the northwest.

FIGURE 5.25

A paleogeographic map based on the lithofacies shown in Figure 5.24.

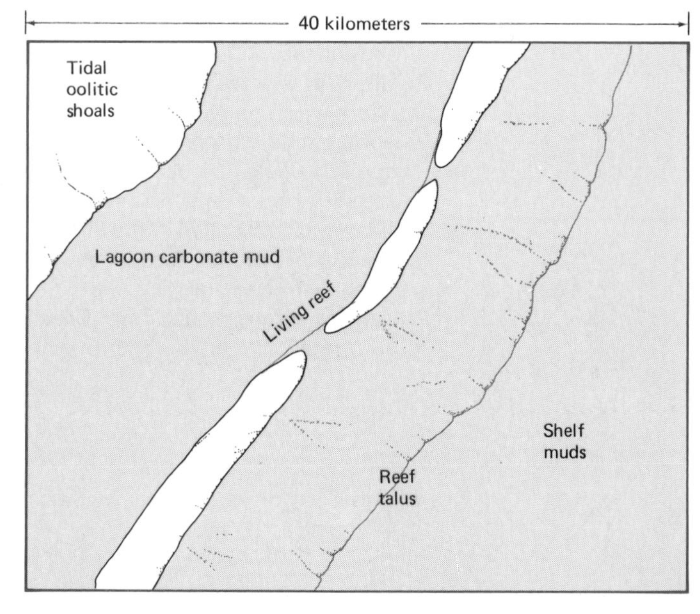

Chapter Summary

The Rock Cycle

This concept helps explain the decreasing abundance of sedimentary rocks with geologic age, for most older rocks have been destroyed and recycled by tectonic activity and erosion.

Stratigraphic Units

Rock units are designated as formations, members, and groups; they are defined solely on the basis of rock type and are the fundamental units used in making geologic maps.

Time-stratigraphic units are more subjective than rock units and include all the rocks deposited during a particular interval of the earth's past; they are recognized primarily by distinctive associations of fossils.

Geologic time units are closely related to time-stratigraphic units but differ in referring *only* to time, rather than to specific rocks deposited during that time.

Unconformity-bounded units: Intervals of continent-wide rock erosion permit the definition of still another type of unit called a sequence.

Biostratigraphy

Paleoecology is the science of interpreting the interactions of ancient sedimentary environments and their contained animal and plant fossils.

Biogeographic provinces can be recognized throughout Phanerozoic time by the distribution of fossil animals and plants.

Fossils and time: Fossils can be used to date the rocks in which they are found because many individual species existed for only a relatively short span of earth history.

Range zones: The geologic time range of a species is influenced by its time of evolutionary origin, time of extinction, and its patterns of migration and distribution during its existence.

Concurrent range zones: Most geologic dating with fossils relies on the overlapping ranges of many fossil species.

Physical Evidence for Time Equivalency

Key beds such as volcanic ash falls can be used to correlate sedimentary sequences.

Position in a transgressive-regressive cycle: Continent-wide or worldwide changes in sea level can also help establish time relations.

Paleomagnetic correlation: The magnetic patterns preserved in ancient rocks can sometimes be used to establish their age relations; particularly useful are sequences reflecting sequential reversals of the magnetic field.

Radiometric calibration of the Phanerozoic time scale: Igneous rocks intrusive into or overlain by fossil-bearing sediments permit the dating in years of Phanerozoic rocks.

Presenting stratigraphic data: Creating maps that show patterns of ancient geography and environments is the ultimate goal of stratigraphic studies.

Important Terms

age
authigenic mineral
bentonite
biogeographic province
biostratigraphic unit
biostratigraphy
concurrent range zone
epoch
eustatic change
group
guide fossil
index fossil
isopach
key bed
magnetic polarity epoch
magnetic reversal
member
paleoecology
paleogeographic map
period
polar wandering
range zone
remanent magnetism
rock cycle
sequence
series
stage
system
type area
type section
zone

Review Questions

1. What does the rock cycle tell us about the survival of rocks of varying ages?
2. Distinguish between rock units, time-stratigraphic units, and geologic time units.
3. On what bases are time-stratigraphic units and geologic time units defined and recognized?
4. Explain the uses of fossils in interpreting past environments and ancient geography.
5. Distinguish between the use of range zones and concurrent range zones as dating techniques.
6. How are key beds, sea-level cycles, and ancient magnetism useful for dating rocks?
7. How are the results of stratigraphic studies usually presented?

Additional Readings

Ager, D. *The Nature of the Stratigraphical Record,* MacMillan, Ltd., London, 1973. *An entertaining presentation of modern stratigraphic ideas.*

Eicher, D. L. *Geologic Time,* Prentice-Hall, Inc., Englewood Cliffs, New Jersey, 1976. *An expanded introduction.*

Garrels, R. M. and F. T. Mackenzie *Evolution of Sedimentary Rocks,* W. W. Norton and Co., New York, 1971. *An advanced survey stressing the rock cycle.*

Harbaugh, J. W. *Stratigraphy and Geologic Time,* W. C. Brown Co., Dubuque, Iowa, 1974. *An introductory survey.*

Matthews, R. K. *Dynamic Stratigraphy,* Prentice-Hall, Inc., Englewood Cliffs, New Jersey, 1974. *An intermediate-level text.*

PLATE TECTONICS

6

One of the most exciting scientific discoveries of recent decades has been that *the earth's present outer skin is divided into seven huge structural plates that move as coherent units, plus about twenty much smaller plates sandwiched between them* (Figure 6.1). The rigid **lithosphere** that incorporates these moving plates includes approximately the outer 100 kilometers of the earth. Thus it comprises the uppermost part of the mantle, as well as all of the crust. The lithosphere rests upon a nonrigid, partly molten zone in the mantle called the **asthenosphere.** No one knows what causes the lithosphere plates to move relative to one another, but it is possible that they are carried along by the slow, plastic flow of the asthenosphere.

Motions of these lithospheric plates produce earthquakes, volcanoes, and certain other geographic features of the modern earth. The discovery of **plate tectonics,** as this subject has come to be called, has provided a unifying theory to explain a number of fundamental earth processes. In this chapter we shall discuss the discovery and the current understanding of plate motions. In succeeding chapters we shall consider the actual plate movements that produced the modern earth.

Discovery of Plate Motions

Continental Drift

Beginning in the early years of this century, geologists suggested various theories describing ancient movements of the continents. Among the first of these was one proposed in 1912 by Alfred Wegener, a German geophysicist. Wegener

FIGURE 6.1 The major structural plates and some selected minor plates. Arrows show the inferred relative motions of the plates. White areas represent the present-day continents.

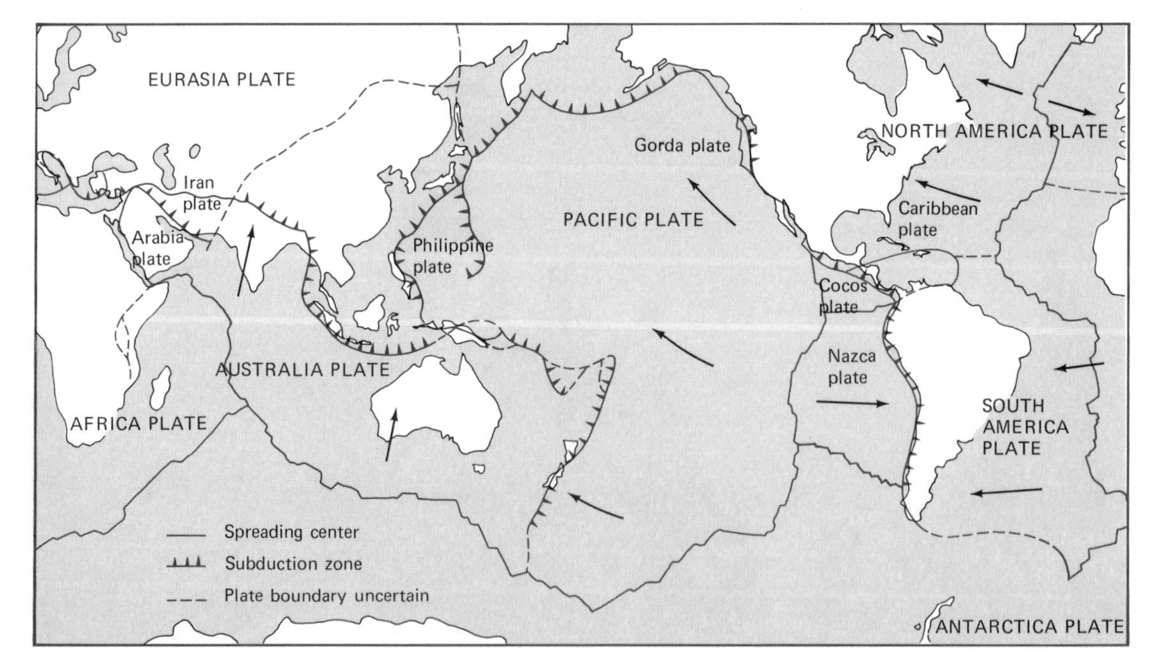

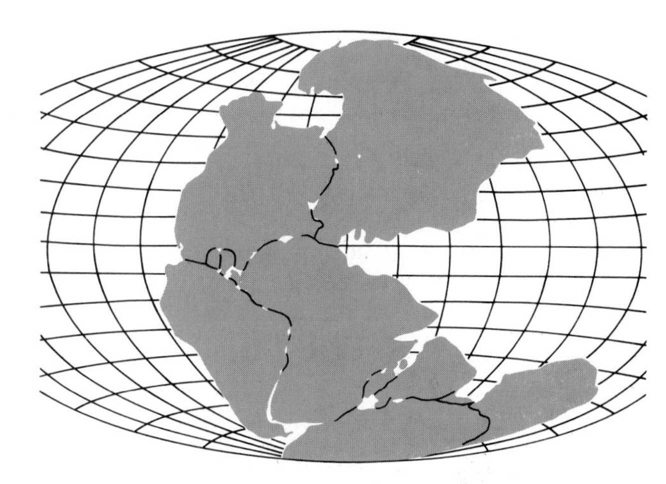

FIGURE 6.2

A reconstruction of the probable relationships of the continents in Triassic time before they drifted into their present configurations. (*Dietz and Holden, 1970*)

was impressed by the jigsaw-puzzlelike fit of the shapes of some of the continents. The most obvious such fit is that between the west coast of Africa and the east coast of North and South America. With suitable juggling, rough fits can be made for other continental margins as well (Figure 6.2). This neat pattern led Wegener to suggest that the continents were originally joined in a single large land mass that subsequently broke apart to create the separate continents we see today. Wegener was unable to suggest a plausible energy source or mechanism for **continental drift**, as his idea came to be called. It was largely dismissed until about 20 years ago, when new evidence derived from study of the earth's magnetic field began to suggest that large-scale continental motions had, indeed, taken place.

When igneous rocks first crystallize, any magnetic, iron-bearing mineral they contain becomes magnetized in the direction of the earth's magnetic field (see Chapter 5). Paleomagnetic studies of continental rocks provided documented evidence that the continents had moved relative to one another and led to a revival of interest in the idea of continental drift (Figure 6.3).

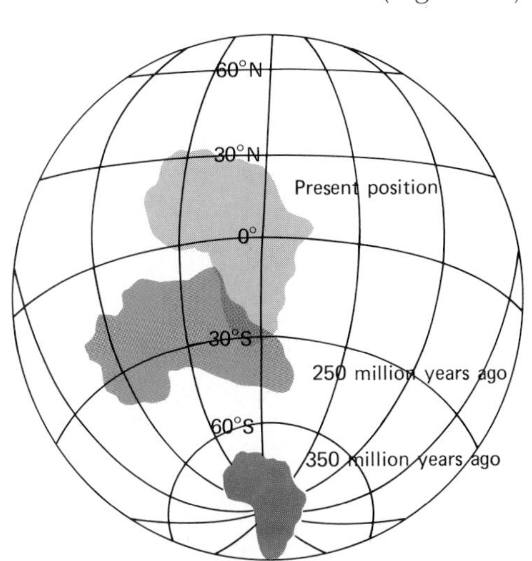

FIGURE 6.3

Paleomagnetic reconstruction of Africa's position relative to the south pole from 350 million years ago to the present. (*Clark, 1971*)

General acceptance of the idea did not come, however, until the mid-1960s, when it was discovered that still more fundamental crustal movements take place beneath the ocean floor.

Ocean-Floor Topography

The most conclusive evidence for large-scale plate motions came not from studies of the exposed continents, but from an improved understanding of the topography and history of the ocean floor.

Because of the ease with which we can observe mountains, hills, and valleys on the continents, most of us are unaware of the extraordinary problems involved in trying to learn whether similar topography exists on the ocean floor. Before 1920 the only method for determining ocean depths or sea-floor topography was to laboriously lower a heavy weight attached to thousands of feet of steel wire over the side of a ship. Even though many such soundings were made, they were so scattered over the vast expanse of the oceans that they provided little evidence for ocean-bottom irregularities. The ocean floors were generally believed to be flat, featureless plains, broken here and there by mountains that rose above the surface as islands.

In the 1920s the first **echo sounders** were developed, which made possible a more rapid determination of ocean depths and bottom topography (Figure 6.4). These instruments measured depths by noting the time required for sound waves to reach the bottom and be reflected back to a shipboard recording device. The first systematic surveys of ocean-bottom topography using these instruments were undertaken in parts of the Pacific Ocean in the early 1940s. The principal result of these early surveys was the discovery of a great many flat-topped mountains, called **guyots**, on the floor of the northern Pacific (Figure 6.5).

The tops of guyots are covered by as much as 1800 meters of water, yet the only known mechanism that could have produced their broad, flat tops is the

FIGURE 6.4

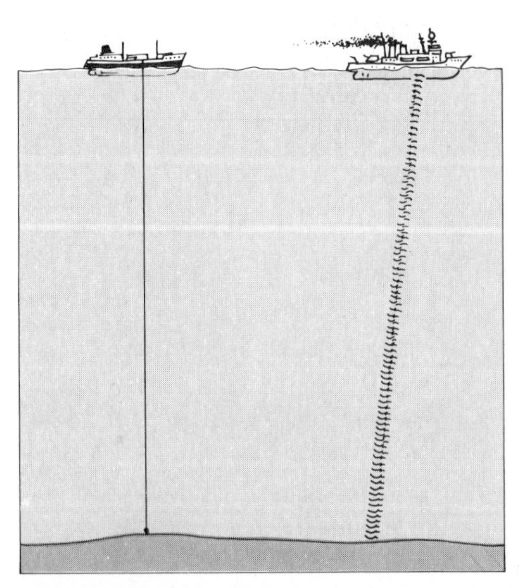

Methods of determining ocean-floor topography. Before the 1920s, the only method was by lowering weighted lines (left); since then, increasingly sophisticated echo sounders (right) have been used to map the ocean floor.

FIGURE 6.5

Idealized drawing of a portion of the Pacific Ocean floor with water removed, showing two flat-topped guyots in the distance. The canyon in the foreground is cut into a guyot, most of which does not show. *(Painting by Chesley Bonestell)*

wearing action of waves at the surface of the sea. Because these flat tops could have been produced only at sea level, either sea level has risen as much as 1800 meters since they were formed or, more probably, the ocean floor on which they stand was somehow lowered by this amount. Guyots thus provided the first suggestion of large-scale movements of the ocean floor.

Since 1945 additional surveys of bottom topography have been undertaken in many areas. The Atlantic, Indian, Arctic, and eastern Pacific Oceans have been intensely surveyed; the other oceans are known in less detail. These surveys have shown that the ocean floor, although relatively flat over large areas, also includes the longest, highest, and most continuous mountain ranges on earth. These ranges make up the **oceanic ridge-rise system**, a worldwide belt of ridgelike mountains and broad plateaus that is of extraordinary importance for understanding movements of the ocean floor (Figure 6.6).

The most intensively studied part of the oceanic ridge-rise system is the Midatlantic Ridge, a chain of mountains lying in almost the exact center of the Atlantic Ocean (Figure 6.7). The existence and general position of the Ridge has been known for many years, because it comes to the surface in several places to make oceanic islands, such as Iceland and the Azores. Furthermore, it has long been known that the Ridge overlies a zone of rather shallow earthquakes. In the 1950s detailed studies of the topography of the Ridge and the distribution of earthquakes under it revealed two important relationships:

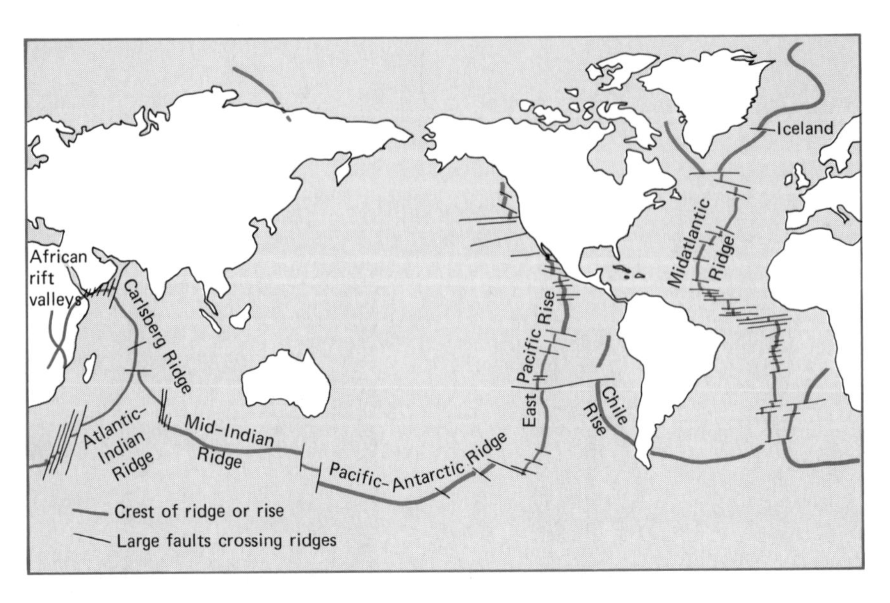

FIGURE 6.6

The oceanic ridge-rise system.

FIGURE 6.7

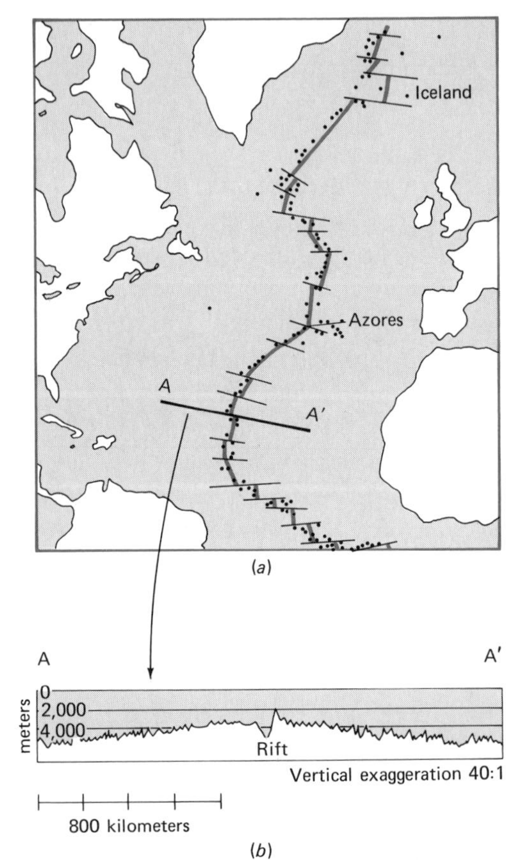

A portion of the Midatlantic Ridge showing (*a*) the location of earthquake epicenters (black dots) and (*b*) the cross section of the rift valley at the crest of the Ridge.

A part of the rift valley of central Iceland. (*Courtesy of Sigurdur Thorarinsson*) FIGURE 6.8

First, although the total width of the Ridge averages several hundred kilometers, the earthquakes are concentrated in a narrow band under only the highest part, or crest. Second, the crest itself did not make up a single long ridge but was, instead, composed of two steep-sided, parallel ridges separated by a deep valley. Similar but smaller and less continuous valleys, called rifts, are also known from the surface of the continents. Rifts form from normal faulting caused by a tension, or *pulling apart,* of the faulted rocks on each side of a valley (Figure 6.8). This immediately suggested that the Midatlantic Ridge might represent a similar, but much larger, zone of crustal tension, in which the Atlantic Ocean floor on both sides of the Ridge was *moving away* from the rifted ridge. The rift valley at the crest of the Ridge and the earthquakes underlying it would mark the line of separation.

Other lines of evidence also pointed to the same conclusion. Where the tops of the Ridge were exposed as islands, the rocks of the Ridge could be sampled. They were found to be basaltic or ultramafic rocks, as would be expected if the Ridge represents a fracture zone in which basaltic lavas poured out from the underlying mantle to build up the ridge. Furthermore, the entire width of the rift valley itself is exposed above the sea in Iceland, where tension cracks and other evidence of movement can be observed directly (Figure 6.8).

Ocean-Floor Magnetism

The discovery of the oceanic ridge-rise system, with its associated earthquakes and central rift valleys, strongly suggested that horizontal movements on a tremendous scale take place in the crustal rocks beneath the sea floor. Proof of this hypothesis of **sea-floor spreading**, as it came to be called, was provided by magnetic studies of the rocks of the ocean floor.

The strength of the magnetic field at the earth's surface is affected by two components: (1) the earth's main magnetic field originates from motions deep in the liquid iron of the core; and (2) lesser magnetism originates from the distribution of magnetic minerals in rocks of the crust. There are no magnetic effects from mantle rocks lying between the crust and the core because the minerals making up the mantle are heated beyond the **Curie point** (the temperature at which they lose their magnetism). After subtracting the large component contributed by the main magnetic field, we can then study differences in the magnetic minerals of crustal rocks by making magnetic surveys of large areas with shipboard or airborne **magnetometers.**

Such surveys have been conducted over land areas for many years and have revealed complex magnetic patterns caused primarily by differences in the amount of the mineral magnetite present in the underlying crustal rocks. The first systematic magnetic surveys of the ocean floor were undertaken off the coast of California in 1955. Surprisingly, they showed patterns that were *far more regular than any from continental areas* (Figure 6.9). In particular, the oceanic magnetic patterns showed a series of extremely long, narrow bands running for hundreds of kilometers approximately parallel to the coastline.

One additional feature of these banded magnetic patterns was of great significance. In certain areas the patterns were broken and offset for hundreds

FIGURE 6.9

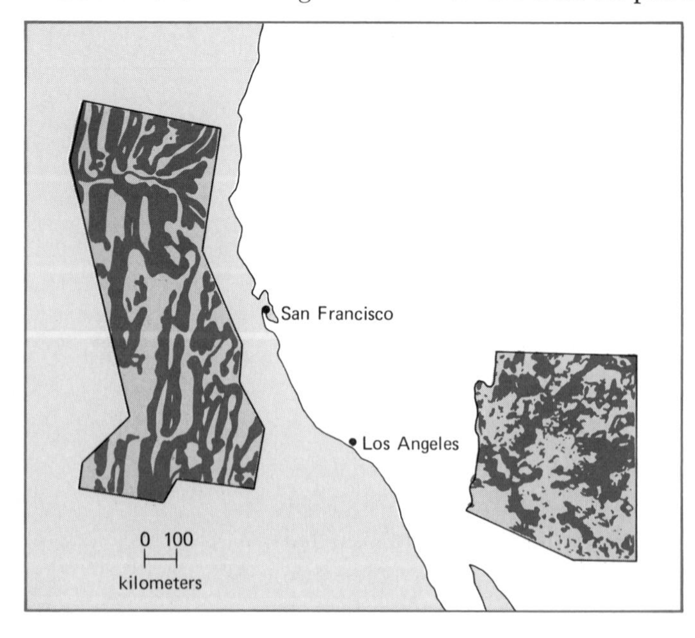

Linear magnetic patterns of oceanic rocks contrast with the irregular magnetic patterns of continental rocks.

of kilometers along huge faults running at approximately right angles to the present coastline. Some of these faults were already known because they make large submarine cliffs discovered previously by echo sounding. The fact that oceanic rocks had moved for hundreds of kilometers along these faults was first shown, however, by the offset patterns of magnetic measurements. Here again was strong evidence for large-scale movements of the ocean floor.

Once the linear oceanic magnetic pattern was discovered, the next question was: What caused it? In continental areas, the much less regular magnetic pattern could be shown by geologic mapping to be caused by differing amounts of magnetic minerals in the underlying rocks. There was, however, another, more likely cause for the regular oceanic patterns. This was related not to differing *amounts* of magnetic minerals, but to differences in their *direction of magnetization.*

Rocks of the ocean floor are almost exclusively basalts containing a large and relatively constant amount of magnetite, which makes them strongly magnetic. When lavas cool to form basalt, this magnetite crystallizes from the liquid lava and eventually cools to the Curie temperature. As this temperature is reached, the magnetic crystals become magnetized in the direction of the earth's magnetic field. As we discovered in Chapter 5, the direction of the earth's magnetic field is not constant but shows regular changes. Lavas forming at different times have different directions of magnetization, reflecting the earth's magnetic field *at the time the lavas cooled.* The original patterns dominate because magnetic minerals permanently retain much of their original magnetic orientation gained on passing the Curie point and do not become completely reoriented to later changes in the earth's field. Fine sedimentary grains of magnetic minerals also become oriented to the earth's magnetic field when they are deposited by water. At about the same time that the puzzlingly regular magnetic patterns were discovered in the oceans, studies of vertical sequences of magnetic mineral orientation in lava flows on land and in samples of sediments from the sea floor showed that the earth's main magnetic field has undergone frequent and rapid *reversals in polarity.* As we saw in Chapter 5, at times in the geologic past the north magnetic pole became the south pole and vice versa (Figure 6.10).

FIGURE 6.10

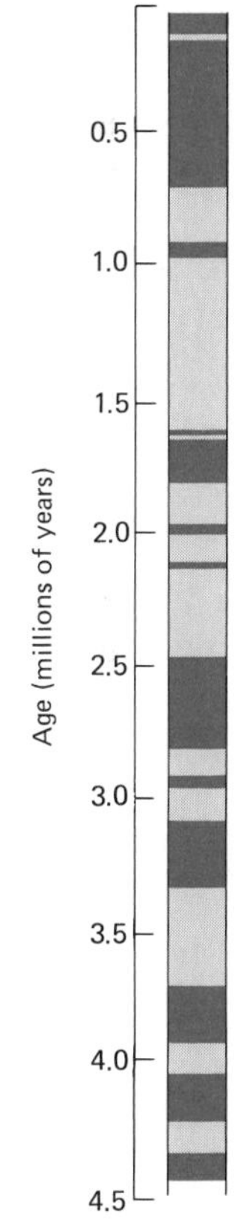

A time scale of paleomagnetic reversals during the last 4.5 million years. Times of normal polarity are dark color.

Soon after the regular oceanic magnetic patterns were discovered, F. J. Vine and D. H. Matthews suggested that they might reflect sudden reversals in the earth's magnetic field in the following way: First, basaltic lavas poured onto the ocean floor from a long crack in the crust, cooled to form basalt, and were then somehow moved away from the crack before new lava was extruded.

If a reversal in the earth's magnetic field took place between lava out-pourings, the result would then be a *series of parallel bands of basalt, each having minerals with a different magnetic orientation* (Figure 6.10). These differences in magnetic direction might account for the regular, parallel patterns of ocean-floor magnetism.

Unfortunately, this idea could not be tested directly because it is impossible to determine the direction of magnetization without having carefully oriented samples of the magnetic rocks. These are extraordinarily difficult to obtain for oceanic basalts covered by thousands of meters of water and hundreds of meters of sediments. Final confirmation that spreading basaltic bands *are* the cause of the oceanic magnetic pattern was obtained, however, from another kind of evidence when, in 1966, the first detailed magnetic analysis was made of the Midatlantic Ridge.

This study showed parallel magnetic patterns similar to those previously found off California, but this time the bands were *identical on each side of the Midatlantic Ridge,* just as would be expected if lavas were pouring from the earthquake rift zone, solidifying into basalt, then being broken along the rift and moving away from it on both sides (Figure 6.11). Furthermore, the width of the parallel magnetic bands closely matched the intervals between reversals in the earth's magnetic field over the past several million years as established independently from magnetic studies of dated sedimentary rocks and lava flows in land-based sections. This discovery indicated that the spreading basaltic rocks of the oceanic crust had acted like a vast magnetic tape, clearly recording changes in the direction of the earth's magnetic field. The evidence shows that basaltic rocks of the oceanic crust rose from the mantle along the oceanic ridge-rise system, solidified, and were then carried horizontally away from the ridges by powerful forces acting from below, within the hotter and less rigid rocks of the mantle. For this reason, the oceanic crust becomes progressively older as you move away from the ridge-rise system.

FIGURE 6.11

An interpretation of symmetrical magnetic patterns on a portion of the Midatlantic Ridge south of Iceland. Lavas from the mantle pour out along the Ridge axis and then move laterally outward on both sides of the Ridge, forming symmetrical bands that are progressively older away from the Ridge.

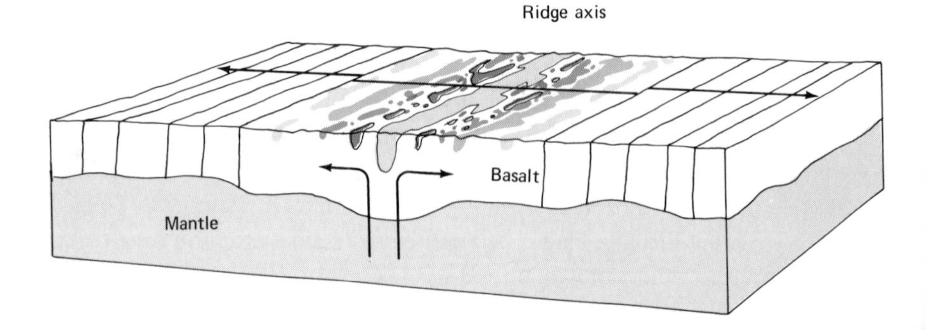

Confirmation of Plate Motions

By the mid-1960s, evidence from the magnetic patterns of ocean-floor rocks had shown conclusively that ocean-floor spreading was taking place. This immediately raised another fundamental question: If new oceanic rocks are being continually created along the ridge-rise system, is the earth increasing in size, or is an equal volume of older rock continually being destroyed to make room for the new materials? At about this same time, the first really precise worldwide map of earthquake locations appeared, providing clues that led to the solution of this question.

Earthquakes and Plate Boundaries

The earthquake map showed belts of medium-strength earthquakes along the oceanic ridge-rise system. These midocean belts were connected to the better-known zones of stronger and generally deeper earthquakes around the margins of the Pacific Ocean. The interconnected earthquake belts defined the boundaries of the seven major plates of moving lithosphere. Further analysis showed that there were three kinds of plate boundaries: (1) those where two plates are moving apart (divergent plate boundaries); (2) those where they are being pushed together (convergent plate boundaries); and (3) those where plates are slipping sideways, one past the other (Figure 6.12).

Regions where plates are actually diverging, called **spreading centers**, coincide with oceanic ridge-rise systems and with large-scale fissures along the margins of continents, for example the Red Sea and the Gulf of California. As two plates move away from one another, the area they vacate is filled in from below with new oceanic crust consisting of basalt flows and ultramafic intrusives, which probably are derived from the underlying asthenosphere. In regions where two plates are converging, older oceanic crust is consumed as

FIGURE 6.12

Idealized diagram of the three types of plate boundary: spreading center, where basaltic crust is added to the plate along oceanic ridges; subduction zone, usually marked by oceanic trenches and deep earthquakes, where the plate plunges deep into the mantle and is destroyed; and transcurrent fault, where only horizontal motions between plates occur.

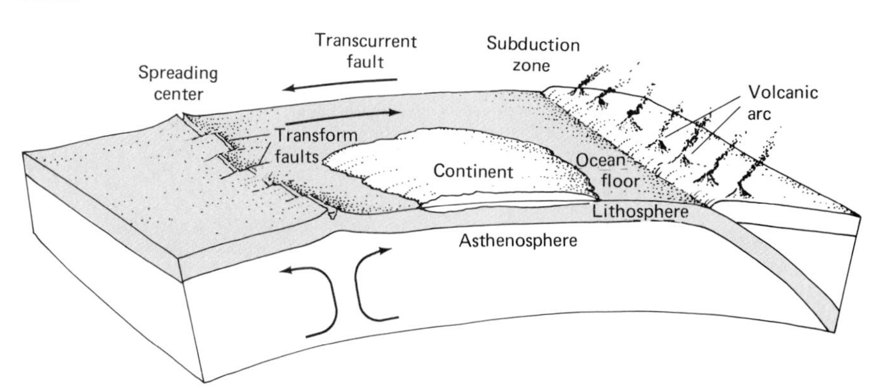

one plate is *subducted* below the other and descends deep into the mantle, where its leading edge is progressively melted, for example in many of the circumpacific deep earthquake zones (Figure 6.12). These **subduction zones** are usually sites of oceanic trenches; they are typically flanked by volcanoes in the form of an *island arc* (such as the Aleutians) or a mainland *volcanic chain* (such as the Andes). The third type of plate boundary, where crustal plates are slipping past one another horizontally, is marked by faults with very large horizontal displacements, called **transcurrent faults.** The well-known San Andreas fault in California is such a transcurrent fault.

Figure 6.1 shows that most spreading centers lie within ocean basins. The reason for this is simple. Spreading centers continually generate new oceanic crust, which is carried away on both sides. Hence, unless a spreading center has formed very recently, it lies within an ocean basin of its own making. At an average separation rate of 4 centimenters per year for 50 million years, a spreading center would produce a 2,000-kilometer-wide ocean between two halves of a continental block *where no ocean existed before* (Figure 6.13). We saw previously that such spreading of former portions of the same continental

FIGURE 6.13

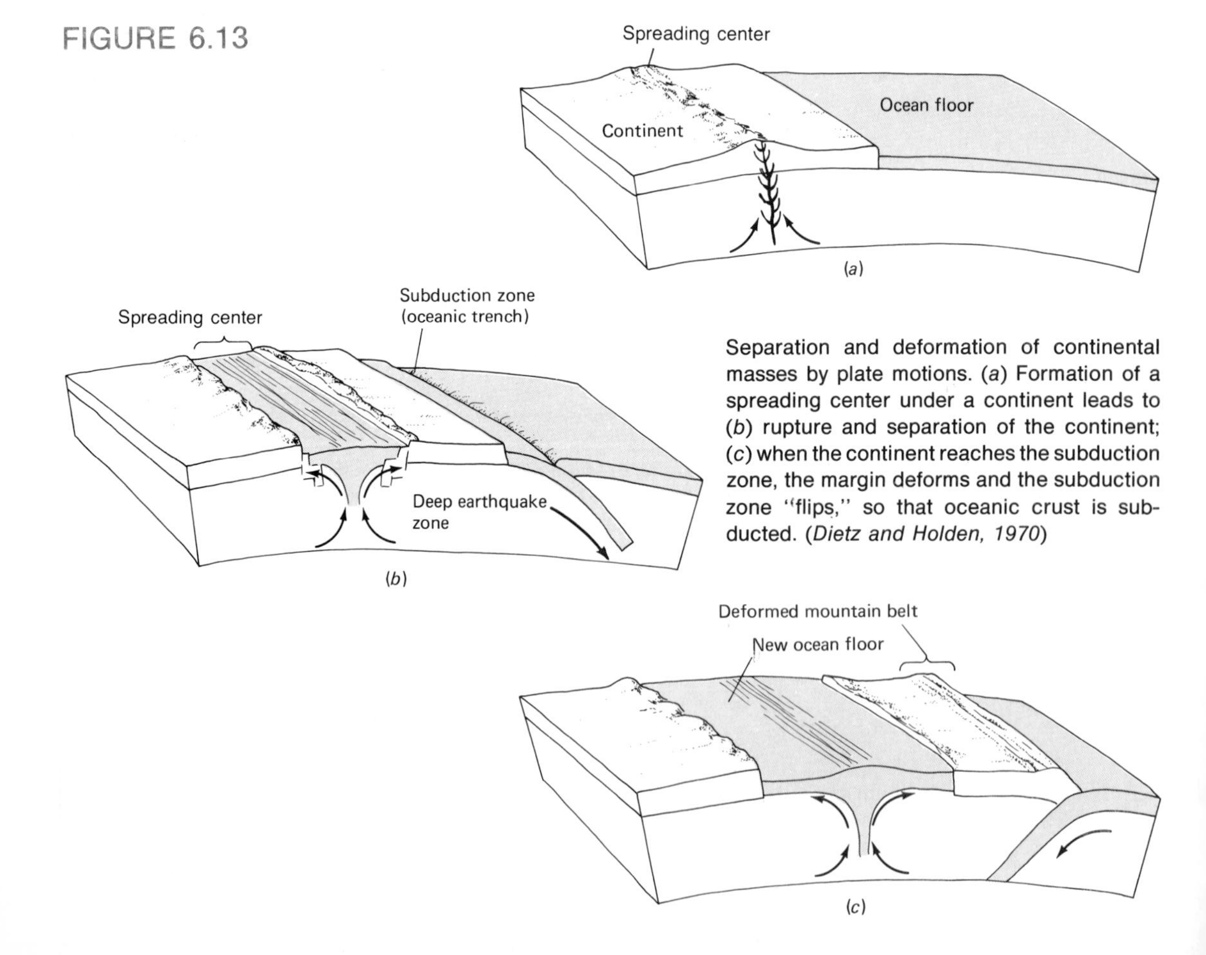

Separation and deformation of continental masses by plate motions. (*a*) Formation of a spreading center under a continent leads to (*b*) rupture and separation of the continent; (*c*) when the continent reaches the subduction zone, the margin deforms and the subduction zone "flips," so that oceanic crust is subducted. (*Dietz and Holden, 1970*)

block had been called continental drift early in this century. Now we realize, of course, that plate motions clearly involve ocean basins as well as continents. Continents are merely thick sialic masses embedded in the lithospheric plates.

Many plate-convergence boundaries coincide with the margins of continents. At these boundaries, it is always the oceanic plate, rather than the continent, that is subducted. When a continent on a subducting plate is carried to the subduction zone, the land mass acts as a buoyant buttress and prevents the plate from being subducted further. If the leading edge is made of oceanic crust, it now becomes the subducted plate and begins to pass *beneath* the continental block and descends into the mantle (Figure 6.13). On the other hand, if both leading edges are made of continental crust, the collision of continental masses tends to stabilize both plates along a **suture zone** between them. Subduction at continental margins causes intense crustal deformation; it is believed to have been responsible for coastal fold mountain chains such as the Andes and Appalachians. Collision of two continents also causes intense crustal deformation; it is believed to have been responsible for interior mountain chains such as the Urals and Himalayas.

Laser Beam Measurements

Much more sophisticated analyses of plate tectonics that involve direct measurement of plate motions are yet to come. During the manned lunar explorations in the early 1970s, small devices designed to reflect laser beams were positioned on the moon. Beams aimed at one of these instruments travel to

FIGURE 6.14

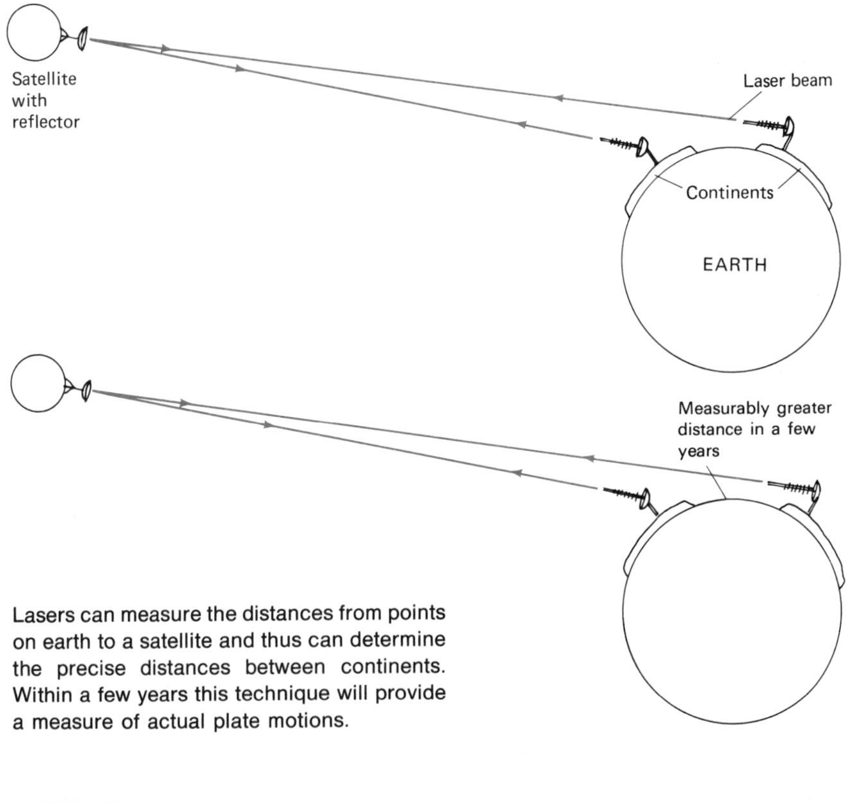

Satellite with reflector

Laser beam

Continents

EARTH

Measurably greater distance in a few years

Lasers can measure the distances from points on earth to a satellite and thus can determine the precise distances between continents. Within a few years this technique will provide a measure of actual plate motions.

the moon station and back again; their purpose is to measure distance, and this they do with incredible precision. We now know the distance from the earth to the moon—or more exactly, from the earth instrument to the lunar instrument—to within 3 centimeters! Simultaneous determinations from two separate continents give their precise distances from the moon. Using these measurements, we can calculate the present distance between the two continents to within a few centimeters. Similar reflectors have been placed on orbiting satellites; accuracy of the distance measurements from them is even greater—to within a few millimeters! These experiments are being conducted now (Figure 6.14). If the North Atlantic Ocean is opening at the rate of 2 centimeters per year, the South Atlantic Ocean at the rate of 4 centimeters per year, and if movement on the Pacific plate is around 5 centimeters per year, the changing distances between surrounding land masses should become apparent in a very few years. We are waiting to see.

The Journey of the Green Turtle

Confirmation of plate motions using lasar-beam reflections may be a long-term experiment by most standards, but the honor of what may be the longest-term experiment of all goes to the green turtle. The green turtle lives its life on the coast of Brazil, very near the equator. Each year, between December and March, it leaves home and swims 2,000 kilometers to tiny Ascension Island (area: 90 square kilometers) in the middle of the Atlantic Ocean. Here the turtle nests and then returns with its young to Brazil. The journey each way takes about two months, but it serves a valuable purpose. Ascension Island has good beaches that are suitable for nesting, and it offers relative freedom from the nest predators that plague similar beaches in South America.

We don't know how the green turtle finds tiny Ascension Island, which is only about 12 kilometers across, in the vast expanse of the Atlantic Ocean, but it may navigate chiefly on the sun. In any case, the navigation feat is astounding. An even more puzzling question, however, is how the green turtle found the tiny, 2,000-kilometer-distant island in the first place, and how the turtle that found it told the other turtles about it, so that the entire species could adapt to nesting on Ascension Island! This scenario would seem to make impossible demands on the process of natural selection. Many geologists now suspect that the turtle's nesting behavior is inherited from a much earlier time, when Brazil and Ascension Island (or more correctly, its midoceanic-ridge predecessors) were a lot closer together.

Mesozoic ancestors of the modern green turtle occur in Brazilian strata. Biologists theorize that, before the South Atlantic opened, these turtles may have migrated between their dwelling grounds and their nesting grounds along the northern coast of what was then a single large continent (Figure 6.15). By 80 million years ago, in the Late Cretaceous, a narrow ocean had opened between Brazil and west Africa in line with the turtles' migration paths. There were probably islands on the nearby Midatlantic Ridge then, as there are now, and many of the turtles found them by simply extending their migration path.

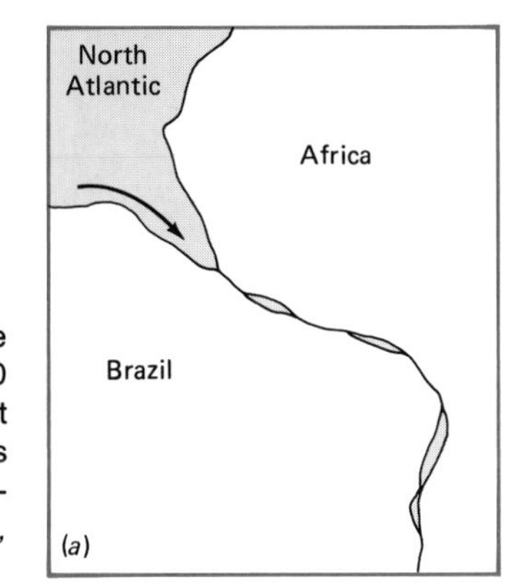

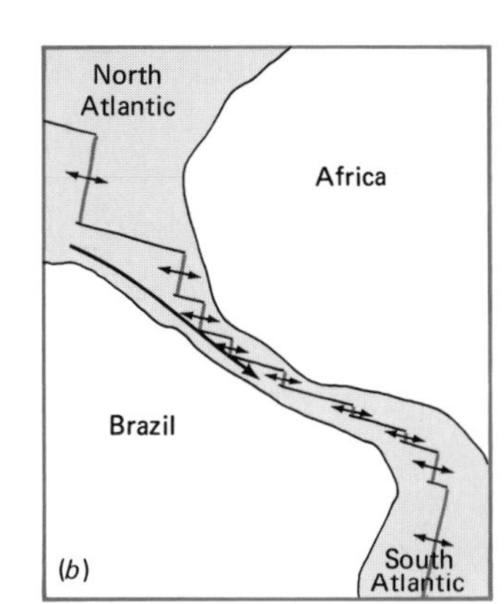

FIGURE 6.15

(*a*) Postulated longshore migration route of the green turtle ancestors prior to 90 million years ago was extended (*b*) about 80 million years ago to offshore islands formed in the rifting that opened the equatorial Atlantic Ocean. (*Carr and Coleman, 1974*)

Eventually, those turtles that nested on the islands survived, but those that nested on the South American shore were selected against, by the rise of mammalian nest predators. As the South Atlantic continued to open through the Cenozoic Era, the islands near the ridge crest moved down the ridge flank and submerged, but new islands formed to replace them (Figure 6.16). For the turtles to make the new islands their nesting grounds, it became necessary for them to extend their travel path, as the once-usable islands slowly submerged beneath the waves. By 70 million years ago, the ancestors of the present-day Ascension Island colony were making seaward breeding migrations of perhaps 300 kilometers; this was extended as new volcanic islands sporadically appeared. Ascension Island, the latest of these, is only about 7 million years old.

This hypothesis places the migration of the modern green turtle in an evolutionary framework in which selective and adaptive processes operate over a vast period of time. If the hypothesis is correct, then the present-day green turtle has inherited a behavior pattern from ancestors that lived many tens of millions of years ago, when the Atlantic was much narrower than it is today. Hence, these animals may be living confirmation of sea-floor spreading.

FIGURE 6.16

After the green turtle's travel path to offshore islands was established, it lengthened throughout Cenozoic time as old islands moving westward sank and new islands appeared on the spreading Midatlantic Ridge. (*Carr and Coleman, 1974*)

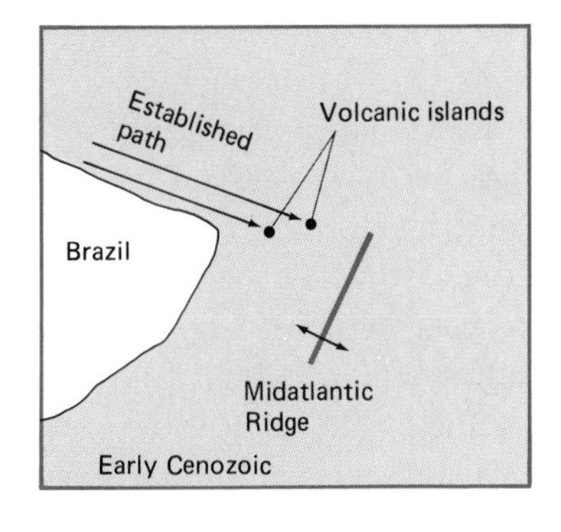

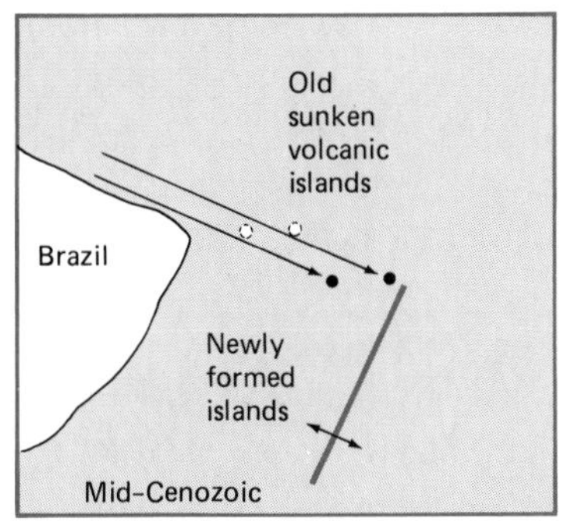

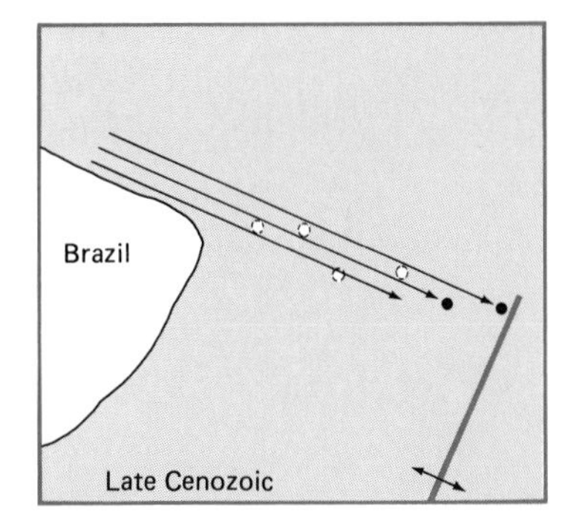

Ancient Plate Motions

The concepts of plate tectonics have revolutionized our interpretations of earth history, especially for the more recent intervals whose record of plate motions is clearly preserved. Unfortunately, critical evidence of ancient movement has been destroyed, particularly in the continually subducted rocks of the ocean floor. In this section we shall review the kinds of clues used to reconstruct ancient plate motions.

Continents and the Ocean Floor

Only oceanic crust is produced at spreading centers and destroyed in subduction zones. Thus, all oceanic crust is geologically young. Figure 6.17 shows the large proportion of existing oceanic crust that has been produced within the past 75 million years. In fact, the oldest rocks known from the ocean basins are mid-Mesozoic in age—less than 200 million years old. Continents, however, persist indefinitely. Surrounded by ocean basins that continually change in shape, continental blocks are sporadically split, transported about, fused together, and generally battered along their margins by plate motions. However, in the protected interiors of continents, as we have seen, very ancient rocks have survived.

A large portion of the ocean floor has now been dated, using inferred positions of the regular oceanic magnetic stripes in the magnetic time scale

FIGURE 6.17

Areas of sea floor created in the last 75 million years (darker color) cover a large part of the ocean basins. In contrast, little continental area has been formed during this time. (*McKenzie, 1972*)

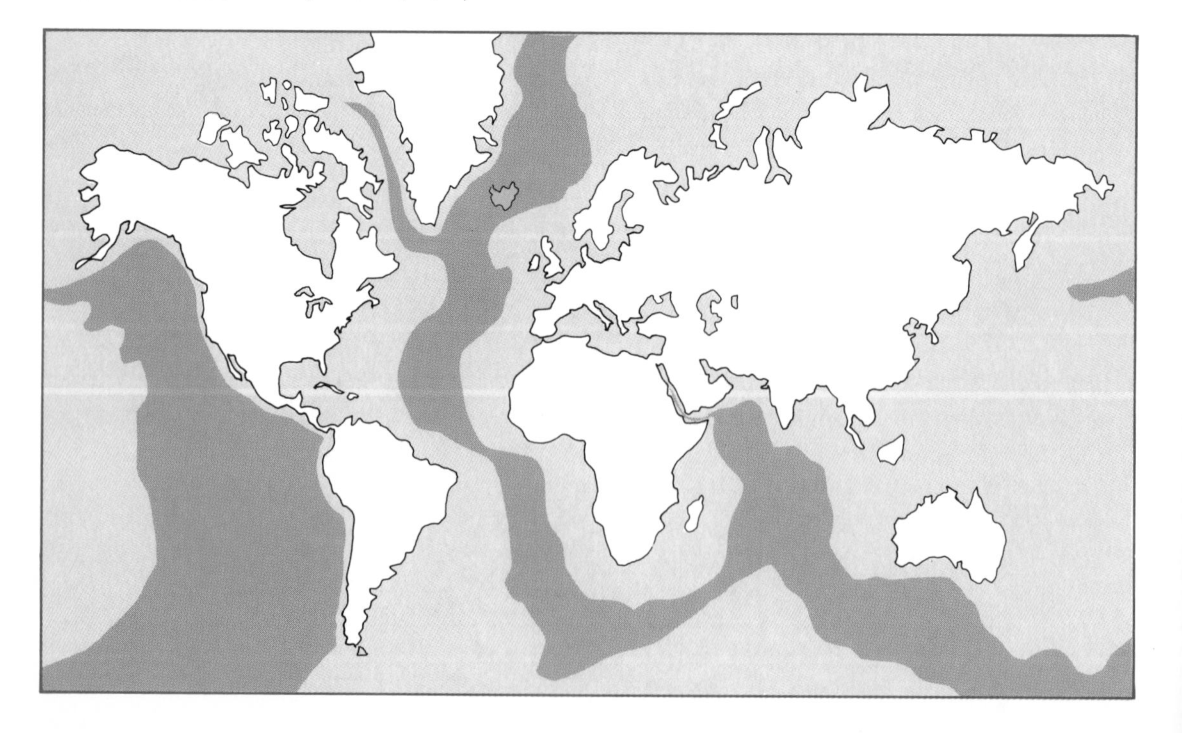

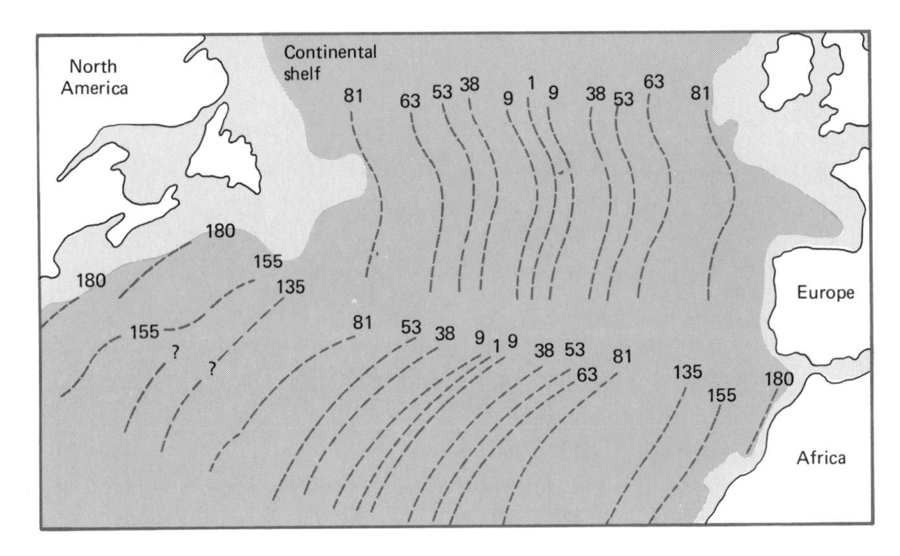

FIGURE 6.18

Ages of magnetic anomalies in the North Atlantic Ocean in millions of years. The numeral 1 marks the Midatlantic Ridge, where the ocean crust is very young. (*Pitman and Talwani 1972*)

(Figure 6.18). These ancient magnetic reversals make possible detailed reconstructions of the continental positions for the past several tens of millions of years. For example, to map the Atlantic Ocean and its bordering continents as they were 38 million years ago (in the Oligocene Epoch), we need only bring the two 38-million-year-old stripes that lie on opposite sides of the Midatlantic Ridge together by moving them parallel to the transform faults in the region. All of the oceanic crust that lies between the 38-million-year-old stripes was formed subsequently. These interpretations are verified by drilling data from the deep ocean, which show that the oldest sedimentary strata that immediately overlie the basaltic crust increase in age with increasing distance from the Ridge. In the area of the inferred 38-million-year-old stripes, the oldest strata have been dated at about 38 million years.

Matching the geology of two diverging continents at their now widely separated margins provides further confirmation that they were once connected, and that the sea floor between them has subsequently spread apart. Figure 6.19 shows Africa and South America fitted together in the familiar predrift reconstruction. In a large area of Northwest Africa, the Precambrian basement rocks have ages around 2 billion years. This Precambrian province is bordered on the east by a province in which the age of the oldest rocks is only 550 million years. The boundary between these provinces heads out to sea. In 1966 a search in Brazil confirmed that the same boundary exists there, *exactly where it was predicted.* This kind of evidence—the matching of age provinces and other geologic features across ocean basins—corroborates the geometric fits based only on shape and thus constitutes important evidence for ancient plate motions.

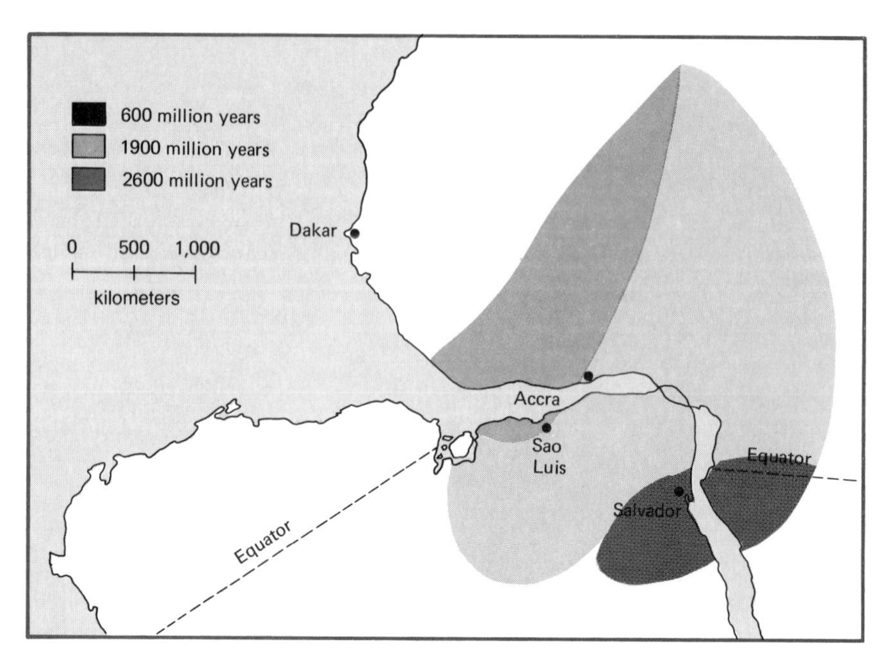

FIGURE 6.19

Areas of South America and Africa containing Precambrian rocks of different ages have been mapped. Boundaries between these areas can be traced directly from one continent to the other when the continents are fitted together as in Figure 6.2. (*Hurley and others, 1967*)

Plate Motions and Mobile Belts

The only surviving evidence for plate motions earlier than about 200 million years is preserved in continental rocks, particularly those that have been deformed by ancient plate movements. In discussing Precambrian history (Chapter 3), we noted the importance of geosynclines and their related belts of folded mountains in the growth of continents. Modern understanding of plate tectonics has greatly clarified the origin and historical significance of these fundamental features of the continents.

Geosynclines, we have seen, are extraordinary thicknesses of sediment that have usually been deformed into folded mountain belts. Geologists had long assumed that geosynclinal sediments accumulated in huge sedimentary troughs where the earth's crust sank for long periods prior to the episode of folding and uplift. From this observation, most geologists drew the conclusion that geosynclines were a necessary forerunner of mountain ranges. They reasoned that there must be a cause-and-effect relationship between geosynclines and mountains—some cycle of subsidence, uplift, and erosion that could explain crustal behavior through time.

A major hitch in the scheme was that, although geosynclinal troughs were supposed to be basic features of the earth's crust, nobody could point to a present-day example of an elongate trough accumulating large amounts of sediment. Geologists questioned whether the side-by-side miogeosynclinal and eugeosynclinal troughs existed only in the earth's past, or whether the very concept of the trough-like shape of the geosyncline was faulty. Perhaps existing geosynclines simply look totally different than expected. If so, where are the modern geosynclines?

Most geologists now believe that modern geosynclinal sequences are accumulating on the quiet margins of continents that are diverging from spreading centers. An example is the Atlantic margin of the United States, which is retreating from the Midatlantic Ridge. A seaward thickening wedge of sedimentary rocks underlies the Atlantic coastal plain and its offshore continuation, the shallow continental shelf. The wedge attains a total thickness of between 3 and 5 kilometers at the outer shelf edge (Figure 6.20). The wedge is composed of well-sorted, shallow-water sediments deposited during the past 150 million years in alluvial plains, in lagoons, along shorelines, and offshore. In thickness and rock type, these deposits resemble those found in ancient miogeosynclinal foldbelts.

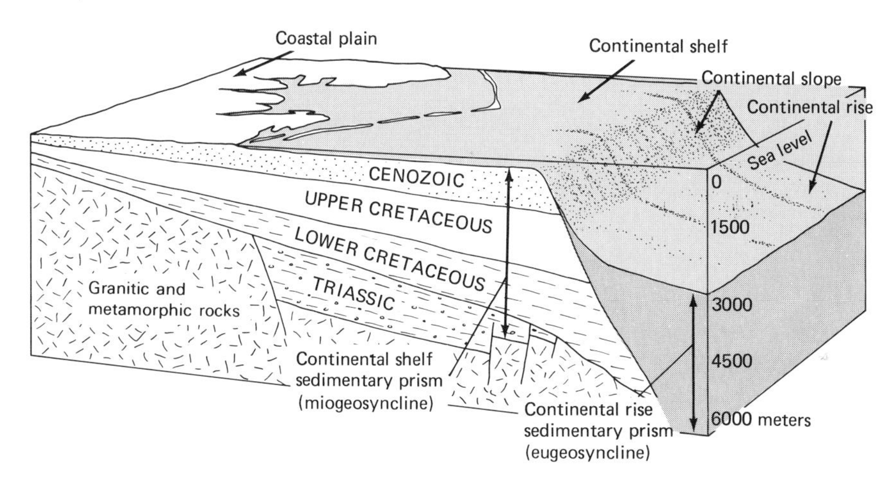

FIGURE 6.20

Accumulating geosynclinal sediments off the Atlantic Coast of the United States. *(Dietz, 1972)*

The Atlantic continental shelf terminates abruptly, marking the continent's margin. Beyond it, the continental slope plunges rapidly into the deep sea. At the base of this slope, however, another important sedimentary assemblage is accumulating, chiefly by turbidity currents that periodically cascade down the submarine canyons and deposit their muddy loads on the subsea fans that lie beyond. The fans coalesce into a broad apron 250 kilometers wide, called the **continental rise** (Figure 6.20). The continental rise parallels the base of the Atlantic continental slope for 2,000 kilometers. Seismic studies reveal that it consists of a lens of turbidites whose maximum thickness is about 10 kilometers. Thus, in thickness and in rock type, the continental-rise sediments resemble those of eugeosynclines, and they are inferred to be modern counterparts. Figure 6.21 shows in block diagram a portion of the Cordilleran Geosyncline as it may have looked in earliest Cambrian time.

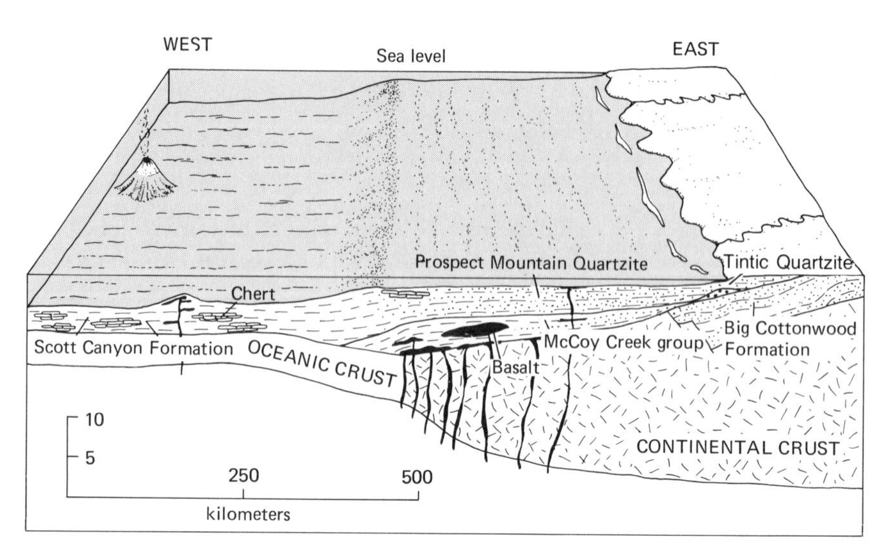

FIGURE 6.21

Block diagram of Upper Precambrian and Lower Cambrian geosynclinal rocks in Nevada and Utah as they appeared in Early Cambrian time. (*Stewart, 1972*)

An interpretation of miogeosynclines as continental shelf accumulations and eugeosynclines as continental rises removes the need to postulate some kind of geosynclinal trough that has no modern counterpart. Ancient geosynclinal deposits appear to have been normal seaward-sloping prisms of sediment, similar to those we can observe today—one accumulating in relatively shallow water on continental crust, and the other accumulating in relatively deep water on oceanic crust. Accordingly, geosynclinal accumulations probably have no formative role in mountain building. Continental margins are deformed by the processes of subduction and continental collision. Geosynclines frequently coincide with fold mountain belts simply because they, too, tend to become localized at continental margins. The old term *geosyncline* (implying a trough-like form) now seems somewhat archaic, and the shortened terms *miogeocline* and *eugeocline* have been proposed as replacements, but have not been widely accepted.

Ancient Convergent Plate Boundaries

How do we recognize subduction zones in the ancient record of folded mountains? Suppose today's plate motions were to reverse, and the Atlantic were to become a closing ocean. If the subduction zone formed along the Atlantic margin of the United States (Figure 6.22), an oceanic trench would be created. The drag of the descending lithospheric plate would slowly crumple the continental-rise sedimentary prism and thrust it landward, forming a eugeosynclinal foldbelt. With continued subduction and thrust faulting, the continental-shelf sedimentary prism would also begin to deform into a parallel miogeosynclinal foldbelt. At depth, melting of the lighter portion of the descending lithospheric plate would produce magmas that would ascend through the overriding plate and give rise to a **volcanic arc.** With continued thickening of the crustal rocks along the Atlantic margin, and with still more melting,

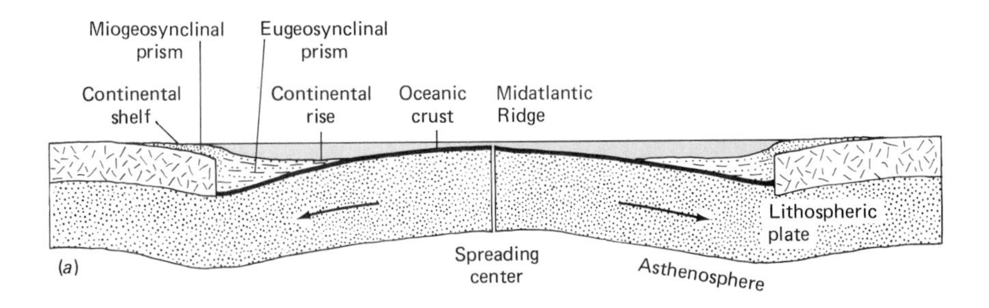

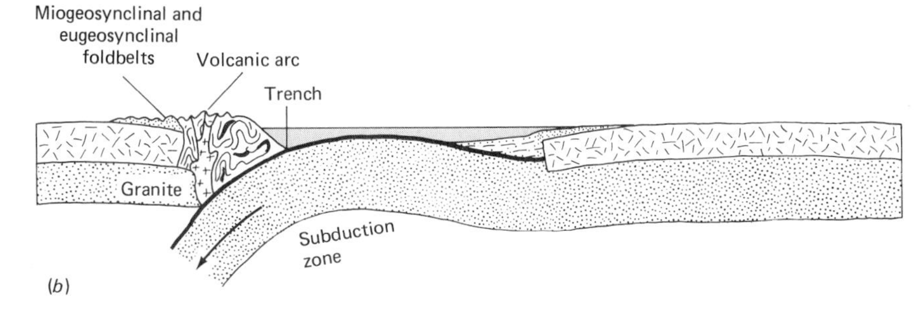

FIGURE 6.22

(*a*) Present-day widening Atlantic Ocean. (*b*) If the Atlantic began to close and the western margin became a subduction zone, the decending lithospheric plate would crumple the continental rise and continental shelf into a eugeosynclinal and miogeosynclinal foldbelt. (*Dietz, 1972*)

FIGURE 6.23

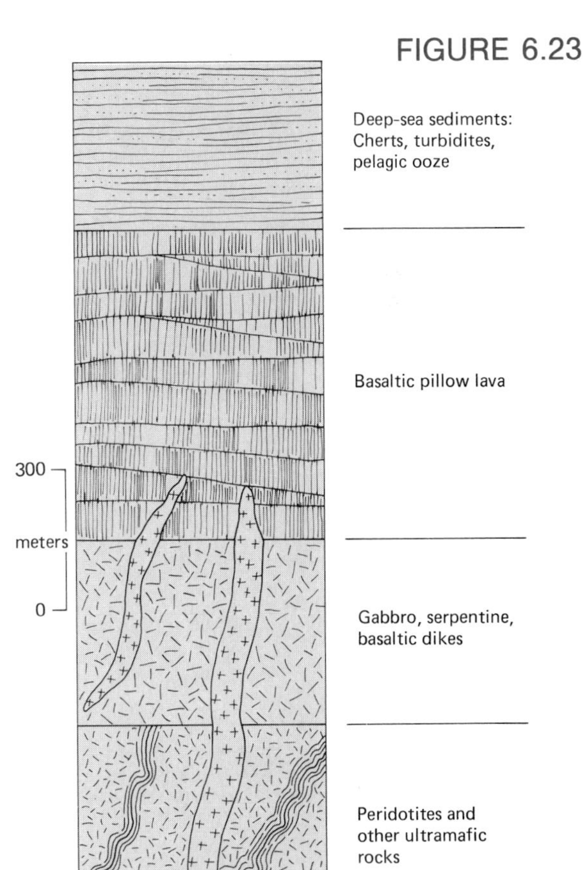

The ophiolite suite is a slice of oceanic lithosphere incorporated in a geosynclinal foldbelt during plate collisions. (*Press and Siever, 1978*)

granitic batholiths would intrude the roots of the foldbelt at depth. Ensuing deformation would be accompanied by vertical uplift, and a young mountain belt would be created at the continental margin.

In this sequence of events, two telltale rock suites would be produced that provide excellent evidence that the foldbelt originated in a subduction zone. These are **ophiolites** and **blueschists.** Together they form much of the **tectonic melange,** the term applied to the intensely contorted and sliced-up outer portion of the eugeosynclinal belt. Ophiolites are suites of distinctive rock types that include, in stratigraphic order, serpentinized peridotite, gabbro and basaltic pillow lavas, and at the top, radiolarian cherts and turbidites (Figure 6.23). These unusual assemblages occur in narrow zones in eugeosynclinal foldbelts; they are believed to be fault slices of oceanic lithosphere that were scraped off the top portion of the descending plate and incorporated into the

FIGURE 6.24 Subduction of the ocean crust that lay between Arabia and Iran during the Cretaceous, and production of an ophiolite belt in the Zagros Mountains. (*Welland and Mitchell, 1977*)

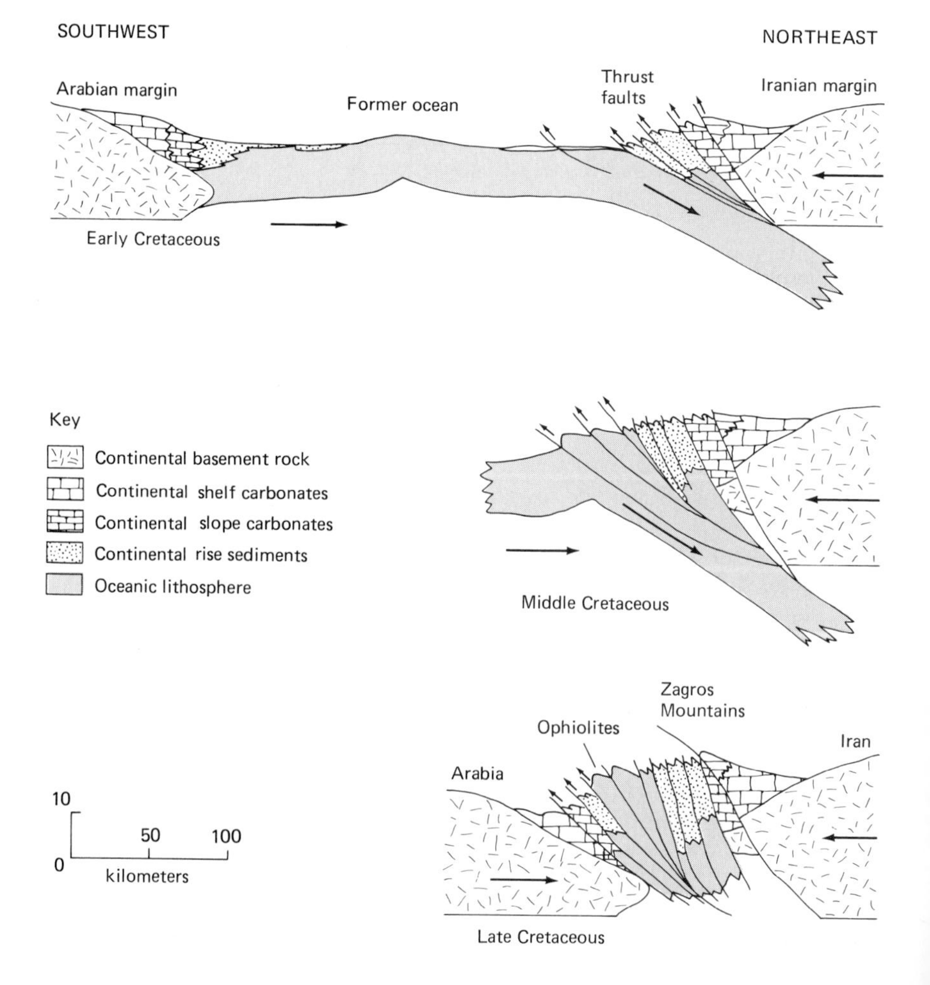

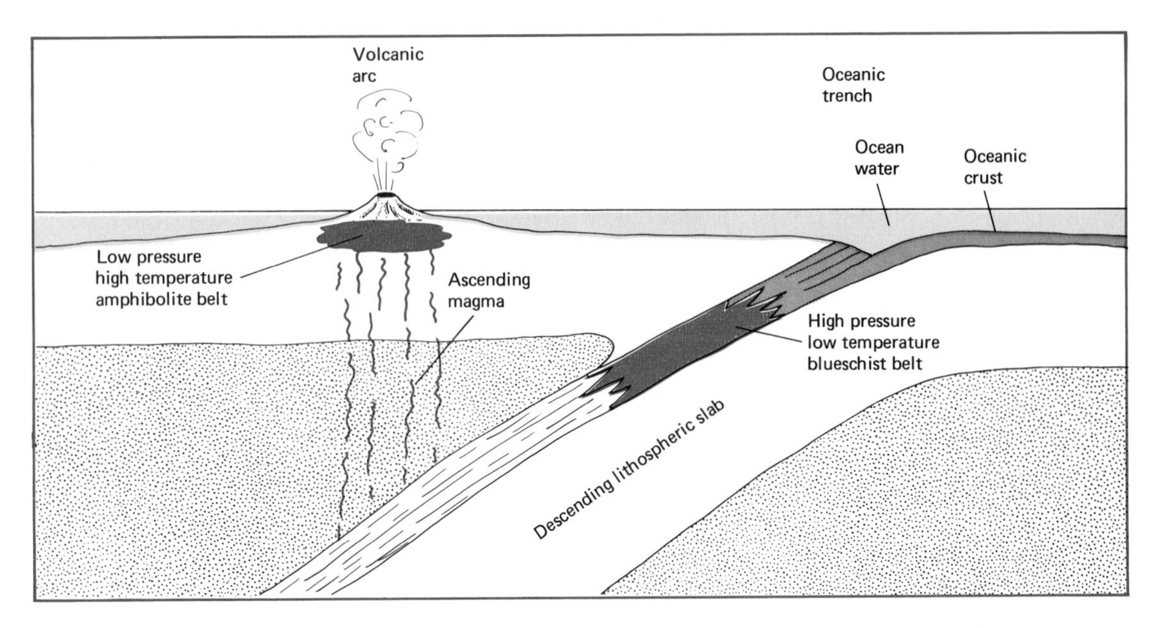

FIGURE 6.25

The two different kinds of metamorphism produced in a subduction zone result in paired metamorphic belts in the rock record of the region that mark the site of the subduction long after it has ceased.

crumpling eugeosynclinal belt (Figure 6.24). Hence, ophiolites are relicts of a former oceanic plate that was otherwise totally consumed by subduction. These unusual and distinctive rock suites typically mark convergent plate boundaries.

Blueschist is a distinctive type of metamorphic rock that forms at very high pressures, yet low temperatures. This metamorphic environment is unique to subduction zones, where relatively cool lithosphere plates descend rapidly into high-pressure environments at depth. The unusual combination of cool temperatures and high pressures produces the distinctive blue metamorphic minerals that give blueschist its name. While the blueschist forms in the melange belt, the zone of ascending hot magmas beneath the adjacent volcanic arc produces a parallel belt of high temperature-high pressure metamorphic rocks called amphibolite. The resulting pair of contrasting metamorphic belts (blueschists and amphibolites) further characterizes the record of ancient subduction zones (Figure 6.25).

Ultimately, with continued shrinking of the ocean, the continents on its opposite sides would meet. The continental blocks might collide head-on (Figure 6.24), or the margin of one might override the other, resulting in even greater crustal thickening. In either case, some oceanic crust would likely be caught between the two plates as they merged; slices of ophiolitic melange would mark the suture zone in the growing foldbelt. In effect, the ophiolites would be all that remained of an ocean that had been consumed between two merging continents. Blueschist metamorphism of the melange rocks would corroborate the mechanism of subduction.

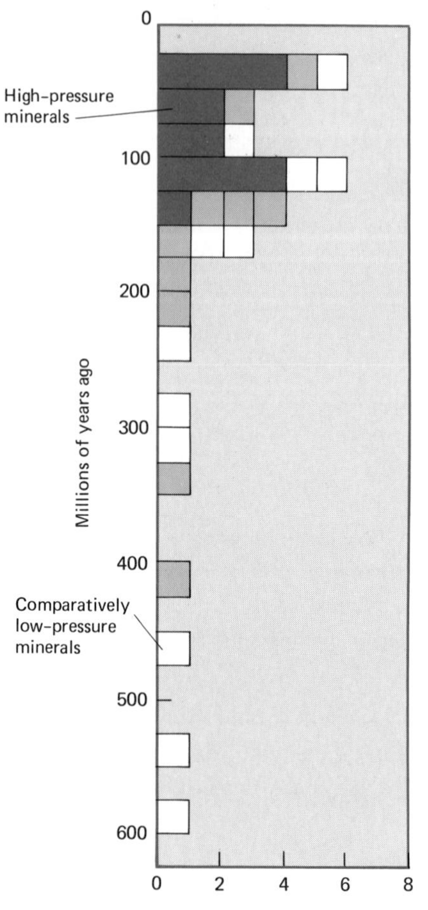

FIGURE 6.26 Histogram showing records of blueschists in the Phanerozoic. The lack of high-pressure blueschist minerals prior to 150 million years ago indicates that temperatures at shallow depths were higher then. (*Ernst, 1972*)

Today, elongate belts of ophiolites and blueschists are prominent features of such young mountain ranges as the Alps, the Himalayas, and the mountains around the margins of the Pacific Ocean. They testify that subduction at convergent plate boundaries was operative as far back as the Jurassic. In Paleozoic mobile belts, ophiolites and blueschists are also common. However, Paleozoic blueschist mineral assemblages show signs of having formed at lower pressures and, hence, at shallower depths than in post-Jurassic time (Figure 6.26). In Proterozoic mobile belts, both blueschists and ophiolites are virtually unknown. This has raised the question of how crustal processes in the Precambrian differed from those of today.

The lower-pressure blueschist environments in the Paleozoic can be explained if the temperature levels beneath subduction zones (Figure 6.25) were higher then than they are today. In other words, at depths sufficient for crystallization of higher-pressure blueschist minerals, the temperatures were too high. The steeper thermal gradient (higher temperatures at shallower depths) in the Paleozoic would be expected, because greater radioactive heat was produced the farther back we go in the geologic past. In the Precambrian, the thermal gradient through the earth's crust and upper mantle must have been so steep that even low-pressure blueschist minerals could not form.

A steeper thermal gradient would also have produced weak, plastic rocks at much shallower depths than today; hence, the asthenosphere would have extended closer to the surface. As a result, the overlying lithosphere plates would have been much thinner (Figure 6.27). Substantially thinner plates might somehow account for the lack of ophiolite assemblages in Precambrian mobile belts. Perhaps the thinner lithosphere plates were less mobile than today's plates, or perhaps their edges were less well defined, so that oceanic crust did not take part in the deformation along subduction zones.

Another possibility is that the mechanism of plate tectonics, as we understand it today, simply didn't operate throughout a large part of the Precambrian. This point of view had serious support for a time, based in part on the grouping of old Precambrian regions when the continents are reassembled in the arrangement they had in the Triassic Period, before they drifted apart. As Figure 6.28 shows, those portions of continents with ages greater than 1,700 million years merge and form what appears to be two discrete entities. The problem is this: If the continents were scattered and moving about independently as portions of separate lithospheric plates during late Precambrian

Higher temperatures at shallower depths in Precambrian time probably produced a shallower asthenosphere and a thinner lithosphere.

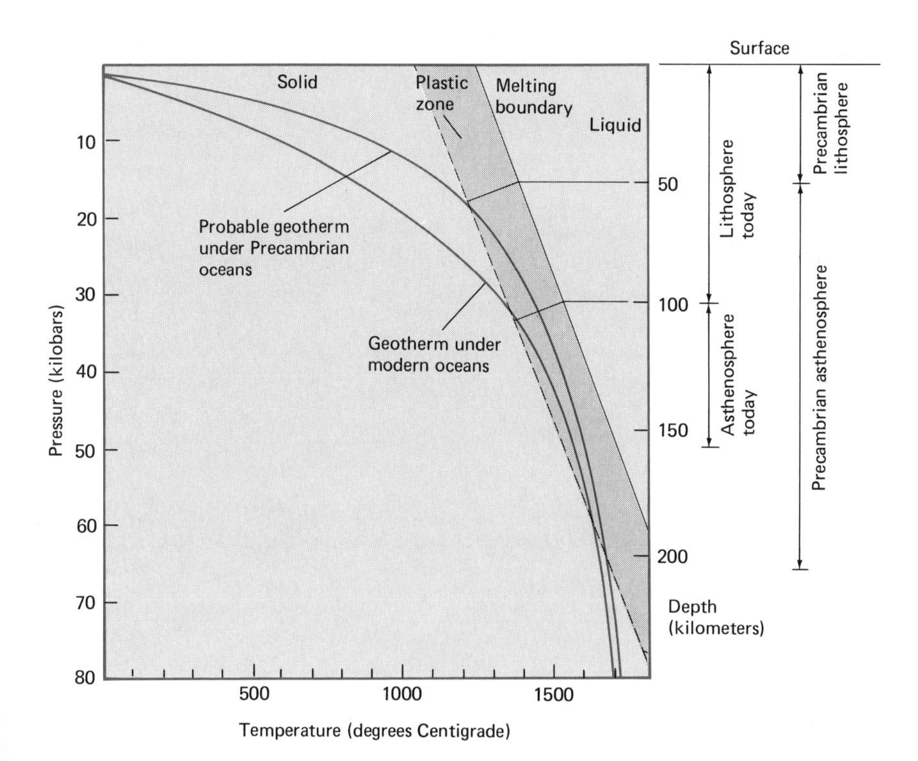

FIGURE 6.27

FIGURE 6.28

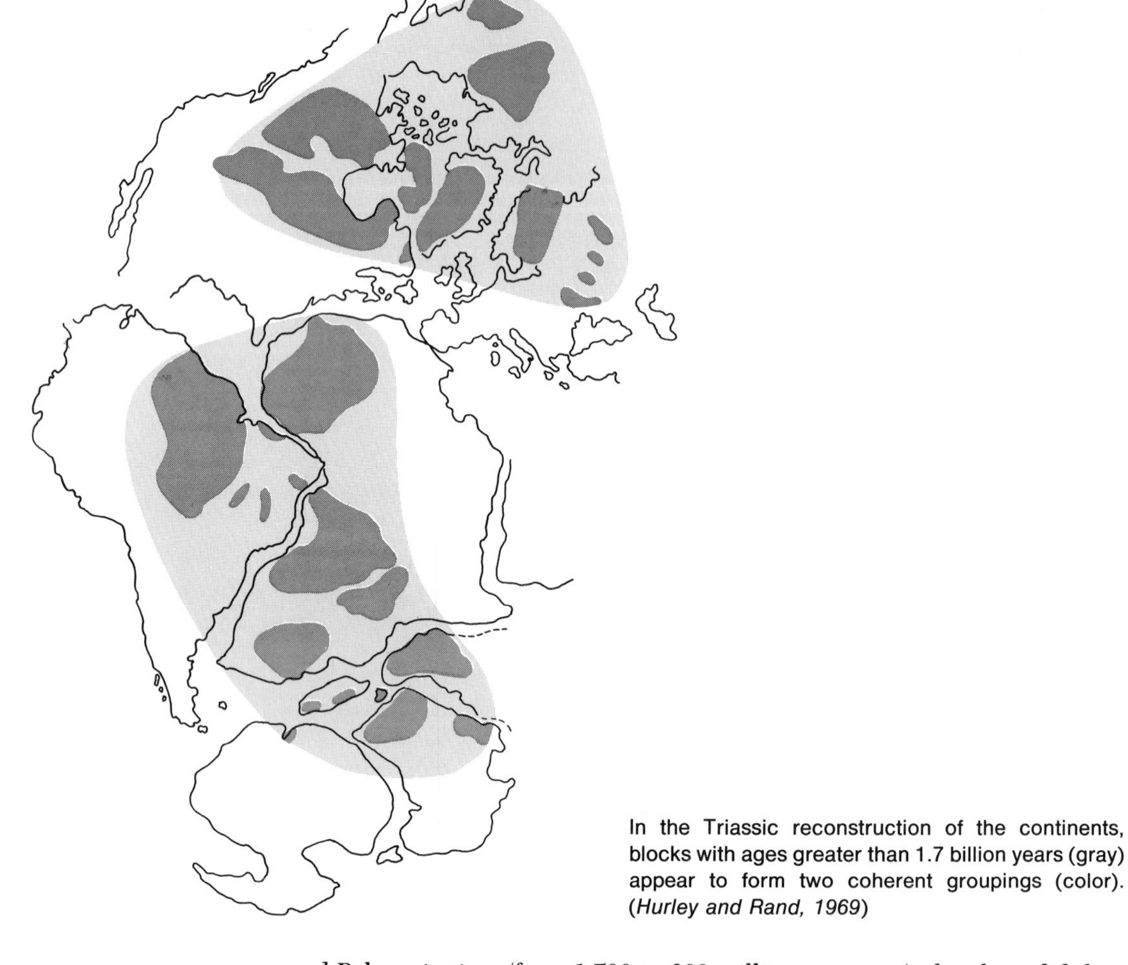

In the Triassic reconstruction of the continents, blocks with ages greater than 1.7 billion years (gray) appear to form two coherent groupings (color). (*Hurley and Rand, 1969*)

and Paleozoic time (from 1,700 to 200 million years ago), then how did they happen to reassemble themselves with this degree of organization in the Triassic? Because this appears improbable, some geologists suggested that, prior to the Triassic, the continents may have been portions of large stationary nuclei around which continental accretion took place up to the time of the great fragmentation in the Mesozoic Era. However, new evidence now strongly supports extensive continental movement during the Paleozoic, which will be discussed in Chapters 7 and 8. As a result, most geologists do not think it likely that the Paleozoic continents were stationary; nevertheless, the groupings in Figure 6.28 have not been explained satisfactorily.

Ancient Divergent Plate Boundaries

The record of ancient subduction zones commands a great deal of attention because subduction zones are believed to be responsible for foldbelts. But what record would ancient spreading centers leave? These might be recognizable today if a continent broke apart across preexisting sedimentary basins, and continental shelf deposits began to accumulate on the newly created continental margin. Just such a framework characterizes the latest Precambrian in western North America. Here strata of latest Precambrian and Early Cambrian age form a generally north-south geosyncline, a portion of which is shown in the block diagram in Figure 6.21. These strata thicken rapidly westward, from a feather edge to 7,500 meters in a distance of only 200 to 300 kilometers (Figure 6.29). This geosyncline is interpreted to represent a trailing continental margin—one that was subsiding as it retreated from a spreading center that lay somewhere to the west. In this same region, older Precambrian rocks of the Belt Supergroup in Montana, the Uinta Mountain Group in Utah, and the Unkar and Chuar Groups in northern Arizona (Figure 6.30), which have ages of 850 to 1,250 million years, had similarly accumulated thicknesses of several thousand meters. However, these strata were not deposited in a north-south geosyncline at all. Instead, they were deposited in discrete depositional basins that appear to be truncated by the younger north-south miogeosyncline. One of these basins (the Belt) actually had a highland source area to the west. The marked change in tectonic framework in the late Proterozoic, following deposition of the Belt Supergroup and equivalent strata, is inferred to represent an ancient divergent plate boundary—the breakup of a larger continent that once extended an unknown distance to the west.

FIGURE 6.29

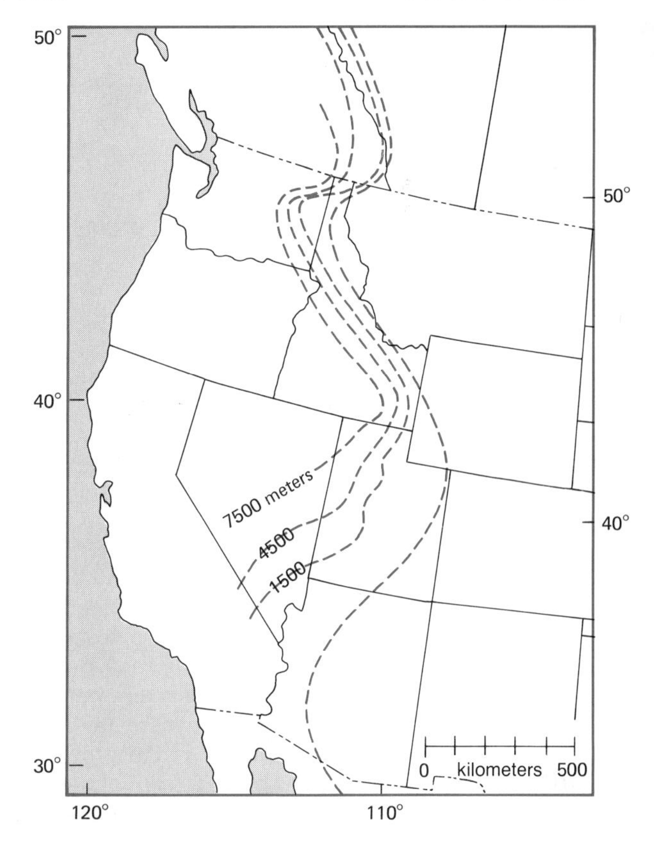

Isopach map of uppermost Precambrian and Lower Cambrian strata in western North America. The geosynclinal thicknesses represent the position of the continental margin at the end of the Precambrian (*Stewart, 1972*)

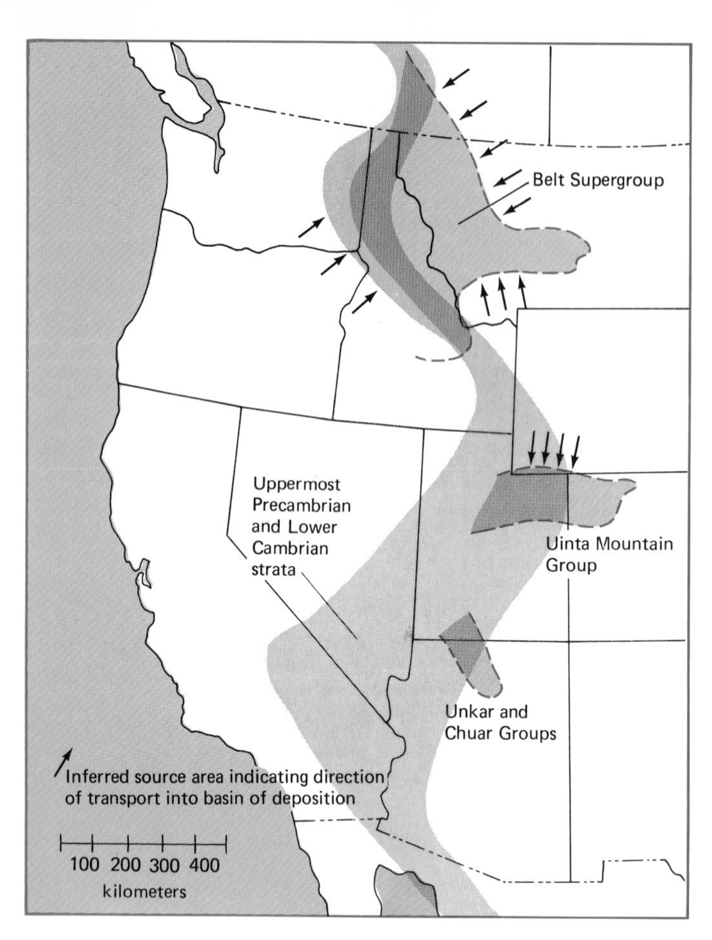

FIGURE 6.30

Contrast in distribution of sedimentary basins of Belt age (850 to 1,250 million years) with distribution of uppermost Precambrian and Lower Cambrian strata (500 to 700 million years). (*Stewart, 1972*)

Chapter Summary

Discovery of Plate Motions

Continental drift has long been suspected, principally because of the complementary shape of many now-separated continental margins.

Ocean-floor topography: The oceans are traversed by a worldwide mountain chain, the oceanic ridge-rise system; rifts along the ridge crests suggest a pulling-apart of the ocean floor.

Ocean-floor magnetism confirmed the movements of the ocean floor by showing linear bands parallel to ridge crests; these bands reflect magnetic reversals preserved in basalts that pour out along the ridge crests and are subsequently moved away by plate motions.

Confirmation of Plate Motions

Earthquakes and plate boundaries: The earth's continuous but narrow zones of earthquakes mark the boundaries of huge crustal plates where new volcanic rocks are created along spreading centers and destroyed as they plunge downward in subduction zones.

Laser beam measurements reflected from the moon will soon provide the first precise measurements of plate motions.

The journey of the green turtle: Breeding on islands of the mid-Atlantic is probably a relict of behavior learned by green turtles millions of years ago when the ocean was much narrower.

Ancient Plate Motions

Continents and the ocean floor: Unlike the rocks of the ocean floor, which are continuously created and destroyed, lighter continental rocks seem to float perpetually at the earth's surface; hence they are generally much older than oceanic rock.

Plate motions and mobile belts: Plate motions earlier than the oldest surviving ocean-floor rocks (about 200 million years) can be inferred from ancient mountain ranges that resulted from earlier continental collisions; such collisions typically deformed the thick wedges of sediments that had previously accumulated along the margins of the continents.

Ancient convergent plate boundaries are typically marked by deformed slices of ancient oceanic rocks (ophiolites and blueschists) preserved in the center of mobile belts; the absence of such rocks from Precambrian deformed belts is an unsolved puzzle.

Ancient divergent plate boundaries can sometimes be recognized where continental rocks broke apart to create new continental margins.

Important Terms

asthenosphere
blueschist
continental drift
continental rise
Curie point
echo sounder
guyot
lithosphere
magnetometer
oceanic ridge-rise system
ophiolite
plate tectonics
sea-floor spreading
spreading center
subduction zone
suture zone
tectonic melange
transcurrent fault
volcanic arc

Review Questions

1. What discoveries concerning the ocean floor first suggested plate motions?
2. How did ocean-floor magnetism provide conclusive evidence of plate motions?
3. What is the relationship between earthquakes and plate motions?
4. What contribution are laser beams expected to make to understanding plate motions?
5. Which is generally older, the continents or the ocean floor? Why?
6. How can former plate boundaries be recognized?

Additional Readings

Condie, K. C. *Plate Tectonics and Crustal Evolution,* Pergamon Press, London, 1976. *An intermediate-level survey.*

Hallam, A. *A Revolution in Earth Science,* Oxford University Press, Oxford, England, 1973. *A popular survey of the growth of the concepts of plate tectonics.*

Sullivan, W. *Continents in Motion: The New Earth Debate,* McGraw-Hill, New York, 1974. *A readable review of the discovery of the plate motions.*

Wilson, J. T. (ed.) *Continents Adrift and Continents Aground,* W. H. Freeman & Company Publishers, San Francisco, 1976. *A collection of important articles originally published in* Scientific American.

Wyllie, P. *How the Earth Works,* John Wiley, New York, 1976. *A well-written textbook stressing geophysics and plate tectonics.*

CAMBRIAN TO DEVONIAN TIME

The Marine Realm Dominant

7

With this chapter, we begin our survey of Phanerozoic history by considering the first one-third of Phanerozoic time—a span of about 200 million years that includes the Cambrian, Ordovician, Silurian, and Devonian Periods. During much of this interval the earth's continents were covered by shallow seas. They deposited a veneer of fossil-bearing sediments, parts of which survive today as sedimentary rocks exposed on every continent. Because of this initial dominance of *marine* rocks and fossils in the Phanerozoic Eon, we shall also use this chapter to examine more closely the life and environments of the oceans.

Early Paleozoic Lands and Seas

The Record of Epicontinental Seas

Although shallow seas had sporadically invaded the continents since the Archean Eon, Cambrian rocks provide the first clear record of continent-wide advance and retreat of the sea. Similar widespread transgressions and regressions of shallow epicontinental seas had occurred in Precambrian time, but they are not widely discernible for two reasons: (1) The lack of abundant fossils makes relative-age determinations difficult; and (2) most of the stratigraphic record of such events was removed during a late Precambrian episode of erosion that was so intense that the continents of that time must have stood higher above sea level than at any time since, with the possible exception of the past few million years. The erosion surface produced at that time is overlain by Cambrian sediments, which record a long, slow transgression of the Cambrian seas onto virtually all the continents.

In most regions the Lower Cambrian contact with the underlying Precambrian rocks is easily recognizable because it is a pronounced unconformity. Along some Cambrian continental margins, as in South Australia and in the Nevada-Utah region, however, deposition of near-shore sediments actually began in latest Precambrian time and continued throughout the Cambrian. In such regions, the Precambrian-Cambrian contact commonly lies within thick sequences of sandstones or other kinds of sedimentary strata, and the only way to locate the boundary is by means of the first appearance of Cambrian fossils.

The great marine transgression in the Cambrian Period began a pattern of continent-wide invasions and retreats of the sea that continued throughout the Phanerozoic. At times, the lands stood high and the marine waters spilled only onto the edges of the continental shelves. At other times, large portions of the continents subsided below sea level and were almost totally covered by shallow seas. It is important to stress that during such times the continents did not subside to oceanic depths, but instead became only slightly submerged platforms, standing high above the surrounding ocean basins.

The cause of such large-scale marine transgressions and regressions is unknown. Conceivably, either continents rise at times and subside at others, or else the capacity of the ocean basins of the world periodically decreases and they overflow. The latter suggestion is favored by the coincidence on several continents of transgressive-regressive episodes. If the oceans' capacity in fact varies, the cause could be variations in the rate of sea-floor spreading or some other cause related to plate motions.

We know that the continents were repeatedly submerged under shallow seas during early Paleozoic time because every present-day continent retains remnants of a formerly more widespread cover of sedimentary rocks deposited beneath these seas. Some of these rocks are exposed today at the earth's surface, where they can be studied directly. The map in Figure 7.1*a* shows in black the present distribution of Cambrian rock exposures in North America; examples of these exposures are shown in Figures 7.1*b* and *c*. Figure 7.1*a* also shows

Cambrian rocks of North America. (*a*) Map showing surface exposures of Cambrian rocks and areas where they are present beneath a covering of younger rocks. (*b*) (top of p. 204) Middle Cambrian Tapeats Sandstone unconformable on Precambrian metamorphic rocks at Bright Angel Trail in Grand Canyon, Arizona. (*c*) (bottom of p. 204) Middle Cambrian Flathead Sandstone on Precambrian granite (lower left) at Rattlesnake Mountain, northern Wyoming.

FIGURE 7.1

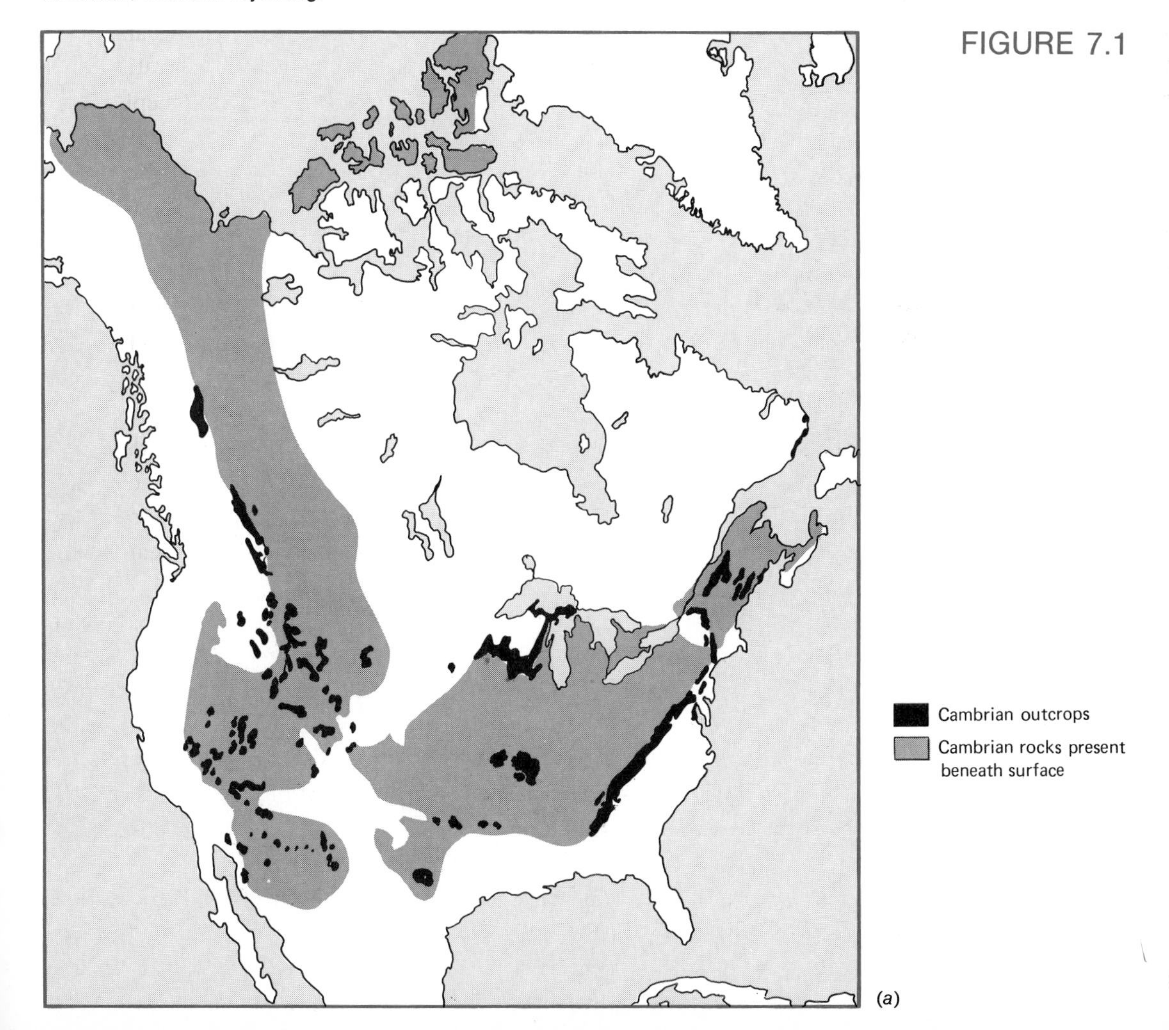

FIGURE 7.1 (*cont.*)

(*b*)

(*c*)

(dark color) the far greater area of Cambrian rocks that lie buried beneath a relatively shallow covering of younger rocks. These subsurface Cambrian deposits can sometimes be studied in mines and more frequently in boreholes that penetrate the overlying cover.

Figure 7.2 provides a worldwide look at exposed and subsurface rocks of Devonian age. Although the largest areas of both surface and subsurface rocks occur in North America and northeastern Europe, widespread Devonian rocks are found on every modern continent. Similar worldwide patterns are found for Cambrian, Ordovician, and Silurian rocks. From studies of these rocks, it is possible to reconstruct many details of early Paleozoic geography.

FIGURE 7.2

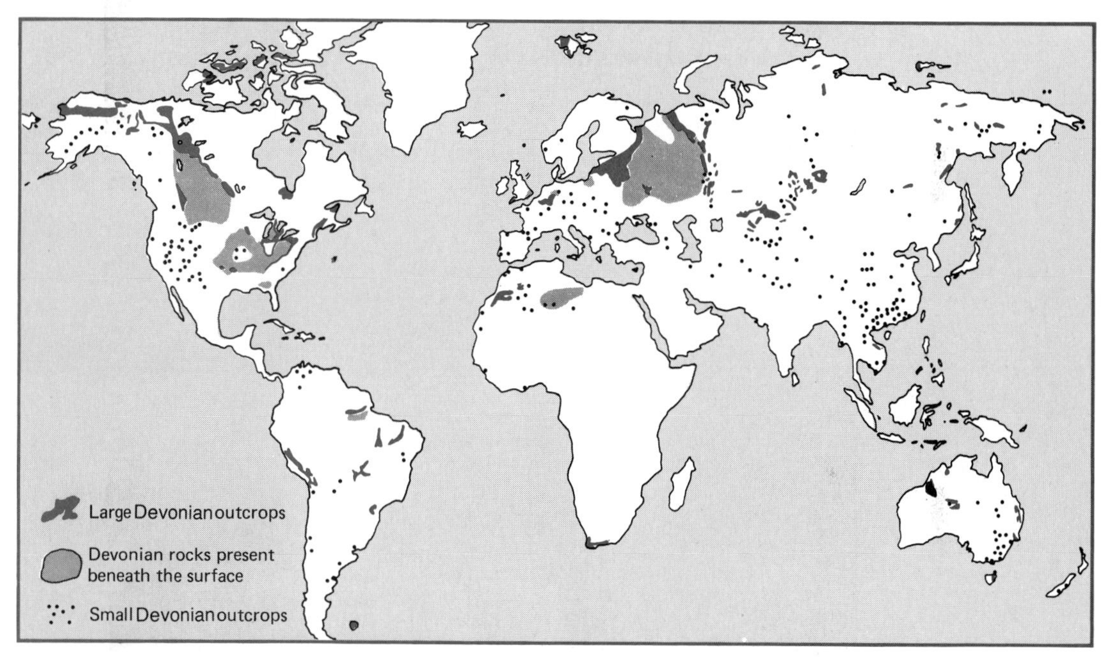

Devonian rocks of the world. (*House, 1971*)

Sedimentary Patterns

Studies show that the distribution of rock types of a single age on any modern continent is not random but shows definite patterns. Figure 7.3, for example, illustrates the patterns shown by the Middle Cambrian rocks in North America. Note that the central portion of the present continent lacks Middle Cambrian deposits and is bounded by a thin belt of near-shore sandstone and shale. This indicates that the central region was a land area undergoing active erosion, rather than sedimentation, during Middle Cambrian time. Erosion of this land area supplied the detrital sands and muds to the bordering shallow seas. Seaward from the sandstone and shale deposits lie belts of marine carbonate rocks that were deposited on the broad continental platform far offshore. These rocks are bounded still farther offshore by belts of shale and limestone that were laid down in relatively deep water at the margin of the platform.

Similar compilations of paleogeographic maps can be made for other continents and for other time intervals. Figure 7.4 shows the present world-wide distribution of Silurian rock types. Note again the dominance of either shallow-water carbonate rocks or shales and sandstones over large areas. The shales and sandstones represent terrigenous clastic material (sedimentary particles derived directly by erosion of the continent), whereas the carbonates were produced in clear water, free from the influx of terrigenous material. These contrasting rock types make up *most* of the marine sedimentary record, not only of early Paleozoic time, but throughout earth history. Later in this

FIGURE 7.3

Sedimentary patterns of North American Middle Cambrian rocks. (*Palmer, 1974*)

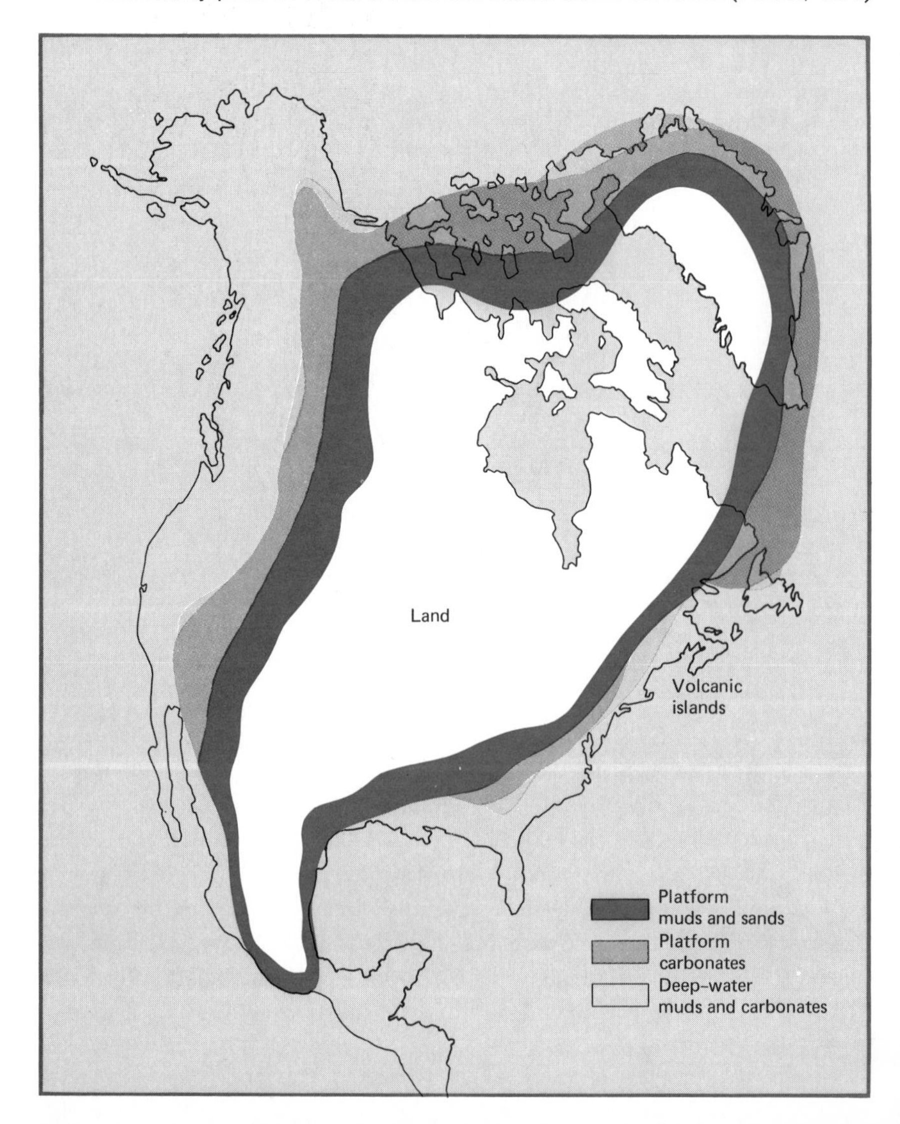

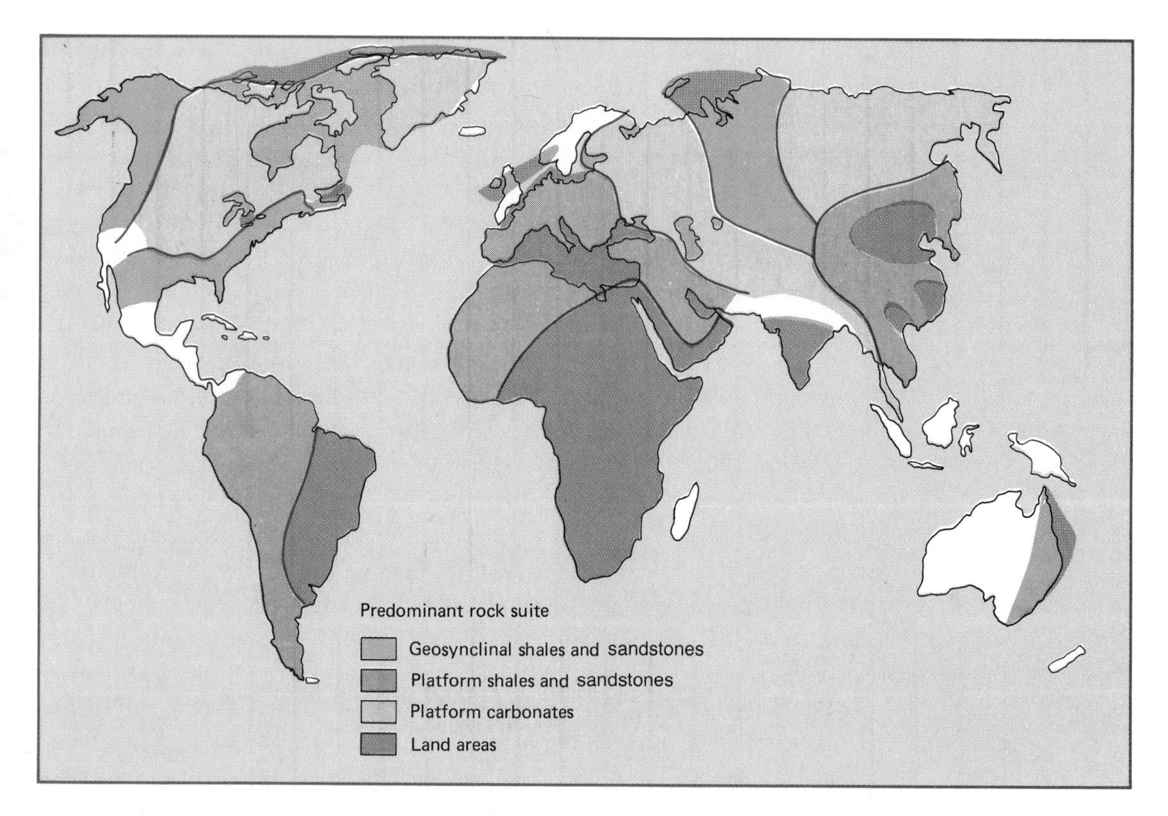

FIGURE 7.4

Worldwide distribution of Silurian rock types. The patterns are inferred from relatively small and scattered areas of Silurian rock exposures. (*Berry, 1973*)

chapter, we shall look more closely at what they tell us of ancient environments. However, we need first consider the role of plate tectonics in producing large-scale changes in paleogeography.

Reconstructions such as Figure 7.4 show the patterns of Paleozoic seas and sediments as they exist today on the *modern* continents. We have seen, in Chapter 6, that the processes of plate tectonics resulted in continual changes in the relative positions of the continents throughout much of the geologic past. In order to understand Silurian geography, for example, we must know not only which part of the present continents were consistently covered by the Silurian seas but also what were the global locations of these continents during Silurian time.

Early Paleozoic Continents and Mountain Belts

Paleozoic Plate Tectonics

When the Paleozoic Era began, several separate continental masses lay scattered about on the earth's surface. Throughout the Paleozoic, these continents slowly drifted together and joined, one by one, so that at the end of the Paleozoic—some 300 million years later—they formed a single huge continent, which we call **Pangaea.** After the Triassic Period, Pangaea gradually broke apart. Its fragments were rearranged during the latter part of the Mesozoic and the Cenozoic Eras to achieve the configuration of continents that exists today. In plate-tectonics terms, Pangaea was *assembled* during the Paleozoic by plate convergence and subduction of the oceanic crust that separated the early Paleozoic continents. In post-Triassic time, it was *disassembled* as a newly defined set of lithospheric plates spread apart along the world's ridge-rise systems, where new oceanic crust formed from upwelling mantle material. Here we shall discuss the evidence for early Paleozoic continental motions and the beginning of the assembly of Pangaea.

We have a far better idea of how the continents were distributed on the earth's surface for times after the Triassic Period than for times before. For times after the Triassic, we can reconstruct continental arrangements not only from positions of ancient magnetic poles, but also by matching up and fitting together present-day continental margins that were formerly contiguous. In addition, we can utilize the rates of separation based upon the age of the oceanic crust that formed in the gradually widening gap between continents (see Figure 6.18). For post-Triassic time, these lines of evidence corroborate one another, enabling us to reconstruct continental positions with a fair degree of confidence.

For times prior to the Triassic, however, we cannot match continental outlines because no divergent boundaries are that old. The outlines of modern continents are no help, because they all originated with the post-Triassic breakup. Furthermore, we cannot reconstruct positions of the continents based on the ages of intervening ocean floor, because none of the present ocean floor was formed before the Triassic. All of the ocean floor that existed in the Paleozoic has been consumed by subduction along convergent plate boundaries, as new crust formed elsewhere along the world's ridge-rise systems to take its place. Consequently, the remaining tools available for tracing the position of the Paleozoic continents are: (1) the positions of ancient magnetic poles, based on rock magnetism; and (2) the distribution of Paleozoic geosynclinal foldbelts, which can commonly be interpreted as the zone of collision between two ancient continental margins. An additional tool, the reconstruction of ancient biogeographic provinces, has been of corroborative value in some cases.

As we saw in Chapter 5, rock paleomagnetism enables us to determine the positions of the north and south poles for times in the geologic past. This record indicates that the continents have continually changed position relative to the poles throughout geologic time. The position of ancient poles alone, however, does not permit us to establish the former positions of a continent. Even if the position of a paleopole is established with confidence, we cannot infer a specific position for the continent at that time. Although we can determine the *orientation* of a continent in relation to the north and south poles and the latitude of the continent, we can never determine the *longitude*. Let us elaborate briefly.

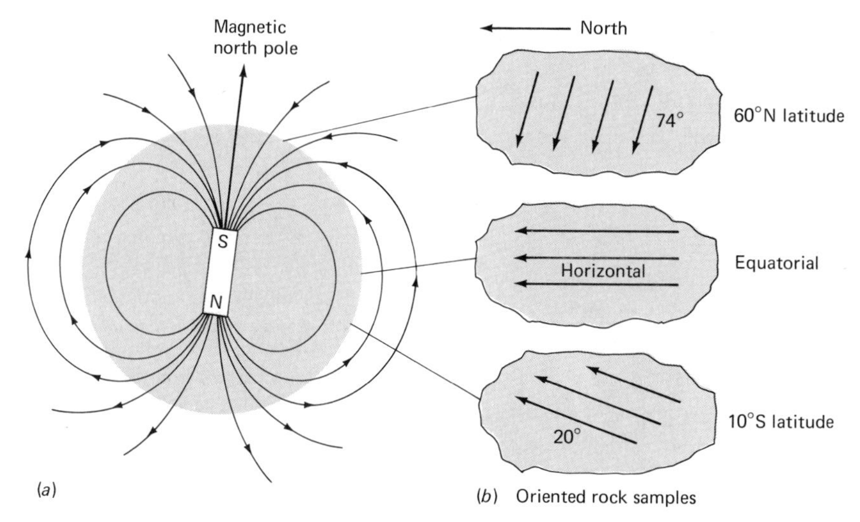

FIGURE 7.5

(*a*) Lines of magnetic force at the earth's surface indicate that the earth has a dipolar magnetic field, as if there were a bar magnet within it. (*b*) Magnetic inclination recorded in rocks from 60 degrees north from the equator and from 20 degrees south from the equator are shown. The arrow points to the north pole.

The direction of a rock's magnetism, considered in terms of a map, establishes the direction of the magnetic poles at the time the rock formed. The **inclination** of the rock's magnetism establishes the latitude at which it formed, because *inclination of magnetic lines of force on the earth's surface increases with latitude.* Figure 7.5 shows that the magnetic lines are horizontal for a rock formed at the equator. For a rock formed at 60° N latitude, the magnetic lines are inclined 74 degrees toward the north pole. For a rock formed at 10° S latitude, they are inclined 20 degrees toward the south pole. The latitude/inclination relationship is summarized in Figure 7.6.

FIGURE 7.6

Inclination of magnetic lines of force as a function of latitude on the earth's surface.

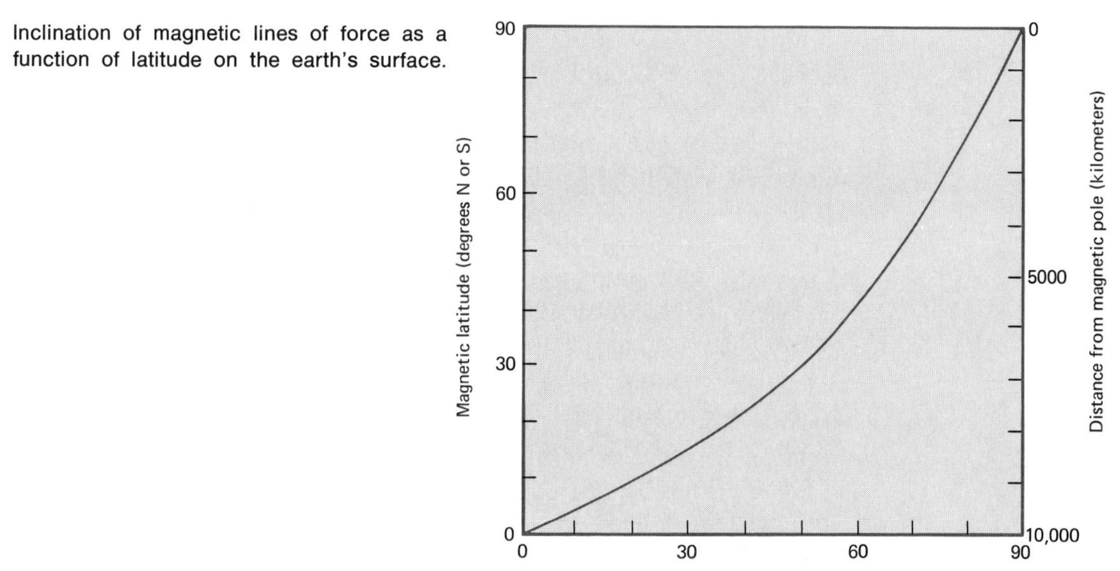

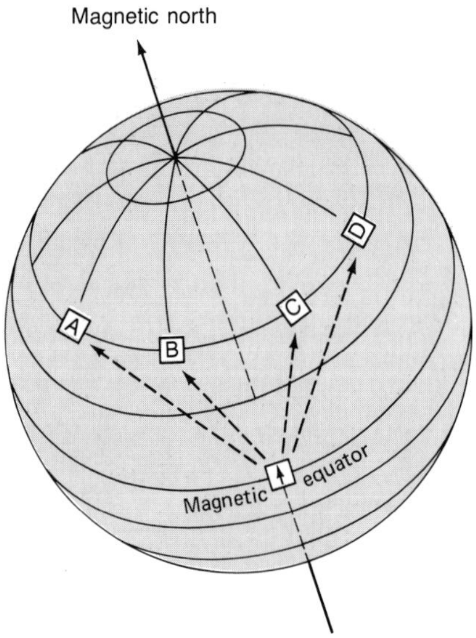

Magnetic north

Magnetic equator

FIGURE 7.7

Magnetic inclination is constant along magnetic latitude. A continent would have the same paleomagnetic record if it moved without rotation from the magnetic equator to any of the four positions shown at 35°N latitude.

At first glance, you might suppose that the exact position at which a rock formed on the earth's surface could be determined from the direction and inclination of its magnetism, but a closer look shows that this is not the case. Consider, for example, Figure 7.7, a perspective view of the earth showing lines of magnetic latitude. As a continent moves from the equator to 35°N latitude, the inclination of the remanent magnetism in the rocks forming on that continent changes from horizontal to about 55 degrees. However, the actual magnetic record in the rocks would be *identical*, whether the continent moved from A to B, C, D or E—all of which are on the same magnetic latitude line. This means that, from paleomagnetic data alone, you cannot tell how far the continents have moved longitudinally in the geologic past. Typically, both latitudinal and longitudinal components are involved in a continent's motion. The latitudinal component can be resolved readily; the longitudinal component, however, is always elusive.

Figure 7.8 shows how the positions of the Paleozoic continents changed between Cambrian and Devonian time. The precise longitudinal spacing of these early Paleozoic continents is very tentative, but we are confident that their relative positions and their latitudes are essentially correct. The Cambrian continents lay at low latitudes. In size and shape, they bore no resemblance to existing continents today. Bounding the continents on the north and south were vast oceans that totally covered the high-latitude regions. Then, in the Ordovician and Silurian, the continent of **Gondwana** swept across the south pole, while Siberia, **Kazakhstania,** and China rearranged themselves at low latitudes. Simultaneously, ancestral North America (labeled **Laurentia**) and ancestral northern Europe and Russia (labeled **Baltica**) rotated counterclockwise and drew close together. Then, in the Devonian, they collided, beginning the assembly of Pangaea. This rapid approach and ultimate collision of Laurentia and Baltica produced an episode of mountain building in the northern Appalachians that has long been known, but has only recently begun to be understood.

Early Paleozoic paleogeography: (*a*) Late Cambrian; (*b*) Middle Ordovician; (*c*) (top of p. 212) Middle Silurian; (*d*) Early Devonian. Mollweide projection showing entire earth surface. (*Scotese, and others, 1979*) The small crosses on the continents represent modern latitudes and longitudes.

FIGURE 7.8

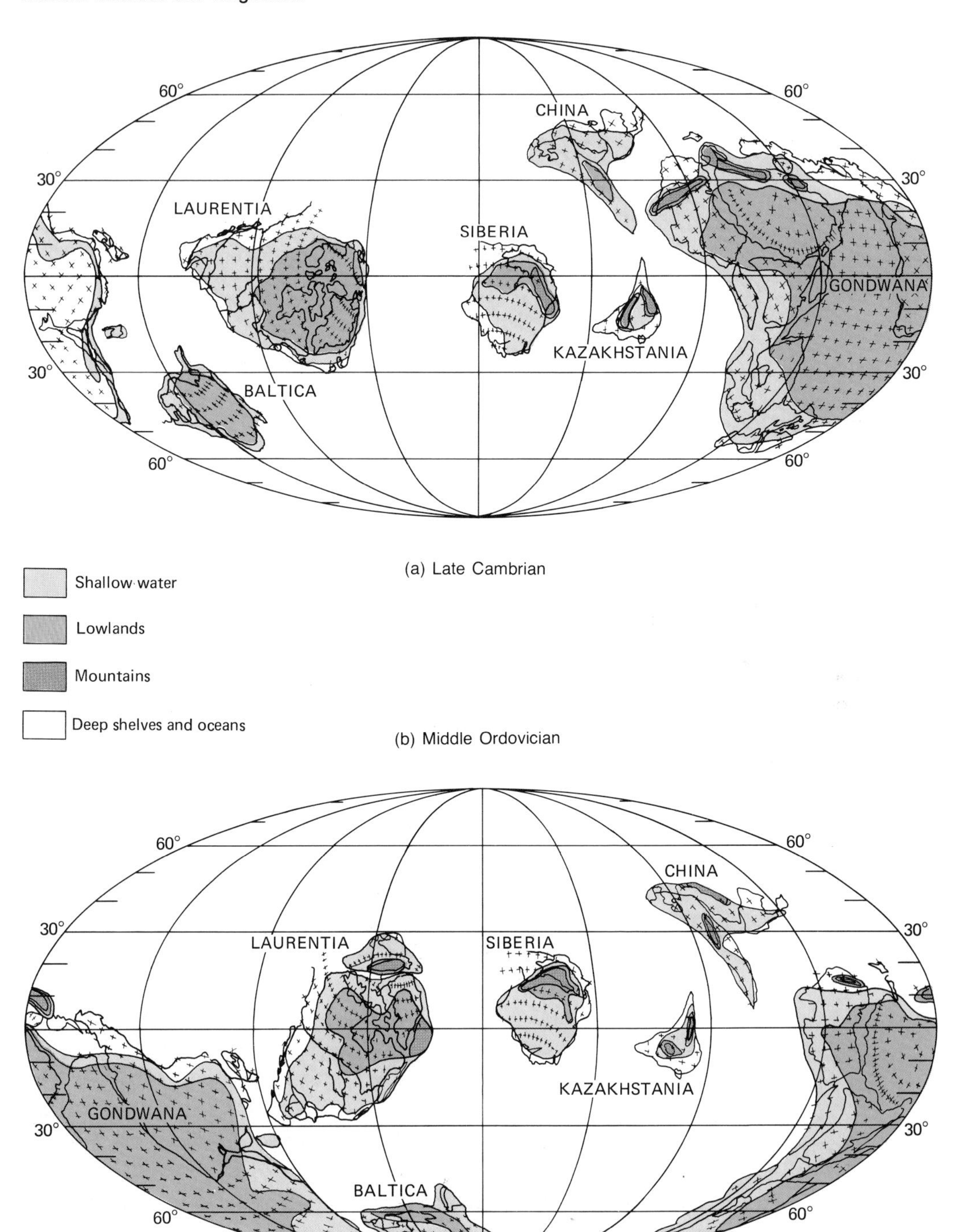

(a) Late Cambrian

(b) Middle Ordovician

FIGURE 7.8 (*cont.*)

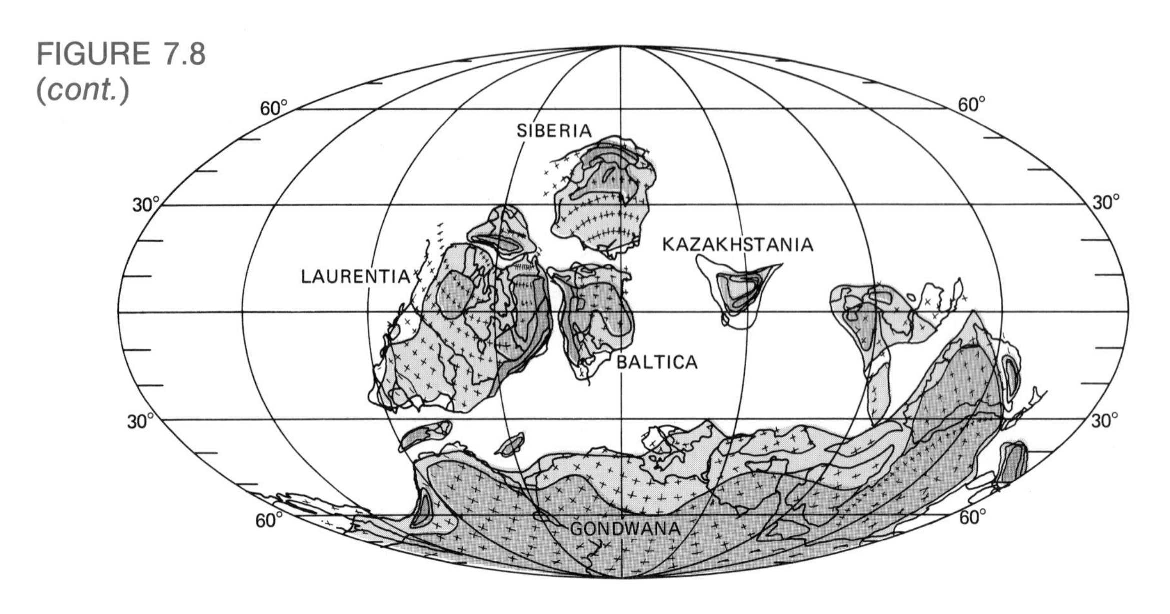

(c) Middle Silurian

(d) Early Devonian

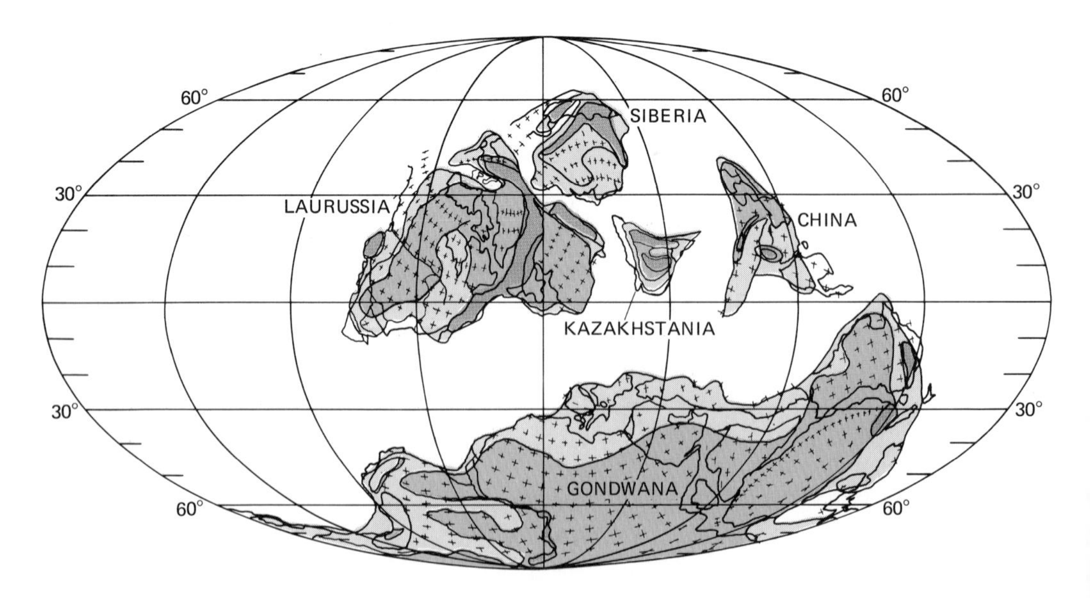

The Northern Appalachian Mobile Belt

During the Cambrian and the early part of the Ordovician, the ancestral North American and European continents lay on opposite sides of a widening ocean. It has been termed the **Protoatlantic Ocean,** but a look at Figure 7.8*a* and *b* shows that it does not resemble the modern Atlantic Ocean at all.

Paleomagnetism and structural and faunal evidence indicate that parts of present-day New England and Canada's maritime provinces were then part of Baltica and much of the southeastern United States was then part of Gondwana. Not only are the names of some of the early Paleozoic continents unfamiliar, but they contained some unfamiliar bits and pieces as well!

In the Middle Ordovician, large-scale thrust faulting took place in the northern Appalachians of the United States and Canada. This was the earliest mountain-building event along the eastern margin of Laurentia. It is named the **Taconic Orogeny** from the Taconic Mountains of easternmost New York state. Somewhat later, in Late Ordovician time, a borderland arose along the continental margin in the same region, and it shed a large volume of detrital sediments westward onto the former continental shelf. In New York and Pennsylvania today, Upper Ordovician shallow-marine shales and limestones may be traced eastward into nonmarine redbeds, chiefly sandstones and shales. The entire sequence represents a delta that built from the east into the shallow epeiric sea. This has been called the Queenston Delta, from the Queenston Formation, which includes many of the deltaic strata (Figure 7.9).

FIGURE 7.9

Upper Ordovician strata in Pennsylvania coarsen eastward toward the highland source area produced in the Taconic Orogeny. Solid lines are time lines. Angular unconformity indicates uplift immediately before deposition of the Juniata Sandstone. This is a reconstruction of the strata at the end of the Ordovician. The line of Section (X - X′) is shown in Figure 7.10.

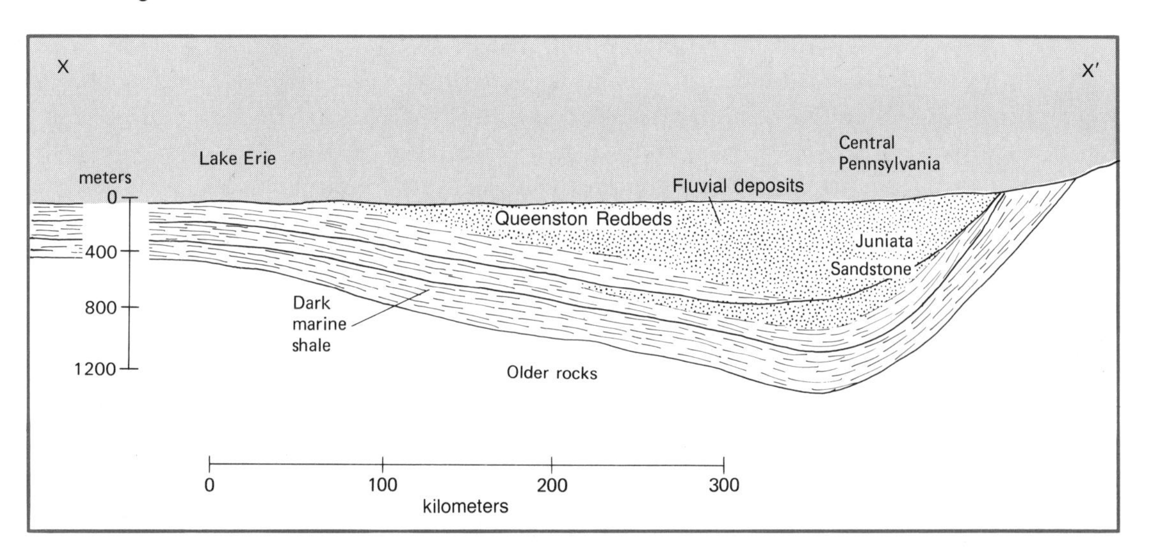

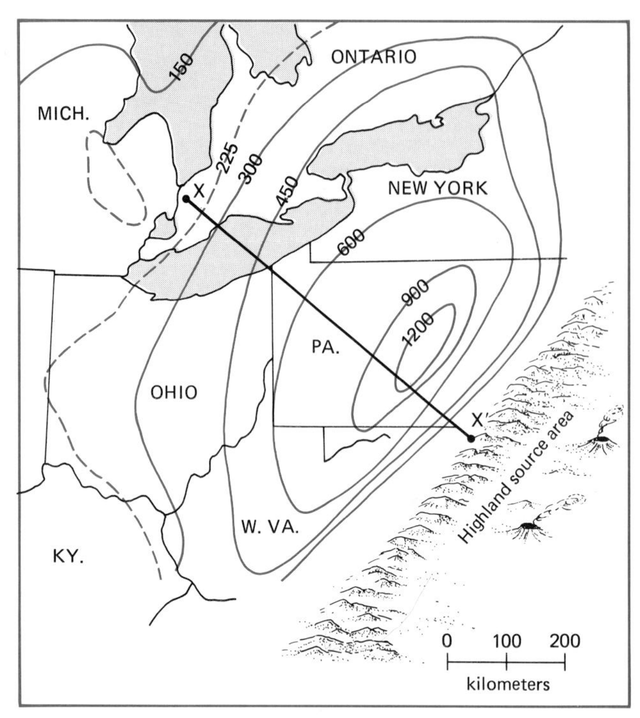

FIGURE 7.10

Isopach map of the Queenston Delta, fluvial and shallow-marine sandstones and shales of Late Ordovician age. X - X′ is the position of the cross section in Figure 7.9. Isopachs are measured in meters.

The Taconic Orogeny that produced these sedimentary patterns marked a fundamental change. What had been a quiet, submerged shelf sloping seaward away from the cratonic interior from which it received small quantities of detritus now *reversed* its slope and began receiving large quantities of detritus from the continental margin—an indication that this area had been uplifted substantially (Figure 7.10). In the northeastern United States, the easternmost outcrops of Ordovician sedimentary rocks contain volcanic debris and interbedded lava flows, strongly suggesting that a volcanic arc had formed along the margin of the continent. This further suggests that the area had become a subduction zone, and that an oceanic trench bordered the volcanic arc on the east. In some areas, such as the Canadian maritime provinces, exotic blocks of Ordovician and Cambrian limestones up to several meters in length occur, engulfed in thick sequences of Ordovician black shale. These are deposits of submarine slumps and debris flows that were triggered on steep slopes by the frequent earthquakes that must have accompanied the tectonic activity.

In the latest Ordovician, the eastern part of New York and Pennsylvania continued to rise. The former deltaic plain was elevated, probably well above sea level, and underwent significant erosion. This produced an unconformity between older Ordovician rocks below and the Silurian or, locally, Devonian strata that overlie them. The succeeding Silurian deposits of this region consist of sandstones and conglomerates that coarsen eastward, reflecting the persistence of highland source areas in the mobile belt.

In both the Canadian maritime provinces and northern Europe, volcanism continued well into the Silurian. In the Late Silurian, the geosynclinal strata in Britain and Scandinavia were strongly folded during the **Caledonian Orogeny**. The volcanism on both the North American and European sides of the Protoatlantic Ocean suggests that volcanic arcs and oceanic trench systems existed on both sides as the Protoatlantic Ocean continued to close (Figure 7.11).

FIGURE 7.11

Suggested evolution of the Taconic-Caledonian-Acadian mobile belt during the opening and closing of the Protoatlantic Ocean basin. (*a*) Late Precambrian: Protoatlantic Ocean opens. (*b*) Cambrian: Protoatlantic continues to widen. (*c*) Mid-Ordovician: Protoatlantic closing as subduction occurs on both sides. (*d*) Devonian: Protoatlantic closes completely. (*Bird and Dewey, 1970*)

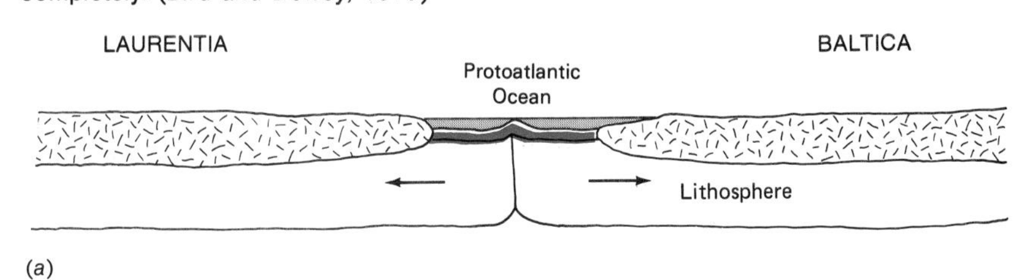

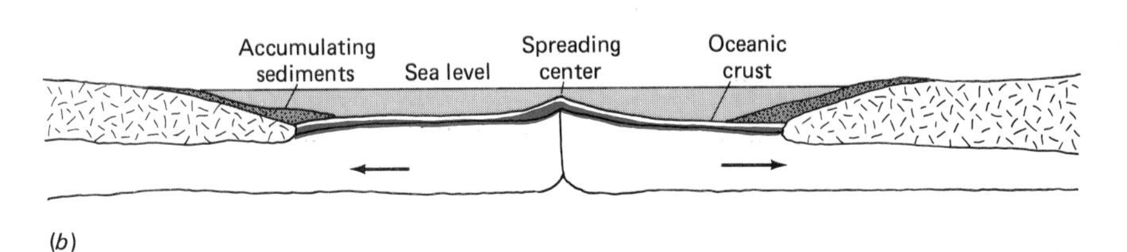

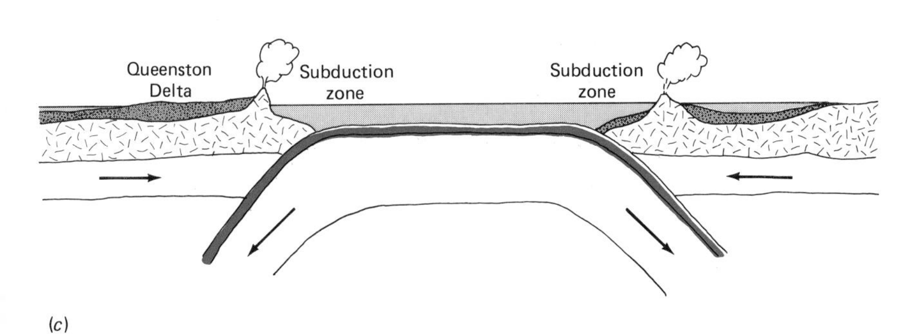

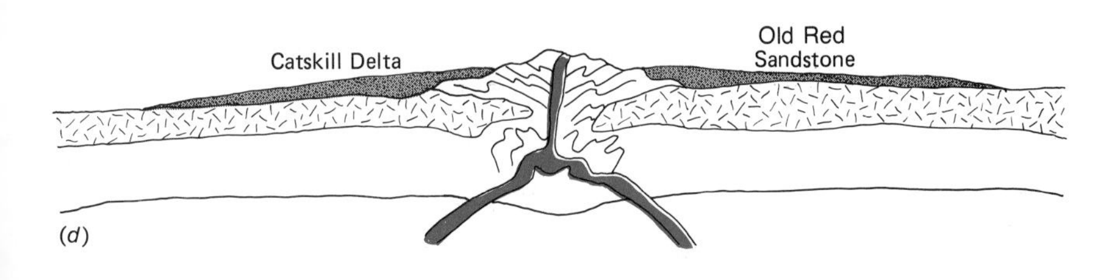

In Middle and Late Devonian time, rising highlands to the east again shed large volumes of sediment westward onto the continental margin of the northeastern United States, the Canadian maritime provinces, and Greenland. Simultaneously, rising highlands west of the European continental margin shed sediments eastward onto Britain and Scandinavia. The resulting sedimentary rocks consist in part of thick sequences of nonmarine redbeds, such as the Old Red Sandstone in Europe and the Catskill Formation in North America. They are composed of sandstones, shales, and conglomerates that accumulated above sea level in vast fluvial plains and deltas.

The similarity of rocks and faunas in both North America and Europe suggests that they were derived from one source area: a huge, elongate mountain belt that formed as ancestral North America and ancestral Europe finally collided (Figure 7.11). The Devonian section in eastern New York and Pennsylvania consists of marine limestones at the base, overlain by near-shore sandstones and shales, and finally by the thick, fluvial redbeds of the Catskill Formation, which thickens and coarsens eastward (Figure 7.12). In the Late Devonian, the Catskill fluvial plain built westward for 500 kilometers behind the regressing epeiric sea (Figure 7.13).

The deposits of the Catskill Delta are similar to those formed earlier by the Ordovician Queenston Delta, but the Catskill deposits are thicker and more widespread, reflecting a higher and more persistent source area to the east. In

FIGURE 7.12

Middle and Upper Devonian strata in southern New York coarsen eastward toward the highland source area formed when the ancestral North American and ancestral European continents collided during the Acadian Orogeny. This is a reconstruction of the strata at the end of the Devonian. The line of section is shown in Figure 7.13.

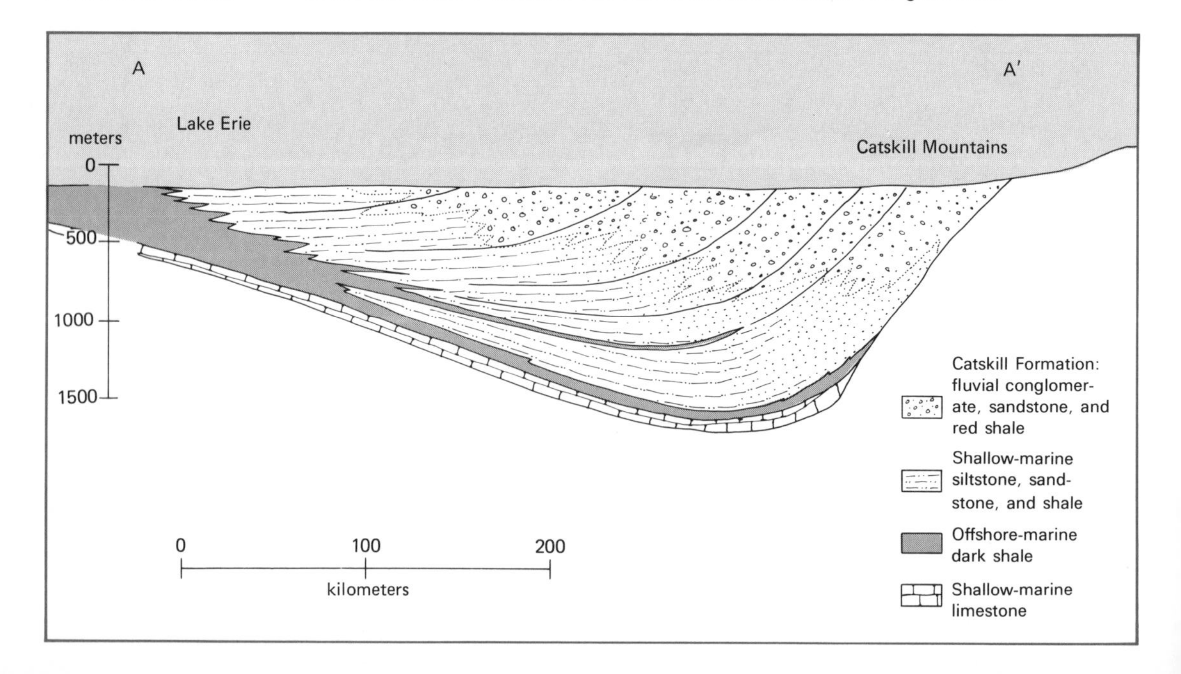

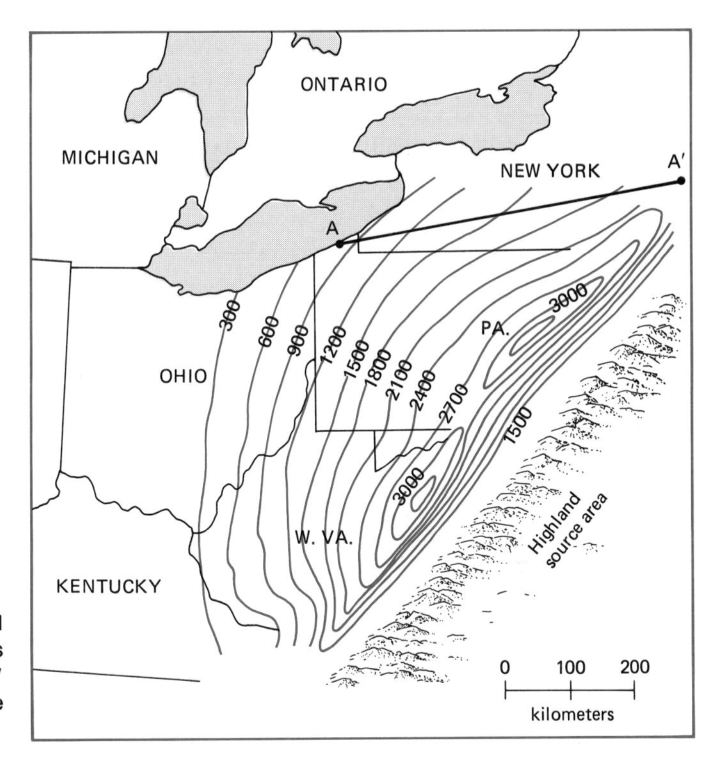

FIGURE 7.13

Isopach map of the Catskill Delta, fluvial and shallow-marine sandstones and shales of Middle and Late Devonian age. A - A′ is the position of the cross section in Figure 7.12. Isopachs are measured in meters.

New England and the adjacent maritime provinces of Canada, regional metamorphism and granitic intrusives of Devonian age testify to the intensity of the mountain-building activity that formed the highland source area. The tectonic episode from this region has been named the **Acadian Orogeny.** Following the Acadian Orogeny, the mobile belt along which ancestral North America and ancestral Europe collided did not immediately become quiescent. It was the site of igneous activity in the ensuing Mississippian Period, and it served intermittently as a source area for large quantities of detrital sediments throughout the remainder of the Paleozoic.

Taconic, Caledonian, and Acadian—we now consider these three orogenies all part of the same grand process: the convergence and ultimate collision of Laurentia and Baltica. Each of these orogenies represents a regional burst of tectonic activity along subduction zones that, with the Acadian event in the Devonian, finally gobbled up the Protoatlantic Ocean altogether. This apparently was the first major collision of northern hemisphere land masses, and with it began the Paleozoic amalgamation of all the continents.

Deciphering Marine Sediments

Mud and sand deposited on the ocean floor normally originate in one of two ways: The first, called **detrital sediment,** is derived from the weathering of preexisting rock *on land,* the fragments being then delivered to the ocean by rivers and streams. Usually quartz sand and various clay minerals dominate such sediments. The second class of sediment is not transported physically from land but is, instead, precipitated from materials dissolved *in ocean water*. Such precipitation can occur either directly or, more commonly, through the life processes of animals and plants. Because these sediments are usually dominated by calcium carbonate (the minerals calcite or aragonite) or by calcium-magnesium carbonate (the mineral dolomite), they are referred to as **carbonate sediments.**

Since much of the early Paleozoic sedimentary record consists of marine detrital and carbonate rocks, an understanding of the environments in which these rocks form is central to unraveling Paleozoic history. In addition, the principles involved in their interpretation apply to both older and younger marine rocks.

Detrital Settings

Shelf Environments Most of the sand and mud delivered to the ocean by streams is deposited on the submerged shelves that surround the continents (Figure 7.14). These continental shelves today have an average width of about 80 kilometers, and they slope gradually seaward to an average water depth of about 180 meters. Beyond them lie the steeper slopes of the continental margins that lead to the deep-ocean floor (Figure 7.15). On the shallow near-shore parts of the shelves, the ocean floor is frequently disturbed by waves and swept by other currents. As a result, sand and gravel can be transported, sorted,

FIGURE 7.14 Present-day continental shelves, the submerged margins of the continents where most detrital sediment accumulates.

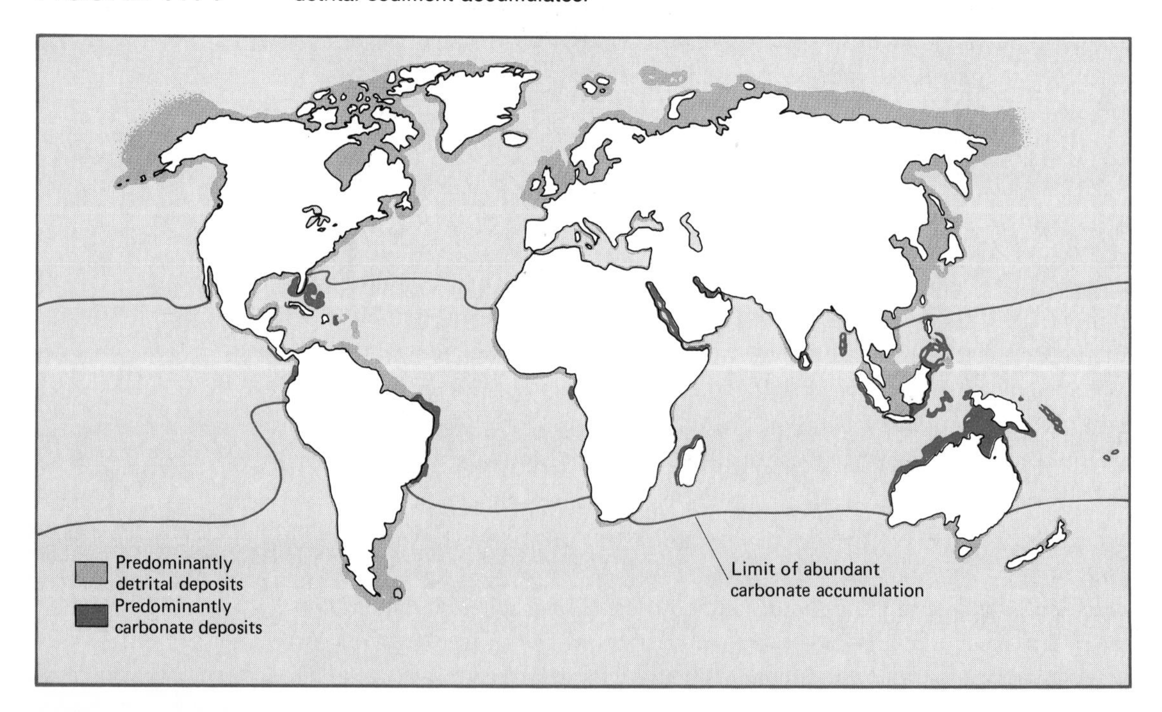

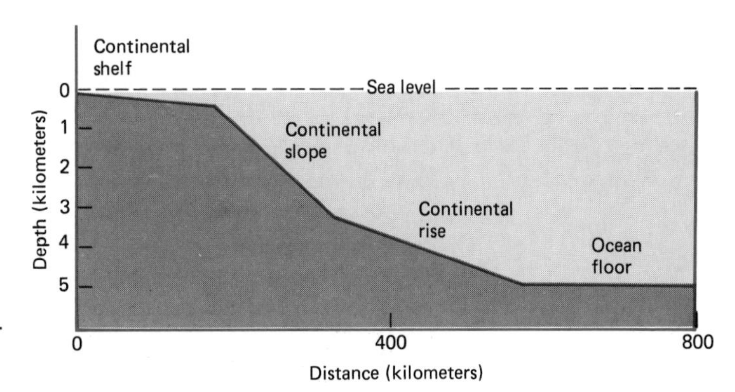

FIGURE 7.15

Topography across a continental margin.

and concentrated in the shallow inner-shelf region. On the deeper outer portions of the shelf, the ocean floor is quiet. Waves rarely disturb the bottom, and current energy is generally so low that only fine silt and clay-sized material can accumulate in this area.

Studies of sedimentary rocks have shown that this pattern of off-shore decrease in particle size was common in the geologic past. Sand and gravel were typically deposited in near-shore regions and graded seaward into finer-grained silt and mud deposits. These deposits are preserved in the geologic record as sandstones that change facies into shales with increasing distance from the ancient shoreline.

Paradoxically, this normal fining-outward sediment pattern does not occur on continental shelves today. Shipboard sampling and underwater observation have shown that *well-sorted sand is the dominant sediment* over much of present-day shelves. It is particularly abundant over the outermost parts of the shelves, where it should be rare or absent! The reason is found in the peculiarities of recent geologic history. The alternating expansion and contraction of continental ice sheets during the Pleistocene caused changes in sea level, as a result of which the shoreline moved back and forth across the continental shelves. Consequently, much of the sediment exposed today on even the outermost parts of the shelves was deposited in high-energy, near-shore environments at times when sea level was much lower. As a result, the shelves today are covered in large part by relict sediments that do not reflect the depth and energy conditions in which they now occur.

Marine life abounds on the continental shelves. Bones, shells, and other hard remains of bottom-dwelling marine animals are common. Shelf sands and silts contain numerous burrowing invertebrates that destroy the layers and generally homogenize the sediments. Details of bedding are lost; all that remains is a gross stratification that is recognizable only by major changes in texture. Exceptions are found in isolated topographic basins on the sea floor of some shelf regions. In these basins there is little current motion, and the oxygen in the water may become depleted as a result of lack of circulation. Such basins are said to be anaerobic or anoxic. Bottom-dwelling shelled invertebrates are absent. Little burrowing takes place and few fossils occur. Organic particles that are transported in are neither consumed nor oxidized, and the usual result is an unfossiliferous, black, evenly laminated mud.

In the vast early Paleozoic epeiric seas, the shale facies that lay offshore of the sand facies was commonly bordered still farther offshore by a carbonate facies. A classic example of this is the record of the Cambrian transgression onto the present North American continent. In the western United States, this record is typified by the stratigraphic sequence in the Grand Canyon (Figure 7.16). Here the Tapeats Sandstone grades upward into the Bright Angel Shale, which, in turn, grades upward into the Muav Limestone. This vertical sequence represents the slow eastward migration of progressively offshore facies. The Tapeats Sandstone is actually a part of a vast blanket of sandstone that, in many parts of the United States, reaches a thickness of 200 meters or more. The sandstone goes by many different names in different areas. Typically it is a mature sandstone—texturally mature in having well-rounded grains, and compositionally mature in consisting of grains of quartz and other stable minerals.

A similar widespread sandstone marks the base of the next great marine transgression in Middle Ordovician time (Figure 7.17). This sandstone, too, goes by different names in different areas, but throughout much of the mid-continent it is called the St. Peter Sandstone. Like the Tapeats, the St. Peter is a mature sandstone. In many areas it consists almost entirely of quartz grains, and it is such a pure quartz sand that it is used widely in the manufacture of glass.

Because they are transgressive deposits, both the Cambrian and the Middle Ordovician sandstones are oldest near the margins of the North American craton and become younger toward the central portion. Exactly how much time the respective transgressions represent is difficult to specify, but each must have required several million years. Hence, these sandstones represent slow and probably sporadic sedimentation over a long period of time. The sand grains that make them up were derived from deeply weathered source areas and were eroded and redeposited many times prior to their final incorporation in the existing sandstone units. In fact, much of the Ordovician St. Peter was probably derived from erosion and redeposition of preexisting Cambrian sandstones.

Both the Cambrian and Ordovician transgressive sandstones represent a variety of near-shore environments. In some areas these have been studied in sufficient detail to permit direct comparison with sands being deposited today. In the ancient sandstones, we can recognize shallow-marine, tidal-flat, and beach environments. In addition, some of the sands are thought to have been deposited as onshore dunes. The dune deposits are closely associated with marine deposits; that is, they are commonly interbedded with marine sand-stones or are traceable laterally into them. However, the very large scale of their cross-bedding and their steep dips are characteristic of eolian (wind-blown) dunes. At some localities, the Ordovician St. Peter Sandstone contains

Cambrian portion of the Grand Canyon section exposed near Bright Angel Trail.

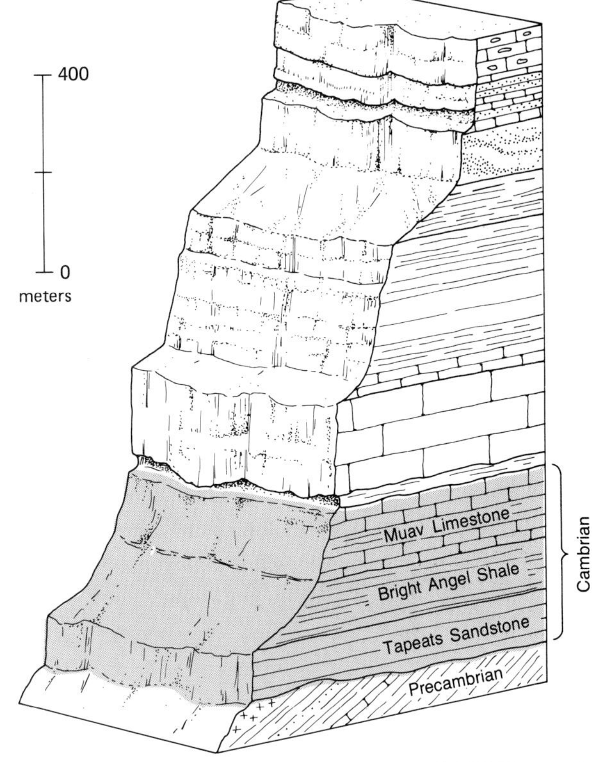

FIGURE 7.16

Middle Ordovician sedimentary facies in a portion of North America.

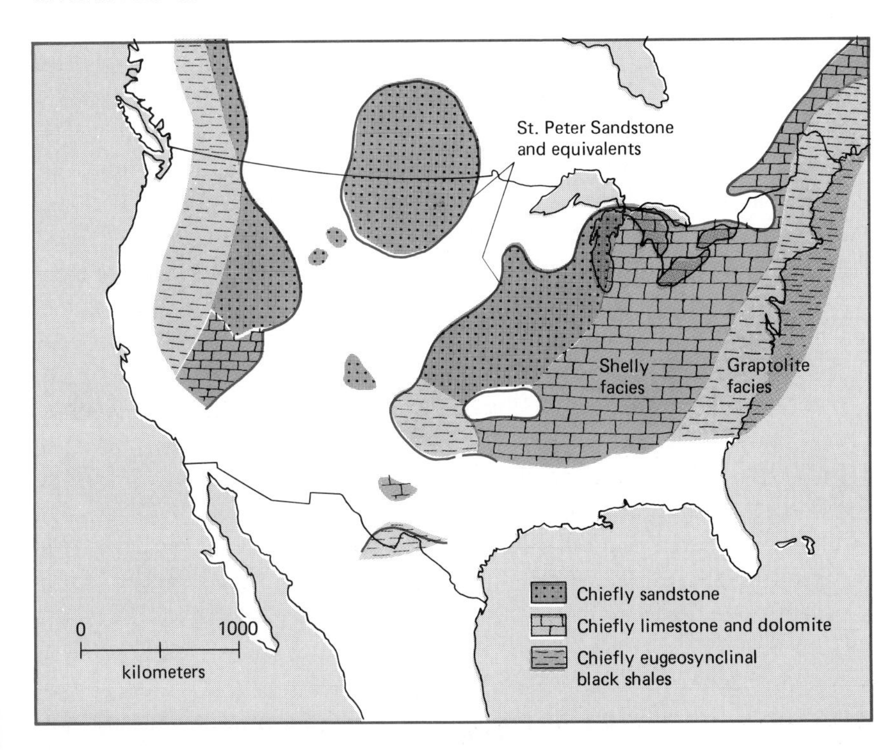

FIGURE 7.17

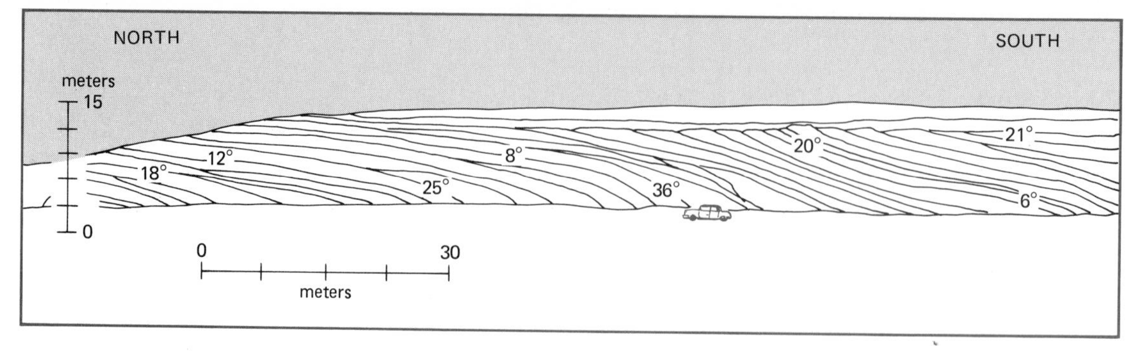

FIGURE 7.18 Cross stratification in the St. Peter Sandstone in a roadcut near Monticello, Wisconsin. The large scale and high dips indicate that the St. Peter here was deposited by large sand dunes. Note the automobile for scale. (*Pryor and Amaral, 1971*)

cross beds 200 meters long that have primary dips of around 30 degrees (Figure 7.18). The sand here was almost certainly deposited on a low coastal plain by migrating sand dunes similar to those transversing certain coastal plains today. For example, on the Gulf Coast of southern Texas and northern Mexico, large sand dunes are migrating inland (Figure 7.19). Here prevailing onshore winds pick up the sand from vast beaches whose sand, in turn, is supplied by marine longshore currents. The mixed shallow-marine, beach, and dune sandstones in the St. Peter probably reflect a similar environmental setting.

FIGURE 7.19 Sand dunes migrating inland near Vera Cruz on the Mexican Gulf coast, seen from an altitude of 3,500 meters. The field of view is about 2 kilometers wide.

Deep Water Environments During the early Paleozoic, well-developed geosynclinal belts flanked the eastern and western margins of the North American craton. Thick miogeosynclinal sequences, chiefly carbonates and clean sandstones, graded seaward into thick eugeosynclinal black shales, greywackes, and locally interbedded volcanic rocks. Greywacke sandstones in the eugeosynclinal settings typically occur in repetitive graded beds that are interpreted as turbidites. In the Ordovician and Silurian, the eugeosynclinal black shales contain **graptolites**, organic-walled planktonic colonial animals that look like shiny black pencil marks on the shale beds. There are few other fossils in these shales, so they have come to be referred to as graptolite facies (see Figure 7.17); carbonate rocks of the miogeosynclines and the stable cratonic platforms that are rich in calcareous shells have been called shelly facies.

Sediments of the eugeosynclines were deposited in deep water, where there was little or no bottom current. Sands were transported into the eugeosynclinal basins by gravity flow and, once deposited, were not reworked. The black shales resulted from the settling out of fine clay particles from suspension. The scarcity of bottom-dwelling life in these eugeosynclinal black shale beds suggests that the sea bottom may have commonly been deficient in oxygen. In the rare cases in which fossils such as trilobites occur in these beds, they are different species from those in the shelly facies, and they are inferred to be deep-water forms.

Graptolites, which are common in eugeosynclinal black shales, did not live on the sea floor but were part of the plankton in the near-surface waters far above. Upon death, their remains settled to the muddy sea floor, where they were buried and preserved. Graptolites evolved rapidly during the Ordovician and Silurian and were carried worldwide by early Paleozoic oceanic currents; hence, they make excellent zone fossils for worldwide correlation. However, correlation between graptolitic and shelly facies, even over a few tens of kilometers, has long been a problem. The carbonates and clean sandstones of the shelly facies are zoned on totally different organisms, generally brachiopods and trilobites. The problem has been resolved, in large part, by rare occurrences of key graptolites in the shelly facies.

Carbonate Settings

Calcium carbonate sediments being deposited today, whether of deep- or shallow-water origin, are derived chiefly from carbonate-secreting organisms. In modern oceans, most carbonate deposition occurs in the deep sea in the form of **pelagic oozes** that accumulate from the calcareous skeletons of tiny planktonic plants and animals that continually rain down from the sunlit waters above. In some regions, however, carbonate deposits are forming today in shallow water, and these deposits provide insights into the much more widespread shallow-water carbonates that formed in the geologic past. During the Paleozoic Era, the open ocean areas contained no calcareous plankton, and carbonate sediments were produced almost exclusively by bottom-dwelling organisms (**benthos**) that proliferated on the shallow floors of the widespread epicontinental seas.

FIGURE 7.20

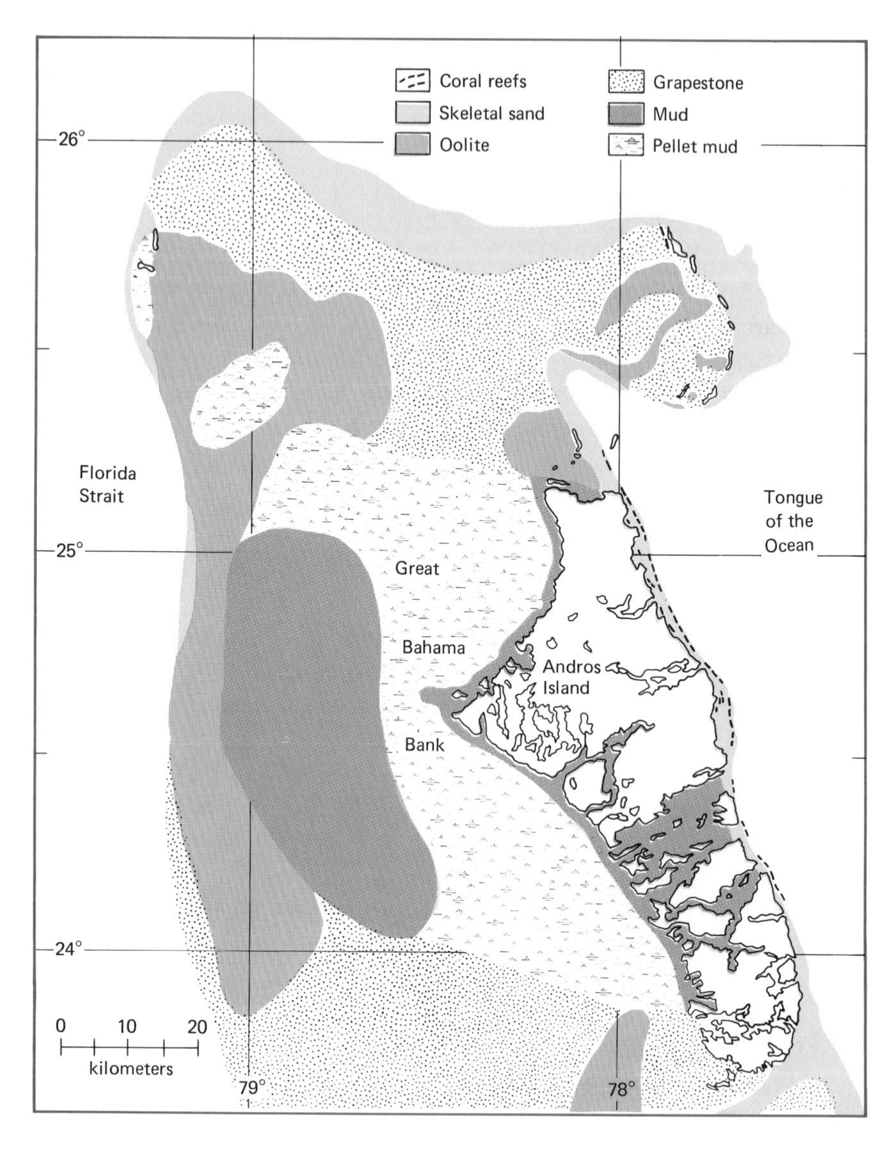

Distribution of the various kinds of calcareous sediments forming today on the Great Bahama Bank. (*Laporte, 1968*)

Carbonate Platform Sediments

Most shallow-water carbonates forming today are found in the tropics, partly because more kinds of carbonate-secreting organisms thrive there than in high latitudes, and partly because, once formed, calcium carbonate is less easily dissolved in warm water than in cold. Nearly all shallow-water carbonate sediments are biogenic (derived from carbonate-secreting organisms), although in special environments some calcium carbonate may be precipitated inorganically. Whether the carbonate sediments are organic or inorganic in origin, perhaps the single most important requisite for the accumulation of large volumes of them in an area is that there be little influx of terrigenous detritus.

Once they have been produced, **carbonate sands** and **carbonate muds** react to water motion in the same way as do land-derived sands and muds. Carbonate sands show cross-bedding and other structures indicative of high-energy water motion. Carbonate muds, on the other hand, tend to accumulate in sheltered or deeper environments where there is less water movement. Among existing areas where shallow-water carbonates are forming, the Bahama Banks is one of the best known. Study of this area has greatly enhanced our understanding of ancient carbonate shelf deposits.

Great Bahama Bank (Figure 7.20) rises from the deep ocean floor 110 kilometers east of Florida. The Bank is a large, shallow platform almost 200 kilometers across, but with water depths generally less than 5 meters. Tidal circulation on the Bank is sluggish, with the result that water salinities and temperatures are generally higher than those of the adjacent deep oceans. The rocks that underlie Great Bahama Bank, as well as the sand and mud sediments being deposited there today, are exclusively carbonates.

Carbonate sands on Great Bahama Bank are of three kinds: **skeletal sand**, **oolite**, and **grapestone**. The skeletal sands consist, not of whole shells of invertebrate animals and plants, but of sand-sized particles derived from the breaking apart of whole shells (Figure 7.21). Most of the skeletal sand on Great Bahama Bank is produced from the skeletons of green algae, **corals**. and other invertebrates that thrive predominantly at the margins of the Bank. When these carbonate-secreting animals and plants die, their skeletons become dis-

FIGURE 7.21

How a green calcareous alga (*Halimeda*) and a stony coral (*Acropora*) contribute skeletal debris to sediments. Note that, depending on the degree of disarticulation and disintegration of the original skeletons, these organisms together produce calcareous sediments of various textures, from coral gravel to lime mud. (*Folk and Robles, 1964*)

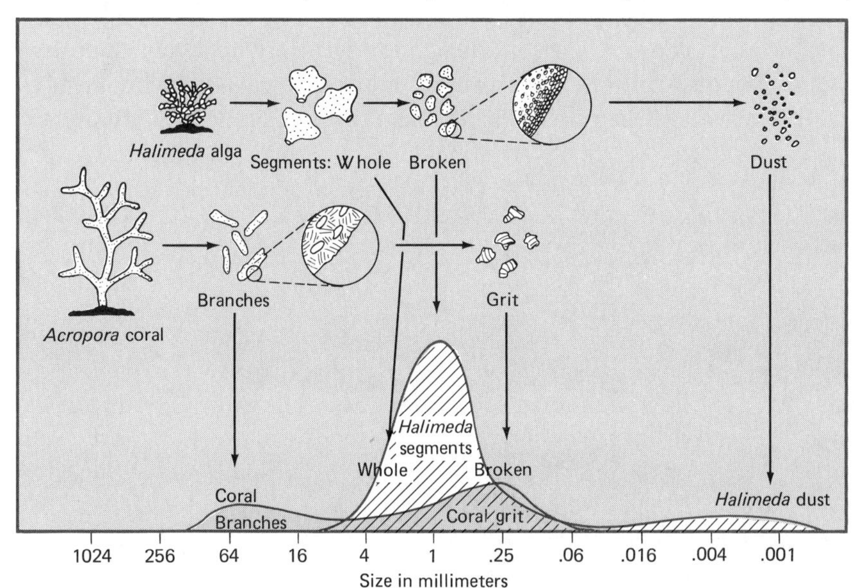

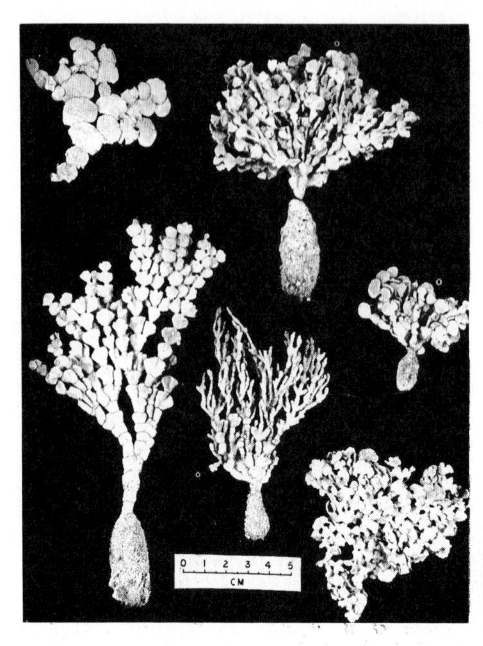

FIGURE 7.22 The green alga *Halimeda,* an important producer of sand-sized calcium carbonate grains in modern-day shallow carbonate environments. (*Blatt, Middleton, and Murray, 1972*)

articulated and are reduced to smaller and smaller particles by mechanical abrasion and by boring organisms. The importance of corals in producing skeletal sand has long been recognized, but the role of algae, which is at least as significant as that of corals, has been fully appreciated only in recent years. On Great Bahama Bank, the algal genus *Halimeda* alone furnishes about 10 percent of the skeletal sand grains (Figure 7.22). Most sedimentologists now feel that algae have been similarly important producers of shallow-water skeletal sand throughout the geologic past.

Oolite sand is characterized by an abundance of spherical grains that are composed of concentric layers of calcium carbonate around a minute central nucleus (Figure 7.23). These grains occur in the high-energy environment of oolite shoals, where they are heaped by tidal currents into underwater ridges.

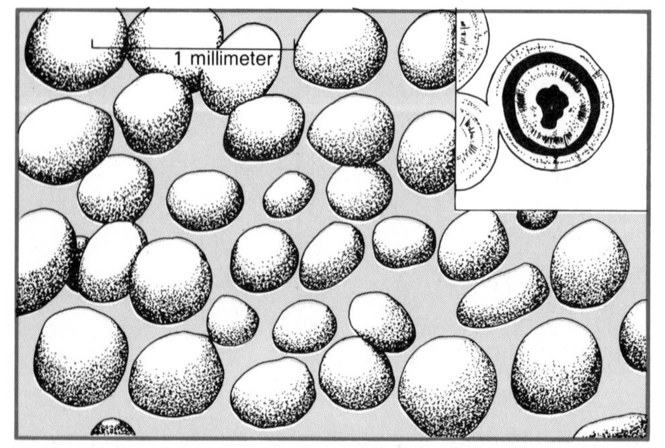

FIGURE 7.23 Oolite grains showing concentric structure (inset).

Aerial view of oolite sand bars on the Bahama Banks. The belt of oolite bars is about 1 kilometer wide.

FIGURE 7.24

The oolite grains are kept in nearly constant motion by the incoming and outgoing tidal currents, which are saturated with calcium carbonate. Few organisms live in this environment because of the unstable substrate. Oolites are accumulating at present in shoal areas near the edge of Great Bahama Bank in a belt 2 to 4 kilometers wide. Commonly, they occur in enormous ripple forms called sand waves (Figure 7.24). Many oolite shoals are actually awash at low tide. Oolitic limestones are common in the geologic record, and the evidence is very good that they formed under similar shallow, high-energy conditions.

Grapestone sediments consist of sand-sized grains that are aggregates of still smaller calcium-carbonate particles cemented together (Figure 7.25). Each grain looks something like a tiny cluster of grapes. Most of the small component particles appear to be oolite grains, but many are not readily identifiable. The Bahamian grapestone occurs in quiet waters that are sporadically agitated. Wave and current energy is sufficient to prevent mud deposition but insufficient to transport sand grains, except occasionally. These generally quiet environments favor cementation of the grains. In some cases grapestone may have originated as eroded fragments of carbonate sands that had been cemented locally into beach rock.

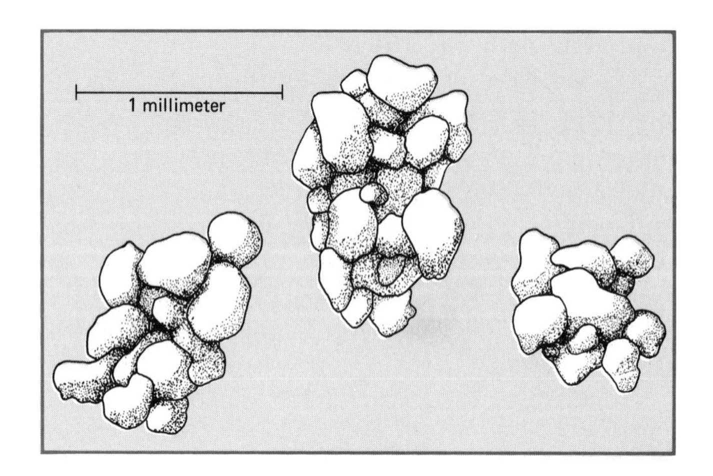

FIGURE 7.25 Grapestone grains.

Carbonate muds consist of silt- and clay-sized carbonate particles (less than 1/8 millimeter). A special variety of mud called **pellet mud** contains numerous ellipsoidal mud pellets that are typically a few millimeters in length. These are actually fecal pellets, produced by burrowing arthropods and molluscs. Pellet mud however, contains few shells. The small pellets that characterize pellet mud are very weakly cohesive. Generally, pellet muds represent the quietest environments of all in the central portion of Great Bahama Bank.

FIGURE 7.26 The marine grass *Thalassia* growing on carbonate mud in Florida Bay. The mound was made by a burrowing shrimp. The scale is 30 centimeters long.

A large quantity of mud-sized carbonate material is mixed in with sand-sized grains in relatively high-energy shallow-marine environments where an underwater grass called *Thalassia* is abundant (Figure 7.26). Although moderately strong currents may sweep such areas, the grass acts as a baffle to create a quiet environment at the sediment interface; here, among the grass blades, silt- and clay-sized particles readily settle out. These particles then become bound by the root network of the grass as they accumulate; hence, they are not easily eroded. Grasses did not exist in the Paleozoic, but dense clusters of algal plants and stalked echinoderms called **crinoids** may have provided similar traps where fine sediments readily accumulated.

The tidal flats that surround some of the numerous low islands on the Bahama Banks are also sites of deposition of carbonate mud. On the **intertidal flats**, which are flooded daily, blue-green algae form temporary mats that trap mud particles, but burrowing organisms continually rework the mud, destroy the laminations, and produce mottling. Shells of some of the molluscs that live in the intertidal flats become a part of the deposit. However, the variety of species in this environment is very limited.

Supratidal flats, which lie still farther landward, are only occasionally covered by seawater during storms or unusually high tides. Here algal mats are the dominant sedimentary agent. Because few burrowing organisms can survive in supratidal environments, the characteristic laminated sediments called stromatolites accumulate and are preserved. Shells are extremely rare. Although burrowers do not disturb supratidal mats, alternate wetting and drying commonly produces mud cracks in them. Magnesium-rich **interstitial waters**, which slowly seep through these finely laminated muds after deposition, commonly alter the tiny grains of calcium carbonate to the mineral *dolomite*. This process of **dolomitization** has occurred on supratidal flats and in other carbonate platform environments throughout geologic time.

The various carbonate facies that are today represented on the Great Bahama Bank occur throughout much of the Phanerozoic rock record. In the Ordovician of northern New York, for example, three formations that constitute the Black River Group represent deepening waters of a shallow Middle Ordovician sea that transgressed from west to east onto the Adirondack Highlands (Figure 7.27). The lowermost unit, the Pamelia Formation, contains numerous stromatolitic and mud-cracked beds, many of which have been dolomitized. The Pamelia represents a wide supratidal mud flat on the eastern shore of the sea. The Pamelia grades upward and westward into the ripple-marked and heavily burrowed limestones of the Lowville Formation. Some beds of the Lowville contain marine invertebrate fossils, but only of a few different kinds. Other beds are mud-cracked and dolomitized. The deposits are believed to represent a dominantly intertidal environment at the sea's edge. The Lowville, in turn, grades upward and also westward into strata of the Chaumont Formation, which consists chiefly of fossiliferous and pelletal limestones with occasional oolitic beds. Here fossils are common, and the

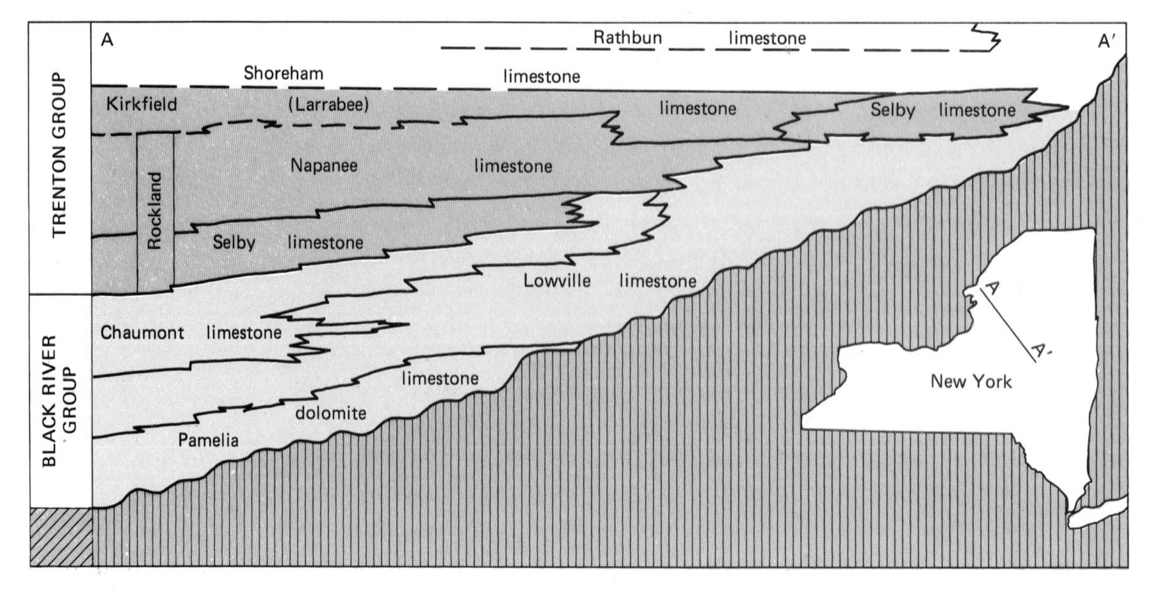

FIGURE 7.27 Stratigraphic relationships of the three transgressive formations of the Middle Ordovician Black River Group in northern New York. (*Textoris, 1968*)

variety is comparatively great. The Chaumont is interpreted to represent a variety of subtidal (permanently submerged) shallow-marine environment, formed offshore from the Lowville.

The model for the deposition of the Black River Group is shown in Figure 7.28. With eastward transgression, the shallow-marine facies (Chaumont Formation) advanced over the intertidal facies (Lowville Formation), which, in turn, transgressed over the supratidal facies (Pamelia Formation). This transgression resulted in the replacement of a low-lying land area by a clear, shallow Middle Ordovician sea.

FIGURE 7.28 Transgression from west to east of an array of carbonate platform environments produced the Black River facies in Figure 7.27. Vertical scale is exaggerated about 100 times. (*Textoris, 1968*)

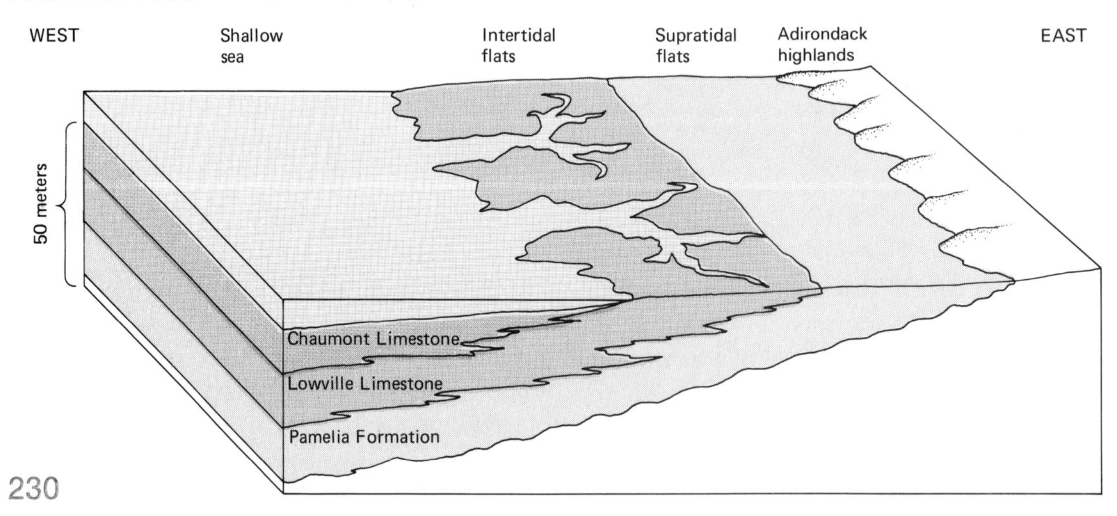

Reefs The marine transgression recorded by the Middle Ordovician Black River Group was one of many that brought shallow seas onto vast expanses of the continental platforms during the early part of the Paleozoic Era. In most regions, the relief of the bordering lands was apparently low, because they shed little or no material into the seas. Beginning in Early Cambrian time, carbonate-secreting marine organisms built massive accumulations on the bottom of such shallow seas that extended to the water surface, where they could intercept the waves. These structures are organic reefs, made of a rigidly cemented mass of calcareous skeletons that has accumulated in place. Today reefs exist widely at latitudes lower than about 30° on shallow-water platforms, along some continental margins, and around volcanic islands (Figure 7.29).

Today and throughout much of geologic history, reef builders have been chiefly corals, other colonial animals, and carbonate-secreting algae. At times, animals other than corals have dominated reef communities, but regardless of which kind of organisms dominated at the time, reefs appear to have always occupied the same ecologic niche that they do today. Reef organisms require a lot of sunlight and clear, agitated water. They grow only on hard substrates where they can anchor themselves firmly. Their upward growth limit is the surface of the sea itself. Reefs flourish in stable conditions where the mean winter temperature of the water is warm—between 27°C and 29°C—and where salinity remains essentially that of the open sea. Reefs favor the windward side of continents and platforms. Corals and algae build strong, firm buttresses that can withstand large waves, but they cannot survive in water clouded by suspended sediment.

FIGURE 7.29

Reef builders today are confined to a narrow belt, mainly between 30°N and 30°S latitude. Within this belt, the greatest number of species occur where the minimum average water temperature is 27°C. (*Newell, 1972*)

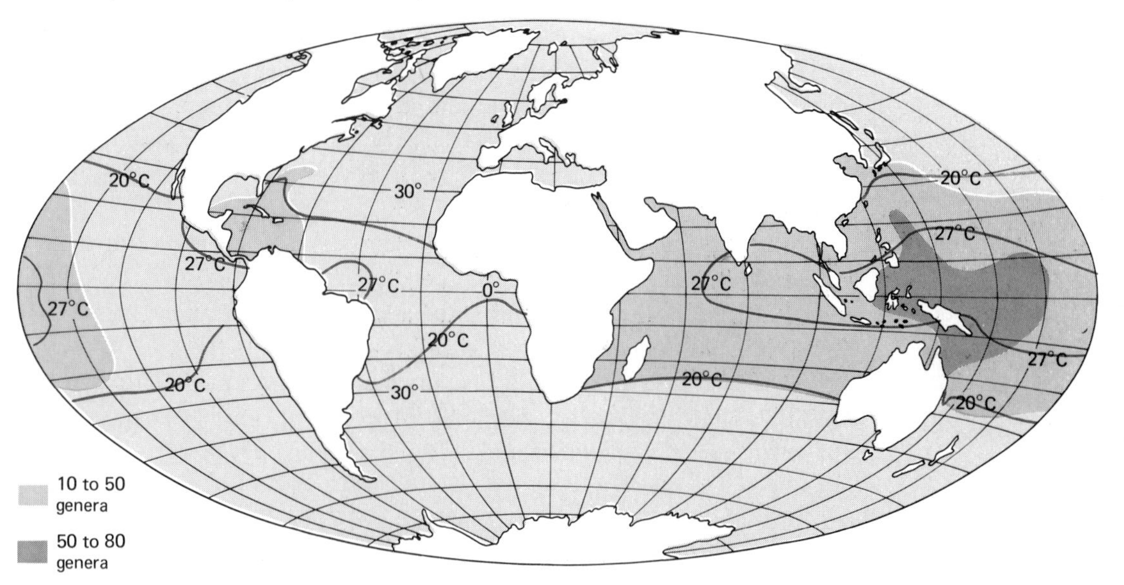

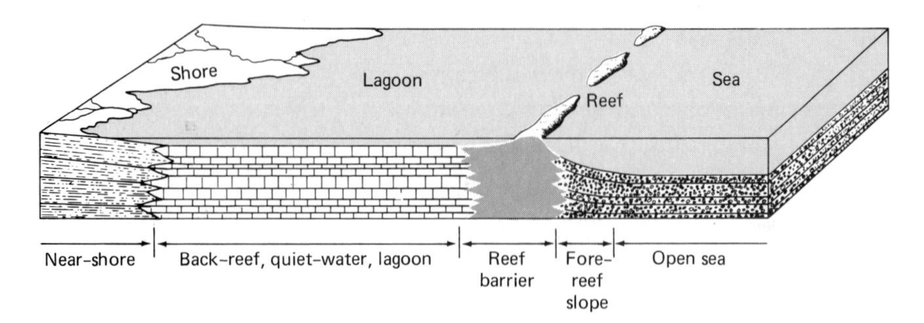

FIGURE 7.30

Ancient reefs often were composed of different kinds of organisms, but they performed much the same structural function as those in modern reefs. Although reefs occur in narrow belts, they influence sedimentation patterns of large areas. (*Laporte, 1968*)

The exact shape of a growing reef is molded by wind, waves, and bottom topography. Most are fairly narrow, not more than a few hundred meters across. They commonly form elongate barriers that grow in a high-energy wave zone between a shallow platform on one side and a comparatively deep marine basin on the other. An exception to the elongate pattern is the small, roughly circular **patch reef**, which may be several meters or more in diameter and which occurs commonly in quiet environments behind a larger barrier reef. Elongate, barrier-type reefs grow most actively on their ocean-facing fore reef (Figure 7.30). Here, where the reef maintains a steep seaward reef front, any skeletal material that becomes loosened moves downslope under the influence of gravity and forms a detrital apron of reef talus. The protected waters behind the reef on the landward or bankward side constitute the back-reef area. This relatively quiet, shallow back reef is called a **lagoon**; typically, it is very broad, compared to the narrow reef zone itself. The back-reef area is commonly a suitable habitat for the growth of sediment-trapping agents such as marine grasses, which cannot survive in the surf zone, and for the growth of fragile species of calcareous algae, which may generate significant quantities of calcareous sand and silt. Thus, reefs not only provide a habitat for sediment-producing organisms, but they also have a profound effect on the sedimentary processes of large regions by absorbing wave energy and by curtailing water circulation.

In the geologic past, when extensive back-reef areas developed under arid conditions on continental platforms, they became important sites of evaporite sedimentation. For example, during the Silurian in the area from Michigan to New York and during the Devonian in western Canada (Figure 7.31), large areas were fringed by reef tracts that acted as barriers to the free exchange of seawater. This caused back-reef evaporites to form as a result of evaporation in the arid climate. Reefs of many ages and in many regions mark similarly profound facies changes.

At times in the geologic past, invertebrate animals other than corals built reefs. Cambrian reefs were made up largely of archaeocyathids, small calcar-

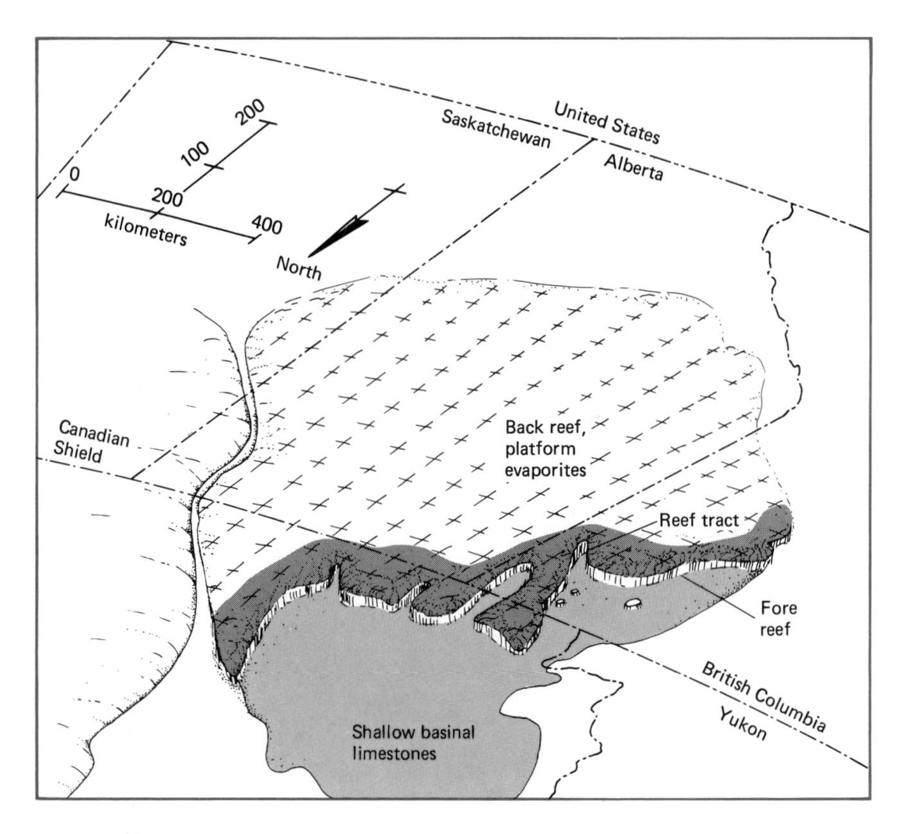

FIGURE 7.31

Looking southeastward across western Canada in a Devonian period. An extensive reef tract controlled regional facies in the Devonian epeiric seaway.

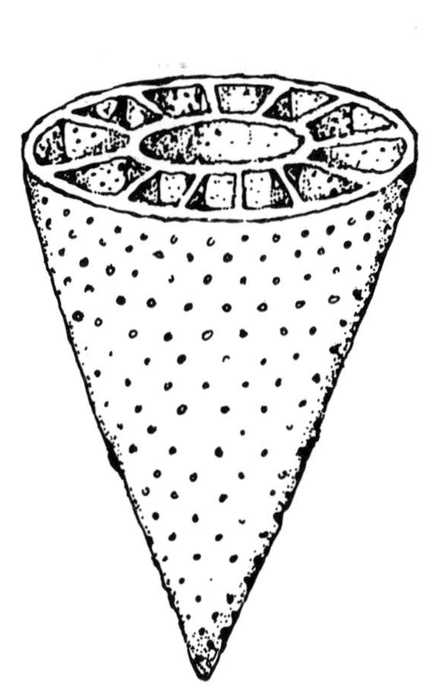

FIGURE 7.32

A Cambrian archaeocyathid, one of the first reef builders.

eous cups shaped like diminutive ice-cream cones whose exact relation to existing invertebrate groups is unknown (Figure 7.32). Following the Cambrian, the first corals appeared. Corals, together with calcareous algae, bryozoans, and stromatoporoids, became the chief reef builders during the Silurian and Devonian. In the late Paleozoic, additional organisms, including green algae, sponges, large foraminifera, brachiopods, and crinoids, became important reef constituents.

Paleozoic reefs are known on nearly all continents; their high-latitude positions today probably were produced by the shifting positions of the continents. Judging by ancient pole positions as inferred from studies of rock magnetism, most reefs of the geologic past appear to have been located at low latitudes—within 30 degrees of the calculated position of the equator—just as they are today. The Devonian reefs of western Canada shown in Figure 7.31, for example, are now at a latitude of about 55°N, but in the Devonian this area was within the tropics (see Figure 7.8).

The Expansion of Marine Life

You will recall from Chapter 4 that the dramatic expansion of animal life that marked the beginning of Cambrian time involved only sea-dwelling, shelled animals, such as trilobites and brachiopods. Such invertebrate animals characterize much of the early Paleozoic fossil record. The first vertebrates, primitive fishes, were not common until Late Silurian time, and large-scale colonization of the land surface by animals and plants first took place during the Devonian Period. For the remainder of the chapter, we shall summarize the life of early Paleozoic seas and explain some of the principles that enable us to understand ocean-dwelling life, especially invertebrate animal life, throughout the Phanerozoic record.

Marine Fossils

Animals and plants that have hard, mineralized shells or skeletons are most easily preserved in the fossil record and therefore make a disproportionate contribution to our knowledge of the history of life in the sea. Of the many chemical compounds that take part in the construction of these skeletons, only three are common: calcium carbonate ($CaCO_3$), silica (SiO_2), and calcium phosphate ($Ca_5(PO_4)OH$).

Calcium carbonate is by far the most abundant and widely distributed skeletal material, occurring in 14 phyla of plants and invertebrate animals. Most shell-bearing animals of the sea make their shells of this material, as do most mineral-depositing algae. Calcium carbonate is deposited by organisms in two crystallographic forms: calcite and aragonite. These two minerals have the same chemical composition, but they differ in the placement of the atoms in the crystal; they are difficult to tell apart without X-ray analysis.

Silica is a much less common skeletal material than calcium carbonate. Two groups of small, unicellular organisms—diatoms and radiolarians—build delicate lacy shells of silica. The only larger organisms with siliceous skeletons are sponges, many of which are supported by a framework of needlelike silica rods. Except for sponges, skeletons of silica are all very tiny; nevertheless, they are common as fossils because they are hard and relatively insoluble.

The third skeletal compound, calcium phosphate, occurs in only three groups, yet it is of extraordinary importance, for it makes up the bones and teeth of vertebrate animals. In addition, two invertebrate groups—brachiopods and arthropods—have some representatives with phosphatic skeletons.

The original silica, calcium carbonate, or calcium phosphate of a shell or skeleton may be preserved without alteration in a fossil. More commonly, secondary changes occur in these materials after they are buried. Porous structures, particularly bone, may fill with silica, calcite, or other minerals deposited from waters in the surrounding sediment; or the original mineral may *recrystallize* to a coarser texture or different structure; or the original mineral may be completely replaced by a different mineral.

The most important replacement process is **silicification,** in which the original shells of calcium carbonate are completely replaced by silica. Silicification is particularly common in fossiliferous limestone that has been exposed to the action of groundwater. In such cases, the silica selectively replaces the fossils, but not the surrounding limestone matrix. It is possible to dissolve the limestone in weak acids to leave only the enclosed siliceous fossils, which are not attacked by the acid (Figure 7.33).

Finally, the original shell or skeletal material may be dissolved and not replaced, leaving only an impression, or **mold,** of the fossil in the surrounding rock (Figure 7.34). These molds can often be as useful as the original shell, for rubber or plaster of paris casts of them can be made that faithfully duplicate the original fossils.

FIGURE 7.33

Silicified Ordovician brachiopod shells, prepared by immersing a block of fossiliferous limestone in weak hydrochloric acid. Part of the limestone block has dissolved away, leaving the shells exposed on the surface.

FIGURE 7.34 Natural molds of Devonian clam shells. The original shell material dissolved soon after burial, leaving only impressions in the surrounding sandstone.

A hard shell or skeleton is the first essential for fossilization, but an equally important consideration for insuring the ultimate preservation of an organism is its life habit. Bottom-dwelling animals and plants, because they commonly live in areas of active sedimentation, are much more likely to be preserved by quick burial than are swimming and floating forms. Large swimming vertebrates are rare as whole, articulated fossils, because their bones and teeth are usually scattered by predators and scavengers after death. On the other hand, the shells of small floating organisms are relatively common fossils, for they tend to sink and accumulate on the bottom in great numbers after death.

Early Paleozoic Animals

Returning now to the early record of marine fossils, we recall from Chapter 4 that the major phylum-level patterns of animals were apparently determined in one evolutionary radiation during late Precambrian and Cambrian time. Since then extensive evolutionary change has occurred *within* the phyla, but few new animal phyla have appeared.

Eight phyla of invertebrate animals have many representatives with mineralized skeletons that are readily preserved in ancient sediments. These phyla are: Sarcodina, Porifera, Coelenterata, Bryozoa, Brachiopoda, Mollusca, Arthropoda, and Echinodermata (Figure 7.35; see also Figure 4.12). Fortunately for our understanding of invertebrate evolution, these eight phyla not only have an excellent fossil record, but they also include the dominant invertebrates in the seas today. Many soft-bodied invertebrate phyla that play minor roles in modern oceans may also have been of relatively minor importance in ancient seas. Only one soft-bodied phylum, the Annelida (segmented worms), is abundant in modern marine invertebrate faunas. The seas have probably been dominated by the eight phyla with mineralized skeletons and the Annelida since early Paleozoic time.

Regrettably, there is no fossil record of the origin of these phyla, for they were already clearly separate and distinct when they first appeared as fossils.

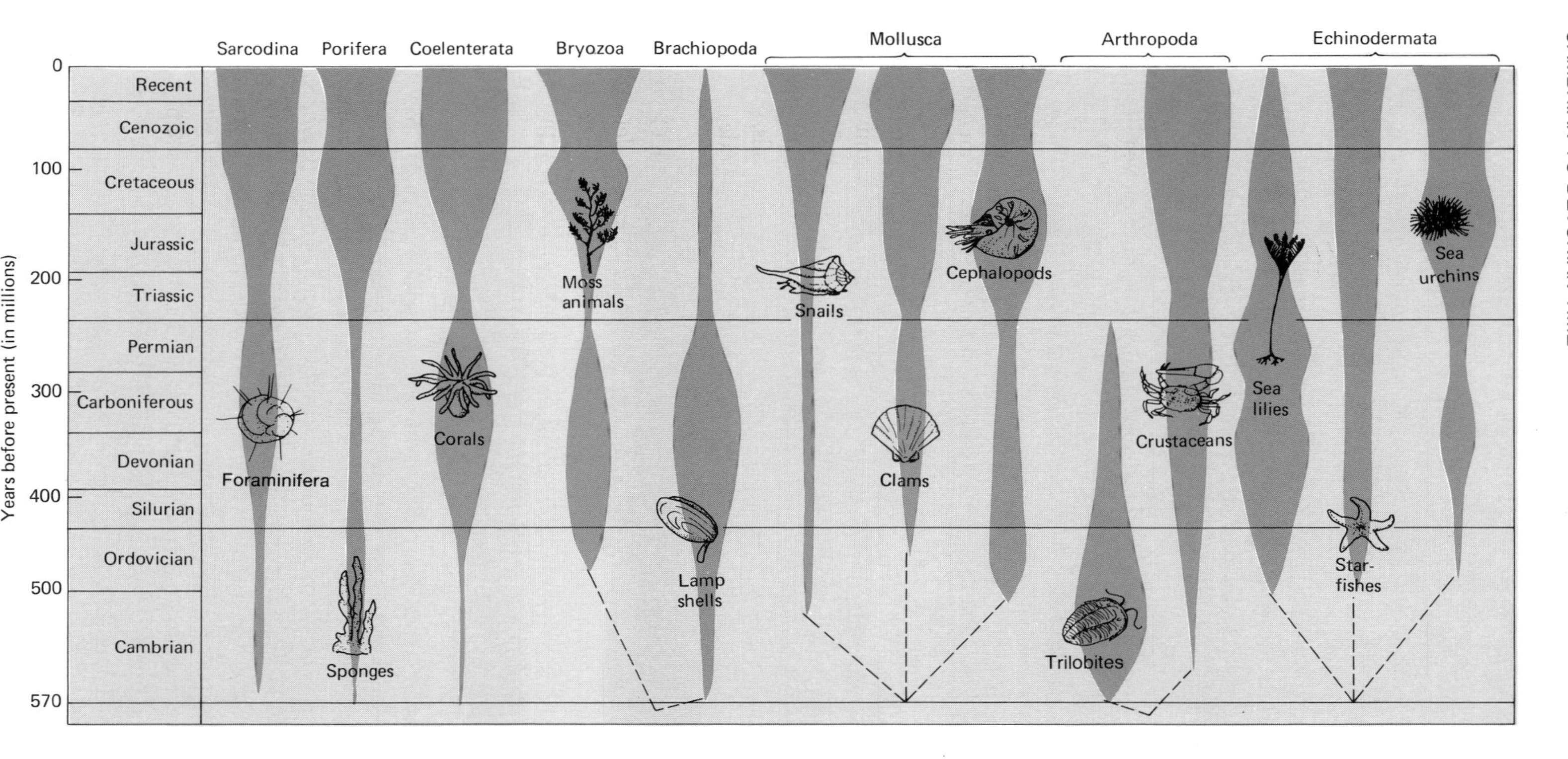

FIGURE 7.35

Geologic record of the principal groups of marine invertebrates. The shapes of the vertical areas indicate the approximate abundance of each group. The dashed lines show the most probable evolutionary relations of the groups.

The ancestors from which they arose either lacked preservable hard parts or were very rare, for they are not preserved in known Cambrian or late Precambrian rocks. Because no intermediate fossils between phyla are known, speculation about their evolutionary relations must depend on indirect evidence, particularly on comparative studies of the anatomy, embryology, and biochemistry of present-day representatives. Such studies have led to conflicting interpretations, but there is reasonable agreement that the Bryozoa and Brachiopoda are more closely related to each other than to the other phyla. Similarly, the Mollusca, Arthropoda, and Annelida appear to be related. The Sarcodina, Porifera, and Coelenterata do not appear to be closely related to any of the other phyla.

If we now look at the next level of organization, the classes within the phyla, we see in Early Cambrian time an example of an evolutionary phenomenon that recurs often in later periods. Even though the modern phyla are clearly differentiated when they first appear as fossils, the modern classes are not. Lower and Middle Cambrian rocks are filled with representatives of primitive classes, most of which are now extinct, having been replaced by the more successful classes that dominate our modern oceans.

Most modern shell-bearing classes first appear in Upper Cambrian and Lower Ordovician rocks. From Ordovician time to the present day, the seas have been dominated by such familiar groups as corals, bryozoans, snails, bivalves, cephalopods, and starfish. In contrast, Cambrian Seas contained great numbers of extinct arthropods called *trilobites; brachiopods,* which still exist but are rare in modern oceans, were also plentiful (Figure 7.35). It has been estimated that 60 percent of Cambrian fossils are trilobites and 30 percent are brachiopods; all other groups make up the remaining 10 percent. Among this 10 percent are several classes of molluscs that are quite unlike our familiar snails, bivalves, and cephalopods, also many unusual echinoderms, and several kinds of soft-bodied arthropods that more closely resemble trilobites than modern crustaceans.

Later History of Marine Invertebrates

Since Ordovician time, many extinctions and evolutionary radiations of invertebrate life have occurred, but most of these involved orders, families, and genera *within* the major classes. For example, note the history of trilobites as shown in Figure 7.35. After Cambrian time the group persisted in decreasing abundance until the close of the Paleozoic Era. Note also that two groups, the brachiopods and crinoids, were far more abundant in Paleozoic seas than they are today; the relative abundance of these groups is the principal difference between Paleozoic and later invertebrate faunas.

Although it is not apparent from a generalized chart such as Figure 7.35, most Paleozoic invertebrates belong to orders and families that either are now extinct or are of minor importance in modern oceans. When we get below the class level of organization, Paleozoic invertebrates are quite distinctive, even though the classes themselves have been the same ever since the Ordovician Period. Still more significantly, replacement of most Paleozoic orders and families by the surviving modern groups of invertebrates took place not gradually, but rather rapidly, in Late Permian and Early Triassic time.

The close of the Paleozoic Era was an extremely critical period for invertebrate evolution. It has been estimated that about *half* of the Permian families of invertebrates are not found in rocks of Triassic age or younger. Among the more important groups that failed to survive the Permian-Triassic crisis were the trilobites and several dominant orders of Paleozoic corals, bryozoans, brachiopods, and crinoids. After Permian time, both crinoids and brachiopods began their decline to their present minor role, although a few kinds continued to flourish during the Mesozoic Era. For some reason, molluscs appear to have been little affected. These massive extinctions were followed by rapid evolutionary radiations of many groups in the Triassic and Jurassic Periods. The principal Mesozoic expansions were among the sarcodines (foraminiferans), sponges, corals, bryozoans, bivalves, crustaceans, and echinoids. (A further discussion on the probable causes of the Permian-Triassic crisis will be found in Chapter 9.)

The Permian-Triassic crisis was the most important in invertebrate history, but a smaller wave of invertebrate extinctions occurred near the close of the Cretaceous Period. Particularly affected were the shelled cephalopods, which had an erratic evolutionary history throughout the Paleozoic and Mesozoic Eras, during which repeated extinctions of entire subgroups were followed by rapid evolutionary expansions of new groups. Shelled cephalopods are represented today by only a single genus, *Nautilus;* on the other hand, unshelled cephalopods (squids and octopuses) expanded to dominance during the Cenozoic Era.

Reconstructing Marine Life

Because so many kinds of early Paleozoic marine life have persisted to the present day, it is often possible to infer the life habits of the early animals by comparing them with their modern relatives. Then, as today, animals and plants living in the sea had three possible modes of life. They could float passively in the water, they could swim actively in the water, or they could live on or within the rocks and sediments of the sea floor. These habits define three great groups of marine organisms: **plankton** (floating organisms), **nekton** (swimming organisms), and **benthos** (bottom-living organisms). Figure 7.36 shows the present-day life habits of the eight principal phyla of shell-bearing marine animals. Note that most are bottom dwellers, the exceptions being swimming cephalopods and such floating forms as tiny foraminiferans and a few very small snails and crustaceans.

Bottom-living animals dominate most shell-bearing fossil assemblages. The mode of life of these animals can be further classified according to their habits

FIGURE 7.36

LIFE HABITS OF THE EIGHT PRINCIPAL PHYLA OF MARINE INVERTEBRATES	
Floating (Plankton)	Sarcodina (foraminiferans, radiolarians) Coelenterata (jellyfishes) Mollusca (pteropod snails) Arthropoda (copepod crustaceans)
Swimming (Nekton)	Mollusca (cephalopods)
Bottom-Living (Benthos)	Sarcodina (foraminiferans) Porifera (sponges) Coelenterata (corals) Bryozoa (bryozoans) Brachiopoda (brachiopods) Mollusca (snails, clams) Arthropoda (trilobites, crustaceans) Echinodermata (sea lillies, starfishes, sea urchins)

of feeding and mobility, as shown in Figure 7.37. Bottom-living marine animals eat in one of four different ways. Two of these are the familiar feeding types found in land animals: plant-eating **herbivores,** which in the sea eat mostly the larger bottom-living algae, and **carnivore-scavengers,** which eat other animals, living and dead. The remaining two kinds of feeding are of great importance in the sea, but are rare among land animals. **Filter-feeders,** a large and important group, strain tiny planktonic organisms and detrital organic matter from the bottom water. Most of these animals create currents of seawater by the beat of tiny hairlike structures; the currents are then passed over or through some sort of straining device that traps and accumulates the tiny food particles, often with the help of a sticky mucus, and transfers them to the mouth. The final kind of feeding adaptation is that of the **sediment-feeders,** or **deposit-feeders,** which take in organic-rich bottom sediments, use part of the organic matter as food, and discharge the undigested sediment particles as feces.

In addition to the four feeding types, bottom-living invertebrates can be divided into four groups on the basis of their living positions and relative mobility. There are **epifaunal animals,** which normally live on top of the rock or sediment, and **infaunal animals,** which have adaptations for living within the rock or sediment of the sea floor. Each of these two categories can be further subdivided into two groups. Epifaunal forms can either be attached to the surface of the rock or be free to move around on the surface. Infaunal forms can be adapted either for burrowing into soft sediment bottoms or boring into hard bottoms of rock or wood.

Figure 7.37 shows the principal fossil groups adapted for each of these habits. Note particularly that most groups that move around on the surface are either herbivores or carnivores-scavengers, whereas most that are attached to the surface are filter-feeders, except for the corals, many of which use stinging tentacles to capture large prey. Most burrowers in soft sediment are sediment-feeders, although some clams living within the sediment are filter-feeders with special tubes or siphons for bringing in the overlying water for filtration. Some snails and crustaceans are infaunal carnivores that creep through the sediment in search of buried filter-feeders and sediment-feeders. In addition, only bivalve molluscs and a few sponges have developed adaptations for boring into hard materials, such as rock or wood. Note also that many of the possible modes of life in the chart are not occupied by *any* animals. As you might expect, there are no epifaunal or boring sediment-feeders, no infaunal or

FIGURE 7.37

LIFE HABITS OF BOTTOM-LIVING MARINE INVERTEBRATES*

		Filter-feeders	**Sediment-feeders**	**Herbivores**	**Carnivores and scavengers**
Epifaunal (living on the surface of the sea floor)	Mobile	Crustaceans		Snails Sea Urchins	Snails Crustaceans Starfishes
	Attached	Sponges Bryozoans Brachiopods Clams Crustaceans Sea Lilies			Corals
Infaunal (living buried in the sea floor)	Burrowing in Soft Sediment	Clams	Clams Sea Urchins		Snails Crustaceans
	Boring in hard rock or wood	Sponges Clams			

*Areas in gray indicate major groups of each feeding type.

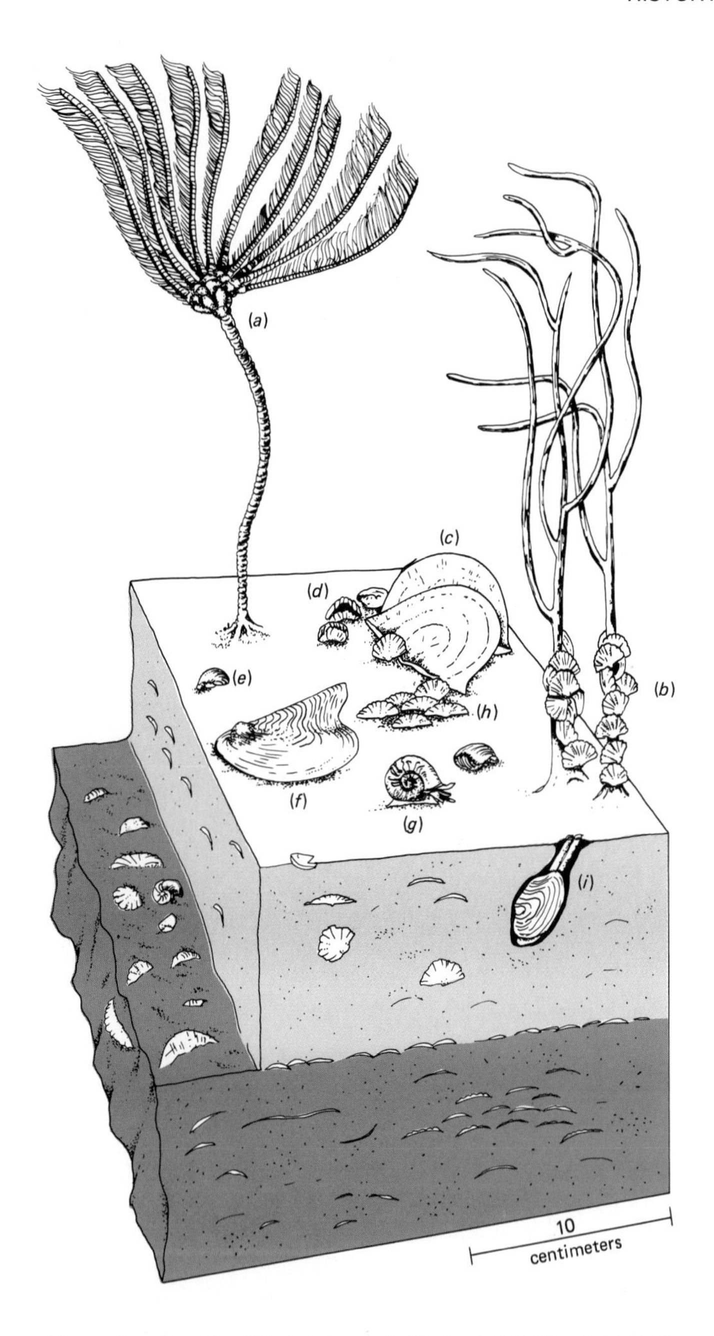

FIGURE 7.38 Reconstruction of a Silurian sandy-bottom community preserved today in western England. The community was dominated by attached epifaunal filter-feeders, including (*a*) sea lilies, (*b*, *c*, *d*, *h*) brachiopods, and (*e*, *f*) clams. Also present were (*g*) a herbivorous snail that probably fed on the seaweed, and (*i*) a sediment-feeding infaunal clam. The darker area below shows fossils as buried in the sediment where they are preserved today in sedimentary rock. (*McKerrow, 1978*)

attached herbivores, and no boring carnivores. Also note that some groups, such as snails, clams, and crustaceans, have adapted to several habits, while others, particularly attached forms, have only a single habit.

Using such data, paleontologists have reconstructed the habits of members of many ancient communities of bottom-dwelling invertebrates. Figure 7.38 shows one such reconstruction from lower Paleozoic rocks of western England. Similar studies have been made for dozens of ancient marine communities from each Phanerozoic period. From those has begun to emerge an understanding of the dynamic interactions of life on the ocean floor over the last 600 million years of earth history.

Chapter Summary

Early Paleozoic Land and Seas

The record of shallow seas: Seas have repeatedly advanced and withdrawn from the continents; many such cycles have occurred, but they were particularly common in Cambrian to Devonian time.

Sedimentary patterns: The sediments deposited on the continents by shallow seas reveal much about early geographic patterns.

Paleozoic plate tectonics: Several widely separated continents of Cambrian time came together in a sequence of complex motions to form a single large continent called Pangaea, by the close of Paleozoic time.

The Northern Appalachian mobile belt: These early plate motions are particularly evident today in the history of northeastern North America, a juncture site between two Paleozoic plates.

Deciphering Marine Sediments

Detrital settings: Sand and mud delivered to the ocean from erosion of the land accumulates both on the shallow submerged shelves of the continents and in the deep ocean beyond; normally the larger sand fragments accumulate near shore, with progressively finer particles occurring on deeper, offshore bottoms.

Carbonate settings: Sand and mud composed primarily of the calcium carbonate fragments of animal and plant skeletons dominate warm shelf environments where there is little influx of land-derived material.

The Expansion of Marine Life

Marine fossils: Many marine animals secrete shells of calcium carbonate, which may become fossilized by burial in sediments; often the original shells are altered by recrystallization, replacement, or dissolution.

Early Paleozoic animals: The eight principal phyla of present-day shell-bearing marine animals originated in early Paleozoic time and have been abundant since Ordovician time.

Later history of marine invertebrates: Most of the changes in marine life since the Ordovician have been in the extinction and radiation of families and orders within the eight principal phyla; extinctions are concentrated in short intervals, the most severe of which ended the Paleozoic Era.

Reconstructing marine life: Comparisons with modern relatives permit the reconstruction of the life habits of ancient marine communities.

Important Terms

Acadian Orogeny
Baltica
benthos
Caledonian Orogeny
carbonate mud
carbonate sand
carnivore-scavenger
coral
crinoid
detrital sediment
dolomitization
epifaunal animal
filter-feeder
Gondwana
grapestone
graptolite
herbivore
inclination (magnetic)
infaunal animal
interstitial waters
intertidal flat
Kazakhstania
lagoon
Laurentia
mold
nekton
oolite
Pangaea
patch reef
pelagic ooze
pellet mud
plankton
Protoatlantic Ocean
sediment-feeder (deposit-feeder)
silicification
skeletal sand
supratidal flat
Taconic Orogeny

Review Questions

1. Outline the changing positions of the continents during early Paleozoic time.
2. Summarize the history of the Northern Appalachian mobile belt. How does it relate to Paleozoic plate motions?
3. What are the principal types of land-derived marine sediments? How might they be recognized in ancient sedimentary rocks?
4. How do carbonate sediments form? Where are they forming in the oceans today?
5. How are shell-bearing animals preserved as fossils?
6. Summarize the principal kinds of early Paleozoic marine animals. What was the later history of these groups?

Additional Readings

Lane, N. G. *Life of the Past,* Charles E. Merrill, Columbus, Ohio, 1978. *Good discussions of the structure and habits of marine animals.*

Laporte, L. F. *Ancient Environments,* Prentice-Hall, Inc., Englewood Cliffs, New Jersey, 1979. *Good introduction stressing marine examples.*

McKerrow, W. S., (ed.) *The Ecology of Fossils,* The MIT Press, Cambridge, Massachusetts, 1978. *Detailed reconstructions of British marine communities of all geologic ages.*

Newell, N. D. "The Evolution of Reefs," *Scientific American,* vol. 226, no. 6, pp. 54–65, 1972. *Excellent review of reefs through geologic time.*

Raup, D. M., and S. M. Stanley *Principles of Paleontology,* W. H. Freeman & Company Publishers, San Francisco, 1978. *A survey of the principles used in interpreting fossil organisms; examples stress marine invertebrates.*

Selley, R. C. *Ancient Sedimentary Environments,* Cornell University Press, Ithaca, New York, 1978. *Intermediate-level text with good treatment of marine environments.*

Stearn, C. W., R. L. Carroll, and T. H. Clark *Geological Evolution of North America,* John Wiley & Sons, New York, 1979. *Good discussion of Paleozoic rocks and Appalachian history.*

Ziegler, A. M., and others "Paleozoic Paleogeography," *Annual Review of Earth and Planetary Sciences,* vol. 7, pp. 473–502, 1979. *Authoritative reconstructions of Paleozoic continental patterns.*

MISSISSIPPIAN TO TRIASSIC TIME

The Expanding Terrestrial World

8

About the middle of the Devonian Period, a vast change began in the earth's geographic patterns. Previously, shallow seas had dominated the continental platforms; the land areas exposed above the oceans, while occasionally extensive, were relatively local and discontinuous. With the Devonian Period began a trend toward increasing emergence of the continental plates above sea level and a consequent restriction of shallow oceans to the margins of the early continents. This trend culminated in Late Permian and Triassic time, when a large proportion of the continental surfaces stood above the surrounding oceans. During this interval, nonmarine environments, rocks, and fossils became increasingly abundant. The interpretation of the preserved record of this expanding terrestrial world will be the principal theme of this chapter.

Late Paleozoic Continents and Mountain Belts

In Chapter 7 we saw that, by the end of the Devonian Period, the continents of Laurentia (ancestral North America) and Baltica (ancestral Europe) had collided, beginning the assembly of Pangaea. Throughout the remainder of the Paleozoic, the separate continental blocks continued to converge on one another. Their paths are documented by a growing body of paleomagnetic data, and their actual collisions are recorded by several Paleozoic foldbelts. From the geosynclinal model outlined in Chapter 6, we infer that geosynclines are produced at continental margins and are then deformed, uplifted, and intruded as these margins become zones of subduction and, finally, zones of continental collision (see Figures 6.22 and 6.24).

Figure 8.1 shows that several Paleozoic mobile belts sliced across the northern hemisphere portion of the supercontinent (**Laurasia**). However, none cut across the southern hemisphere portion (Gondwana), although they did form a nearly continuous belt around it. Hence the subcontinent of Gondwana appears to have existed as an entity *throughout* the Paleozoic Era, whereas Laurasia is revealed as a mosaic of several continental blocks assembled *during* the Paleozoic.

The Paleozoic mobile belts that cut across Laurasia contain thick, highly deformed sequences of Paleozoic sedimentary rocks. They also contain ophiolites of Paleozoic age, which are inferred to be remnants of the ocean basins that once lay between the continental blocks. It appears that, prior to the construction of Pangaea, these continental blocks were individual entities and possibly widely separated. The available paleomagnetic evidence supports this interpretation. The foldbelts themselves were deformed, not only at the time the continents actually collided, but also for a considerable time prior to the collisions.

Because crustal deformation begins before the actual collision, the timing of continental collisions is sometimes difficult to establish. In cases where the suture zone between once-separated continental blocks is marked by linear belts of ophiolites, these can usually be dated. However, the ages of the ophiolites themselves indicate only when the particular piece of ocean floor that they represent formed, not when the collision occurred. The merging of two continental blocks results in the disappearance of the intervening ocean basin, the uplift of new source areas, and an influx of terrigenous sediments into adjacent sedimentary basins. These events have a marked effect on regional sedimentary patterns, which provide us with the best clues to the actual timing of the collision.

FIGURE 8.1

The Triassic supercontinent of Pangaea showing its two major subcontinents, Laurasia and Gondwana. The Paleozoic mobile belts show how Pangaea was assembled. (*Seyfert and Sirkin, 1973*)

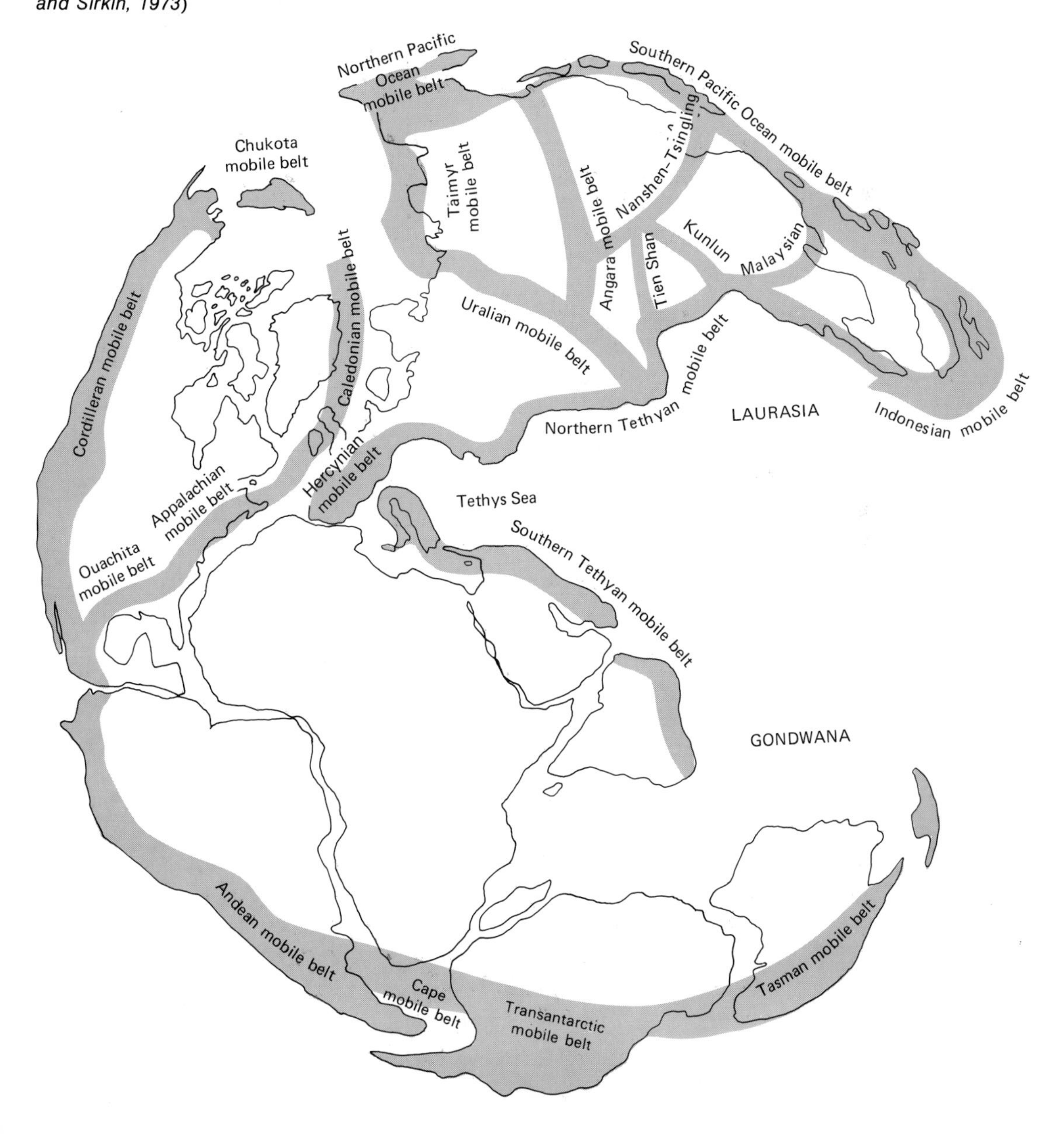

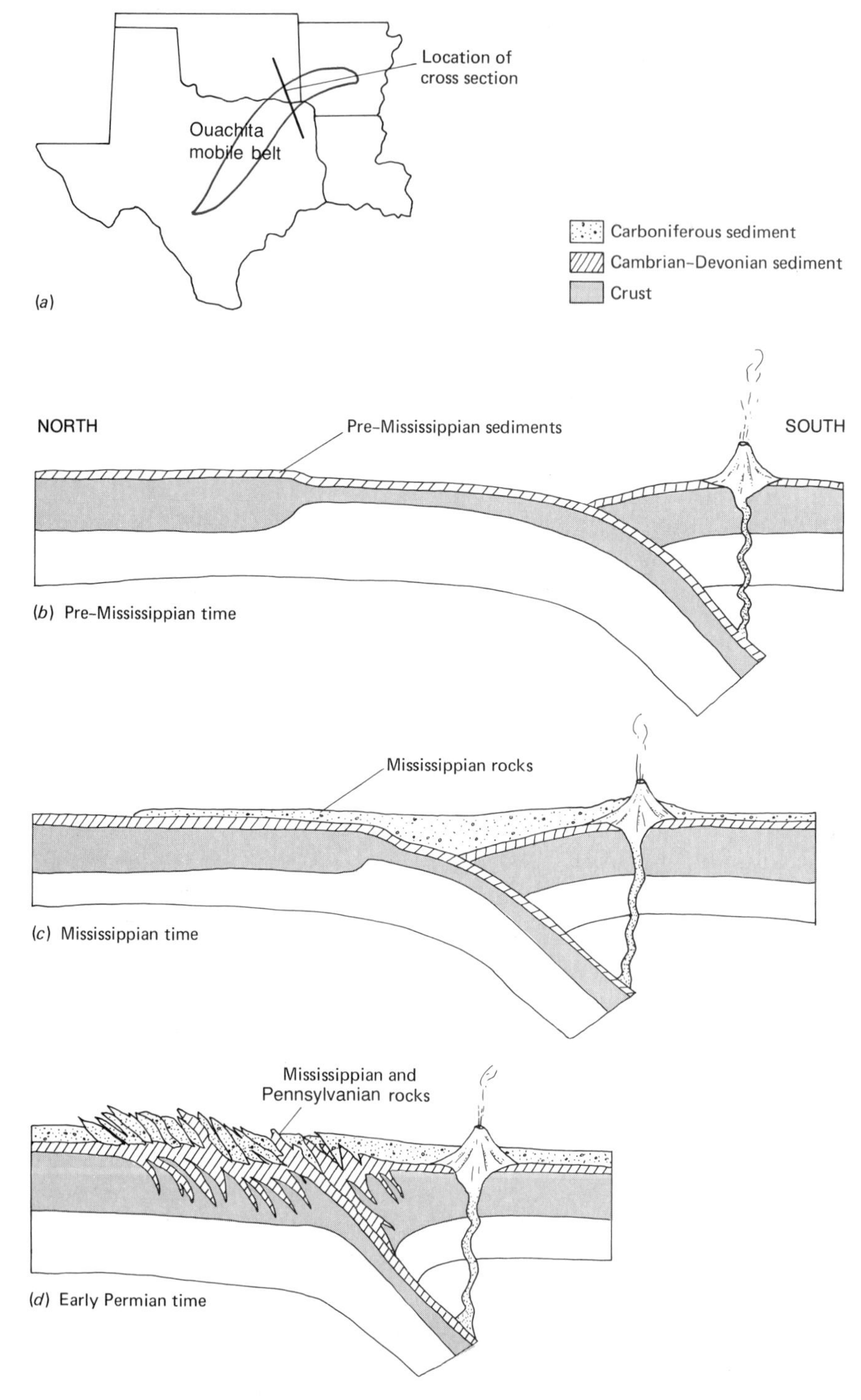

FIGURE 8.2 A continent-to-continent collision probably built the Ouachita Mountains in the late Paleozoic. (*Wickham, Roeder, and Briggs, 1976*)

The orogenies that accompanied the converging of the Paleozoic northern hemisphere continents apparently occurred in two major episodes. In Chapter 7 we reviewed the first episode, deformation of the northern Appalachian mobile belt, which culminated in the Devonian Period. Here we shall discuss the second orogenic episode, which culminated in the Permian. We begin with a regional example, the **Ouachita mobile belt,** at the southern margin of the ancestral North American continent, and conclude with a summary of the worldwide **Hercynian Orogeny,** of which the Ouachita mobile belt was a part.

The Ouachita Mobile Belt

The Ouachita Geosyncline in Oklahoma and Arkansas lay at the southern margin of the Paleozoic North American continent (Figure 8.2*a*). Like the northern Appalachian Geosyncline, the Ouachita Geosyncline represents a convergent margin; but unlike the northern Appalachian Geosyncline during its early history, the Ouachita Geosyncline was generally without volcanism and orogeny. This probably indicates that the southern margin of the Paleozoic North American continent was not bordered by a subduction zone. During the Ordovician, Silurian, and Devonian Periods, while considerable orogenic activity occurred along the eastern margin of the continent, the Ouachita Geosyncline quietly accumulated several thousand meters of limestone, sandstone, and shale (Figure 8.2*b*). Finally, in Mississippian time, the Ouachita Geosyncline began to receive sediments from a volcanic source area that lay to the south (Figure 8.2*c*). Increasing quantities of **volcanic tuffs** in the Mississippian shales suggest an encroaching volcanic arc and subduction zone that bordered an approaching continent.

In the Pennsylvanian Period, boulders derived from orogenic uplift of the outer part of the Ouachita Geosyncline were transported northward by submarine slumps and debris flows along the steep scarps produced by the northward thrust-faulting of huge crustal slabs that accompanied the encroaching subduction front (Figure 8.2*d*). In Late Pennsylvanian and Early Permian time, thrusting and folding continued as a result of regional compressive stresses, and the Ouachita Geosyncline became a fold-mountain chain, much like that produced in the northern Appalachian Geosyncline earlier in the Paleozoic.

The southern continental mass with which the Ouachita Geosyncline collided was apparently South America. Venezuela is the site of much late Paleozoic igneous and metamorphic activity, and this region apparently formed the southern border of the Ouachita mobile belt.

The Hercynian Orogeny

The Ouachita Geosyncline was connected on the southwest with the Marathon Geosyncline in west Texas and on the east with the Appalachian Geosyncline (Figure 8.3). The Appalachian Geosyncline, in turn, was continuous with the Hercynian Geosyncline of southern Europe (see Figure 8.1). This elongate, composite belt marked the southern boundary of the continent of **Laurussia,** which combined the ancestral European and North American continents

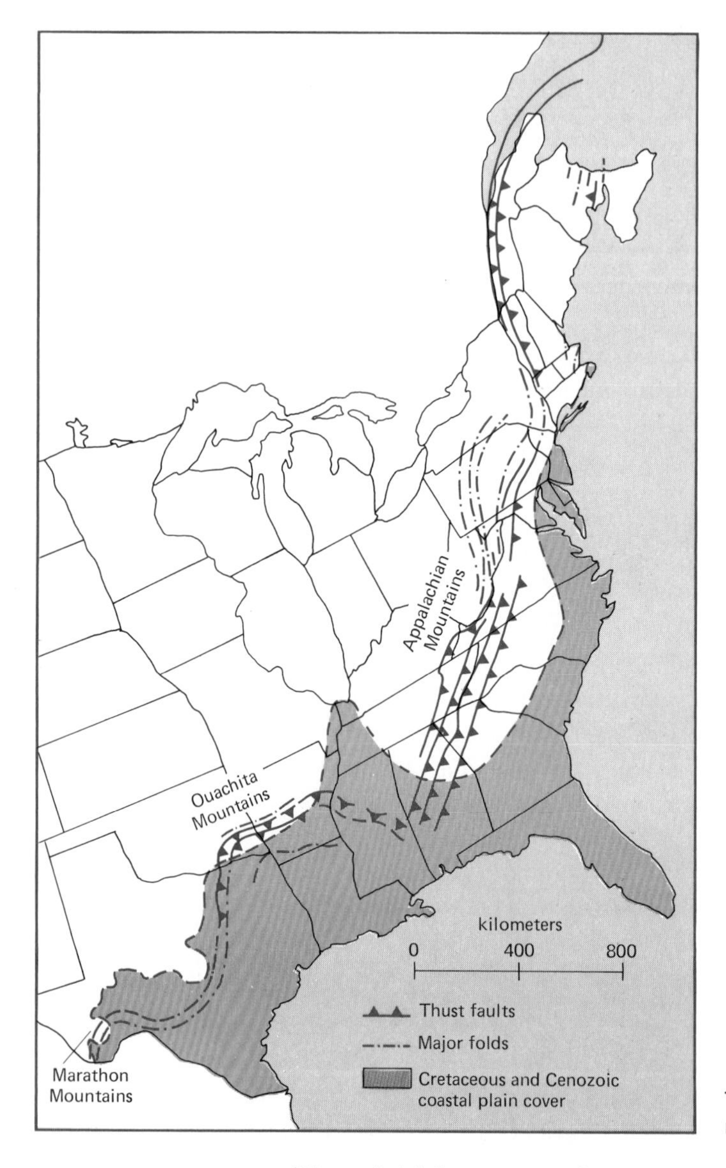

FIGURE 8.3

The Marathon, Ouachita, and Appalachian mobile belts in North America.

(Figure 8.4*a*). Its entire southern margin was strongly folded and thrust-faulted as a result of the collision, beginning in the Pennsylvanian, with the Gondwana continent (Figure 8.4*b*). Prior to this collision, the southeastern border of Laurussia extended across the Carolinas. Much of what is now the southeastern United States was then a part of Gondwana.

Beginning at about this time and continuing throughout the remainder of the Paleozoic, numerous geosynclinal belts in Asia were similarly deformed and uplifted into mountain ranges. This widespread late Paleozoic tectonic activity, which took place from about 300 million to 220 million years ago, is known throughout much of the world as the Hercynian Orogeny. In the eastern United States, it is called the **Appalachian Orogeny.**

FIGURE 8.4

Late Paleozoic paleogeography: (*a*) Middle Mississippian; (*b*) Middle Pennsylvanian; and (*c*) (top of p. 254) Late Permian. Mollweide projection showing entire earth surface. (*Scotese, and others, 1979*) The small crosses on the continents represent modern latitudes and longitudes.

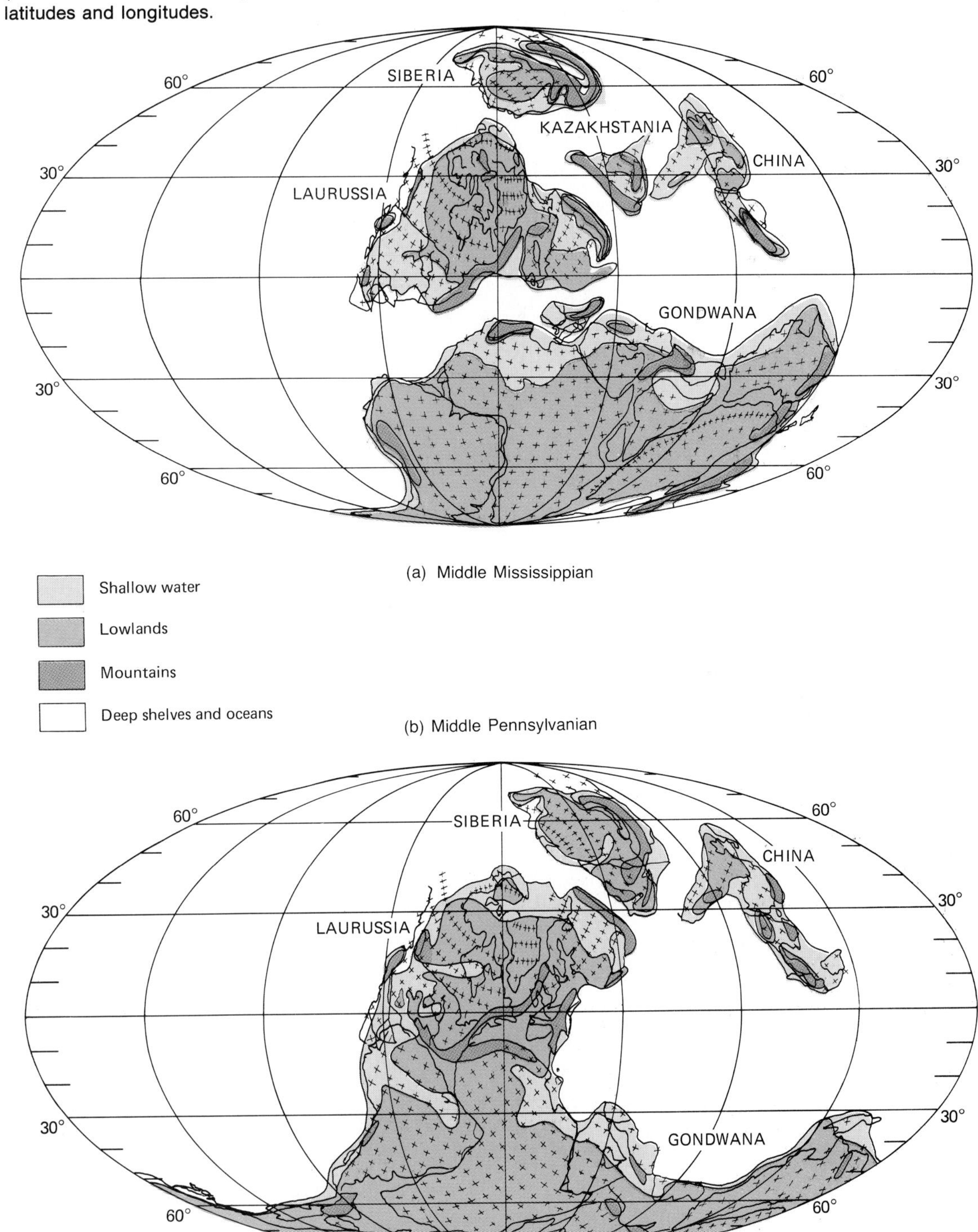

(a) Middle Mississippian

(b) Middle Pennsylvanian

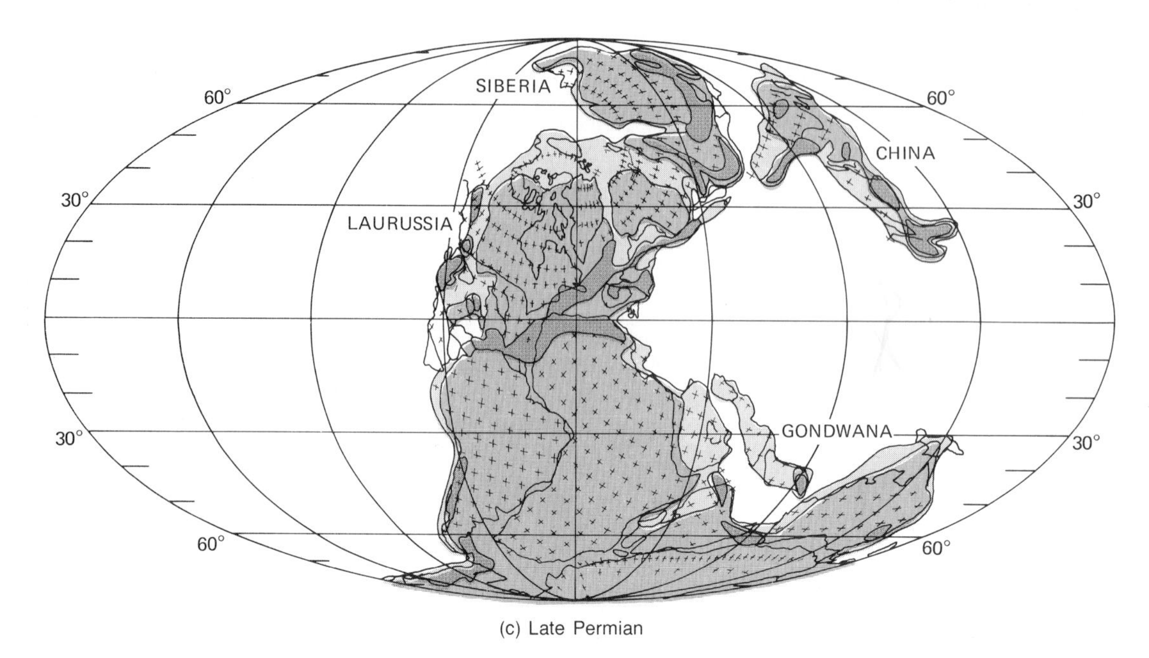

(c) Late Permian

FIGURE 8.4 (*cont.*)

The Hercynian Orogeny represents the joining of the northern hemisphere continents with Gondwana to form the supercontinent of Pangaea. Figure 8.4 shows how Pangaea gradually grew during this episode. By the end of the Devonian, as we have seen, ancestral North America (Laurentia in Figure 7.8) and ancestral northern Europe and Russia (Baltica in Figure 7.8) had merged, forming the Laurussia of Figure 8.4. By Mississippian time, at least four separate continental blocks, which now form Asia and a part of Europe, lay close together in the northern hemisphere. Beside Laurussia, these included Siberia, Kazakhstania, and China. In the Pennsylvanian Period Kazakhstania joined Siberia; Laurussia moved southward, and its southern margin collided with Gondwana along the Marathon-Ouachita-Appalachian-Hercynian geosynclinal trend (Figure 8.4*b*).

At about the same time, numerous uplifts were occurring throughout a vast region of the American West, from Texas to Idaho (Figure 8.5). These uplifts have long since been worn away by erosion, but some of them must have been imposing mountain ranges at the time, because they were important source areas that greatly influenced regional sedimentary facies. In Colorado, for example, the Uncompahgre and Ancestral Front Range uplifts shed voluminous quantities of arkosic (feldspar-rich) gravels and sands into adjacent basins (Figure 8.6). These Pennsylvanian mountains appear to have been simple fault-bounded blocks that may have been produced mainly by **tensional forces.** We do not know how they are related to the folded and thrust-faulted geosynclinal foldbelts to the south and east, which were produced mainly by **compressional forces.** The Pennsylvanian **block mountains** appear to have been similar to the block mountains of the present-day Basin and Range province, which formed in a setting of regional tensional forces.

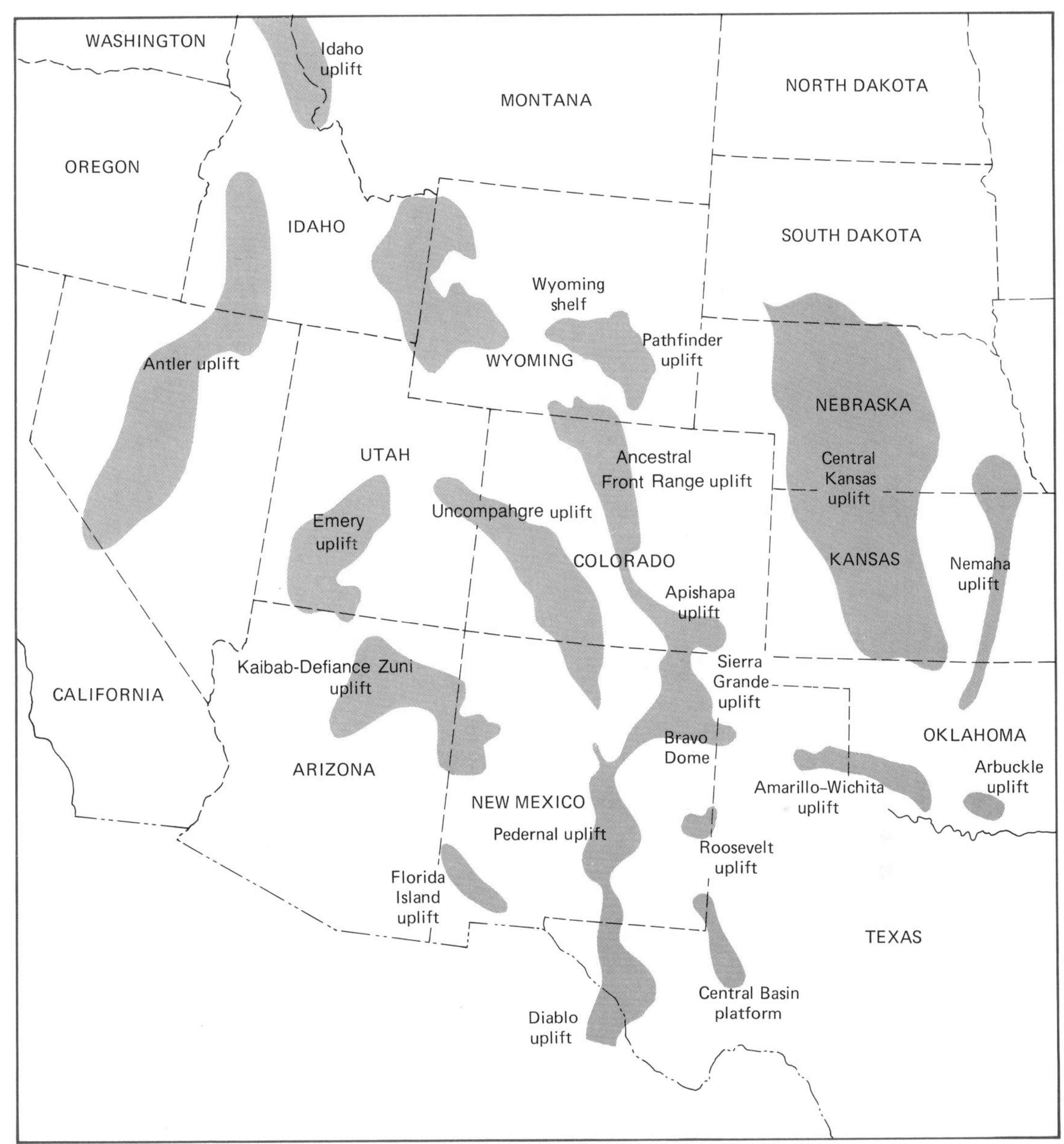

Pennsylvanian uplifts in the western United States. (*McKee and others, 1975*) FIGURE 8.5

In Early Permian time, Siberia with the attached Kazakhstan continental block moved southward to join the Laurussia-Gondwana mass, and the supercontinent of Pangaea was taking shape (Figure 8.4*c*). By this time China was moving westward toward the remainder of the world's continents to complete the enormous Pangaea land mass. In the Triassic Period, the mobile belt between China and the former Kazakhstan block at the northeast extremity of Pangaea underwent considerable deformation, which is interpreted to mean that the final throes of the collision were still taking place at that time.

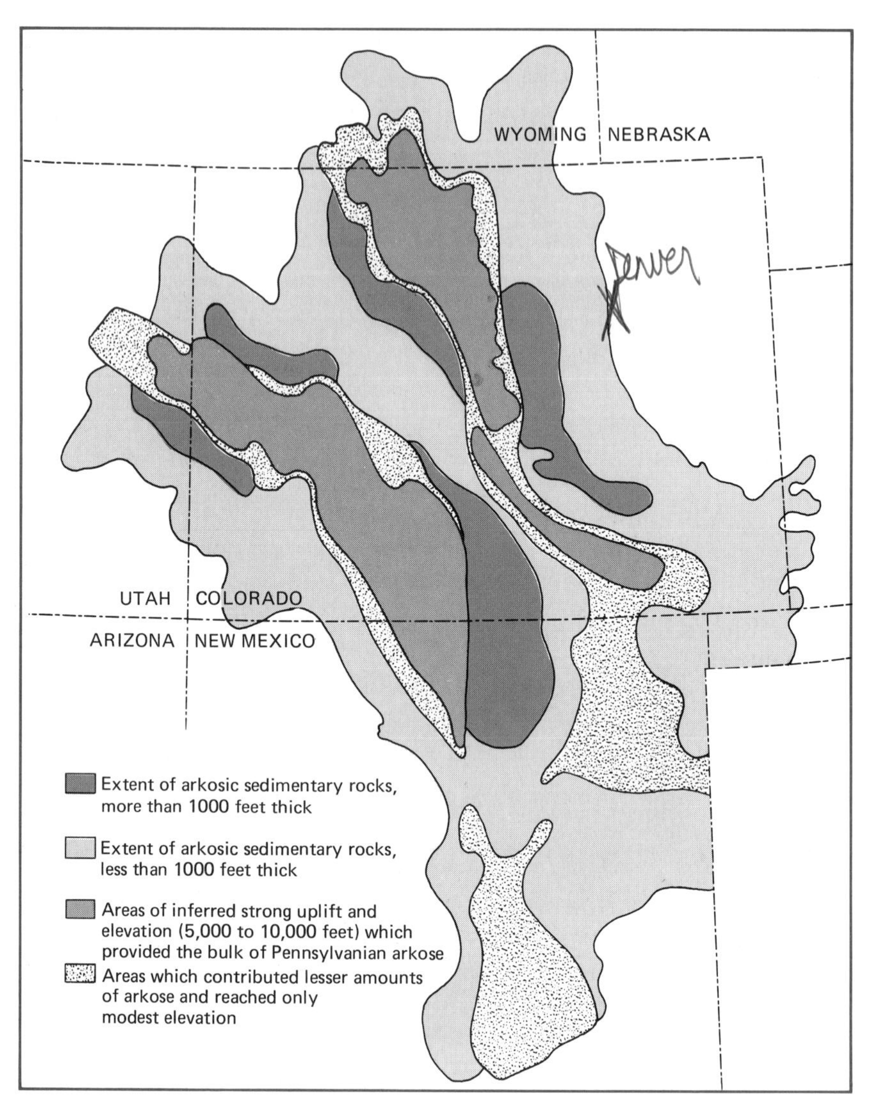

FIGURE 8.6

The Pennsylvanian Uncompahgre and Ancestral Front Range uplifts in Colorado shed huge quantities of arkosic gravels and sands into adjacent sedimentary basins. (*Mallory, 1972*)

Terrestrial Sediments and Environments

In the Mississippian Period, shallow seas covered large areas of the continents. Like the early Paleozoic seas, they left a widespread record of diverse marine deposits. During the Pennsylvanian and Permian, however, the continents underwent sporadic but progressive emergence, and gradually marginal-marine and continental environments became dominant. By the end of the Paleozoic, normal marine environments had virtually disappeared from the continental platforms.

The late Paleozoic worldwide emergence of land masses was probably related to the convergence and amalgamation of all the earth's continents. Perhaps the assembly of Pangaea was accompanied by an overall thickening of sialic crust, with the result that the effects of **isostasy** caused the continents to stand higher above the ocean basins than they had previously. Or perhaps the frequency of continental collisions during the late Paleozoic somehow slowed down the plate motions and the rate at which new ocean crust was created. Reduced rates of sea-floor spreading might, in turn, have decreased the total volume of oceanic ridges and thus increased the capacity of the ocean basins. As a result, sea level would have been lower, and the continents would have stood relatively higher.

Whatever the cause, the late Paleozoic record has an abundance of thick, widespread, continental and marginal-marine strata, including deposits of streams and flood plains, coal swamps, deserts, and even glaciers. We shall now look at some examples of these deposits.

Fluvial Environments

Fluvial deposits are produced by sediment-laden streams. The water comes from rain or snow, and the sediment comes from the weathering of surficial materials in the stream's drainage basin. There are two major kinds of streams: *meandering streams,* which occupy a single channel, and *braided streams,* which follow a multitude of channels (Figure 8.7). Braided streams have an overabundance of sediment, much of which is coarse sand or gravel. This is transported dominantly as **bed load**. Meandering streams carry a larger proportion of silt and clay, which is transported mainly as **suspended load.** Braided streams have higher gradients and larger, more rapid fluctuations than do meandering streams in the amounts of water they carry.

Meandering streams produce a distinctive stratigraphic sequence, commonly referred to as a *fining-upward sequence.* This sequence results directly from the meandering of the channel. It includes the vertical sequence of sediments and sedimentary structures that can be observed from the bottom of the channel, where energy is high, to the point bar at the inside of the meander curve, where energy is low (Figure 8.8).

Braided streams may well have been relatively more important in the geologic past, before the land was widely covered by vegetation. In arid regions today, where precipitation is *less* than about 30 centimeters per year, the amount of sediment carried out of an area by its streams is proportional to the amount of precipitation that falls. In such regions, the streams tend to be braided. The runoff of sudden rains sporadically transports large quantities of sediment from the drainage basin; the sediment consists mostly of bed load.

In regions today that receive *more* than about 30 centimeters of rainfall, the amount of sediment that streams carry out of an area is strongly influenced by vegetation. Not only does vegetative cover markedly inhibit the quantity of runoff, but it also greatly slows the rate of runoff, thereby smoothing out the

(a)

FIGURE 8.7

The two basic types of streams: (*a*) The Yazoo River in Mississippi, a large meandering river (road shows scale). (*Courtesy of Frank Beck*); (*b*) the Knik River in Alaska, a braided river about 3 kilometers wide. (*Courtesy of W. C. Bradley*)

(b)

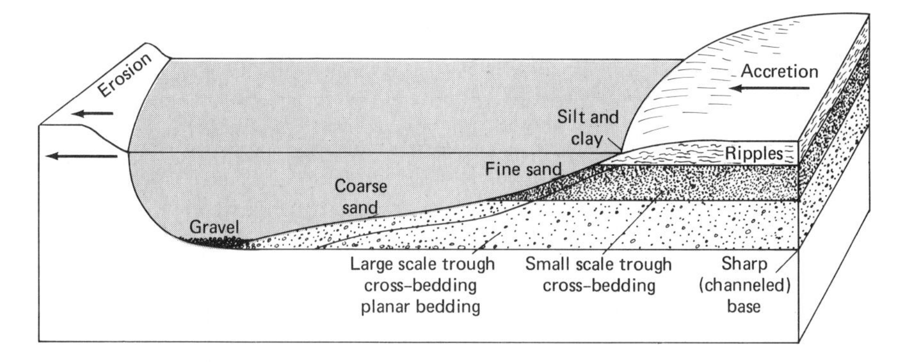

The basic fining-upward unit of meandering stream deposition. FIGURE 8.8

effects of sporadic precipitation. Vegetation also very effectively protects soil from erosion. Above 30 centimeters of precipitation per year, a further increase in precipitation actually causes the amount of sediment carried away by streams in a given period of time to *decline* because of the correspondingly greater vegetative cover (Figure 8.9, Curve C). In humid regions, the retention of weathered surface material as soil permits long and thorough weathering, so that when the detritus is finally carried away by streams, it has weathered to clay- and silt-sized fragments. These weathered fragments are generally carried as suspended load by meandering streams.

Prior to the Silurian Period, there was no land vegetation. All the lands on earth, regardless of the quantity of precipitation they received, must have appeared the same as today's deserts. Even in the late Paleozoic, the primitive types of upland vegetation then extant were probably ineffective in stabilizing hillslopes. With any increase in rainfall, there must have been a progressive increase in sediment carried by the streams (Figure 8.9). The nearly bare Paleozoic hillslopes probably lost material as rapidly as it weathered to a size that could be transported. Fine sediment was practically nonexistent, and streams probably carried a comparatively large proportion of their sediment as bed load. In regions of high rainfall, braided streams in unconfined channels temporarily occupied the entire floor of alluvial valleys. Periodic floods would have produced widespread, sheetlike deposits of conglomerates and sandstones that contained far less fine-grained material than would flood deposits today. Hence, interpretations of such deposits must be made cautiously.

After land plants appeared in the Silurian, first in the moist lowlands and later in upland areas, sediment yield for a given amount of rainfall began to decline (Figure 8.9, Curve B). The trend probably accelerated after flowering plants first appeared in the Cretaceous Period. After the appearance of grasses in the mid-Cenozoic (Figure 8.9, Curve C), sediment yield for a given amount of rainfall achieved modern values; stream deposits of late Cenozoic age are probably directly comparable with stream deposits forming today.

FIGURE 8.9

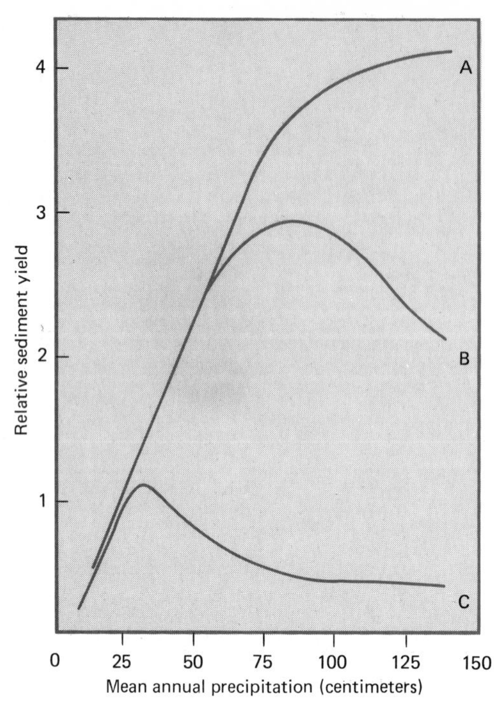

Relationship between quantity of precipitation in an area and its relative sediment yield. Curve A represents the time prior to the appearance of any land vegetation, Curve B represents late Paleozoic time, and Curve C, modern time. (*Schumm, 1968*)

Coal

Conditions were ideal for the formation of coal during the Pennsylvanian Period. The huge coal fields of the midwestern and eastern United States (Figure 8.10), northern Europe, and Asia are all of this age. In recognition of this, the term *Carboniferous System* was coined early and is applied in Europe and much of the world for strata that, in the United States, are placed in the Pennsylvanian System and the underlying Mississippian System.

FIGURE 8.10

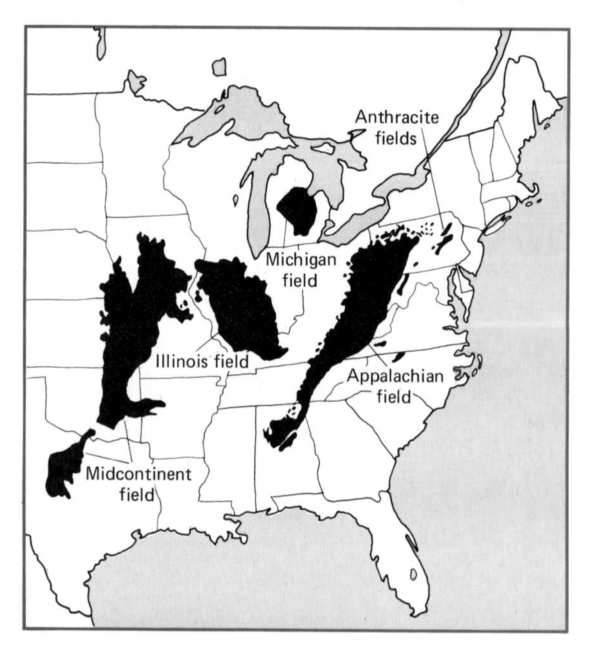

Coal fields of Pennsylvanian age in the central and eastern United States.

Reconstruction of a Pennsylvanian coal swamp. (*Part of a mural by Rudolph Zallinger in the Yale Peabody Museum*)

FIGURE 8.11

Coal comes from peat. The Pennsylvanian peats accumulated in extensive swamps that covered vast coastal regions. These swamps contained many so-called scale trees (**lycopsids**) and jointed-stemmed plants (**sphenopsids**), (Figure 8.11). Ferns made up much of the undergrowth. Why did these plants produce so much coal during Pennsylvanian time? First of all, many of the large swamp-loving land plants had evolved only a short time before and simply were not available previously to make coal. Even more important, however, were the special climatic and tectonic conditions that combined to produce widespread moist lowlands.

Swamps require high rainfall, at least seasonally, and flat regions where drainage is poor. In the Pennsylvanian, large areas received abundant rainfall because epeiric seas were still widespread enough to act as important sources of moisture for adjacent lands. Huge areas of flat, poorly drained lowlands formed in the interior portions of the Pennsylvanian continents, chiefly on large delta systems. The large volume of sediments that built the deltas came from the erosion of the extensive uplands that were produced by Pennsylvanian mountain-building.

The abundance of Pennsylvanian coal is due not only to favorable climate and to the huge expanse of swamp environments, but also to repeated fluctuations of the coastline, which caused coal-forming conditions to recur again and again. Throughout the world, wherever Pennsylvanian coals are found, numerous important coal beds and dozens of lesser ones commonly occur in succession. This striking *cyclicity* of the coal-bearing sequences is indicated by the term **cyclothem** which is applied to each of the repeated rock sequences that make up the cycles. The cause of the cycles has been a topic of much speculation.

Cyclothems

In the Illinois coal basin cyclothems are very well developed. During the Pennsylvanian, this area was a vast, low coastal plain that was bordered on the east by the newly uplifted Appalachian Mountains and on the west by a large epeiric sea that occupied the interior of the continent. Periodically the coastal plain was occupied by streams flowing westward from the mountains and periodically it was invaded by the sea. The streams brought large quantities of sediments to the edge of the sea and deposited them as vast deltas.

FIGURE 8.12

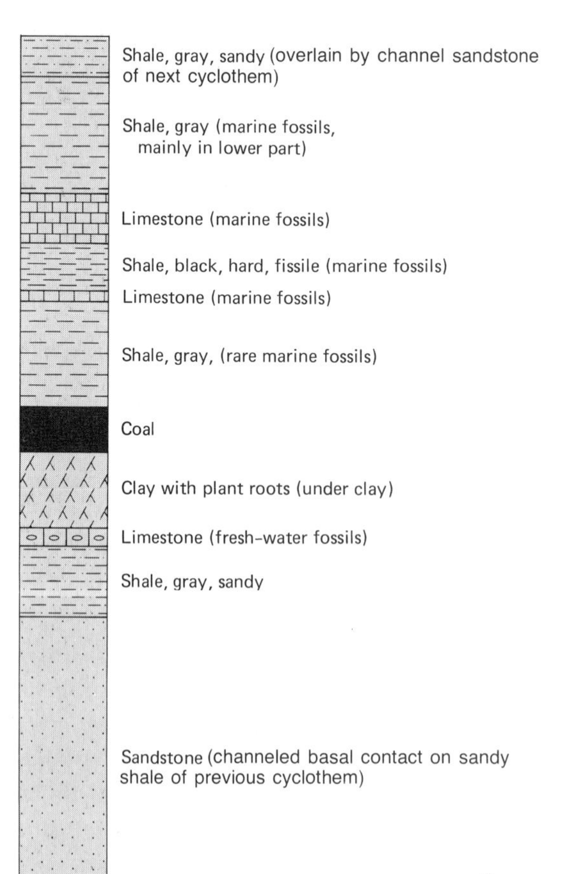

Sequence in an ideal Pennsylvanian cyclothem in the Illinois region.

A typical cyclothem in this region (Figure 8.12) is between 10 and 20 meters thick. It begins with cross-bedded sandstone that rests on a channeled surface on the underlying beds. These basal sandstones are interpreted as river channel deposits. The shale and limestone strata immediately above are interpreted as deposits of flood plains and lakes. The coal in the middle of the cyclothem apparently formed in vast coastal swamps full of lush vegetation that typically covered tens of thousands of square kilometers. Plant material that accumulated on the swamp floor was protected from oxidation by stagnant waters and rapid burial. With time, large quantities of the organic material were compacted by the weight of overlying sediments to form peat and, later, coal. The upper part of the cyclothem generally contains fossiliferous marine shales and limestones. They indicate that the sea transgressed from the west, changing the coastal swamp into a shallow-marine shelf. The limestone in the upper portion of the cyclothem apparently represents the peak of marine transgression.

During Pennsylvanian time, the Illinois region must have been exceedingly flat and very near sea level, so that a small change in sea level would produce a very large change in the location of the coastline. The marine beds at the top of each cycle are overlain by fluvial sandstones that mark the beginning of the next overlying cyclothem. The approximately 50 Pennsylvanian cyclothems in the Illinois basin were produced in about 50 million years. This constitutes a record of unusually frequent transgressive-regressive cycles. Coal was produced in most of the cycles, and this is the key to its abundance in the Pennsylvanian.

Some of the Pennsylvanian transgressive-regressive cycles may not have resulted from actual changes in sea level. Instead, they may have been generated locally within the depositional basin—regressing when deltas grew temporally seaward, and transgressing when river distributary systems shifted from one coastal area to another. This point of view is supported by two observations: (1) The ideal cyclothem shown in Figure 8.12 is rarely achieved; most lack one or more of the units. (2) Different areas contain different numbers of cyclothems; for example, while some 50 occur in the Illinois basin, about 90 have been recognized in West Virginia. Clearly some of the cyclothems are not basin-wide. Other cyclothems, however, extend over vast distances throughout much of the central and eastern United States, and these were probably caused by numerous eustatic sea-level changes.

We do not know what caused these unusually rapid sea-level changes. Another relatively unusual event, continental glaciation, was occurring at the time in the southern hemisphere, and the two may have been related. During the comparatively recent Pleistocene ice ages, continental ice sheets did not simply form only once, occupy the continents for the duration of the Pleistocene, and then disappear. Instead, the ice sheets expanded and contracted many times, producing sea-level fluctuations of about 200 meters. A similar expansion and contraction of the late Paleozoic ice sheets may have produced the eustatic sea-level changes necessary for the cyclothems. On a coastal plain of low relief, a 200-meter sea-level change would cause the sea to transgress and regress over a distance of several hundred kilometers.

Deserts and Evaporites

Deserts are, by definition, dry areas where evaporation greatly exceeds precipitation. They are commonly, but not necessarily, hot lands. Sand dune deposits and redbeds are considered to be **climate-sensitive rock** types suggestive of ancient desert conditions. Although both can form locally in regions of considerable rainfall, their abundance and wide distribution in the Permian and Triassic probably indicate widespread dry climates. Evaporite deposits—chiefly calcium sulphate (the minerals gypsum and anhydrite) and sodium chloride (the mineral halite)—which are associated with many of the sand dune and redbed deposits, confirm that the climates were warm and arid. This supports the inference that the surrounding lands were, indeed, deserts.

Evaporites are unique among sedimentary deposits in that they require no physical transport of sedimentary particles from afar, nor do they need to be produced by living organisms at the site of deposition. Instead, they precipitate directly and sometimes rapidly from seawater when it evaporates nearly to dryness. Gypsum begins to precipitate after 80 percent of the seawater has evaporated; halite begins to precipitate after 90 percent is evaporated; bittern salts (salts of magnesium and potassium) precipitate only after about 99 percent of the seawater has evaporated. Marine evaporite deposits require an area with an arid climate and impeded circulation with the open sea.

Evaporite conditions occurred repeatedly in Paleozoic epeiric seas. For example, in the Pennsylvanian Paradox Basin of southern Utah, even as coal swamps flourished under humid conditions in the midcontinent region, about 1,500 meters of halite and bittern salts accumulated, reflecting the extreme aridity of the region (Figure 8.13). These salts occur in thick beds separated by

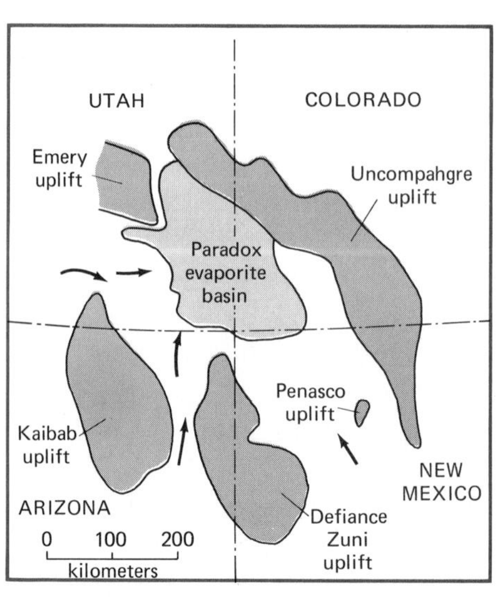

FIGURE 8.13 Paradox evaporite basin during Middle Pennsylvanian time showing bordering land areas and probable inlets for normal marine water. (*Hite and Cater, 1972*)

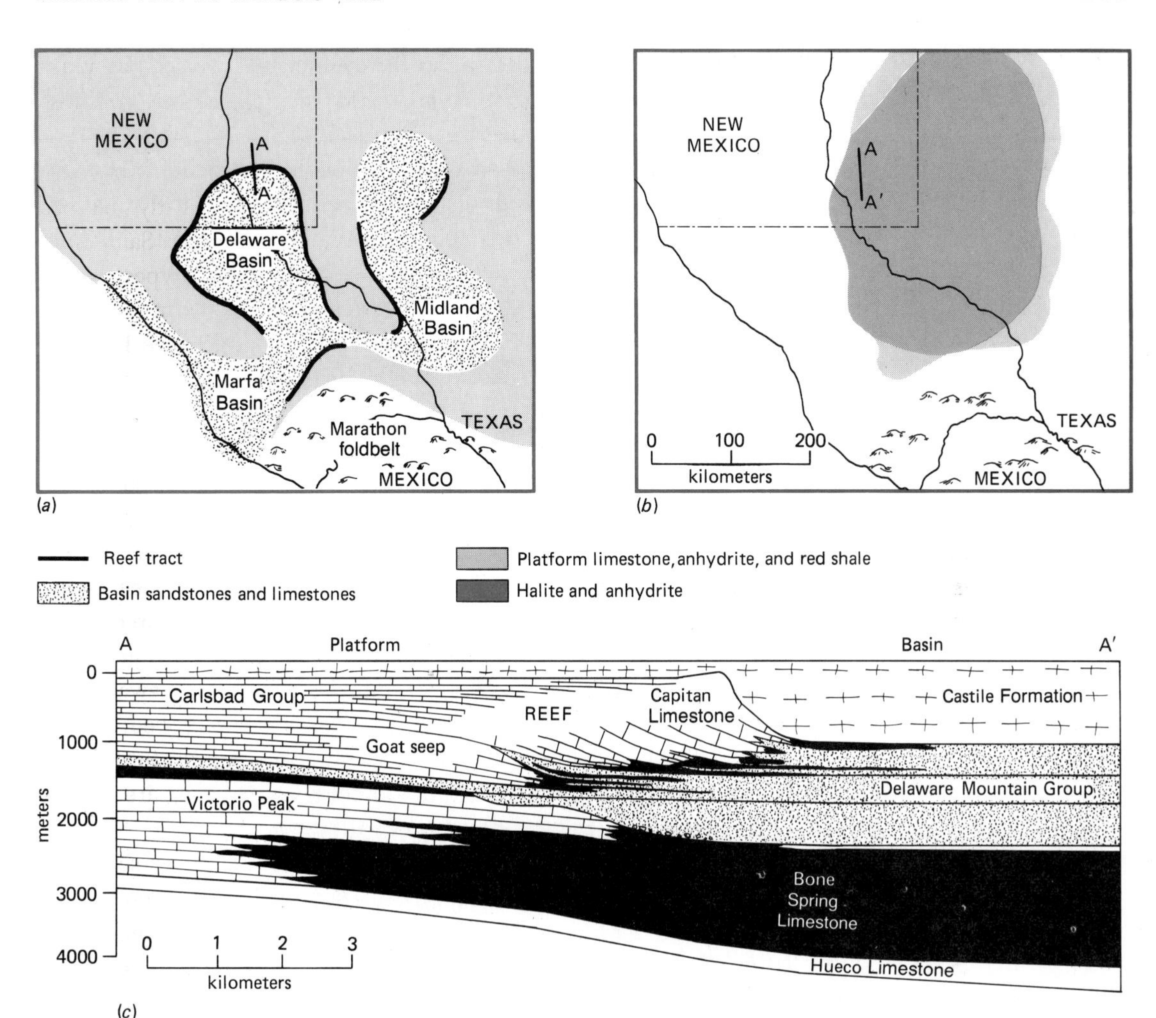

FIGURE 8.14

Permian of west Texas. (*a*) Middle part of Permian: well-demarked basins largely flanked by barrier reefs and surrounded by shallow-marine platforms. (*b*) Latest Permian: salt and anhydrite fill basins and blanket part of surrounding region. (*c*) Cross section A - A′ through basin margin at end of Permian time. (*King, 1948*)

beds of black shale. The regular alternation of shale and evaporites may be the result of the same cyclic sea-level changes that, at that same time, were causing the coal cyclothems in the midcontinent area.

During the ensuing Permian Period, humid, coal-producing climates gradually disappeared altogether from low latitudes. As aridity became more and more widespread, salt and gypsum deposits accumulated in shallow restricted seas throughout the world. Some of the most famous deposits occur in deep basins in west Texas near the western margin of the Pangaean supercontinent. These basins formed early in the Permian. Throughout most of the period, they were sites of marine-sand and carbonate deposition. Extensive barrier reefs (the Goat Seep and Capitan Limestones of Figure 8.14*c*) grew around their margins. Then in latest Permian time, circulation was curtailed between the basins and the open ocean. The basins, which by that time were 600 meters deep, filled up in less than a half-million years with anhydrite and halite. The anhydrite and halite beds, which constitute the Castile Formation, covered the former reef tracts and much of the adjacent platform (Figure 8.14*c*). Geologists estimate that a net evaporation of about 3 meters of water per year was required to produce these evaporite deposits. By comparison, the most arid area in North America today—Death Valley, California—has only a slightly greater net evaporation of about 3.5 meters per year.

If the 600 meters of west Texas evaporites had formed by simple evaporation of a standing body of seawater of normal salinity (35 parts per thousand), the water would have had to be about 800 kilometers deep! This, of course, is unreasonable, and we conclude that all the seawater did not simply enter the basin once and then evaporate. Instead, as it evaporated, new water continually entered the basin through a limited opening, possibly a shallow sill. This provided an effective barrier to circulation, so that little water was able to leave the basin except by evaporation (Figure 8.15).

For evaporite deposits that did not accumulate in clear-cut basins, the cause of impeded circulation is not so apparent. In Europe, the Permian Zechstein Formation, a widespread and long-known source of salt, accumulated in a sea so vast and so shallow that there may not have been a need for a physical barrier to prevent the exchange of water with open-ocean regions. The arid climate of the time removed the water from this "evaporite pan" as fast as it flowed in.

FIGURE 8.15

Silled basin model for evaporite sedimentation by direct precipitation from seawater. Vertical scale much exaggerated.

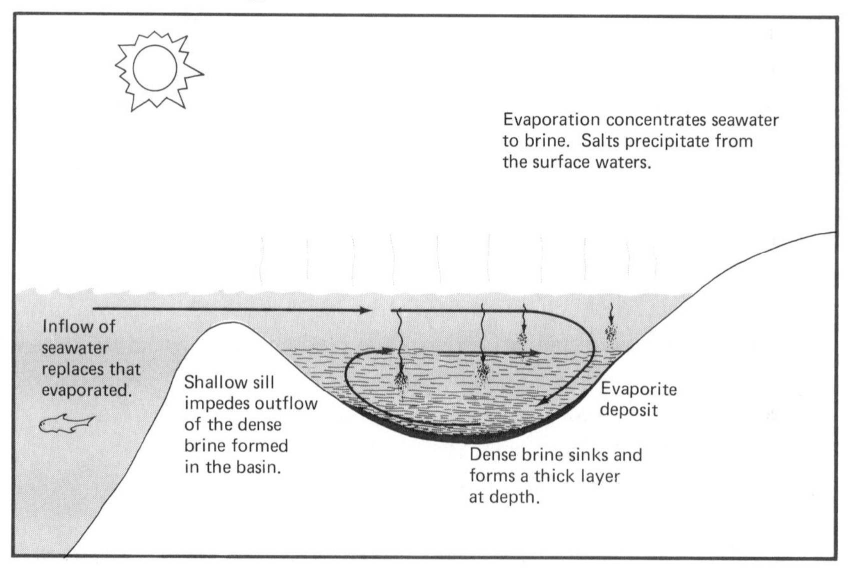

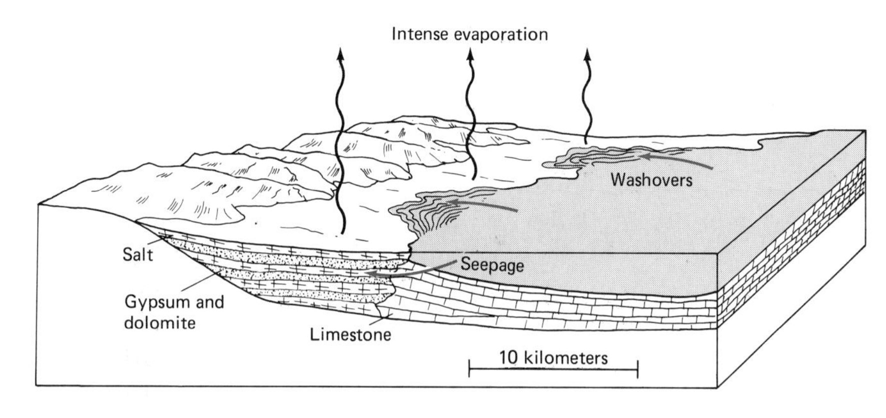

Evaporites form on arid supratidal flats from seawater supplied from occasional washovers and by slow landward seepage into the sediments.

FIGURE 8.16

Some present-day evaporite deposits that resemble those of the Zechstein are forming in extremely arid regions on vast coastal flats that are actually above high-tide level in the supratidal zone. Widespread gypsum and halite deposits forming in the Persian Gulf today, for example, precipitate on broad supratidal flats from Gulf waters that are occasionally carried onto the flats by storms or that seep into the sediments laterally and then escape upward by capillary action (Figure 8.16).

The Redbed Problem

Redbeds occur throughout the geologic record from the middle part of the Precambrian to the Pleistocene. In the upper Paleozoic and lower Mesozoic, extensive sequences of red sandstones, shales, and conglomerates of terrestrial and marginal-marine origin are particularly widespread. These rocks lend spectacular color to many regions where they crop out, particularly in arid climatic regimes.

The red coloring in both ancient redbeds and modern red soils is caused by ferric iron oxide, usually the mineral hematite. What is the source of all the iron that colors the voluminous redbed deposits in the geologic record? Actually not much iron is necessary! As little as 0.1 percent iron pigment may make a rock intensely red if the iron is thoroughly oxidized. Most drab-colored sedimentary rocks, in fact, contain as much iron as do brick-red rocks. What is unique about redbeds, then, is not the quantity of iron, but its completely oxidized state. Once formed, ferric iron oxide is stable. It survives in the chemical environments that commonly exist in areas of sediment accumulation.

Until recent years, most geologists thought that redbeds formed from red detritus eroded from areas of red soils. Because such soils are common today in tropical regions of high rainfall, redbeds were widely believed to be indicators of such climates. However, recent studies have produced three lines of evidence against this simplistic interpretation. First, many redbeds occur in association with evaporites, which necessarily form in arid climates (Figure 8.17); redbeds are virtually nonexistent in coal cyclothems, which form in humid climates. Second, most stream detritus weathered from tropical soils in Central and South America *is not red* but is commonly yellow or brown. This

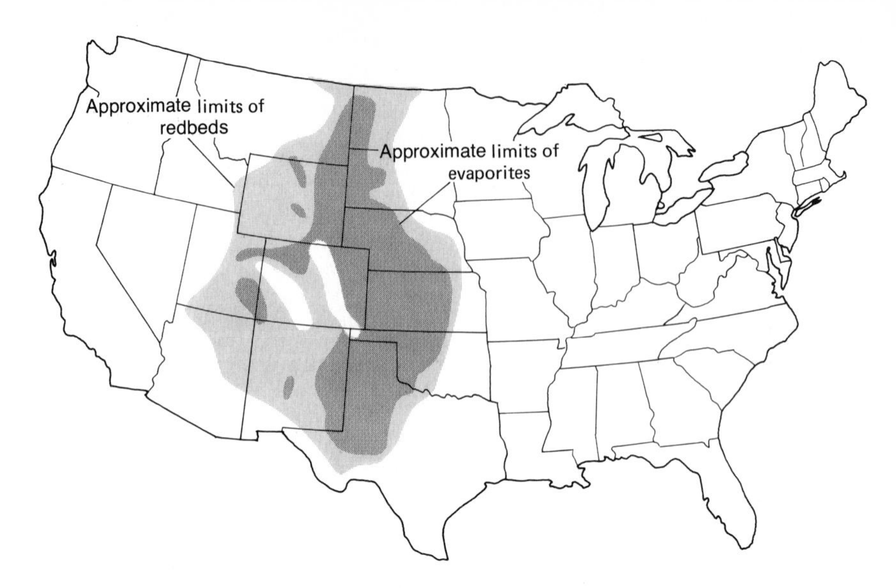

FIGURE 8.17 The area occupied by late Paleozoic redbeds in the western interior of the United States is nearly identical to that occupied by evaporites of the same age. (*Walker, 1967*)

suggests that the red iron oxides in the soils are either reduced in fluvial environments, or else that they are masked by larger volumes of nonred sediments. In any case, modern sediments produced from red soils appear to lose their red color prior to lithification. Third, careful studies of ancient sandstones suggest that the red ferric iron oxide is largely **diagenetic** in origin; that is, it results from chemical alteration of dark, iron-bearing mineral grains within the sediment some time after it was deposited. The red pigment is not uniformly distributed throughout the sandstone, but is concentrated around dissolved, iron-bearing mineral grains (Figure 8.18).

Today red desert soils and alluvial-fan deposits are forming in desert environments in northeastern Baja California and the southwestern United States. This region has been arid at least since Pliocene time. The red soils have clearly formed in place and have not been transported in from a distant source area, nor can they be the product of a former humid climate, for they contain immature clays and carbonates, which form only under arid conditions. Recent soils and newly deposited sediments in Baja California and the southwestern United States are gray, reflecting the color of the granitic source rock. The older soils, however, show a progressive reddening, accompanied by a decrease in their contained iron-bearing mineral grains.

Redbeds can and do form in arid environments today. This does not mean that redbeds form *only* in arid regions. Similar oxidative diagenetic environments may also prevail in humid regions. In southern Puerto Rico, alluvial sediments below the water table appear to be gradually changing from yellow and brown to red. They are saturated by alkaline waters, which provide a favorable diagenetic environment for the formation of hematite. Etched hornblende grains in these sediments have clearly undergone solution, showing that these, and probably other unstable mineral grains as well, are breaking down and releasing iron. Given enough time, this iron will probably oxidize to form the red mineral hematite. Possibly these sediments in the humid tropics will become red with time, in the same way as do those in desert environments.

In summary, redbeds are not believed to indicate particular climates. Interpretations of the climate in which ancient redbeds formed will need to be based on associated sedimentary minerals, sedimentary structures, and diagnostic fossils. Redbeds that are associated with evaporites probably indicate arid climates; those associated with fossil tropical floras or with coal beds probably formed in humid regions. In order for redbeds to form, it now appears that unstable iron-bearing grains must be present, and that the chemical environment within the deposit must promote the alteration of these grains and the oxidation of the iron.

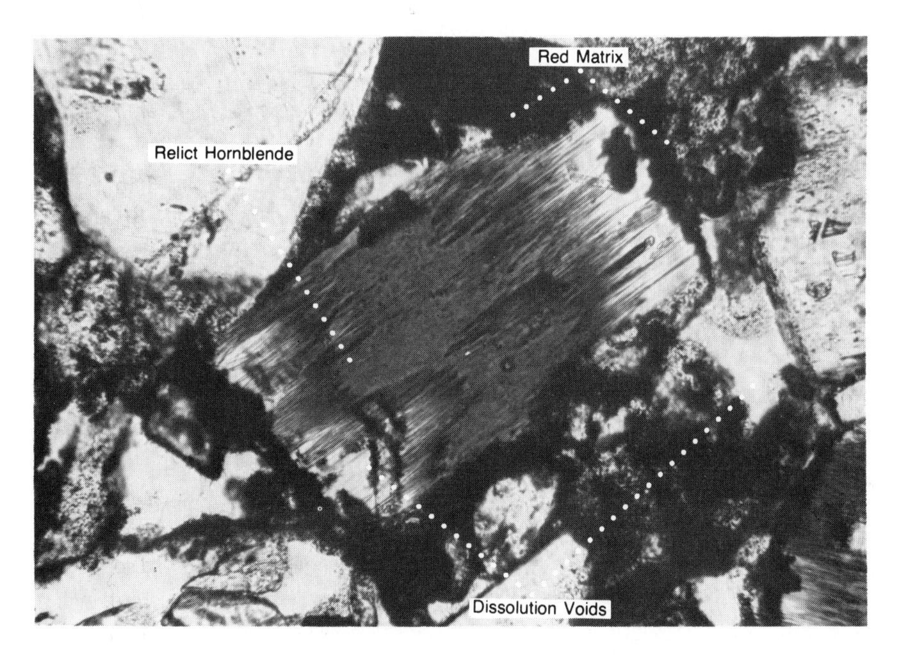

FIGURE 8.18

Thin section of a hornblende grain oxidizing to red pigment. Width of photograph is 0.6 millimeters. (*Courtesy of T. R. Walker*)

Paleozoic Glaciers

Some of the most interesting Paleozoic rocks occur chiefly in the southern hemisphere and are unstratified deposits of boulder-laden mudstones that are inferred to be *tillites*—lithified till that was deposited by Paleozoic continental glaciers. These deposits contain strongly faceted and striated pebbles and cobbles that were ground across the bedrock while held fast in the ice. The bedrock on which these deposits rest is itself polished, scratched, and deeply grooved in places as a result of the slow passage of massive, pebble-laden ice. Other lines of evidence that support the glacial origin for these and associated deposits include: (1) large boulders weighing several tons that were transported for hundreds of kilometers; and (2) scattered boulders in adjacent marine shales that are inferred to be dropstones—the deposits of melting icebergs in shallow seas that bordered the glaciated lands. These glacial deposits cover vast areas of all the present southern hemisphere continents that once constituted late Paleozoic Gondwana. The ice sheets appear to have centered around the Paleozoic south pole as it migrated from north to south across Africa and then to Antarctica during the Paleozoic (Figure 8.19).

The earliest Paleozoic glacial deposits that we know about were deposited during the Ordovician Period in what is now the Sahara Desert region of North Africa. Grooves and striations in the bedrock immediately below the coarse Ordovician tillites indicate that the glaciers moved dominantly northward (Figure 8.20). Tillites of Middle Devonian age occur in western Argentina, which at that time lay due south of the Sahara. Widespread tillites of Pennsylvanian and Permian age occur throughout the southern part of South America. In South Africa, too, a thick Late Pennsylvanian tillite, called the Dwyka Tillite, covers many thousands of square kilometers. Similarly extensive tillites that lithologically resemble the Dwyka also occur in Antarctica, India, and Australia (Figure 8.19)

Studies of the orientation of glacial striations indicate that a huge ice sheet originated in the late Paleozoic from two large centers, one in southwestern Africa and the other in eastern Antarctica (Figure 8.19). Both centers are near the Permo-Pennsylvanian south pole, as determined from rock paleomagnetism. The wide distribution of the glacial deposits and the radial pattern of the

FIGURE 8.19 Path of the south pole across Gondwana during the Paleozoic (dashed line) and the inferred distribution of Permo-Pennsylvanian ice sheets. Arrows show direction of ice flow determined from glacial pavements.

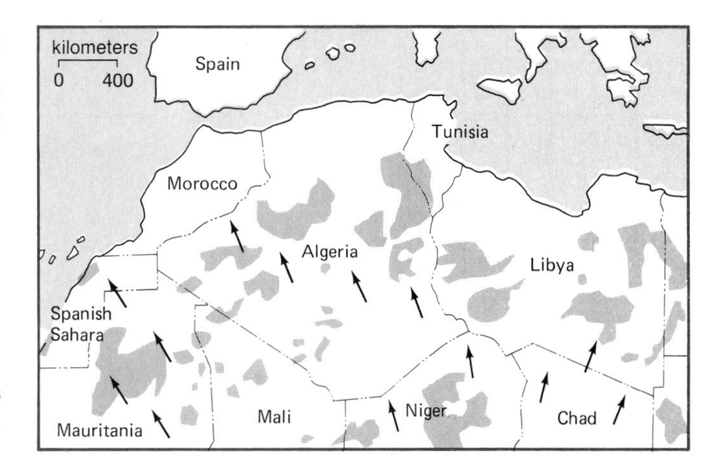

FIGURE 8.20 Striations and grooves in the bedrock immediately below Ordovician tillites reveal the direction of flow (arrows) of a great ice sheet that occupied the Sahara Desert region. Darker areas are sand-covered today. (*Sullivan, 1974*)

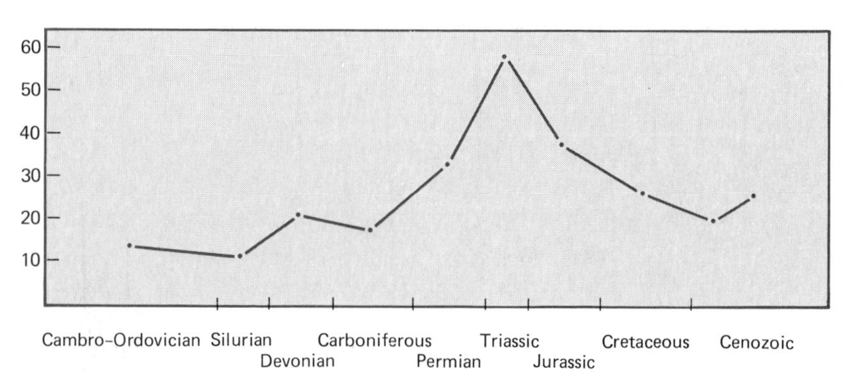

FIGURE 8.21 Areas covered by evaporite deposits throughout the Phanerozoic. (*Gordon, 1975*)

ice movement indicate continental glaciation, rather than alpine glaciation. Glaciers apparently extended to within 30 degrees of the equator at the time. In much more recent geologic history, the Pleistocene ice sheets reached to within 40 degrees of the equator. Thus, the extent of late Paleozoic and Pleistocene glaciation is comparable.

Climates and Continents

In Triassic time, the assembly of the continents that had begun in the Devonian was completed. This was a unique time in Phanerozoic history. On one side of the earth lay Pangaea, a single enormous continent, most of it in the southern hemisphere, covering about one-quarter of the earth's total surface. Over the remainder of the earth lay only ocean—a vast single expanse of water punctuated perhaps by an occasional volcanic island, but uninterrupted by any significant continental masses. By now Pangaea was largely emergent. As the individual continents assembled, the shallow seas, which had been so widespread in the early Paleozoic, gradually withdrew. An increasing proportion of sedimentary rocks were deposited in continental, rather than in marine environments.

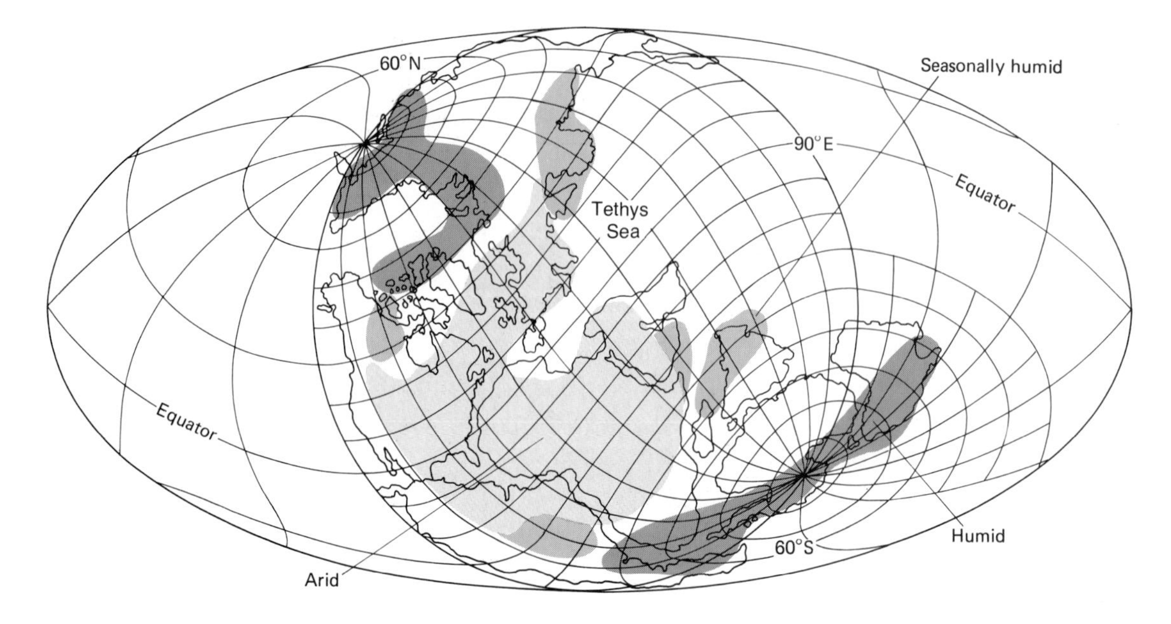

FIGURE 8.22 Triassic climatic regimes based on distribution of evaporites and coal (oblique Mollweide projection). (*Robinson, 1971*)

The emergence of the continents, together with the distinct separation of continent and ocean on the earth's surface, had a considerable effect on climate. In general, dry climates are made by large land areas, and wet climates occur near, or downwind from, large water bodies. If we were to try to arrange the earth's continents to provide the greatest possible aridity, we would put them all together and drain off the **epicontinental seas.** These are precisely the conditions that appear to have evolved in Late Permian and Triassic time. Our expectation of dry climates is borne out by the deposits of widespread evaporites, redbeds, and sand dunes that are conspicuous in the Permian and Triassic Systems in North and South America, Europe, and Africa. The extent of the continental platforms that were covered by evaporites increased dramatically as the continents completed their assembly and reached an all-time peak in the Triassic (Figure 8.21). Following the Triassic, as the continents broke apart and became increasingly submerged, evaporites rapidly decreased in abundance.

Humid conditions, in contrast, are indicated by coal. The peat from which coal forms requires abundant vegetation, which, in turn, requires a good water supply, at least seasonally. Coal can form in both cool and warm climates, but its presence indicates fairly high rainfall and poor drainage.

Figure 8.22 shows the inferred distribution of climatic regimes in the Triassic, based on the distribution of the climate-sensitive rocks, coal and evaporites. Pangaea at this time was bisected by the equator into two approximately equal halves that stretched from the equator to very high latitudes, and that formed a chevron shape to the east around the huge Tethys Sea. Abundant evaporites, redbeds, and dune deposits occur in the low and middle latitudes, indicating arid climates over vast regions. Coal occurs only at high latitudes, marking these as belts of moist climates. Between these extremes, particularly on the lands bordering the Tethys Sea, lay belts of highly seasonal (monsoon) rains. This pattern of climatic distribution is one that would be expected if the global wind circulation during the Triassic were similar to that of today.

Life on Land

In earlier chapters we saw that life originated in the sea. Most phyla, particularly those of primitive plants and invertebrate animals, are still found primarily in the oceans. Not long after the great proliferation of marine life in early Paleozoic time, the first plants and animals began to colonize the lifeless surface of the land, for we find the oldest land fossils in Silurian rocks.

The problems faced by early sea-dwelling life in making this transition to land were formidable. In the sea animals and plants have an inexhaustible supply of water, but on land they must obtain water from rain, streams, soil, or the food they eat. To prevent loss by evaporation of this water, they must have tough, watertight coverings, such as the waxy surface of leaves or the skin of reptiles and mammals. Land animals also require special structures for breathing the oxygen of the air, rather than absorbing oxygen from the surrounding water. Reproduction presents still other problems. In the sea, organisms normally deposit eggs and sperm directly into the water for fertilization. On land, special adaptations are required to prevent the drying out of these delicate structures. In spite of these difficulties, the advantages of colonizing the land were enormous. For plants, there was vast, unoccupied space with direct sunlight for photosynthesis. For animals, there was the abundant free oxygen from the atmosphere and, after the spread of land plants, an almost limitless supply of food.

Most organisms probably made the transition from sea to land by way of the fresh waters of rivers and lakes; that is, they first became adapted to water that lacked the dissolved salts of the sea, and only then developed further adaptations for life out of water. Although the transition from salt to fresh water did not present the complexities of water supply and reproduction on land, for animals it involved an almost equally serious problem—that of maintaining the balance of salts in their body fluids. The blood and other body fluids of all animals contain about the same dissolved salts as does seawater—an amazing fact that, in itself, suggests that life originated in the sea. Because of this similarity, marine animals have no difficulty in maintaining the proper balance of body salts. However, fresh water animals and land-dwelling animals must develop special organs and expend a great deal of energy to prevent a fatal dilution of their body fluids. For this reason, even the transition from the salty sea to fresh-water rivers and lakes presented a complex adaptive problem.

More groups have overcome the problem of salt balance in fresh water than have developed the additional adaptations required for life in the open air.

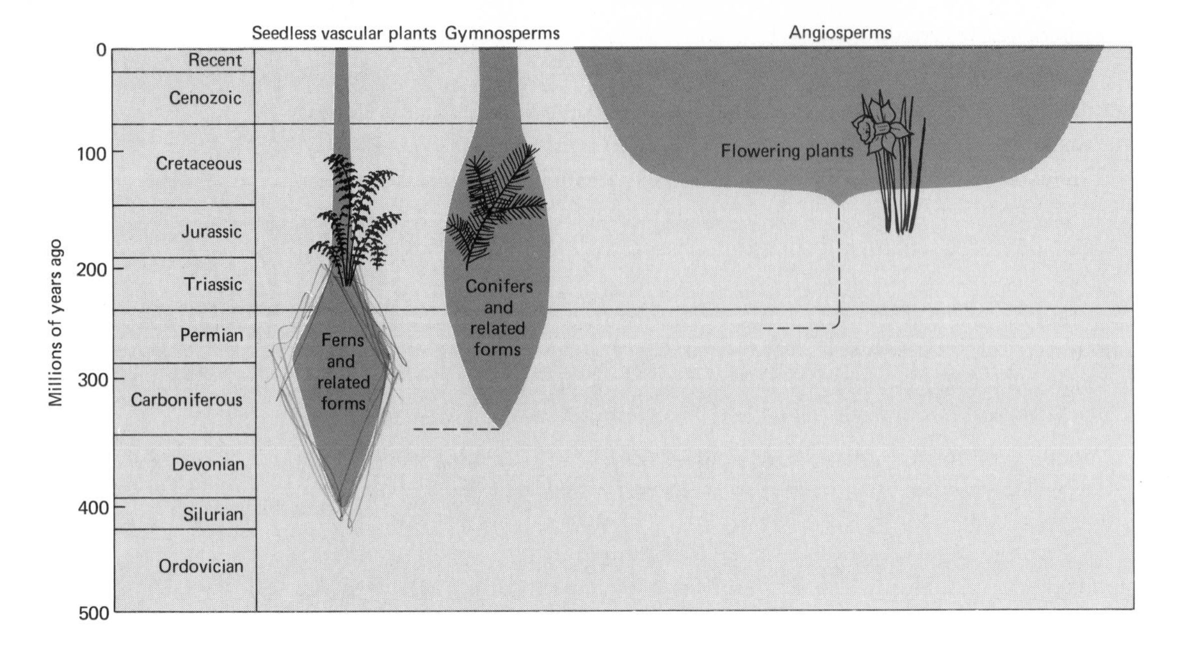

FIGURE 8.23 Evolutionary history of the tracheophytes. The dashed lines show the most probable evolutionary relations of the groups; the width of the vertical areas indicates the approximate abundance of each group.

About ten major phyla have representatives adapted to fresh water, but only four have members that are fully adapted to life on dry land. These four are: Tracheophyta (vascular plants), Mollusca (some snails), Arthropoda (arachnids, insects), and Chordata (reptiles, birds, mammals) (see Figure 4.12).

Land Plants

Simple, water-dwelling algae are the most probable ancestors for the phylum Tracheophyta, which includes all the dominant land-dwelling plants, such as ferns, conifers, and flowering plants. Since the first **tracheophytes** arose from these aquatic ancestors in Silurian time, the phylum has gone through three great periods of evolutionary expansion, each leading to progressively more advanced groups of land plants (Figure 8.23). The three groups differ primarily in the efficiency of their adaptations for reproduction on land.

The first great expansion led to ferns and related plants, which were to dominate the land through much of the Paleozoic Era. These groups had no seeds and lacked fully effective mechanisms for protecting the sperm from drying out; they were, therefore, still restricted to environments that were relatively moist.

Beginning in the Carboniferous Period, these early seedless plants were gradually replaced by conifers and other groups that are known informally as **gymnosperms.** Gymnosperms differed from their predecessors by developing seeds and pollen, which are primarily adaptations to protect the sperm and insure successful reproduction in dry environments. These adaptations were highly advantageous, and by Triassic time, gymnosperms had replaced seedless plants as the dominant element of the land flora. Gymnosperms were the principal large land plants through much of the Mesozoic Era, and today they still make up great forests of pine, spruce, and fir trees.

They were to be overshadowed, however, by a third and final group of land plants, one that developed another adaptation for still more successful reproduction on land—the flower and its enclosed, covered seed. Fossil flowering plants first appeared in Lower Cretaceous rocks and then went through a rapid radiation, so that by Late Cretaceous time they were dominant over much of the land surface. Today, aside from ferns and conifers, most familiar land plants belong to this group.

The Rise of Vertebrates

Although invertebrate animals, particularly insects, spiders, and snails, have been extremely successful in adapting to life on land, it is the history of terrestrial vertebrate animals—fresh-water fishes, amphibians, reptiles, birds, and mammals—that is of greatest interest to us. Not only are vertebrates more familiar, for we ourselves are vertebrates, but they also have a much more complete fossil record.

Vertebrates certainly arose from some sort of invertebrate ancestor, but, as is so often the case, the exact ancestral group is uncertain. The fossil record provides no clues, because the earliest fossil vertebrates, fragments of primitive fish found in Cambrian rocks, were already fully differentiated from their invertebrate ancestors. Once differentiated, however, vertebrates have an unusually detailed historical record.

The earliest vertebrates, from which all others arose, were primitive fishes; indeed, four of the eight classes of vertebrates are fishes. Today, as in the past, fishes far outnumber land dwellers, both in number of species and of individuals (Figure 8.24). However, most of the evolutionary history of fishes is a side issue in the mainstream of vertebrate development that leads to land-dwelling amphibians, reptiles, birds, and mammals.

FIGURE 8.24

Evolutionary history of fishes and amphibians. The dashed lines show the most probable evolutionary relations of the groups; the width of the vertical areas indicates the approximate abundance of each group.

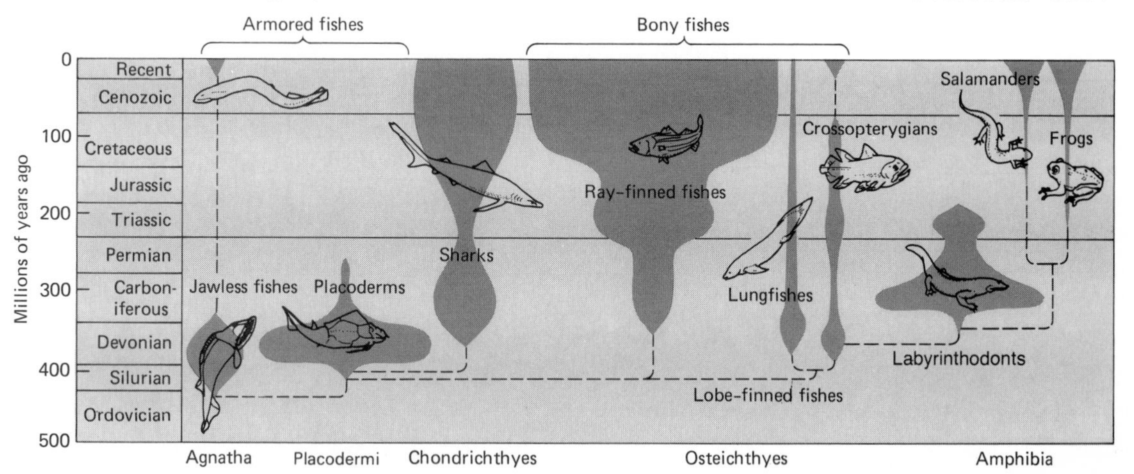

Although the first fragments of fossil fishes are found in Upper Cambrian rocks, remains of fishes are rare and poorly preserved until the end of the Silurian Period. Then fishes began an explosive evolutionary radiation that lasted throughout the succeeding Devonian Period. Just as with the invertebrates of the preceding Cambrian and Ordovician Periods, fishes first went through a stage of evolutionary experimentation that lasted from Silurian to Middle Devonian time. This phase of fish evolution was dominated by two primitive and now mostly extinct classes, Agnatha and Placodermi. From these early experimental groups arose the two dominant classes of modern fishes, the Chondrichthyes (sharks) and Osteichthyes (bony fishes). These groups rapidly replaced the earlier classes in Late Devonian time and have been dominant ever since.

Note that most of the evolutionary expansion and replacement of the fish classes took place during the Devonian Period, a key time in fish evolution. For this reason, the Devonian Period is sometimes called the Age of Fishes, even though fossil fishes are much less common than invertebrate animals in Devonian rocks.

The First Amphibians

The Devonian Period was not only a critical time in the evolutionary history of fishes, but it was also the time when vertebrates first made the transition to land. The oldest fossil amphibians are found in Upper Devonian rocks of Greenland and eastern Canada. In adapting to life on land, these early vertebrates faced the same problems of reproduction, water-retention, and oxygen-respiration that were earlier solved in land-dwelling plants and invertebrate animals. In addition, they faced a unique problem of *locomotion* on land. Plants, of course, do not move around. The first land-dwelling invertebrates were arthropods and snails that developed from ancestors already adapted for moving on the solid surface of the sea floor by means of legs or a muscular foot. Fishes, on the other hand, were adapted to a swimming mode of life; a profound modification was required of their land-dwelling descendants.

FIGURE 8.25 Arrangement of fin bones in (*a*) typical lobe-finned and (*b*) ray-finned fishes. In the lobe-fins, the muscles extend into the fin, permitting greater control and flexibility of movement. Such fins developed into the limbs of land-dwelling amphibians.

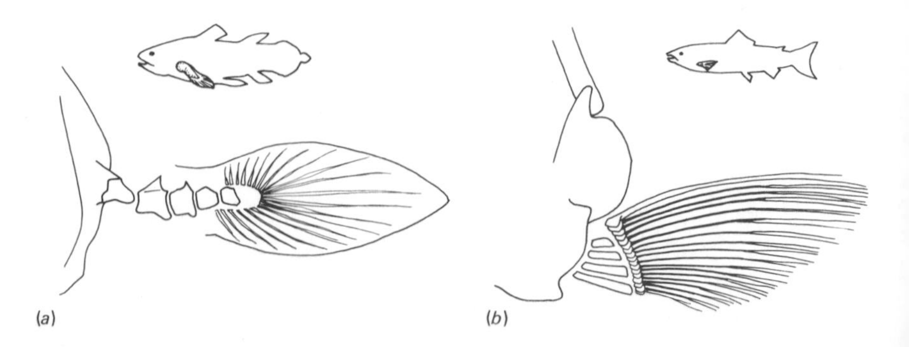

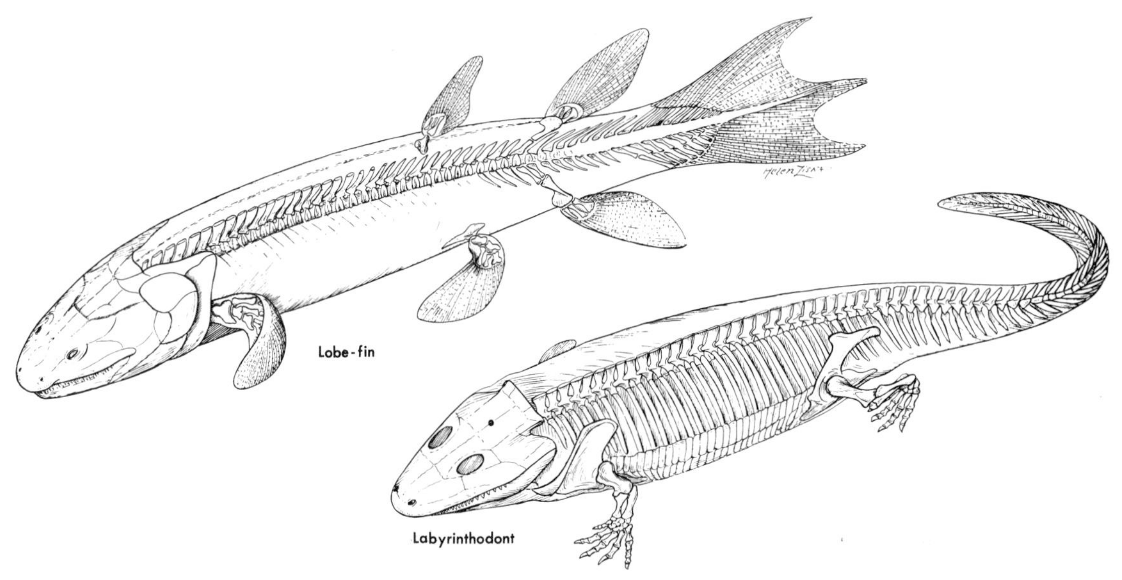

FIGURE 8.26

Devonian lobe-finned crossopterygian fishes had labyrinthodont amphibian descendents. (*Courtesy of the American Museum of Natural History*)

It was the potential for developing a means of locomotion on land that distinguished one group, the early *lobe-finned fishes,* from their *ray-finned* relatives (Figure 8.25). The bone arrangement of lobe-finned fishes permitted the fins to move freely at their point of attachment to the body; in addition, the muscles extended *into* the fin to allow precise movements. Ray-finned fishes, on the other hand, lacked this flexible bone and muscle arrangement. The lobe-finned pattern, with its complex muscles, was ideally suited for development into elongated, flexible limbs to support and move the animal on land.

This occurred in one group of lobe-finned fishes, the **crossopterygians** (Figure 8.26). Indeed, the skull and skeleton of Devonian crossopterygians are almost identical with those of the earliest fossil amphibians, except for the modification of the lower fins into stubby limbs for walking on land. The similarities are so close that the crossopterygian-amphibian transition is one of the best documented in the entire fossil record.

Although amphibians solved the problem of locomotion on land, they did not conquer another difficulty of land life—that of reproduction. All amphibians (except for some very specialized present-day forms) must return to the water to reproduce, for their delicate eggs quickly dry out and die if exposed to the air. It remained for their descendants, the reptiles, to solve the reproduction problem by developing eggs with tough outer coverings to prevent drying.

Surviving modern amphibians (frogs, toads, newts, salamanders, and their relatives) originated in Triassic and Jurassic time as specialized descendants of early amphibians called **labyrinthodonts,** which were common during the Pennsylvanian Period (Figure 8.27). Labyrinthodonts appeared more rep-

FIGURE 8.27

Cacops, an early Permian labyrinthodont amphibian. These amphibians had short but very powerful limbs and probably were completely land-dwelling as adults, although the young were probably water dwellers. (*Robert Bakker reconstruction*)

tilelike than their modern relatives and, in general, looked something like fat, stubby-nosed alligators. The habits of many of them were probably alligator-like as well, for they appear to have spent much of their lives in water and along the banks of rivers and lakes. Some developed into large animals 3 meters long. They were still abundant in the Permian Period but were beginning to decline, as their more successful offspring, the reptiles, expanded. By the close of the Triassic Period, the labyrinthodonts were extinct.

Chapter Summary

Late Paleozoic Continents and Mountain Belts

The Ouachita mobile belt, exposed today in Oklahoma and Arkansas, is an example of a deformed continental margin of late Paleozoic age.

The Hercynian Orogeny is the name used for the late Paleozoic deformations that occurred along the margins of two large continents, Laurussia and Gondwana, assembled from smaller units; in Triassic time these joined to form a single continent called Pangaea.

Terrestrial Sediments and Environments

Fluvial environments: Sediment-laden streams and rivers have distinctive patterns of sand and mud deposition that can be recognized in ancient sedimentary rocks.

Coal forms from the accumulation of land-plant remains in swamps; coal-forming swamps were especially widespread on the continents of Pennsylvanian time.

Cyclothems: Coal often occurs in repetitious sequences of sediment caused by repeated flooding of low-lying coastal areas.

Deserts and evaporites: Dry regions have characteristic patterns of sediments deposited in periodically evaporated lakes or oceanic lagoons, wind-blown dunes, and alluvial fans; such deposits were especially common in Permian and Triassic time.

The redbed problem: Not all red sandstones and shales indicate desert conditions, for they are known to form today in humid regions as well.

Paleozoic glaciers: Vast ice sheets repeatedly covered the present southern hemisphere continents, which were assembled in late Paleozoic time.

Climate and continents: The widespread dry climates of Permian and Triassic time were largely the result of the assembly of the continents into Pangaea.

Life on Land

Land plants: Ferns and their seedless relatives dominated the land in Devonian time but were largely replaced during late Paleozoic time by seed-bearing conifers and related groups.

The rise of vertebrates: The oldest primitive fishes occur in Cambrian rocks; the dominant modern groups arose in Devonian time.

The first amphibians: In Late Devonian time, the first land-dwelling vertebrates arose from fishes through modification of their muscular fins into limbs for locomotion.

Important Terms

Appalachian Orogeny
bed load
block mountain
climate-sensitive rock
compressional force
crossopterygian
cyclothem
diagenesis
epicontinental sea
gymnosperm
Hercynian Orogeny
isostasy
labyrinthodont
Laurasia
Laurussia
lycopsid
Ouachita mobile belt
sphenopsid
suspended load
tensional force
tracheophyte
volcanic tuff

Review Questions

1 Where were the principal zones of deformation during the Hercynian Orogeny?
2 Summarize the history of the Ouachita mobile belt.
3 What are the principal environments of sediment deposition on land?
4 What are redbeds, and how are they formed?
5 Outline the evolutionary history of land plants.
6 What were the first vertebrate animals? How and when did land-dwelling vertebrates originate?

Additional Readings

Banks, H. P. *Evolution and Plants of the Past,* Wadsworth, Belmont, California, 1970. *An excellent intermediate-level text.*

Burk, C. A., and C. L. Drake (eds.) *The Geology of Continental Margins,* Springer-Verlag, New York, 1974. *A collection of review papers.*

Friedman, G. M., and J. E. Sanders *Principles of Sedimentology,* John Wiley, New York, 1978. *Advanced text with good discussions of terrestrial sediments.*

Rigby, J. K., and W. K. Hamblin *Recognition of Ancient Sedimentary Environments,* Society of Economic Paleontologists and Mineralogists, 1972. *An advanced review.*

Stahl, B. J. *Vertebrate History: Problems in Evolution,* McGraw-Hill, New York, 1974. *A readable intermediate-level text.*

Windley, B. F. *The Evolving Continents,* John Wiley, New York, 1977. *An advanced review stressing tectonic history.*

THE BREAKUP OF PANGAEA AND EARTH REORGANIZATION

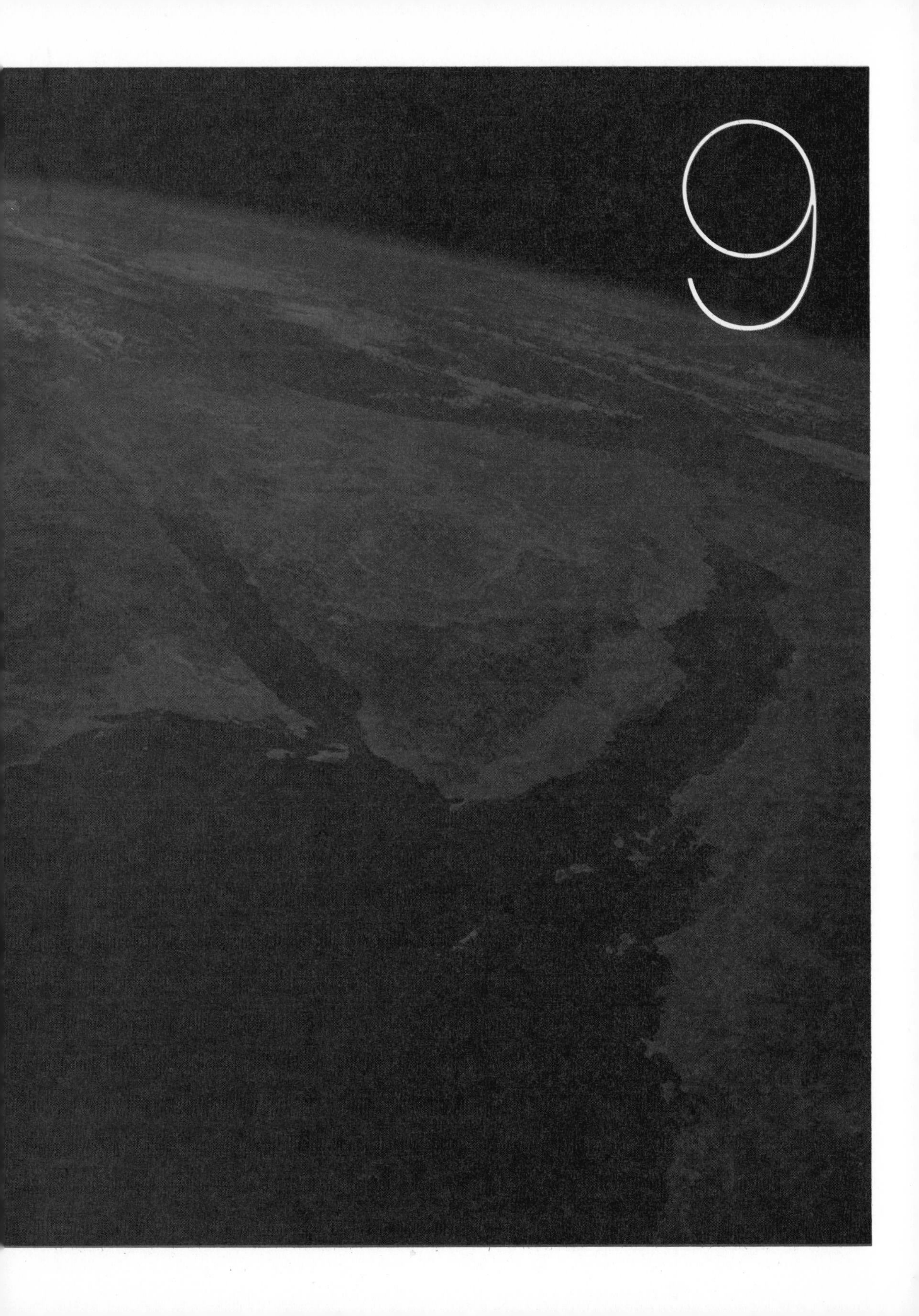
9

Our modern world had its beginning late in the Triassic Period—about 200 million years ago—when the supercontinent of Pangaea began to break apart. This breakup, involving wholesale motions of huge crustal plates, can now be deciphered in considerable detail from several kinds of evidence. In this chapter we shall examine this evidence and discuss the patterns of change in the earth's geography and life that followed the breakup of Pangaea.

Post-Triassic Plate Motions

We saw in Chapter 8 that in Triassic time the forerunners of today's major land masses lay joined together in one great continent, Pangaea. Most of that huge continent lay above sea level throughout the period, for most Triassic sedimentary rocks, like those of the latest Paleozoic, were deposited in terrestrial environments. The evidence for the timing of the Mesozoic breakup of Pangaea and the subsequent drifting apart of the continents is provided in part by paleomagnetism and in part by dating of the sea floor through **magnetic anomalies.** This evidence is confirmed by the radiometric ages of lava flows that spread out along the boundaries of the newly created ocean basins.

Figure 9.1 shows that, in the middle of the Triassic Period, the earth's continents were nearly equally distributed between the northern and southern hemispheres; today two-thirds of all land lies north of the equator. To achieve their present distribution, the present southern hemisphere continents (excepting Antarctica) have moved toward the equator, away from the high-latitude positions that permitted their widespread glaciation in the late Paleozoic. The northern hemisphere continents have moved northward, away from the equator.

The Moving Continents

Just as the continental collisions that formed Pangaea left compressive structures (foldbelts) to mark the sites of former continental margins, the breakup of Pangaea left tensional structures (**fault-block basins** or **rift valleys**) along the newly formed continental margins to testify to their origin. For example, prior to the actual opening of the North Atlantic Ocean, a broad belt of rift valleys was produced by the initial pulling apart of the continents. One of these valleys was finally torn open to produce the new ocean. Others survive on both sides of the Atlantic as a series of down-faulted basins that are filled with Triassic sediments. In the United States, these Triassic fault basins extend southward from Connecticut into the Carolinas (Figure 9.2). Basaltic magma welled up through cracks formed by the rifting and produced sills and lava flows within the accumulating Triassic strata.

Initially the Atlantic Ocean rifting occurred along a nearly east-west line not far north of the Triassic equator (Figure 9.1*a*). After 20 million years, Laurasia had separated from Gondwana and rotated clockwise, as shown in Figure 9.1*b*. Late Triassic and Early Jurassic evaporite deposits formed in the

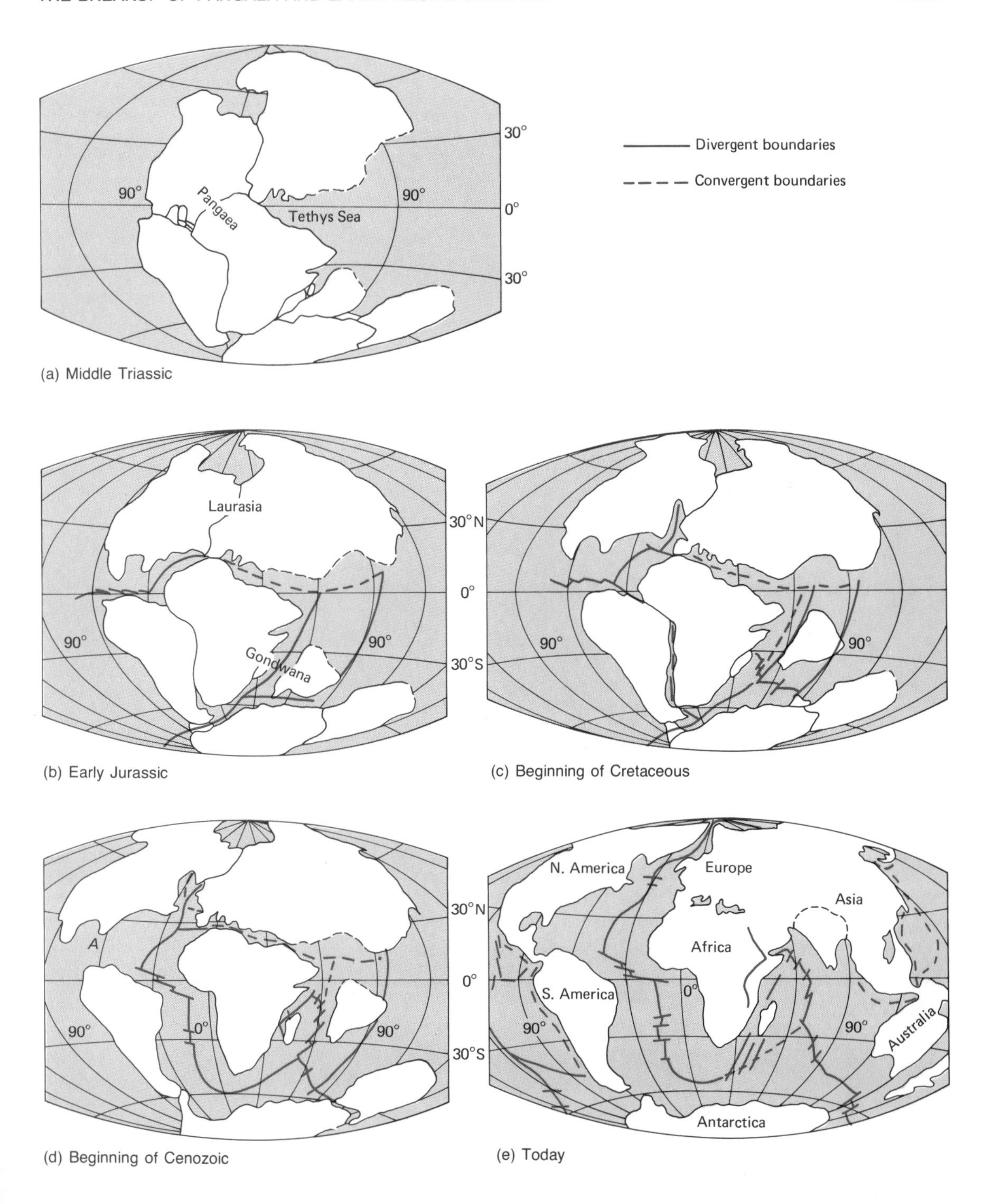

FIGURE 9.1

Disruption and dispersal of the supercontinent of Pangaea during the past 200 million years: (*a*) Middle Triassic; (*b*) Early Jurassic; (*c*) beginning of Cretaceous; (*d*) beginning of Cenozoic; and (*e*) today. (*Dietz and Holden, 1970*)

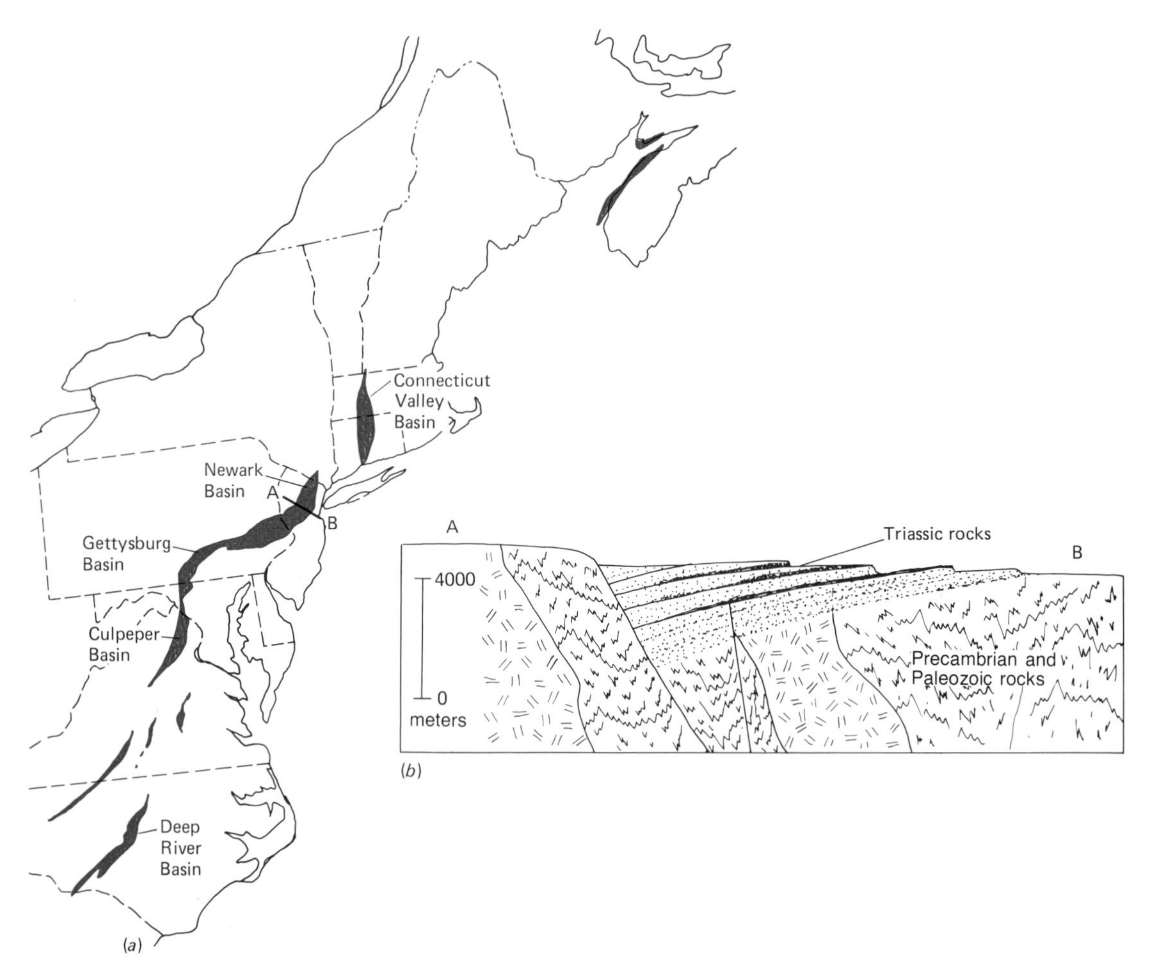

FIGURE 9.2 (*a*) Map of exposed Triassic block-fault basins along the Atlantic coast of North America. Others lie buried beneath coastal plains states. These basins, which formed from tensional forces during the opening of the Atlantic Ocean, contain thick terrestrial sedimentary rocks, basalt flows, and sills of Triassic age. (*b*) Cross section A-B shows the Newark Basin of New Jersey as an example.

barely opened North Atlantic where water from the Tethyan Gulf flowed into a narrow seaway that must have been much like the Red Sea today (Figure 9.3). Similar evaporites formed in what is now the Gulf of Mexico, but which in the earliest Jurassic must have been a restricted embayment fed by water from the Pacific Ocean. Still later, in Early Cretaceous time, South America and Africa began to separate (Figure 9.1*c*), creating still another narrow evaporite basin, which was fed by water from the southern ocean. Today, the salts produced there underlie the South American and African shelf areas indicated in Figure 9.3. In all of these regions, the salt occurs as **diapirs** (salt domes), which have been locally squeezed upward through a thick sequence of younger strata (Figure 9.4). These diapirs have created numerous oil traps at depth, many of which are now being exploited.

FIGURE 9.3

Shaded areas contain abundant salt diapirs, indicating the extent of massive salt deposition by oceanic sources during rifting in the Jurassic and Early Cretaceous. (*Burke, 1975*)

Thick salt deposits and accompanying diapirs of Early Cretaceous age along the western margin of the Atlantic Ocean off Brazil. (*Kumar and Gamboa, 1979*)

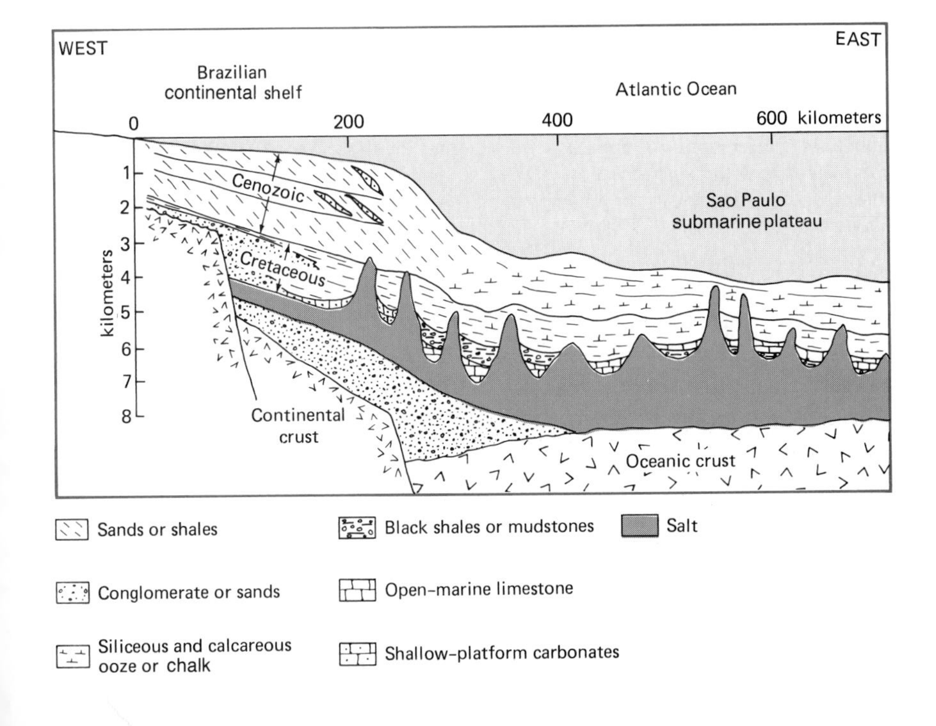

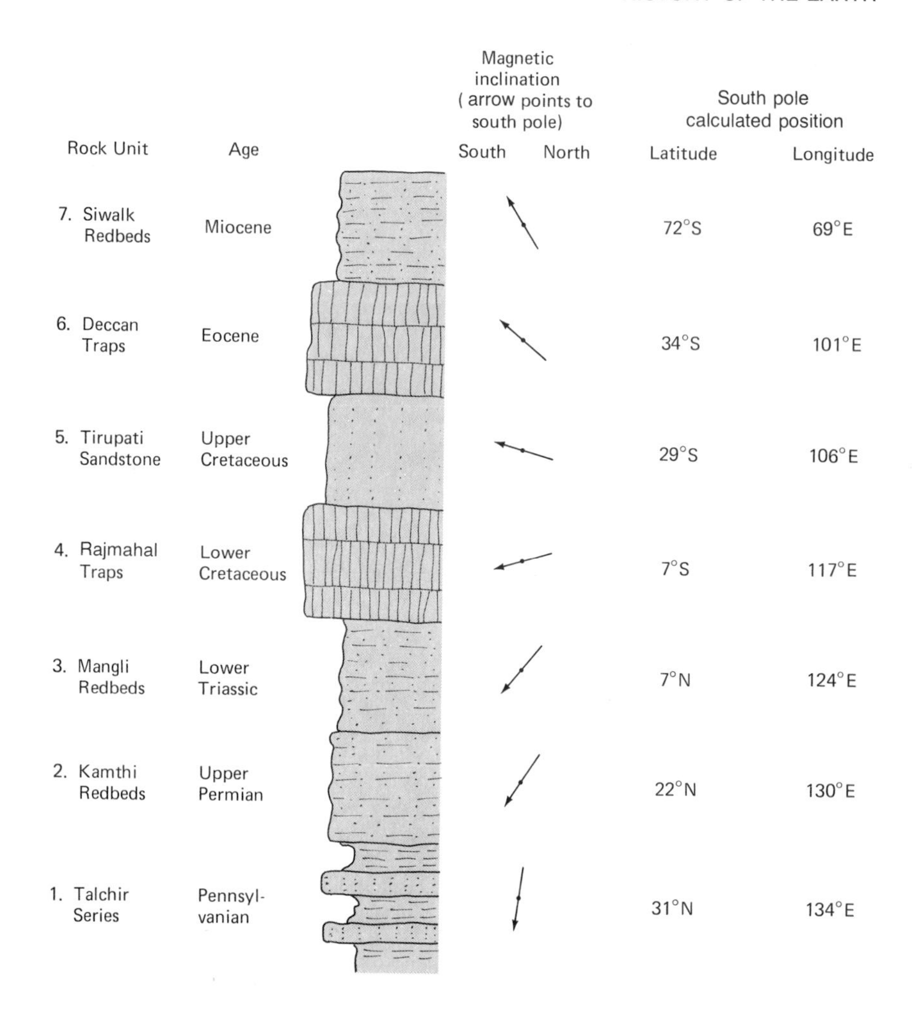

FIGURE 9.5 India's long northward travel path is documented by the magnetic inclination of rocks since the late Paleozoic. Section is diagrammatic. Magnetic deviation, which is not shown, helps determine south pole positions. These paleopoles are mapped in Figure 9.6. (*Wensink, 1973*)

By this time (Figure 9.1*c*), India had separated from Antarctica and was moving northward toward Asia, as the Indian Ocean opened behind. Simultaneously, eastern Laurasia and Africa were converging, as the Tethys Sea closed. By the end of Cretaceous time (Figure 9.1*d*), the South Atlantic had widened and become continuous with the North Atlantic, as the two American continents migrated westward. North America had broken away from the remainder of Laurasia along a rift immediately west of Greenland. Shortly thereafter, in the early Cenozoic, the rifting shifted to the east side of Greenland, where divergence has continued to the present time.

Finally, during Cenozoic time, Australia broke away from Antarctica and moved rapidly northward. North and South America became connected by an isthmus built largely of volcanic eruptions. The Mediterranean Sea was formed by the motions of several microplates on the site of the western part of the former Tethyan seaway, and the Red Sea was produced by rifting, as Arabia split away from Africa. Within this broad framework of worldwide plate motions following the Triassic breakup of Pangaea, many regional details have been worked out. Here we shall briefly review some selected examples.

India's Paleomagnetic Track Record

India provides an excellent example of tracking a particular continent by the paleomagnetism of its exposed rocks. Most of the rocks in this case come from the region of the Deccan Plateau in south central India. Here a thick succession of lava flows and sedimentary strata of Mesozoic and Cenozoic age rests on late Paleozoic and Triassic rocks. Several samples from each rock unit in the succession were analyzed and dated so that the paleopole positions could be calculated with precision. Over time, the actual rock magnetism showed a progressive change in inclination: from a steep dip to the south, to level, to a gentle dip to the north (Figure 9.5). These results indicate a systematic north-

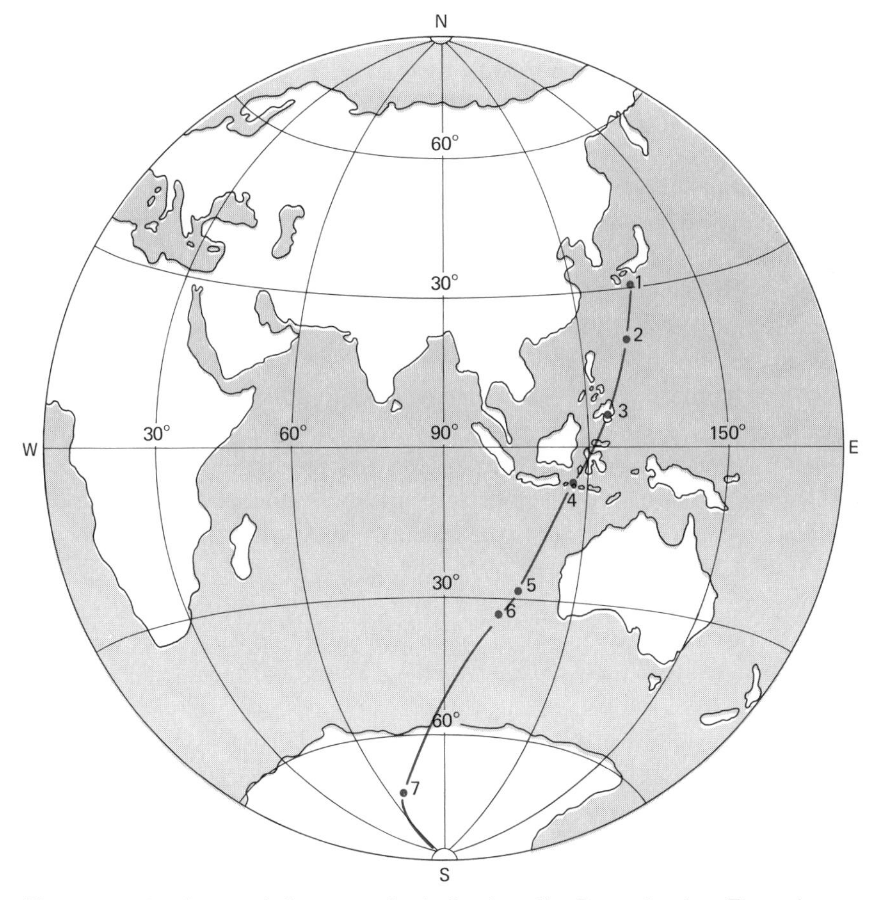

FIGURE 9.6

The apparent polar-wandering curve for India since the Pennsylvanian. The poles are numbered to correspond to Figure 9.5.

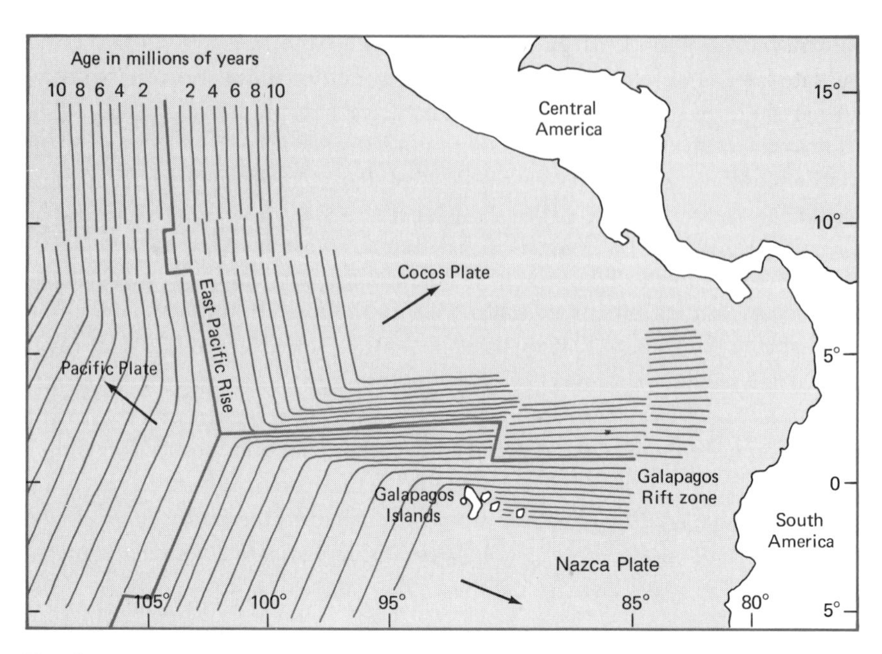

FIGURE 9.7 The Galapagos triple junction formed by intersection of three spreading zones. Seafloor ages are shown schematically, but spacing is an approximation of spreading rates. (*Hey and others, 1972*)

ward change in latitude. The inclination and directional data together indicate that India moved from near the south pole, all the way across the equator, and into the northern hemisphere since the Paleozoic.

Relative to the present position of the Indian subcontinent, the south pole appears to have moved from a position near Japan to its present position far south on Antarctica. This result can be depicted in the **polar-wandering curve** shown in Figure 9.6. However, no one believes that the south magnetic pole actually crossed latitude lines in this way. Instead, if the pole maintained its present position, as is believed, then India must have moved northward about 120 degrees, from a high southern latitude to a low northern latitude. Magnetic anomaly stripes and age-dating of the Indian Ocean corroborate the northward migration of India.

Triple Junctions

The movements of the earth's lithospheric plates illustrated in Figure 9.1 show many places where three separate plates converge. Although these **triple junctions** are limited to small areas, their development and evolution influence plate motions and exert important controls on regional geologic history.

A few triple junctions are stable and maintain the same configuration for a long time. One example is the Galapagos triple junction at the boundary of the Pacific, Cocos, and Nazca plates (Figure 9.7). The Cocos and Nazca plates are spreading from each other, and both are spreading from the Pacific plate. The Galapagos triple junction occurs as three ridges converge and has apparently maintained a similar configuration throughout much of Cenozoic time.

Unlike the Galapagos example, most triple junctions are unstable and migrate laterally as plate boundaries change position about them. For example, at the boundary of the Pacific and North American plates, a complex

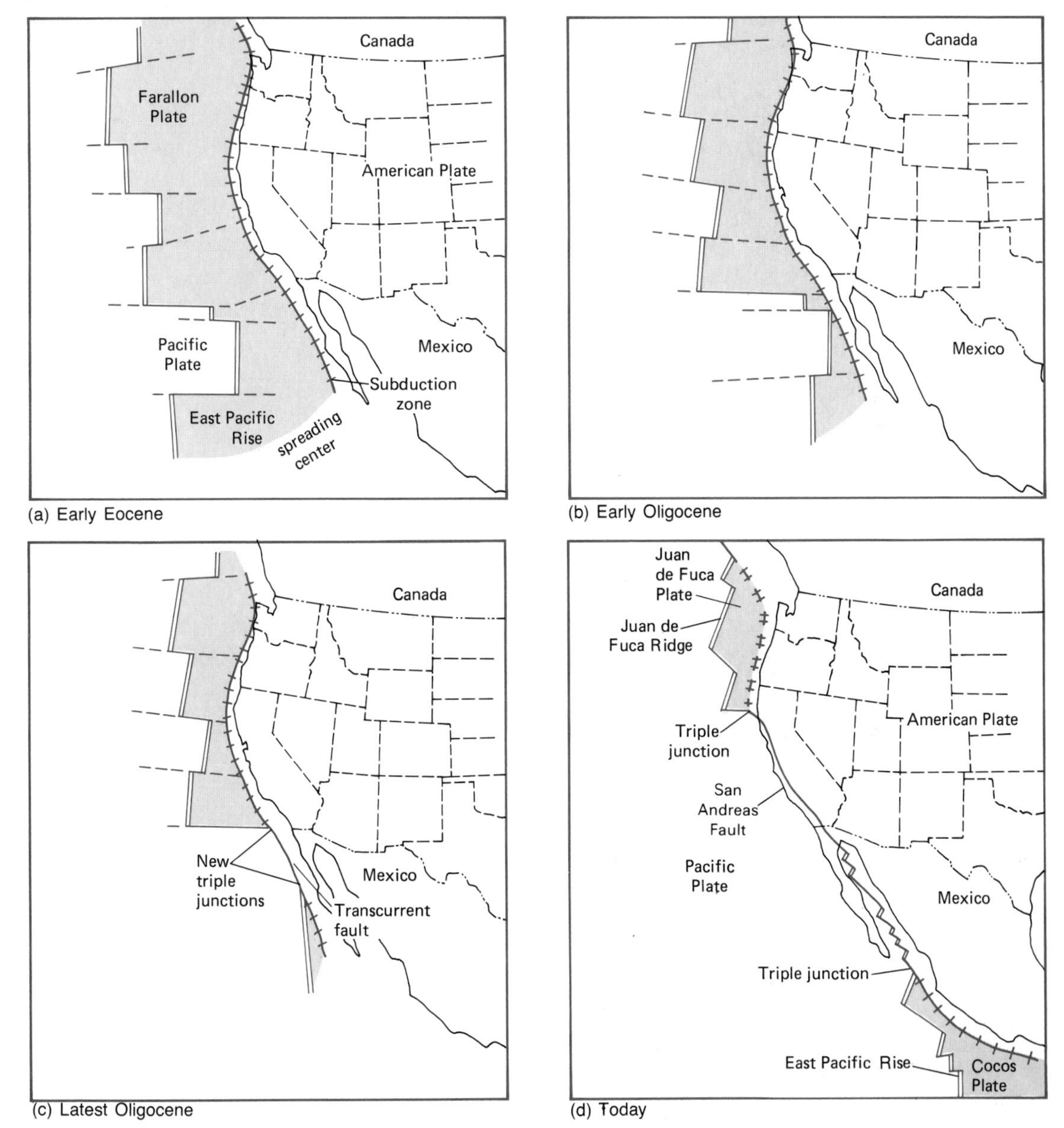

FIGURE 9.8

During the Cenozoic, the once-huge Farallon plate was largely subducted along the western margin of North America. As the East Pacific Rise encountered the North American plate, the subduction zone became a transcurrent fault. (*a*) Early Eocene: the East Pacific Rise nears the American plate. (*b*) Early Oligocene: the Pacific plate first meets the American plate at a subduction zone. (*c*) Latest Oligocene: triple junctions form and migrate laterally along subduction zone. (*d*) Today: the Pacific and American plates are in extensive contact; most of the subduction zone is replaced by the San Andreas Fault. (*Lipman and others, 1971*)

history is recorded in the magnetic stripes on the floor of the eastern Pacific Ocean. In the area of California and Baja California, the northwest-trending San Andreas fault system, which many geologists believe to be a huge transcurrent fault, marks the boundary between the Pacific plate and the North American plate. Off Washington and Oregon, the Pacific plate is separated from the North American plate by the small Juan de Fuca plate; south of Baja California, it is separated from the North American plate by the Cocos plate (Figure 9.8*d*). Both the Juan de Fuca and Cocos plates bound the North American plate at subduction zones, and both bound the Pacific plate at actively spreading ridges that are sliced up by large transform faults.

The Juan de Fuca Ridge and East Pacific Rise are active spreading centers; consequently, the age of the ocean crust increases both westward and eastward from these features. Where the Pacific plate contacts the North American plate, the oceanic crust ranges in age from 6 million years on the south to 29 million years on the north, and the age everywhere increases systematically westward.

Apparently this oceanic crust was produced on the East Pacific Rise at a time when the Rise was situated well offshore from North America and extended northward to join the Juan de Fuca Ridge (Figure 9.8*a*). The huge oceanic plate that existed east of the rise has been named the **Farallon plate;** the boundary between the Farallon plate and the North American plate was a subduction zone. Today the Juan de Fuca plate on the north and the Cocos plate on the south are all that remain of this enormous plate; their present-day velocities and directions of movement can be extrapolated backwards in time to reconstruct the Farallon plate's history.

During early Cenozoic time, the Farallon plate was moving toward the subduction zone that bordered California and Baja California (Figures 9.8*a* and *b*). The East Pacific Rise was also moving toward the subduction zone, but at a slower rate. Wherever a segment of the Rise encountered the subduction zone at the western edge of the North American plate, the Farallon plate disappeared and the Pacific plate came into contact with the North American plate (Figure 9.8*c*). As the Farallon plate was replaced by the Pacific plate, the plate motion immediately west of the North American plate boundary changed from easterly to northerly. As a result, the subduction zone was changed into a transform fault along migrating triple junctions. Figures 9.8 *c* and *d* show how the triple junctions followed the shrinking fragments of the Farallon plate both to the north and to the south. Today's long transcurrent-fault complex along the coast of California and Baja California was thus inherited from this collision.

Marginal Basins

During the subduction of the Farallon plate beneath California and Baja California, the trench was probably flanked on the side of the North American plate by volcanoes, either in the form of an island arc or as a mainland volcanic chain. Today the subduction zones along the west coasts of North and South America are bordered only by mainland volcanoes; the same was likely true of

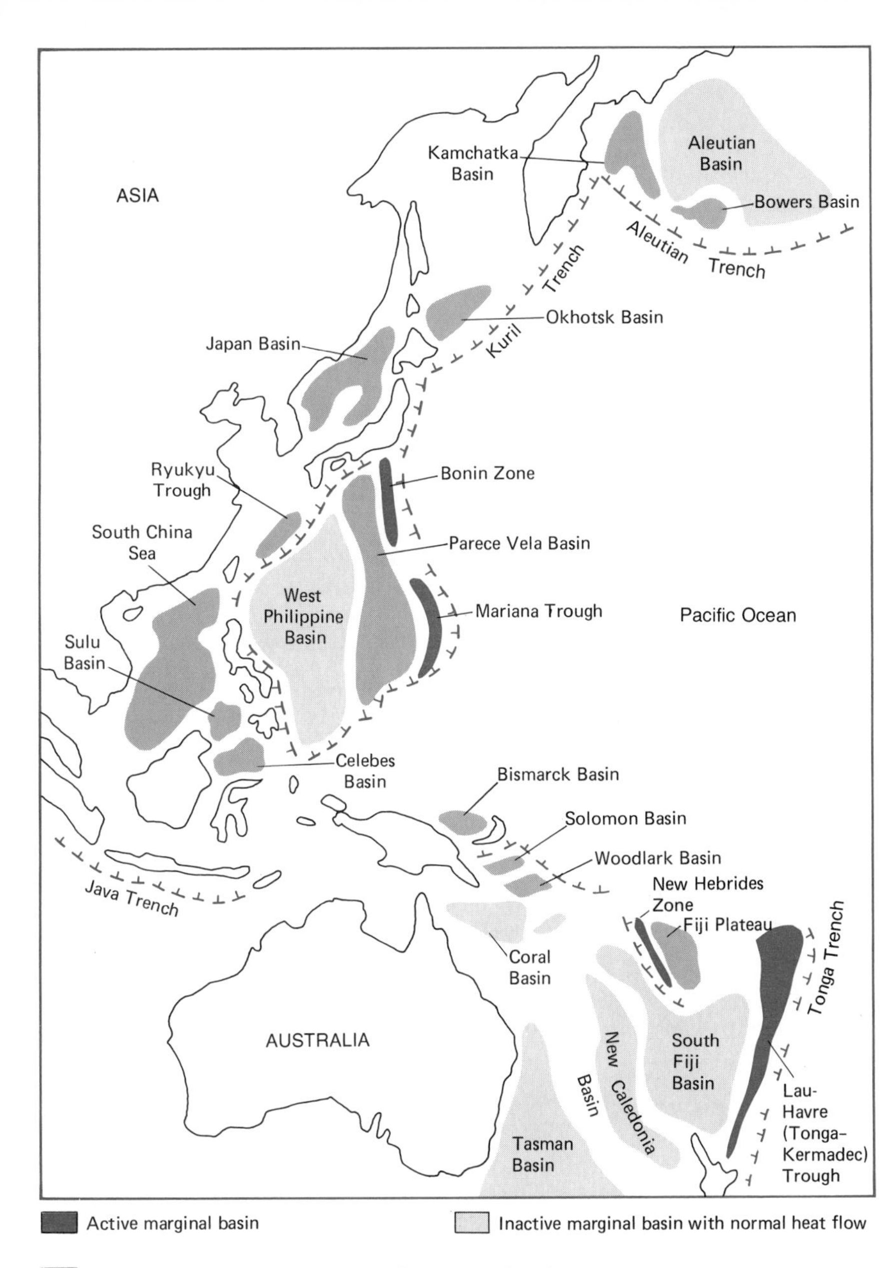

FIGURE 9.9

Distribution of marginal basins in the western Pacific. (*Karig, 1974*)

the Farallon plate. In the western and northern Pacific, by contrast, the subducting Pacific plate is bordered, not by mainland volcanic chains, but by island arcs. Behind these island arcs lie small basins covered by the ocean, known as **marginal basins.** These basins are situated between the arc and either the mainland or other island arcs (Figure 9.9). Such marginal basins lie immediately adjacent to zones of pronounced plate convergence, where much of the subduction on the earth's surface is taking place. Yet they appear to have formed as a result of crustal extension, and not as a result of compression.

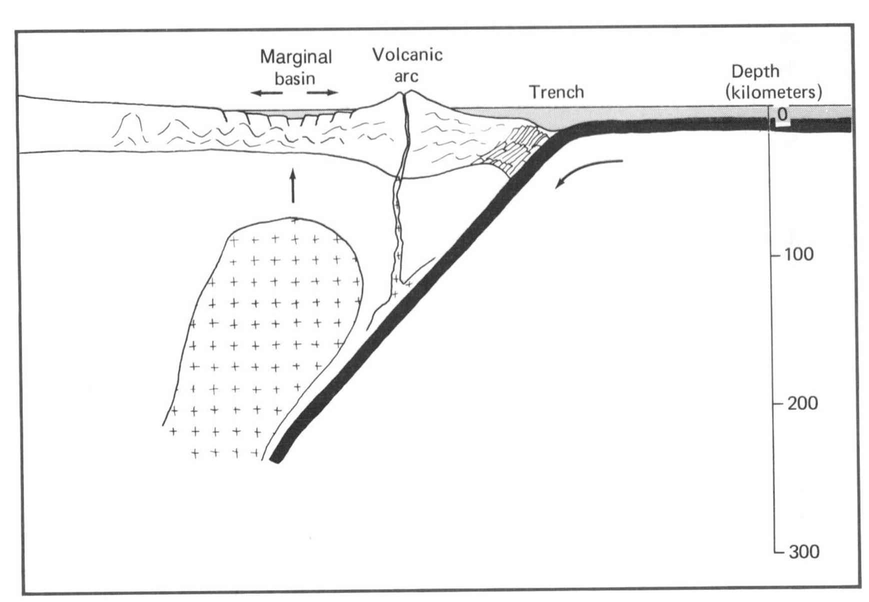

FIGURE 9.10

A rising dome of hot mantle material below a marginal basin as a possible cause of extension. (*Karig, 1971*)

This raises a fundamental question: What processes could cause an extension in a region where major lithospheric plates are meeting head on? The evidence for such extension is clear. The oceanic crust on the floor of the basins is broken into ridges and troughs characteristic of rifting. Those rifts which are active are widening by the addition of new oceanic crust along the middle of the basin. The marginal basins contain linear magnetic anomalies similar to those of the major oceanic plates, which are believed to reflect the emplacement of belts of new crust during extension.

Some geologists believe that the marginal basins are expanding in response to the forceful injection of basaltic magma that is differentiated from a rising dome of hot mantle material (Figure 9.10). The mantle is believed to be heated by friction at the top of the descending lithospheric plate, after which it rises buoyantly from deep within the mantle below the marginal basin. This rise of very hot mantle material would also explain the high heat flow observed in marginal basins. Although this explanation is possibly correct, it is by no means proven, and the origin of marginal basins remains a fundamental problem of plate tectonics.

Some Consequences of the Breakup of Pangaea

The plate movements that converted the Pangaean supercontinent into the present-day earth had far-reaching effects on the earth's rocks, minerals, climates, and life. For the remainder of this chapter, we shall examine some large-scale aspects of this reorganization before turning, in Chapters 10 and 11, to the smaller-scale events of post-Triassic history.

Hot Spots

Scattered around the globe are more than 100 small regions of isolated volcanic activity known to geologists as **hot spots.** Unlike most volcanoes, they are not necessarily found at plate boundaries; many occur in the middle of lithospheric plates, far from centers of seismic activity. Hot spots are certainly generated in the mantle. They may be the surface manifestations of rising plumes of hot mantle material, which ascend plastically from below the asthenosphere and intrude into the crust. The plumes apparently remain stationary relative to the earth's interior. Thus, hot spots generate chains of volcanoes as the lithospheric plates move over them. The Hawaiian Islands, which are thousands of kilometers away from the nearest plate boundary, are an excellent example of a volcanic island chain produced in this way.

The Hawaiian Islands were apparently formed as the Pacific plate moved northwestward over a single source of lava. The plate thus carried northwestward a trail of volcanoes of increasing age. At the western end of the Hawaiian Islands, a line of submerged mountains, the Emperor Seamounts, extends to the north (Figure 9.11). The entire Hawaiian-Emperor system

FIGURE 9.11

The northwest part of the Hawaiian Island chain consists largely of submerged seamounts, which turn northward to become the Emperor Seamount chain. These seamounts and islands, which are all of volcanic origin, formed from the northwest to the southeast as the Pacific plate passed over a hot spot.

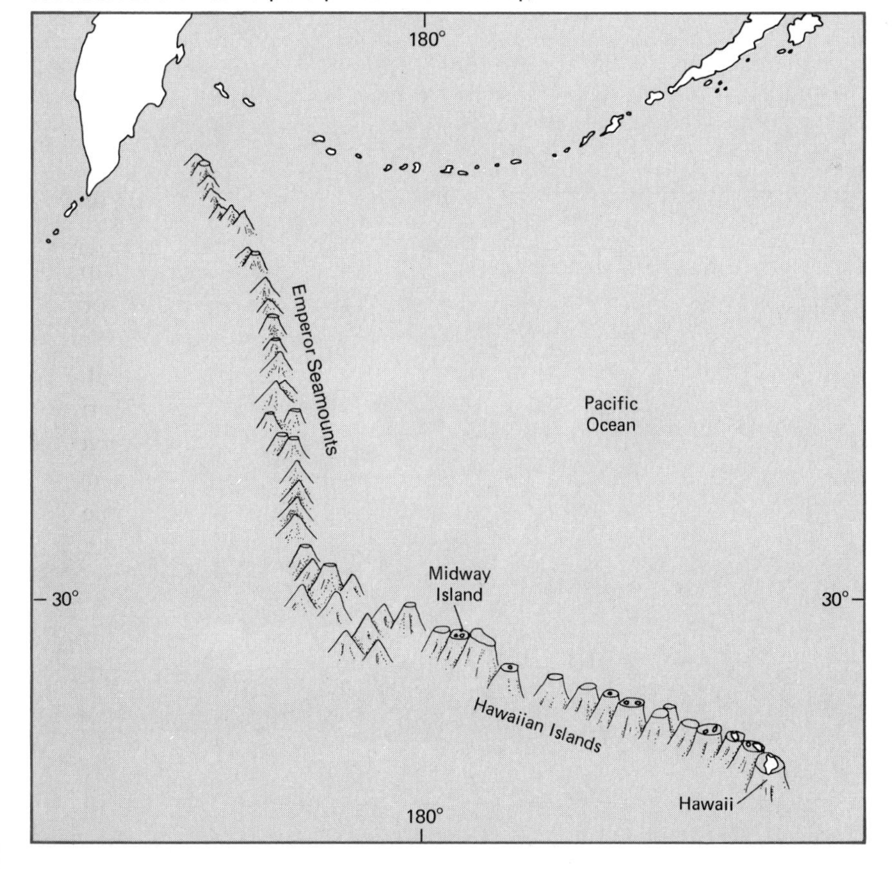

appears to be a single chain that is not straight because the direction of movement of the Pacific plate has changed. Near the bend, the oldest of the Hawaiian Islands are 40 million years old. The Emperor Seamounts continue the age sequence without interruption, beginning near the bend with an age of 40 million years, and continuing to an age of about 80 million years where the chain ends at the Kamchatka Peninsula. Two other chains of islands on the Pacific plate whose trends parallel that of the Hawaiian chain are the islands of the Austral Ridge and the Tuamotu Ridge (Figure 9.12). Like the Hawaiian Islands, the most recent activity in these chains is near the eastern end. Toward the northwest, both chains become older and both have an abrupt bend to a more northerly direction. These similarities provide good evidence that all three chains were generated as a result of the same plate motion.

Similar chains of islands and seamounts in other oceans are also due to hot spots. In the South Atlantic, the Walvis Ridge, extending seaward from Africa, and the Rio Grande Ridge, extending seaward from South America, were generated by a hot spot that coincides with the active volcanic island of Tristan da Cunha at the western limit of the Walvis Ridge (Figure 9.13*c*). The two ridges must have begun to form as soon as the South Atlantic opened; that is, the volcanic center that is now at the site of Tristan da Cunha was already active, and it lay on the newly opened rift. Volcanoes at this hot spot were built on both sides of the ridge and were rafted away by the spreading plates. As the

FIGURE 9.12

The Austral and Tuamotu volcanic island chains have the same pattern as the Hawaiian chain and were probably generated by separate hot spots below the Pacific plate. The dashed lines are submerged seamounts that are not well dated. (*Burke and Wilson, 1976*)

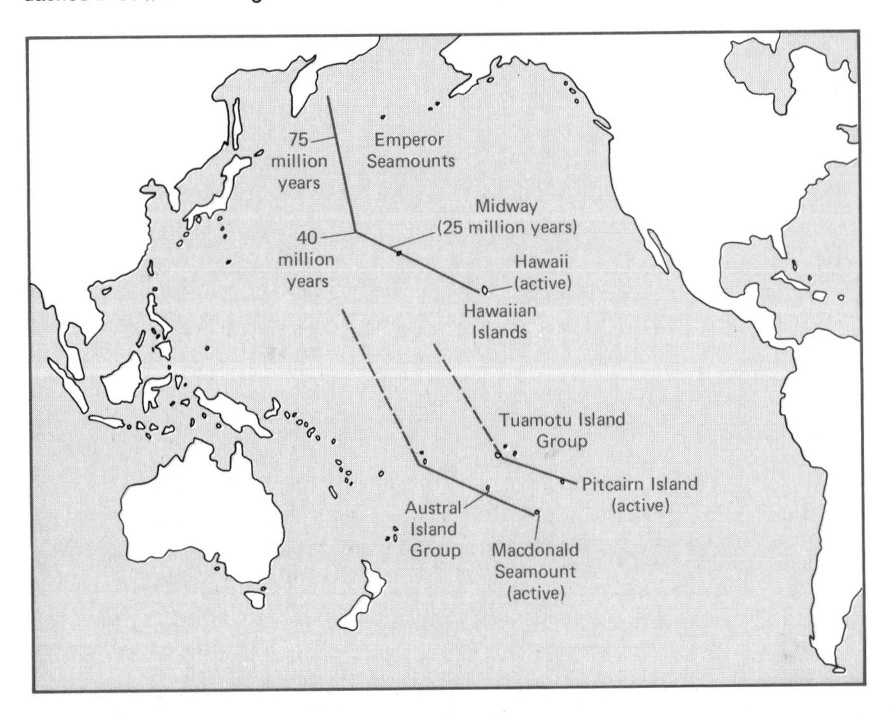

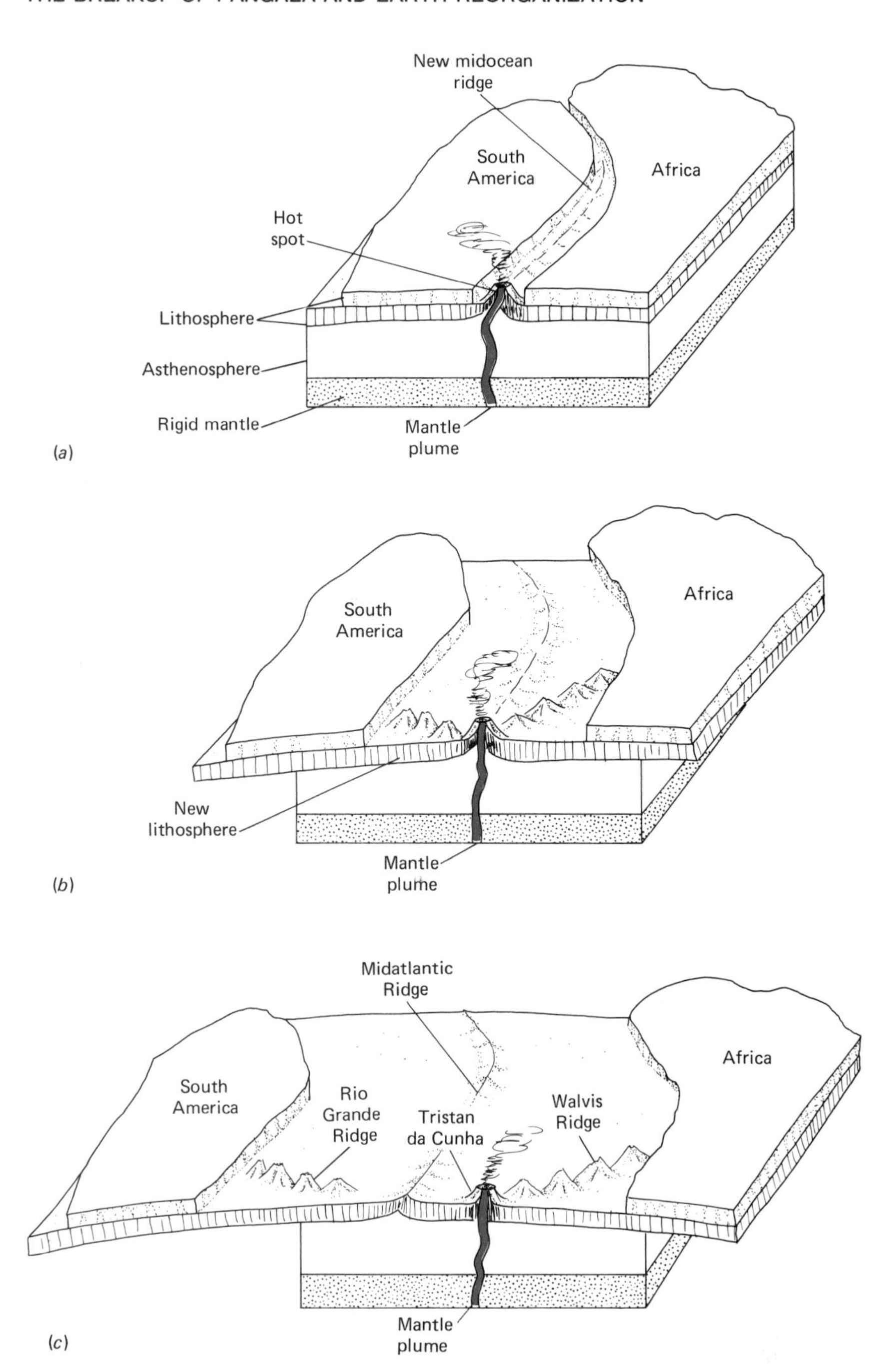

FIGURE 9.13

(*a*) Rising mantle plume about 90 million years ago forms a hot spot at new Midatlantic Ridge between Africa and South America. Volcanoes erupt over the plume. (*b*) Until 30 million years ago, the Midatlantic Ridge was stable relative to the deep, rigid mantle. The hot spot remained at the Ridge, forming new volcanoes as old ones were carried outward in both directions. (*c*) After 30 million years ago, the African plate became stable relative to the earth's deep mantle. The Ridge then migrated westward, leaving the hot spot behind.

plates separated, they also moved slightly northward, and the continuing volcanism formed a V-shaped pair of tracks. Today, however, the seaward end of the Rio Grande Ridge is inactive, and it is separated from the Midatlantic Ridge by a gap equivalent to 30 million years of sea-floor spreading (Figure 9.13*c*). The island of Tristan da Cunha lies a similar distance *east* of the present Midatlantic Ridge on oceanic crust about 30 million years old, and it has been a volcanic center since about that time.

It appears that about 30 million years ago the African plate, which had been moving eastward, became stationary with respect to the earth's mantle. The Midatlantic Ridge, which previously was stationary with respect to the mantle, began to move west. The hot spot was left behind on the stationary African plate. It could no longer produce a ridge; instead, its successive lava flows simply piled up in one place. Since lava was no longer deposited on the American plate, the Rio Grande Ridge was also terminated. Corroborative evidence that the African plate came to rest relative to the earth's mantle comes from the African continent itself, where other hot spots appear to have remained stationary for the past 30 million years.

Aulacogens

When a continent first begins to split apart on a newly formed oceanic ridge system, the initial spreading center characteristically consists of a series of domes produced by the upwelling of deep mantle material. The individual domes then develop large-scale tensional faults (rifts), typically in a three-armed pattern. Two of these arms widen progressively to become part of the new spreading center. The third arm, however, fails to widen further and remains behind as a fairly straight rift, or **aulacogen**, that extends directly from the newly formed continental margin onto one of the continental fragments produced by the breakup (Figure 9.14). These fault-bounded troughs are generally several tens of kilometers wide and up to several hundred kilometers long. They persist for tens, and in some cases hundreds, of millions of years, during which time they receive thick sequences of sediment. Aulacogens are distinctive in lying nearly perpendicular to continental margins or geosynclinal foldbelts.

FIGURE 9.14 Development of an aulacogen from a three-armed rift system. (*Burke and Wilson, 1976*)

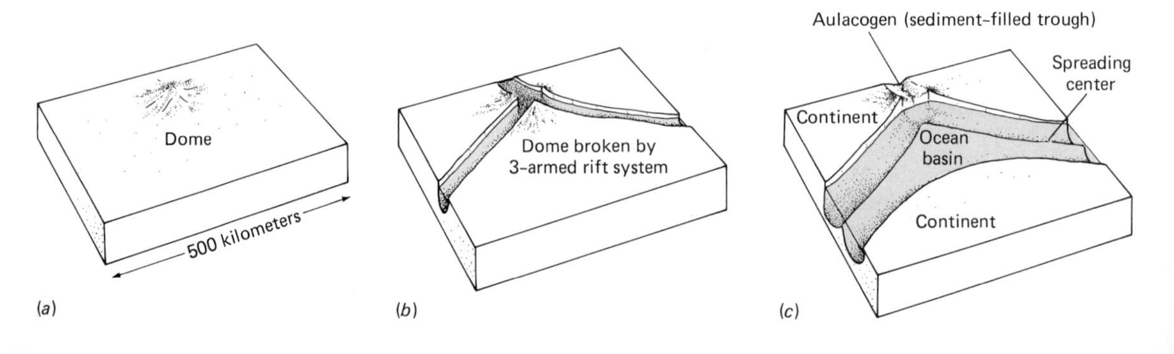

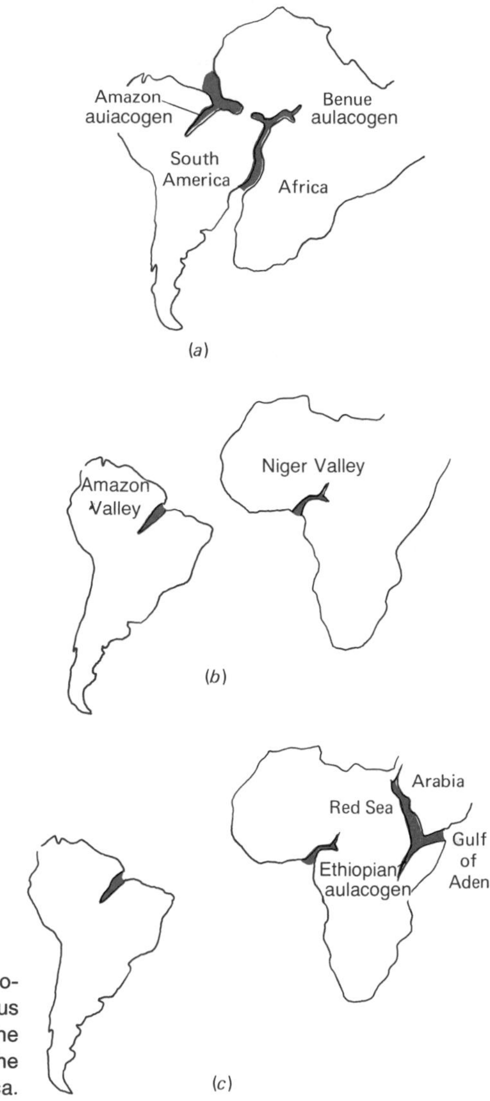

FIGURE 9.15

(*a*, *b*) The Amazon and Benue aulacogens formed during the Cretaceous opening of the South Atlantic. (*c*) The Ethiopian aulacogen formed during the Miocene splitting of Arabia from Africa.

Several elongate sedimentary basins on the margins of continents bordering the Atlantic Ocean have been identified as aulacogens whose origin dates from the opening of the Atlantic. Aulacogens have probably been important in localizing some of the major streams, such as the Amazon in South America and the Niger in Africa (Figures 9.15*a* and *b*). An example of a recently formed three-armed rift system lies at the tip of the Arabian Peninsula, where the Red Sea and the Gulf of Aden mark a spreading center that dates from the Miocene (Figure 9.15*c*). The Ethiopian rift trends southwestward, away from the widening ocean. It has ceased to enlarge further, but it is a site of much sediment accumulation and is considered an active aulacogen.

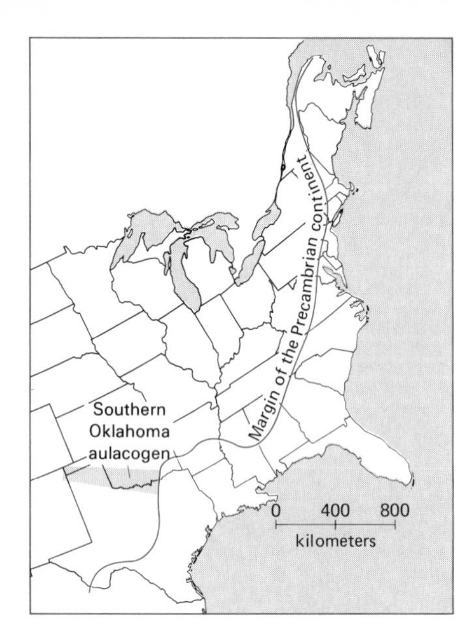

The Southern Oklahoma aulacogen was initiated in a late Precambrian continental breakup. In the late Paleozoic, it was deformed when a continent from the south collided with Paleozoic North America during the Hercynian Orogeny.

FIGURE 9.16

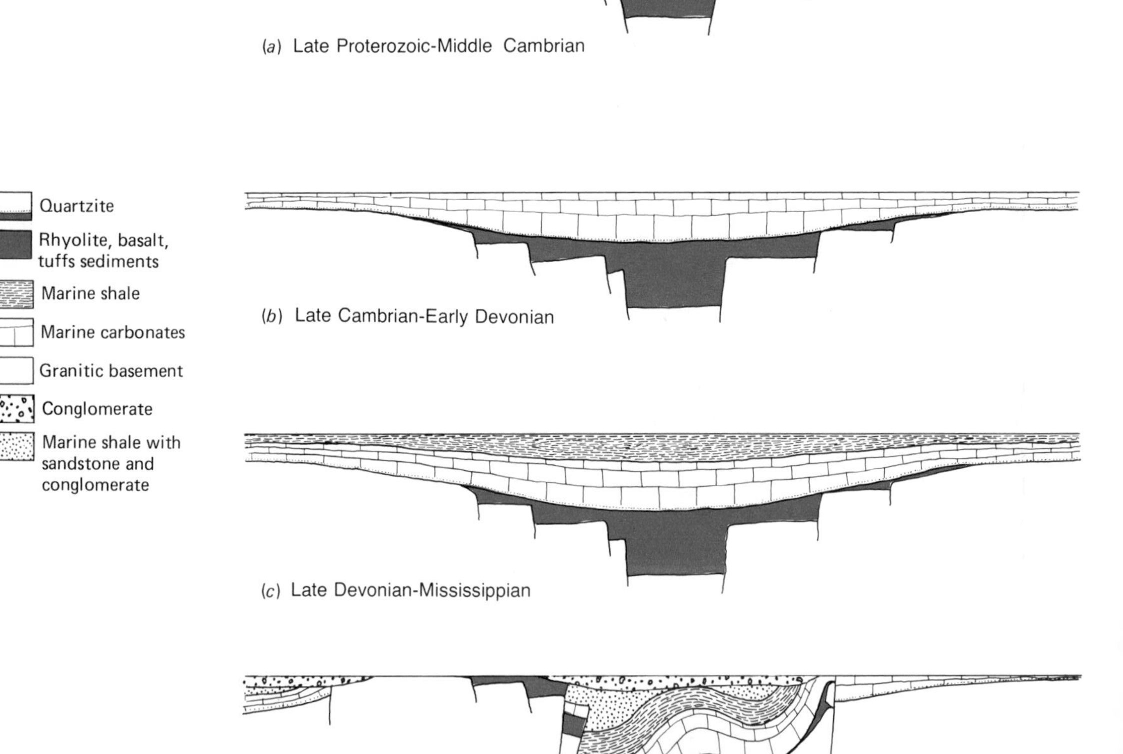

The Southern Oklahoma aulacogen begins with (*a*) block faulting and volcanism, (*b*, *c*) evolves into a broad unfaulted downwarp, and (*d*) ends with compressional stage of folding, faulting, and continental sedimentation. Location is shown in Figure 9.16. (*Hoffman and others, 1974*)

FIGURE 9.17

Aulacogens have been recognized far back in geologic history. The Oklahoma aulacogen (Figure 9.16), which lies adjacent to the Ouachita mobile belt, was apparently initiated in late Precambrian time, when the southern Oklahoma region flanked a new spreading center (Figure 9.17). This aulacogen was deformed in the late Paleozoic during the continental collision that produced the Ouachita folding. Two Precambrian aulacogens developed on the western margin of the Canadian Shield about 2 billion years ago when this area, too, was apparently a trailing continental margin at the boundary of a widening ocean. These features have been used to argue that plate motions like those in effect today were operating some 2 billion years ago.

Climatic Change

We saw in Chapter 8 that desert climates prevailed over much of the earth's land surface during Triassic time. This was a result of the continents being assembled into a single mass that straddled the equator. Triassic temperatures in the tropics were probably not much warmer than today's, but temperatures at the high latitudes were substantially warmer; that is, the *temperature gradient* from the tropics to the poles in Triassic time was somewhat gentler than it is today (Figure 9.18).

Estimates of the Triassic temperature gradient are based on the distribution of land and sea in the northern hemisphere. High latitudes dominated by seas are warmer than high latitudes dominated by land; this is because oceans absorb 90 percent of the solar radiation that falls on them, but continents absorb only about half; if they are snow-covered, 30 percent. The remainder is reflected back into space and is lost. Therefore, from the former positions of the continents, we can estimate the probable reflectivity of various portions of the earth's surface. These estimates plus rates of heat transfer by the ocean and atmosphere form the basis for the northern hemisphere **isotherms** (lines of equal temperature) during Triassic time (Figure 9.18*a*). As Figure 9.18 shows, the calculated Triassic temperature gradient for the northern hemisphere had a range of 20°C, whereas the temperature gradient today has a range of 41°C.

From Triassic time until the present, global temperature gradients increased as the continents in the northern hemisphere gradually moved northward to their present positions and displaced high-latitude oceans. Figure 9.19 summarizes the calculated temperature changes for the northern hemisphere following the breakup of Pangaea. From Triassic to the present time, the mean annual temperatures for the entire northern hemisphere decreased by only 3°C, but the mean *winter* temperatures for the 10-degree band of latitude between 60°N and 70°N decreased by 20°C. At this latitude (which at that time was only water) Triassic winter temperatures were about the same as today's summer temperatures. Seasonality on land increased greatly from the Triassic to the present as continents migrated to high latitudes.

As temperatures at the high latitudes decreased, circulation of the ocean and atmosphere increased because of the greater temperature difference between tropics and poles. (This difference is the driving force for the cir-

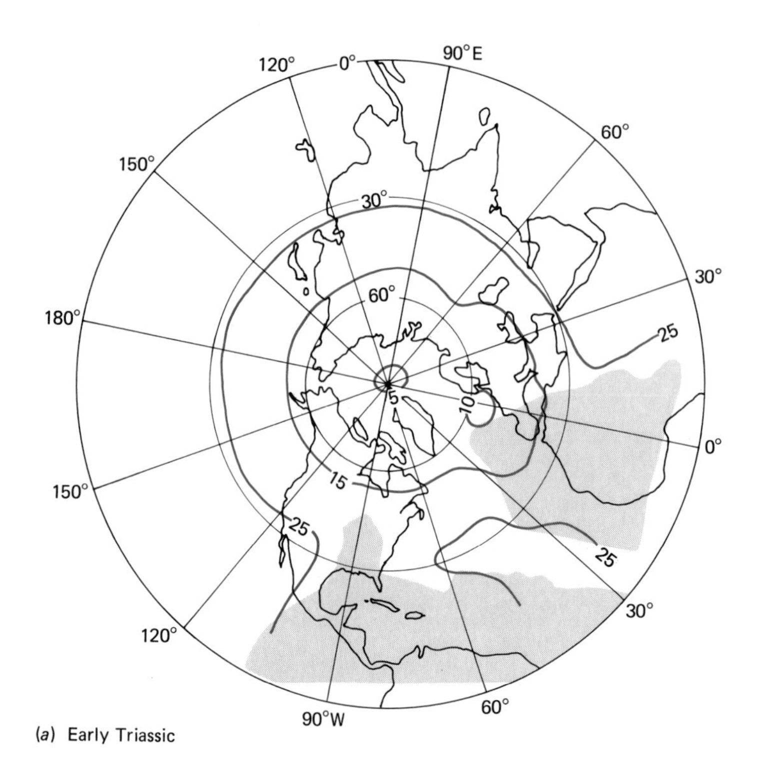

(a) Early Triassic

FIGURE 9.18

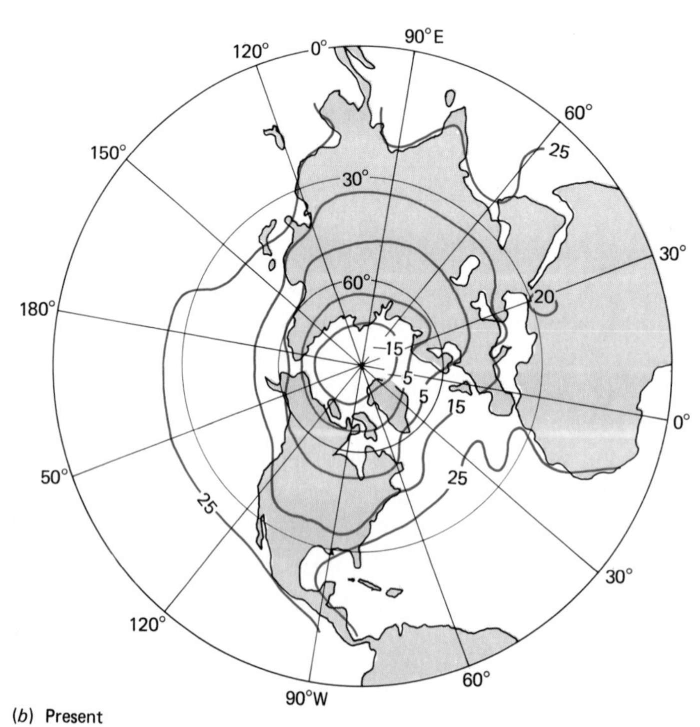

(b) Present

North pole views of the earth showing (a) northern hemisphere isotherms computed for the Early Triassic in increments of 10° C. Shaded areas are reconstructed positions of Triassic continents superimposed on present geography. (b) Northern hemisphere isotherms computed for the present. (*Donn and Shaw, 1977*)

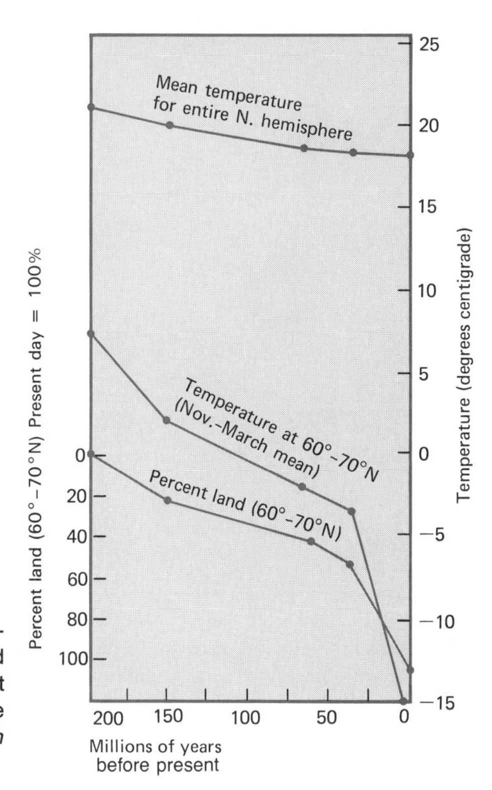

FIGURE 9.19

Increasing land in high northern latitudes during the last 200 million years greatly increased the northern hemisphere temperature gradient but changed the mean temperature of the entire northern hemisphere very little. (*Donn and Shaw, 1977*)

culation of the oceans and atmosphere.) During the Triassic, the global winds must have been lighter and the ocean currents weaker than they are today. One consequence of the lower rate of ocean circulation in the Triassic was probably much lower organic productivity. Today the productivity of the oceans depends greatly on upwelling, the vertical circulation that recycles nutrients from depth and brings them to the light zone, where they can be used by planktonic algae. In the Triassic, when overall circulation, including vertical circulation, was a lot less than today's, productivity of the seas must also have been less.

The global cooling that began after the Triassic breakup of Pangaea culminated in the late Cenozoic with large-scale glaciation, which has profoundly shaped our modern world. The causes of the glaciation are not known, and this subject will be reviewed in Chapter 11. Here we wish only to point out that it may not be necessary to call upon profound changes—such as alteration of the earth's orbit, variation in the composition of the atmosphere, or decrease in the energy output of the sun—in order to explain the global cooling that gave rise to the glaciation. Much of this cooling since Triassic time can be explained by plate motions acting to change the configurations of the land and seas.

Petroleum

During a comparatively brief time span of only about 30 million years in the middle of the Cretaceous Period, more than half of the world's known petroleum reserves were formed. About 70 percent of the middle Cretaceous petroleum accumulated in one relatively small region around the Persian Gulf.

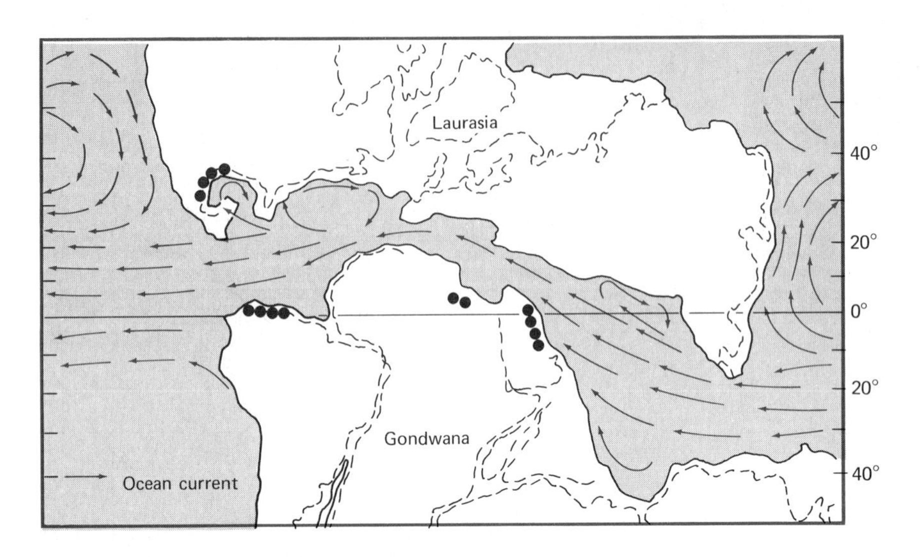

FIGURE 9.20

Possible relationship of the continents during a period of rapid plate motion in the middle Cretaceous showing a continuous Tethys-Atlantic seaway. Dots are major oil basins of the Persian Gulf, Libya, and the Americas. (*Irving and others, 1974*)

A large proportion of the remainder accumulated in another relatively small region of the Americas, between Venezuela and the Gulf of Mexico. Exceptionally rapid plate motions were largely responsible for preserving this great amount of petroleum during so brief a time.

By middle Cretaceous time the North Atlantic Ocean had opened substantially and had effectively extended the Tethys Sea 5,000 kilometers to the west, where it joined the Pacific. Thus a vast, low-latitude seaway now separated Laurasia on the north from Gondwana on the south (Figure 9.20). Along the continental margins that were retreating from the spreading centers in the seaway, large areas subsided, creating sedimentary basins. Among these basins were the Gulf Coast of the United States and Mexico, the Maracaibo Basin of Venezuela, the Sirte Basin of Lybia, and the Persian Gulf (Figure 9.20). The petroleum in these areas was derived from organic matter that accumulated in fine-grained marine sediments, which are referred to as **source beds.** In addition, sandstones, reefs, and other porous **reservoir rocks** from which petroleum may be produced were deposited in the vast sedimentary basins.

The phenomenal Cretaceous petroleum reserves owe their existence largely to extraordinary conditions of preservation of organic-rich sediments. Recent deep-sea drilling in the Atlantic Ocean encountered widespread middle Cretaceous black shales that are exceptionally rich in organic material. Today sediments similarly rich in organic material are forming only in isolated regions of the sea where the water contains little or no oxygen. We infer that, during the middle Cretaceous, huge regions of the oceans must have been anoxic, perhaps because the water circulated very slowly as a consequence of the weaker temperature gradients.

Much of the organic material was produced in the sea by planktonic microorganisms. In the Tethys seaway, the middle Cretaceous may have been a time of unusually great production of these organisms. Today the greatest sustained planktonic productivity occurs in tropical seas that are rich in nutrients, chiefly phosphorus and nitrogen compounds. The Tethys seaway had opened wide by the middle of the Cretaceous, and this may have altered circulation in the low-latitude oceans in a way that increased their productivity. By the middle of the Cretaceous, two groups of microorganisms (the planktonic foraminifera and the calcareous nannoplankton) were beginning to diversify widely for the first time. These new groups were perhaps able to exploit the nutrients in the planktonic environment more efficiently than had their predecessors. Thus the evolutionary diversification of these groups may have resulted in an increase in the overall planktonic productivity of the seas.

Large quantities of nutrients were buried with the organic material and needed to be replenished constantly from the land. Perhaps the extraordinary supply of these nutrients can be attributed to high sediment yields from rugged terrains and to high levels of volcanic activity at spreading centers and subduction zones, all of which could be attributed to high rates of sea-floor spreading and plate motions.

FIGURE 9.21

Relationship of large copper and molybdenum deposits to Mesozoic and Cenozoic subduction zones. (*Eimon, 1970*)

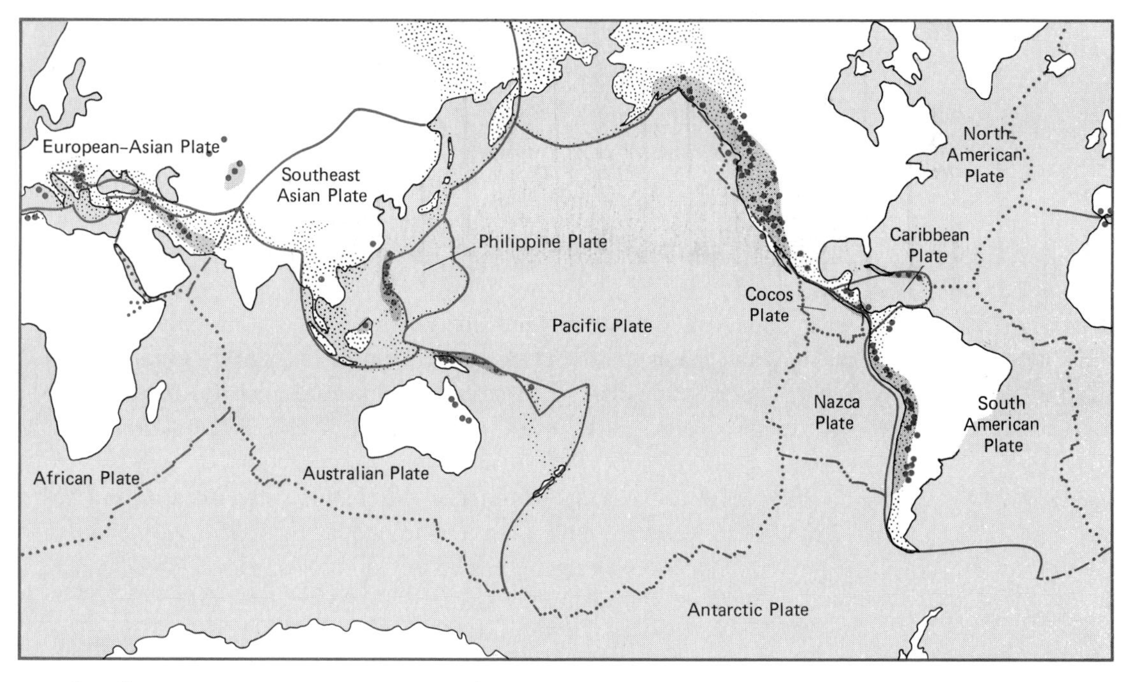

Mineral Deposits

Plate motions have implications for the origin not only of petroleum, but also for metallic mineral deposits. Much of our supply of copper, lead, zinc, gold, and silver is produced from **hydrothermal** concentrations, in which the ore minerals have been precipitated from metal-rich solutions. Until the discovery of plate motions, the origin of such deposits was uncertain, but in recent years it has become clear that most occur along formerly active plate margins.

Such deposits have been discovered to be forming today in deep basins at the bottom of the Red Sea, an incipient spreading center. There, direct sampling shows that volcanic magmas from below interact with extremely salty seawater at the bottom of the basins to produce the characteristic minerals of hydrothermal deposits. Many productive ore deposits appear to have originated in a similar fashion during the Mesozoic breakup of Pangaea, and still others appear to have originated above subduction zones. Figure 9.21, for example, shows the relation of plate boundaries to large-scale deposits of copper ore. Such knowledge, in turn, has helped to guide the search for new sources of these important minerals.

Reorganization of the Living World

Accompanying the dramatic late Phanerozoic reorganization of the continents were some equally striking changes in the living world. Many of the details of these changes are described elsewhere, for example, the Permian-Triassic extinction of many kinds of marine animals (Chapter 7), or the Mesozoic expansion of flowering plants on land (Chapter 10). There is, however, a broader perspective in which these changes can be viewed. It concerns not the details of individual kinds of animals and plants, but rather the overall changes in the complexity and diversity of the entire living world, or *biosphere.* In this section we shall view these broad patterns of life history before turning, in Chapters 10 and 11, to a more specific examination of late Phanerozoic life.

Patterns of Biological Diversity

The simplest and, in many ways, most instructive way of looking at overall patterns and changes in the living world is simply to count the number of *kinds* of fossil animals and plants that existed at some particular time in the geologic past. Such counts can be made at any level of classification—from species through genera, families, orders, and classes, to entire phyla.

Figure 9.22 shows such counts for all the common phyla, orders, and classes of shell-bearing marine invertebrates throughout the entire span of Phanerozoic time. Note that the number of phyla, the highest level of organization, has been almost constant; as we noted in Chapter 7, most animal phyla originated in the initial Cambrian radiation of metazoan life. Unlike the number of phyla, the number of orders and classes of marine invertebrates peaked in early Paleozoic time, declined sharply to Triassic time, and has been nearly constant ever since.

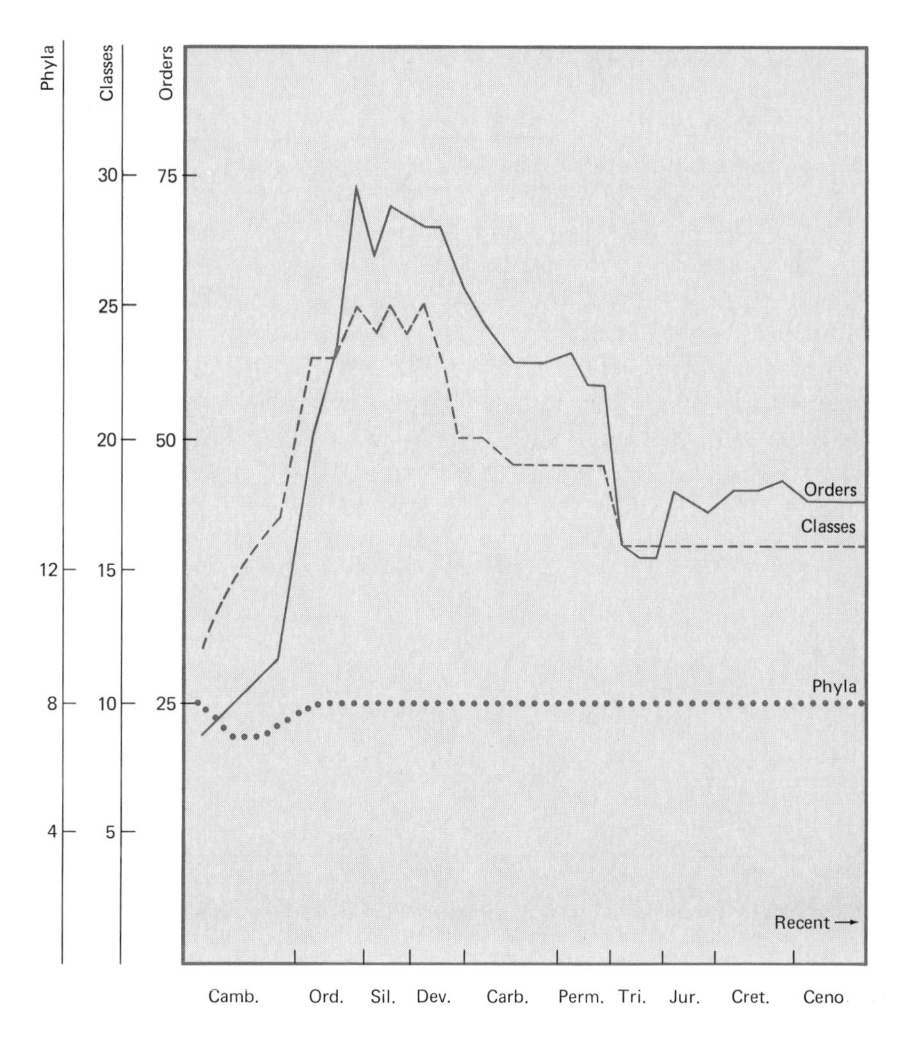

FIGURE 9.22

Diversity levels of phyla, classes, and orders of well-skeletonized marine invertebrates during Phanerozoic time. (*Valentine, 1969*)

Dropping down to the next level, that of families, we see an entirely different pattern emerging (Figure 9.23). The Paleozoic rise and decline seen before remains, but a still sharper rise in diversity begins in the Triassic Period and continues through Mesozoic and Cenozoic time to the present day. Counts at still lower levels (genera and species) are less reliable because the fossil record at these levels is less complete, but counts for some abundantly preserved groups, such as the foraminifera (Figure 9.24), show an even more dramatic rise in generic diversity after Triassic time.

Broadening our perspective still further, Figure 9.25 shows a plot of family-level diversity for the *entire* living world throughout Phanerozoic time. Note that the dramatic post-Triassic rise in diversity already seen in marine

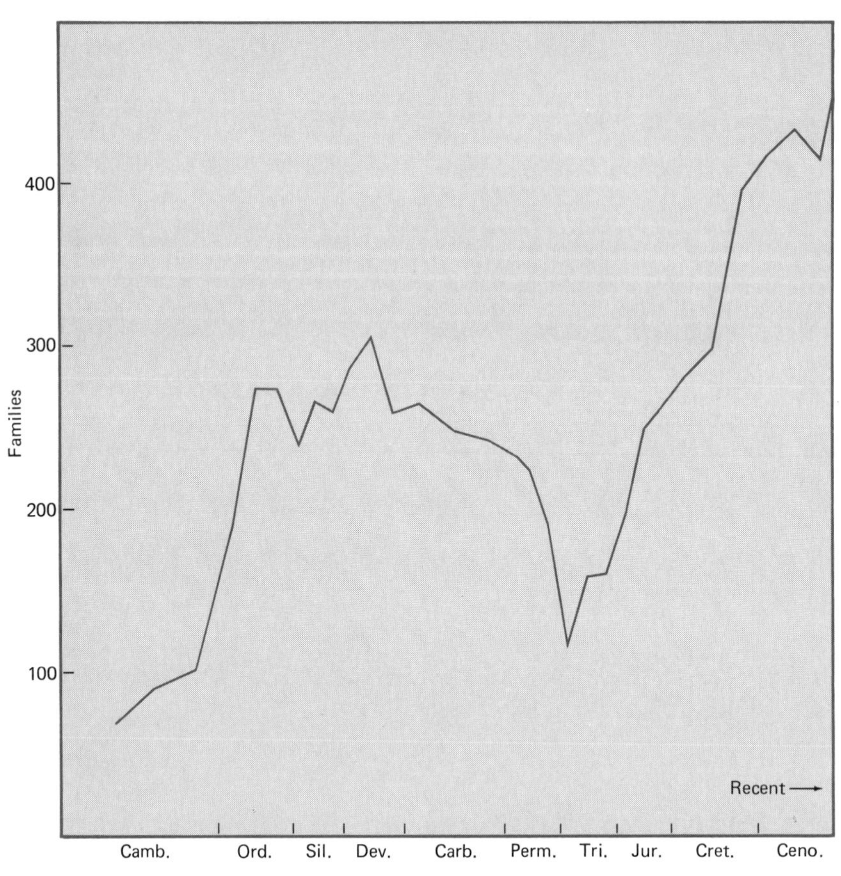

FIGURE 9.23 Diversity levels of families of well-skeletonized marine invertebrates during Phanerozoic time. (*Valentine, 1969*)

invertebrates is mirrored throughout the well-fossilized living world. Note also that unlike ocean-dwelling animals, which show their maximum diversity today, plants and land animals both reached their maxima in late Mesozoic or early Cenozoic time and have since declined. Note especially that the enormous Mesozoic increase in animal and plant families coincides closely with the breakup of the Pangaean continent, with which it is almost certainly interrelated.

(top of p. 307) Diversity levels of genera of fossil foraminifera during Phanerozoic time. (*Valentine, 1969*)

(bottom of p. 307) Standing diversities of Phanerozoic fossil families. (*McAlester, 1973*)

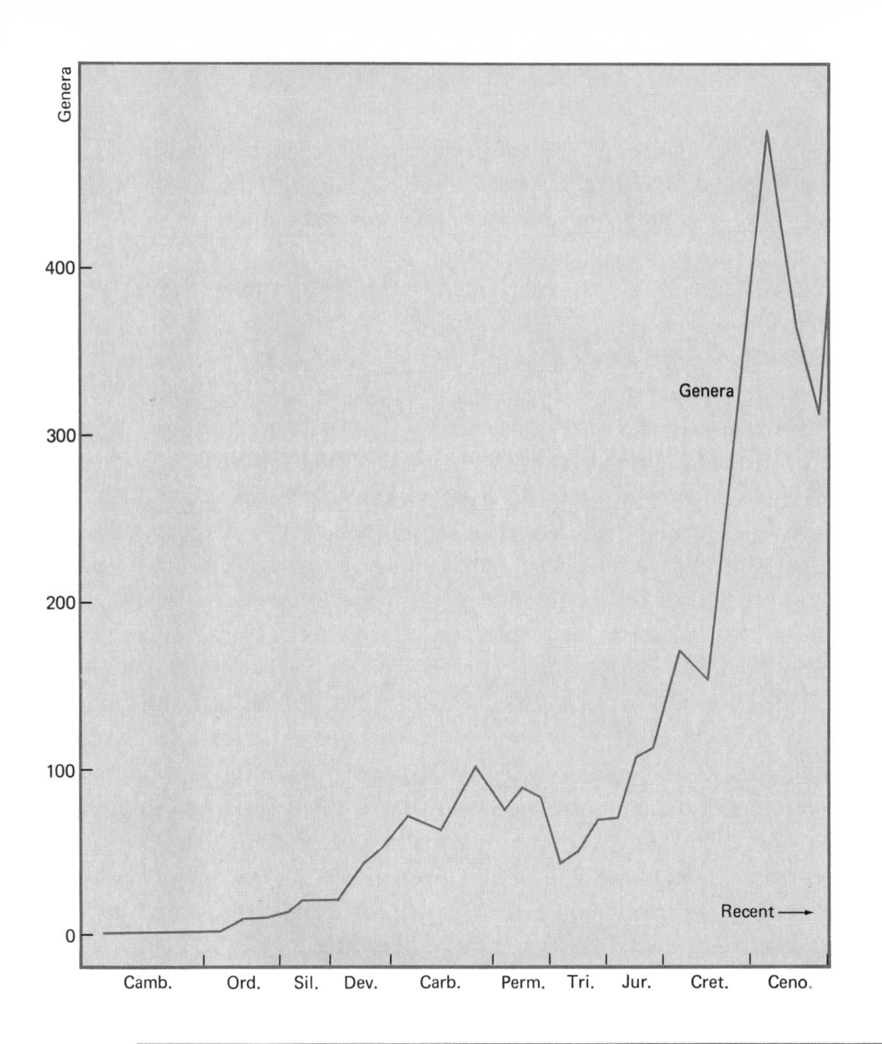

FIGURE 9.24

FIGURE 9.25

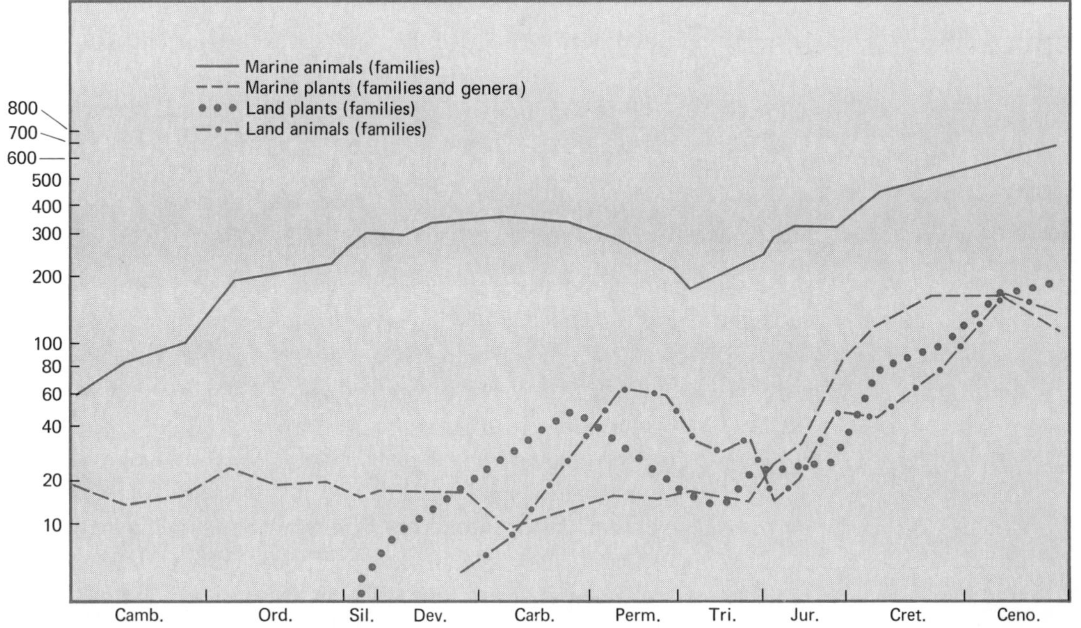

Mechanisms of Diversity Change

Before considering more closely the possible interactions between continental fragmentation and the living world, we need to look at patterns of change in diversity as they relate to evolutionary radiations and extinctions.

The counts shown above in Figures 9.22 to 9.25 reflect what is known as **standing diversity,** that is, the actual numbers of living families, orders, etc. on the earth at any one interval. Such counts do not, however, show the *turnover* of the living world, as some groups become extinct and others arise during evolutionary radiations. For example, if 90 percent of the families of land plants became extinct in one time interval, but were replaced by an equal number of newly evolved families in the next time interval, the standing diversity in each interval would be the same. It would appear as a straight line on graphs such as Figures 9.22 to 9.25, despite the large-scale extinctions and radiations. For this reason, changes in standing diversity occur only when the number of radiations differs from the number of extinctions over two or more time intervals. Standing diversity plots thus show only the broadest patterns of change in the living world.

A somewhat more detailed look at diversity change is given in Figure 9.26, which shows the *percentage* increase or decrease (expansions or extinctions) in animal and plant families throughout Phanerozoic time. Note how periods of extinction and expansion tend to occur at the same time for most of the components of the living world. In particular, note that the dramatic extinctions of Permian and Late Triassic time were followed by strong Jurassic and Cretaceous evolutionary radiations. These events caused the sharp Mesozoic rise in standing diversity seen in Figures 9.22 to 9.25.

Many hypotheses have been advanced to explain these parallel expansions and contractions of the living world. Evolutionary expansions normally follow periods of extinction and can most logically be viewed as a refilling of environments vacated during the preceding extinction. A more difficult problem lies in explaining periods of wholesale extinction.

All theories that seek to explain widespread extinctions call on profound, worldwide environmental changes. Among the most commonly suggested are increased radiation from space, large-scale changes in climate, or variations in atmospheric composition.

An increase of high-energy particles from space (cosmic radiation) might occur during intervals when the earth's magnetic field, which normally deflects such particles, is reversing. Since these cosmic particles can be absorbed by a thin layer of water, they should affect only land-dwelling organisms, not bottom-dwelling life in the sea. However, Figure 9.26 shows that such marine life tends to decline at the same times as life on land.

Perhaps the most commonly advanced cause of extinctions is large-scale changes in climate. Since most animals and plants are adapted to a rather narrow range of temperature variations, severe changes in climate thus might be expected to lead to wholesale extinction. The difficulty with this concept is that although climatic changes normally shift the position of climatic zones, they do not eliminate them completely. Because such climatic changes take

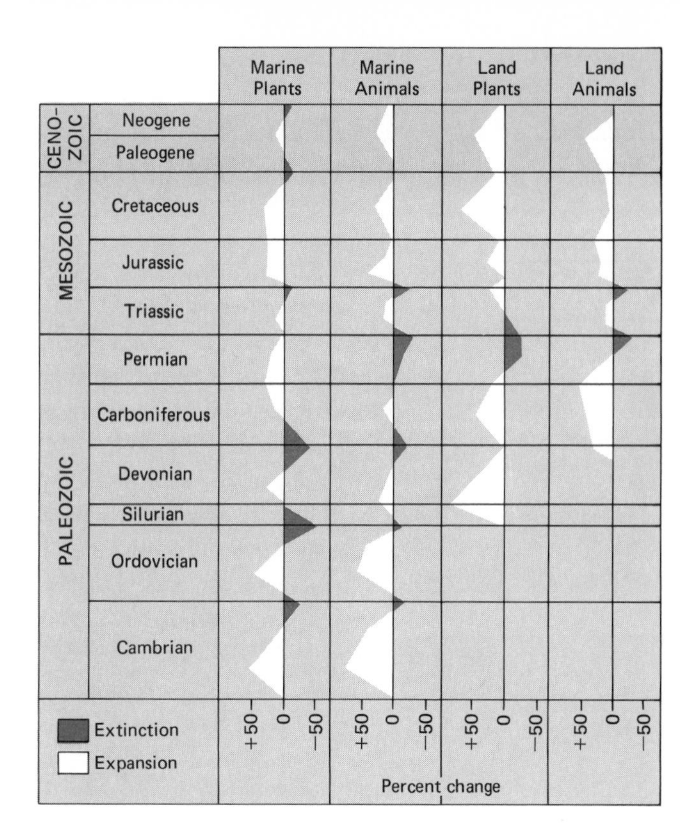

FIGURE 9.26

Phanerozoic expansions and extinctions of life. The curves show the net expansions or extinctions of animal and plant families during each Phanerozoic period. Paleozoic intervals of general extinction occurred near the end of the Cambrian, Ordovician, Devonian, and Permian periods; each was followed by a major expansion. (*McAlester, 1973*).

place rather slowly, the inhabitants of a zone might have gradually moved with the changing position of the zones. For this reason, many geologists feel that climatic changes have served principally to alter the ranges and distributions of animals and plants, rather than to cause wholesale extinctions.

Still another hypothesis attempts to relate large-scale extinctions to changes in the composition of the atmosphere. We saw in Chapter 4 that a slow Precambrian build-up in atmospheric oxygen may have been a possible cause for the dramatic Cambrian expansion of metazoan life. Similarly, changes in the rate of production of oxygen by green plants or the depletion of oxygen by weathering of rocks and by combination with reduced volcanic gases might have led to changes in the amount of atmospheric oxygen during post-Cambrian time. Such changes, if severe, might have led to the extinction of many sensitive, oxygen-dependent organisms, both animals and plants. There is, however, little independent evidence of such large-scale Phanerozoic fluctuations in atmospheric composition.

Still another hypothesis concerning evolutionary expansion and extinction is particularly appealing in relating the sharp Mesozoic rise in family diversity to continental fragmentation. In theory, when continents coalesce to form one large land mass, there should be relatively few barriers to the migration of animals and plants, whether on the continent itself or on the interconnected shallow seas surrounding it. Under such conditions, there might be competition for food or living space among formerly isolated organisms, leading to the extinction of many groups. Conversely, during times of continental fragmentation, independent evolutionary change could occur on each continental mass, leading to a sharp worldwide increase in standing diversity.

In addition to this simple effect of continental isolation on diversity, there is the more complex question of the relation of climate and continental position. Today latitudinal position affects the extent of animal and plant diversity: Tropical regions typically have far greater numbers of animal and plant species, genera, and families than do adjacent temperate regions. Temperate regions, in turn, are far more diverse than adjacent polar regions. Because the earth has almost certainly always shown some equator-to-pole temperature variations, it is likely that ancient continents in tropical latitudes had greater diversities than those in polar regions. In addition, single continents extending over a range of climatic belts, such as present-day Africa and the Americas, have even greater total diversities, because separate associations of animals and plants develop in each belt to form distinct provinces.

Such considerations have led to the appealing suggestion that the Mesozoic breakup of Pangaea might be the *direct* cause of the dramatic Mesozoic rise in the number of families of animals and plants. It is tempting to speculate further that the mid-Paleozoic rise in diversity similarly reflects increasing latitudinal distribution of continents. The dramatic late Paleozoic extinctions might thus correspond to the assembly of these continents into the Pangaean supercontinent.

Chapter Summary

Post-Triassic Plate Motions

The moving continents: The Triassic breakup of Pangaea left tensional rift structures along the margins of the newly separated continents; thick evaporite deposits formed in the narrow seaways between.

India's paleomagnetic track record indicates a progressive northward movement from a starting point near the south pole.

Triple junctions are found where three plates converge; they are sites of complex motions and interactions.

Marginal basins lie behind island arcs along some subducting plate margins; they show tensional features that are difficult to explain in zones where plates are converging head on.

Some Consequences of the Breakup of Pangaea

Hot spots are isolated volcanic regions away from plate margins; they probably reflect rising plumes of hot material from the underlying mantle.

Aulacogens are deep rifted troughs formed at right angles to newly opened continental margins; ancient examples are evidence of former plate margins.

Climatic change: Progressively cooler and more seasonal climates resulted from the breakup of Pangaea and the general northerly movement of the newly formed continents.

Petroleum: Much of the earth's petroleum originated from a mid-Cretaceous expansion of oceanic plankton in a large equatorial seaway.

Mineral deposits: Much of our supply of copper, lead, zinc, and other metals originated from the interactions of volcanic gases and ocean water along post-Pangaean plate boundaries

Reorganization of the Living World

Patterns of biological diversity: The numbers of families, genera, and species of fossil life show a dramatic post-Triassic rise.

Mechanisms of diversity change: The history of life reflects periodic intervals of widespread extinction followed by evolutionary expansion; some of these might have been caused directly by the assembly and breakup of continental masses.

Important Terms

aulacogen
diapir
Farallon plate
fault-block basin
hot spot
hydrothermal
isotherm
magnetic anomaly
marginal basin
polar-wandering curve
reservoir rocks
rift valley
source bed
standing diversity
triple junction

Review Questions

1. Summarize the history of the narrow ocean basins formed by the initial breakup of Pangaea.
2. How do triple junctions and marginal basins originate from plate motions?
3. What are hot spots, and what do they suggest about the earth's deep interior?
4. What effects did the breakup of Pangaea have on world climates? Why?
5. How have plate motions led to the origin of certain valuable mineral concentrations?
6. How might the breakup of Pangaea have affected the diversity of the living world?

Additional Readings

Bonatti, E. "The Origin of Metal Deposits in the Oceanic Lithosphere," *Scientific American,* vol. 238, no. 2, pp. 54-61, 1978. *Good review of latest ideas on the origin of hydrothermal ores.*

Burke, K. "Aulacogens and Continental Breakup," *Annual Reviews of Earth and Planetary Sciences,* vol. 5, pp. 371–96, 1977. *An advanced review.*

Burke, K. C. and J. T. Wilson "Hot Spots on the Earth's Surface," *Scientific American,* vol. 235, no. 2, pp. 46–57, 1976. *Reviews recent ideas on these unusual volcanic features.*

Dietz, R. S. and J. C. Holden "The Breakup of Pangaea," *Scientific American,* vol. 223, no. 4, pp. 30–41, 1970. *A much-quoted popular summary.*

Marsh, B. D. "Island-Arc Volcanism," *American Scientist,* vol. 67, pp. 161–72, 1979. *Intermediate-level review of marginal basins and related features.*

Sclater, J. G. and C. Tapscott "The History of the Atlantic," *Scientific American,* vol. 240, no. 6, pp. 156–74, 1979. *Up-to-date summary of this important phase of the breakup of Pangaea.*

Toksoz, M. N. "The Subduction of the Lithosphere," *Scientific American,* vol. 233, no. 5, pp. 89–98, 1975. *Good nontechnical survey of marginal basins.*

Valentine, J. W. *Evolutionary Paleoecology of the Marine Biosphere,* Prentice-Hall, Inc., Englewood Cliffs, New Jersey, 1973. *Advanced discussions of biotic diversity changes through time.*

JURASSIC TO PLIOCENE TIME

Progressive Modernization of the Earth

10

Evolving Landscapes

When the Mesozoic Era began, the continents were emergent, much as they are today. As the Triassic Period progressed, however, epeiric seas began to occupy portions of what had previously been dry land. Then, during the ensuing Jurassic, shallow seas advanced widely over western North America and also over parts of Europe, Asia, South America, and Africa (Figure 10.1). For the first time since the late Paleozoic, a substantial portion of the continental platforms was covered by a thin layer of water. Because continental patterns had changed markedly since the Paleozoic, the configurations of the Mesozoic epeiric seas bore little resemblance to their predecessors. Also, the faunas that accompanied the Mesozoic seas onto the continental platforms were significantly different from those of the Paleozoic, reflecting the profound evolutionary radiations of many invertebrate groups following the Late Permian extinctions. Widespread Jurassic limestones, shales, and sandstones found in northern Europe contain abundant fossils; these strata were among the first anywhere to be mapped and their orderly sequence of faunas to be perceived as distinct zones.

At their maximum extent, near the end of the period, the Jurassic seas covered about one-fourth of the total area of the continents (Figure 10.2). Then they regressed temporarily, but early in the Cretaceous Period they again advanced widely, and during portions of the Late Cretaceous they covered a greater area of the continental platforms than they have at any time since the Paleozoic (Figure 10.3). As the end of the Cretaceous approached, the seas began a large-scale retreat from the continents. Numerous minor marine

Distribution of land and sea in part of North America in Late Jurassic time. (*After U.S. Geological Survey Map I-175*)

FIGURE 10.1

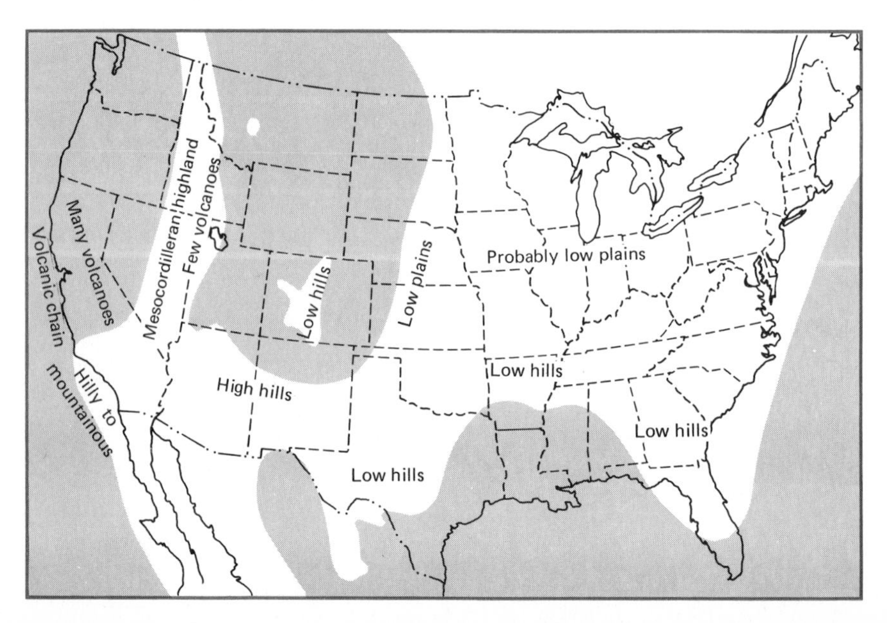

FIGURE 10.2

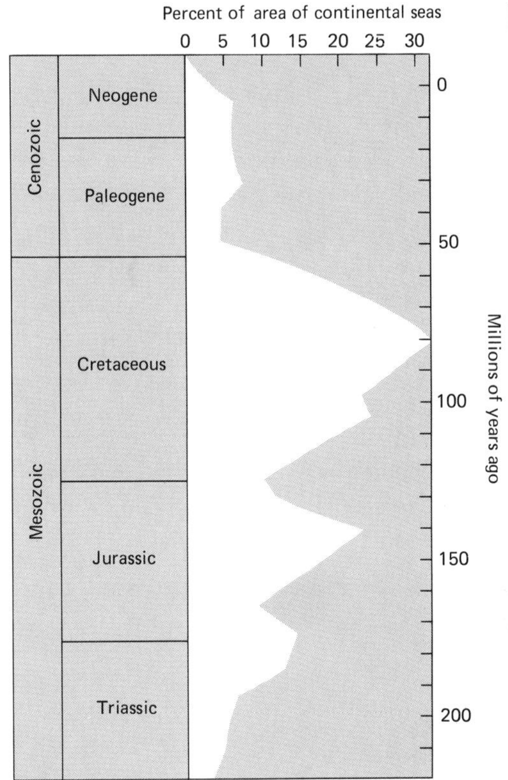

Proportion of North America covered by shallow seas since Triassic time. (*A. G. Johnson, 1971*)

incursions occurred during the Cenozoic, but the continents have generally tended to become increasingly emergent until the present day (Figure 10.2).

During the Jurassic, Cretaceous, and Cenozoic, shallow-marine strata were deposited widely, but many sediments also accumulated in marginal, near-shore environments and on adjacent low-lying continental areas as well. The seas that occupied much of the Jurassic and Cretaceous continental platforms provided a source of moisture to the adjacent lands, and the arid climates that had prevailed widely during the Triassic were replaced by more humid conditions. Evaporites and eolian sandstones indicate that arid conditions still existed locally, but during Jurassic and Cretaceous time many regions that were formerly deserts acquired moist climates.

FIGURE 10.3

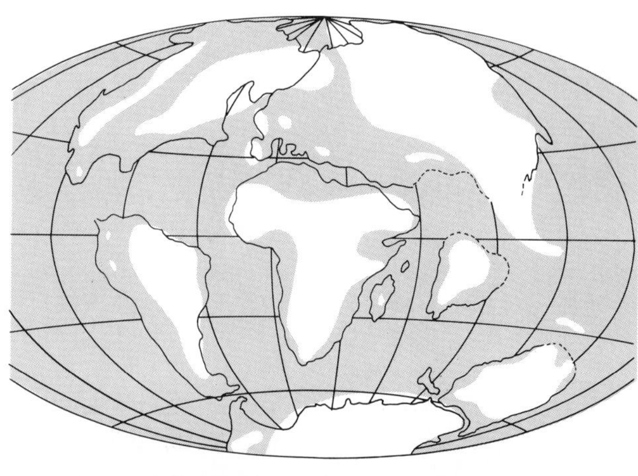

Extent of Cretaceous seas during the widest coverage of the continents in Late Cretaceous time.

Between Jurassic time and the present, as shallow seas periodically transgressed and regressed from the continental platforms, the deep-ocean basins changed shape markedly, as we have seen, some growing by sea-floor spreading and others shrinking. All parts of the deep-ocean basins have received sediments sporadically since they formed, so that the oldest stratum on a given portion of oceanic basaltic crust is a good approximation of the age of the underlying basaltic crust itself. Remarkably, nowhere in the ocean basins have sedimentary rocks older than Jurassic been found. Several hundred deep holes have now been drilled in the ocean basins, so the likelihood of finding significantly older rocks anywhere in the ocean basins appears to be small. Virtually all the oceanic crust that was produced prior to the Jurassic has apparently been subducted and thereby returned to the mantle. Jurassic and Cretaceous strata in the deep sea thus provide our earliest records of undisturbed open oceanic enviornments.

Tectonic Patterns

Present-day earthquake activity outlines the dynamic margins of today's lithospheric plates (Figure 10.4). These patterns evolved from plate motions that actually began during the initial Mesozoic breakup of Pangaea. One such margin marks the site of the former Tethyan seaway and extends from the Alps, across southern Asia, to the Himalayas; another is the Cordilleran belt that parallels the western boundary of the Americas.

FIGURE 10.4 Locations of epicenters of all earthquakes recorded from 1961 to 1969. (*National Earthquake Information Center, U.S. Department of Commerce*)

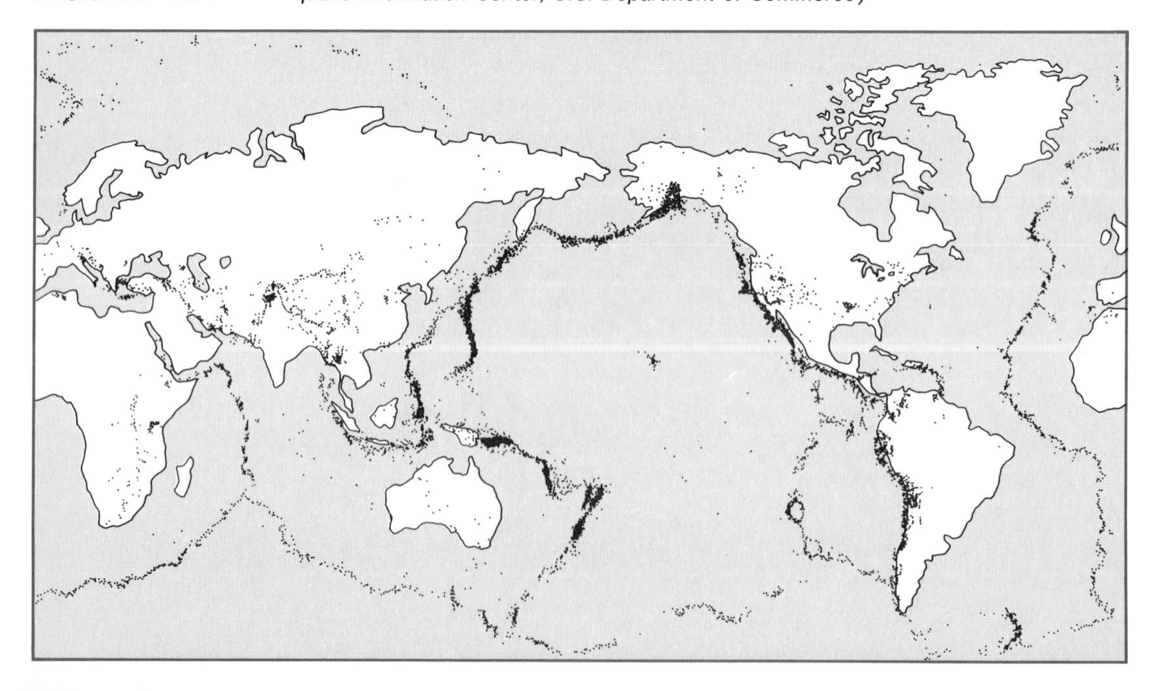

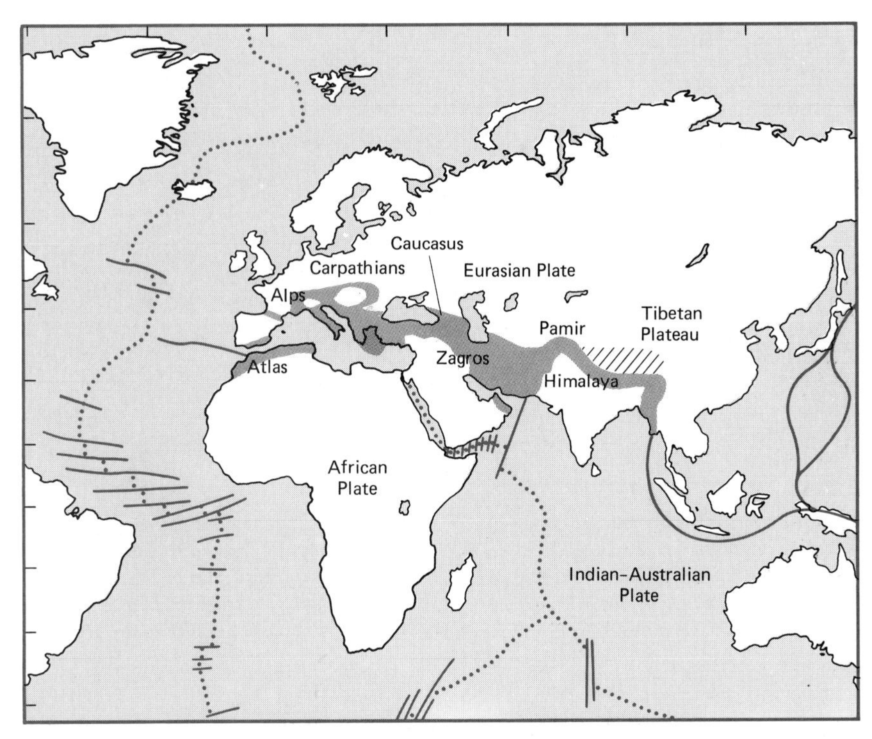

FIGURE 10.5

The Alpine-Himalayan mountain belt marks the collision of the African and Indian-Australian plates on the south with the Eurasian plate on the north.

The Alps and Himalayas During the Triassic, the Tethys Sea lay across the tropics and formed an enormous oceanic gulf between Laurasia on the north and Gondwana on the south (see Figure 9.1). The warm waters that covered its bordering platforms favored the deposition of carbonate sediments. From Triassic to Oligocene time, as the Tethys closed, thick sequences of limestone and dolomite accumulated; these strata contain rich and diverse invertebrate faunas and numerous reefs. The close of the Tethys Sea during the Mesozoic and early Cenozoic not only altered world geography, but also altered global patterns of oceanic circulation and climate. Where the Tethys was consumed, the colliding plates formed a huge chain of mountains, with the Atlas and the Alps on the west and the Himalayas on the east (Figure 10.5). This collision zone is still seismically active, as Figure 10.4 shows.

In Oligocene and Miocene time, large-scale folding and overthrusting occurred at the boundary between the African and Eurasian plates. During this episode, huge crustal slabs were transported northward several tens of kilometers, some of them rolling over their frontal lobes like huge caterpillar tractor treads as they moved. These nappes, as they are called, piled up on the southern margin of the Eurasian continent, where today they constitute a significant part of the geologic framework of the Alps (Figure 10.6). Since Miocene time, the Alps have risen to their great height as a result of isostatic

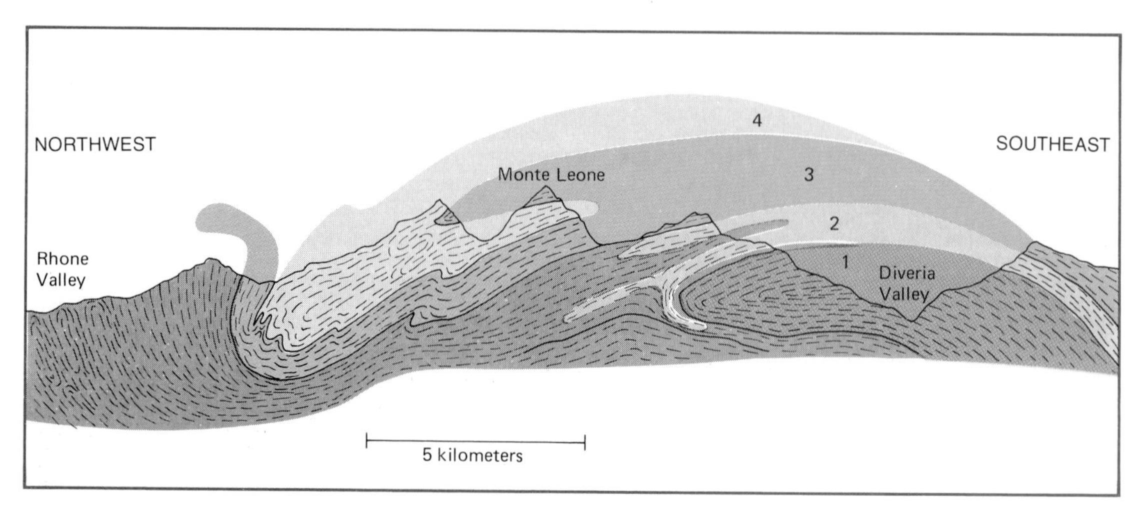

FIGURE 10.6 Cross section through part of the Alps on the Swiss-Italian border showing (1) the Antigorio Nappe, (2) the Lebendum Nappe, (3) the Monte Leone Nappe, and (4) the Great St. Bernard Nappe, which were thrust northward during the Alpine Orogeny. The rocks are chiefly metamorphosed sediments of late Paleozoic and Mesozoic age. The vertical scale is not exaggerated.

Microcontinents between the converging African and Eurasian plates during Paleocene time. Later collisions here produced the Alps. (*Dewey and others, 1973*)

FIGURE 10.7

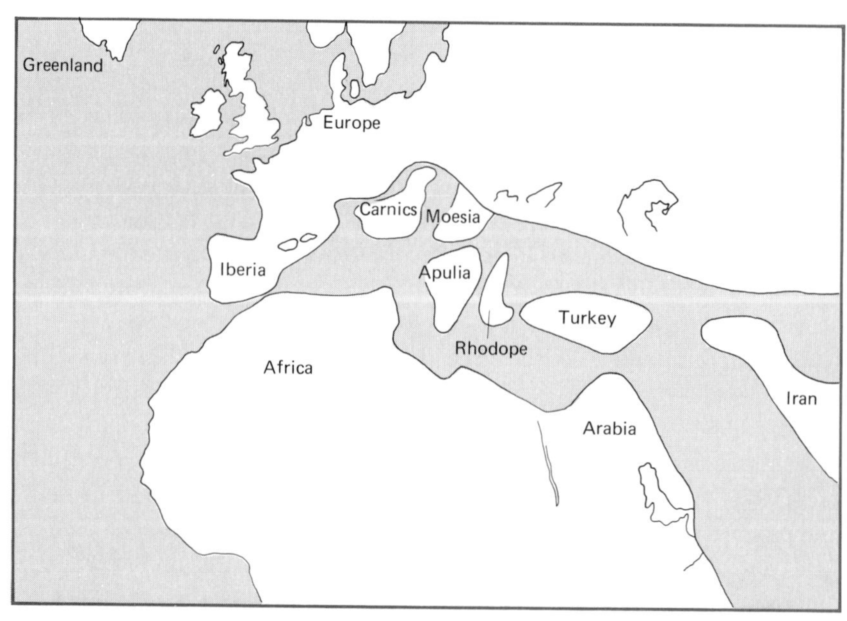

uplift. The present-day Alps are thus a culmination of a long episode of subduction of the ocean crust that once floored the Tethys Sea, followed by the collision of the Eurasian and African continents.

The process was complex, because the northern boundary of the African continent became fragmented into numerous smaller blocks, sometimes called **microcontinents**, each separated by oceanic crust (Figure 10.7). Today numerous ophiolite belts that range in age from Jurassic to Eocene constitute the only remains of the oceanic crust that once separated these microcontinents (Figure 10.8). These ophiolite belts lie chiefly in the Alpine belt *north* of the Mediterranean Sea, indicating that the Mediterranean is not a simple remnant of the old Tethyan seaway, as was once supposed. Rather, the Mediterranean opened up behind the small continental blocks that were actually part of the leading edge of the African plate and that now make up the Italian and Ionian Peninsulas and Turkey.

The highest mountain range on earth, the Himalayas, extends for 2,400 kilometers along the boundary between the Indian subcontinent and the remainder of Asia. The Himalayas form a sharp, northward-protruding spur where they encounter the Pamir Mountains, a massive range in central Asia (Figure 10.5). Immediately east of the Pamirs lies the Tibetan high plateau, which is the largest concentration of mass above sea level on the earth's surface.

Geologists now believe that the Himalayan Range and the very high terrain immediately to the north actually formed from the collision of India with the remainder of Asia, with concomitant thickening of the continental crust of that region. The great vertical relief of the Himalayas and the Tibetan Plateau, equalled nowhere in the world, reflects a virtual doubling of crustal thickness

Distribution of the ophiolites believed to be relict ocean floor and related igneous and metamorphic rocks mark ancient subduction zones that evolved into the Alpine system. Age of activity ranges from Jurassic to Miocene. (*Dewey and others, 1973*)

FIGURE 10.8

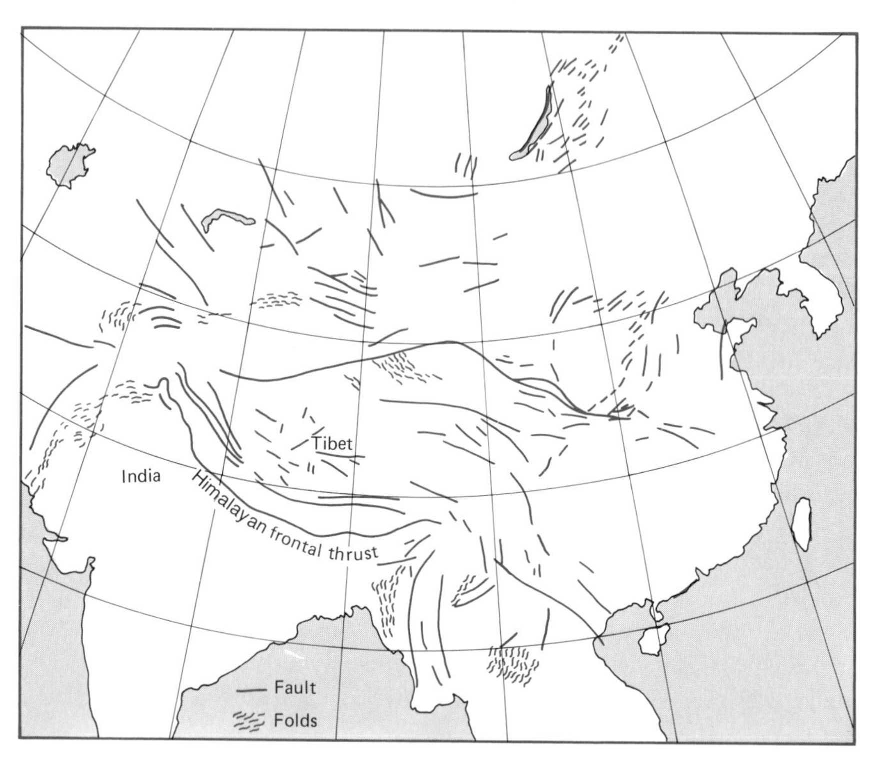

FIGURE 10.9

Faults and folds believed to have been caused by the continuing collision of the Indian and Eurasian plates. (*Molnar and Tapponnier, 1977*)

that was perhaps caused by accordionlike crustal shortening or, less likely, by the entire Tibetan Plateau being underthrust by the northern part of the Indian subcontinent. The Himalayas are very young mountains, and even today they are being actively uplifted, as earthquakes throughout the region attest (Figure 10.4). Fault systems extend from the Himalayas far to the north and east (Figure 10.9). Analyses of movements along these faults suggest that China is actually being shoved eastward, out of the way of the advancing Indian plate.

The Indian continent probably first collided with the Eurasian continent in the Middle Eocene. At this time a large number of Asiatic land animals appear for the first time in the fossil record of the Indian peninsula, indicating that a land connection had been established. Actual deformation of the Himalayan belt began a few million years later in the Oligocene, after the entire ocean basin that once lay between India and Eurasia had disappeared, and the two continents actually met along their full length. Subsequently, the extraordinarily thick continental crust beneath the Himalayas and the Tibetan Plateau was produced, and the region was strongly elevated. The details of the collision itself are still poorly understood. An ophiolite belt called the Indus suture north of the Precambrian crystalline rocks that form the axis of the Himalayas (Figure 10.9) indicates that this was an important zone of collision. But some geologists believe that the Himalayas originated through the interaction of one or more microcontinents between the Indian and Eurasian plates, similar to the way in which the Alps developed between Africa and Eurasia (Figure 10.7).

Cordilleran Region Prior to its Triassic breakup, Pangaea was bordered on the west, along what is now North and South America, by an oceanic trench. When the American continental masses broke away from the remainder of Pangaea and moved westward, they overrode the trench, and their western margins underwent a great deal of volcanism, folding, and faulting. In North America, this tectonic activity produced a highland within the Cordilleran geosyncline, which, in the Late Triassic, began to furnish sediments both westward toward the eugosyncline, as well as eastward toward the central part of the continent. This **Mesocordilleran highland** (see Figure 10.1) persisted into the Jurassic; by the latest Jurassic, the highland had grown into a very long chain of sporadic hills and mountains from Alaska to Central America.

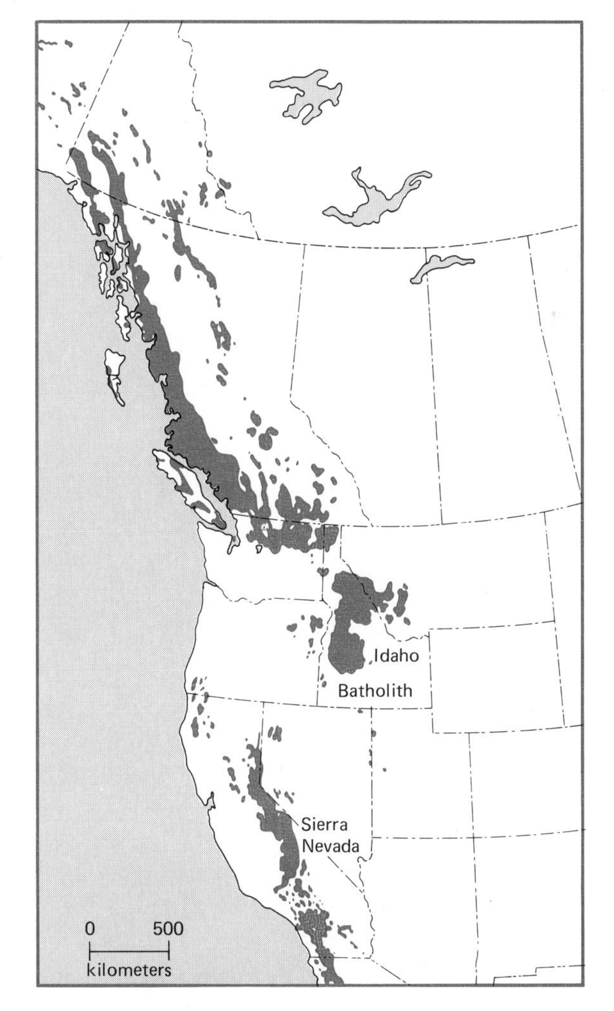

FIGURE 10.10

Distribution of Jurassic and Cretaceous granitic batholiths in western North America.

Meanwhile, volcanic activity continued in the eugeosyncline that lay west of the Mesocordilleran highland. This area was first intruded by granitic rocks in the Late Triassic. In the Late Jurassic and Early Cretaceous, during the **Nevadan Orogeny**, huge volumes of granitic material were generated at depth above the subduction zone created by the oceanic plate that was underthrusting the North American continent from the west. These granitic masses ascended as huge batholiths that today include the Sierra Nevada of California and even larger batholiths in British Columbia (Figure 10.10).

FIGURE 10.11

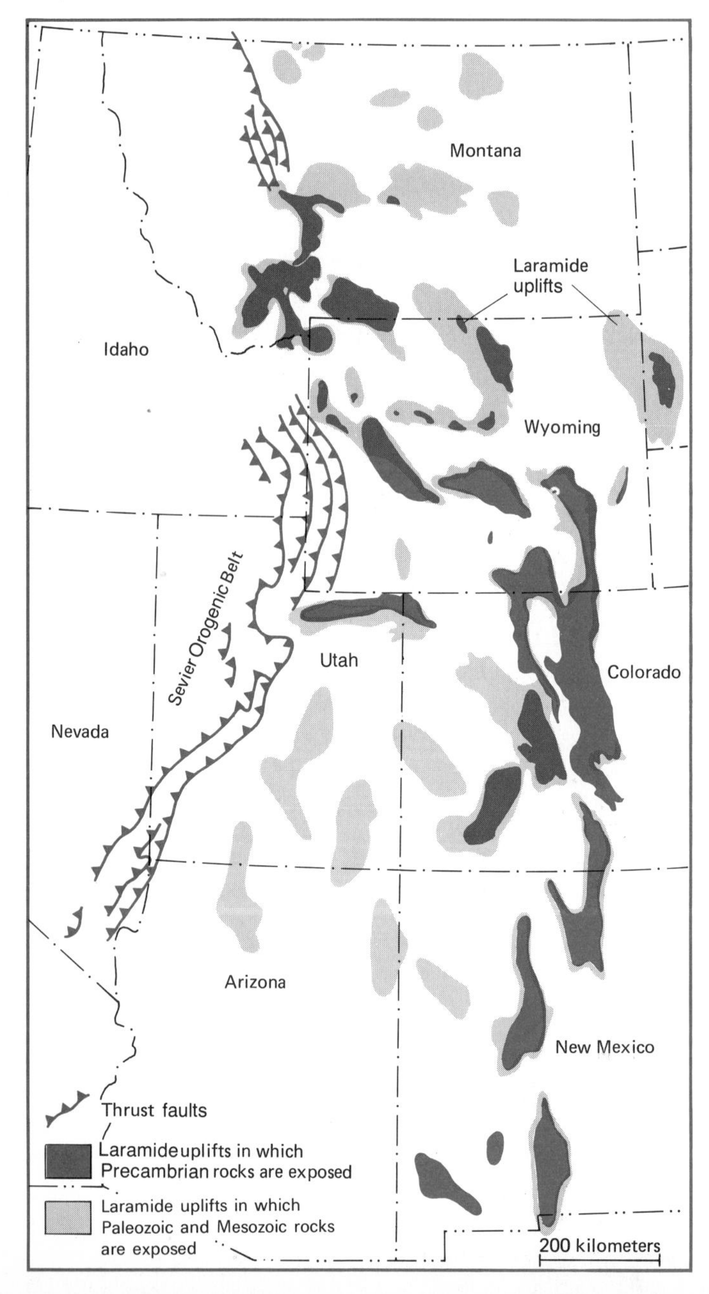

Areas of Sevier and Laramide Orogenies.

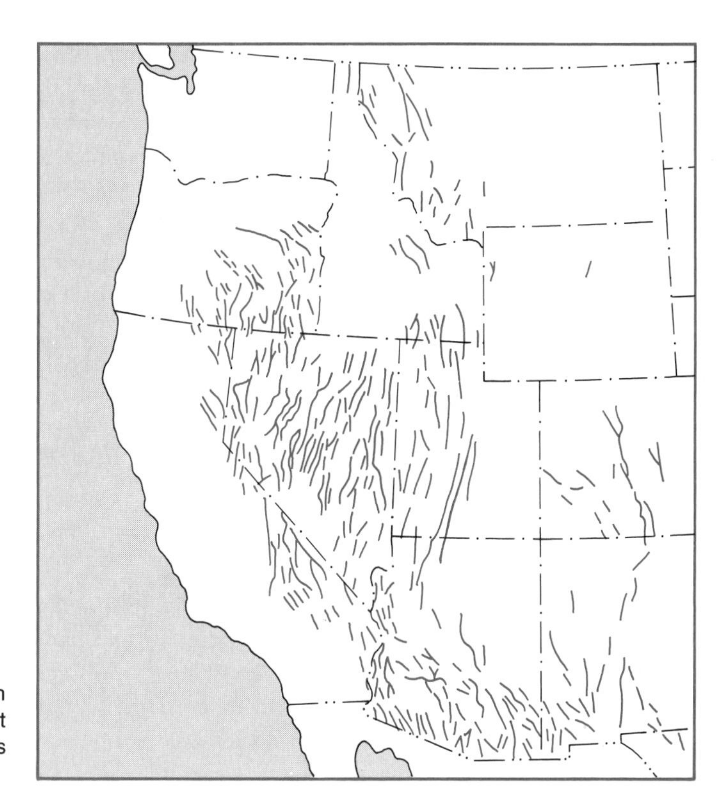

Block faults (heavy lines) occurred in response to tensional forces throughout a wide area of the western United States during Miocene time. (*Gilluly, 1963*)

FIGURE 10.12

In Early Cretaceous time, the Mesocordilleran highland broadened and shed large volumes of clastic sediments into adjacent oceans, both westward into what is now the California region and eastward into the western interior of the United States. Along the eastern margin of the Mesocordilleran highland, from southern Nevada to western Montana, rocks from the west moved many kilometers eastward on huge overthrust faults (Figure 10.11). This Early Cretaceous episode of deformation is known as the **Sevier Orogeny.** In the latest Cretaceous and early Cenozoic, widespread vertical faulting far east of the overthrust belt produced block mountains from New Mexico northward to Canada (Figure 10.11). This episode is known as the **Laramide Orogeny.** Uplifts that began in both these orogenies have evolved into the present-day Rocky Mountains. The Rockies actually consist of many isolated uplifts, and the basins between these began to fill up with sediments weathered from the rising mountains early in the Cenozoic. During these orogenies, widespread mineralization produced many of the deposits of metallic ore that occur throughout the Rocky Mountain region.

Much later, about 15 to 20 million years ago during Miocene time, the entire Basin and Range region of Nevada and adjacent states underwent large-scale block faulting as a result of new tensional, or pull-apart, forces (Figure 10.12). Almost simultaneously throughout the Basin and Range province, andesitic volcanism gave way to basaltic volcanism, underscoring the fundamental change in tectonic style. This change apparently coincided with the collision of the East Pacific Rise with the subduction zone that bordered the western margin of the continent discussed in Chapter 9. Shortly thereafter, pull-apart forces also caused the Baja California peninsula to separate from mainland Mexico, as the Gulf of California began to widen.

Precisely how the pull-apart forces developed after the East Pacific Rise collided with the subduction zone is not clear. They may have evolved as a by-product of large-scale transcurrent motion on the San Andreas fault system. Alternatively, they may have resulted from regional doming of deep crustal material that had been subducted below the Basin and Range Province. Whatever their cause, the tensional forces also permitted the uplift of the Sierra Nevada at that time. The Sierra Nevada batholith had formed during the Mesozoic but was somehow held down, possibly by regional compressional forces, so long as Pacific Ocean crust was being subducted. When subduction ceased in the Miocene, these forces relaxed and the Sierra Nevada block rose isostatically.

Volcanoes were active in much of the western United States throughout the Cenozoic Era. In the Eocene, andesitic volcanism occurred in the San Juan Mountains of southwestern Colorado and in the Absaroka Mountains and Yellowstone Plateau of northwestern Wyoming. In these areas, the thick accumulations of volcanic rocks consist partly of lava flows, but mostly of sandstones and breccias that were derived directly from the slopes of the volcanoes and transported to intervening low areas by streams and mud flows. Locally, volcanism continued into the Oligocene Epoch. The Miocene, however, was the time of the most extensive volcanism. During this epoch, the Columbia Plateau and much of the Snake River Plain were formed by vast basaltic outpourings (Figure 10.13). Somewhat later, in the Pliocene, the Cascade Mountains were built and were capped by andesitic volcanism in the Pleistocene. This episode has been called the **Cascadian Orogeny.**

FIGURE 10.13

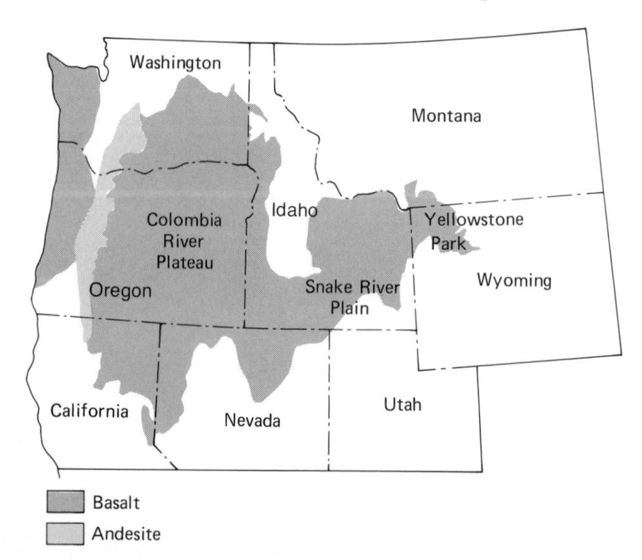

Cenozoic volcanic rocks cover a large region of the northwestern United States. (*Gilluly, 1963*)

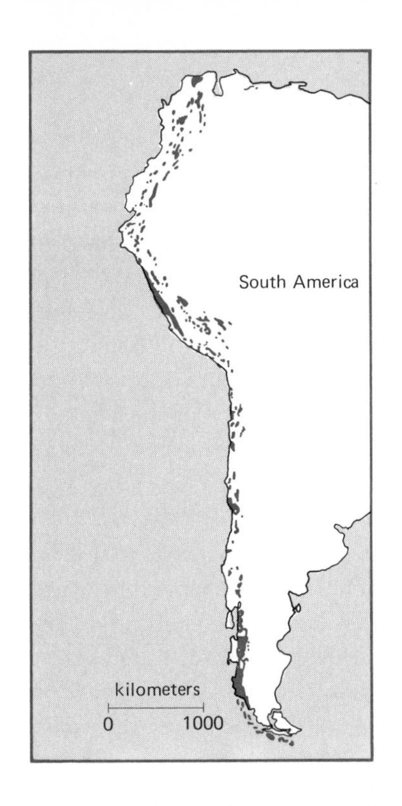

FIGURE 10.14

Granitic batholiths of Cretaceous age in South America.

The Snake River plain terminates on the northeast in the Yellowstone Park area of northwestern Wyoming (Figure 10.13). Northeastward, along the Snake River-Yellowstone trend, volcanism becomes progressively younger; in the Yellowstone area, large-scale volcanic activity has occurred as recently as 70,000 years ago. A batholith, in part still molten, underlies Yellowstone Park, and igneous activity will probably recur there in the future. Some geologists have suggested that the linear Snake River-Yellowstone trend may have been formed by a deep mantle hot spot, similar to that believed to have formed the Hawaiian Islands, as the North American plate drifted slowly westward over it.

The Cordillera of South America had a history similar to that of North America. Volcanism and folding occurred widely along the geosynclinal western margin of the South American continent during the Triassic, possibly as a side effect of the initial stages of separation of North America from Gondwana. A second period of deformation in the Late Jurassic may have been related to an increase in subduction activity as South America began to separate from Africa. The most intense episode of deformation and metamorphism began late in the Early Cretaceous and culminated in the Late Cretaceous and early Cenozoic, at the same time as the Laramide Orogeny was taking place in North America. During this episode, massive amounts of magma intruded and crystallized at depth to form enormous batholiths that are exposed along the western flanks of the Andes (Figure 10.14). The volcanic superstructure of the present-day high Andes began to form during the Miocene, about 15 million years ago. Since that time, numerous large andesite volcanoes have been built in the region. Some of the Andean volcanoes rise to an elevation of well over 6,000 meters and are still active.

Tectonic Control of Sedimentary Environments

Areas of uplift tend to become sources of clastic sediments. These are shed into adjacent areas of relative subsidence that, regardless of their shape, are termed **sedimentary basins.** We can learn about the timing and magnitude of ancient uplifts, which may have long since disappeared, by studying the sedimentary strata that accumulated in basins adjacent to them. The following discussion will briefly examine three examples from the Mesozoic and Cenozoic sedimentary record. The first will refer again briefly to the Alps and some of the strata that record their evolution; the next two examples come from the western interior of the United States, where tectonic patterns led to thick sequences of strata that contain valuable fossil fuels.

Flysch and Molasse

In the Alps, the first stages of orogeny in a given area are typically marked by the appearance of dark, evenly bedded marine sandstones in areas that had previously received only limestones, shales, or bedded cherts, which are indicative of quiet, relatively deep marine environments. In the Alps, these distinctive sandstones, derived from newly created source areas, are referred to as **flysch.** Most flysch deposits are graded and contain **sole markings;** they are inferred to be marine turbidites. The Alpine flysch ranges in age from Late Cretaceous to early Cenozoic. Similar deposits mark the beginnings of the orogeny in the Himalayas and in numerous other mobile belts of many ages.

Overlying the Alpine flysch are thick sequences of light colored, irregularly stratified conglomerates, highly cross-bedded sandstones, and some shale and coal. These sedimentary rocks, which are clearly of nonmarine origin, reflect the final uplift of the Alps above sea level. They are referred to as **molasse,** and they represent the late stages of the mountain-building cycle. The Alpine molasse is mid-Cenozoic in age. Similar deposits mark the late stages of orogeny in other mobile belts as well.

Delta and Barrier Island Systems

In the western interior of the United States, east of the Mesocordilleran highland, a thick, widespread sequence of sedimentary rocks, aggregating about 4 billion cubic kilometers, accumulated during the Cretaceous Period. The sediments themselves were derived almost exclusively from the west—the site of the Mesocordilleran highland (Figure 10.15). This accumulation on the eastern side of the Mesocordilleran highland alone, represents the removal of about 8 kilometers of material from the Mesocordilleran land mass. The greatest thickness of strata accumulated close to the fluctuating western shoreline of the Cretaceous sea (Figure 10.16). The deposits consist of intertonguing marine and marginal-marine strata, and they include thick, extensive coal beds, which originated as thick peat deposits in coastal marshes on large **deltas** and behind extensive **barrier islands.** Cretaceous and early Cenozoic coals from this region make up a large portion of the United States' substantial coal reserves.

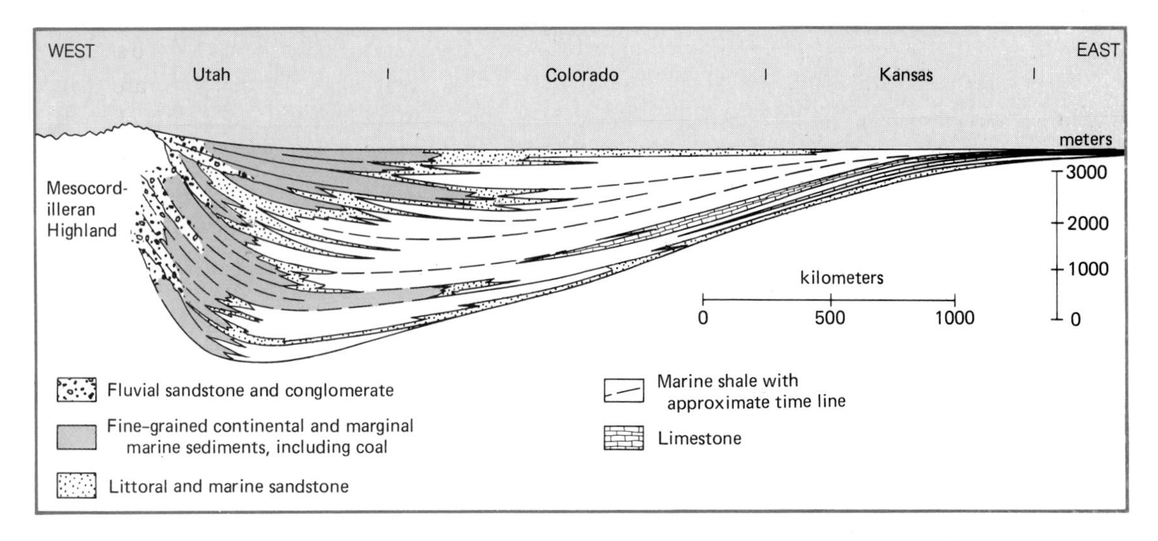

FIGURE 10.15

The thick section of Cretaceous sedimentary rocks in the western interior of the United States coarsens westward toward the source area. (*King, 1959*)

FIGURE 10.16

Thickness in meters of Upper Cretaceous sedimentary rocks in the western interior of the United States. (*Eardley, 1964*)

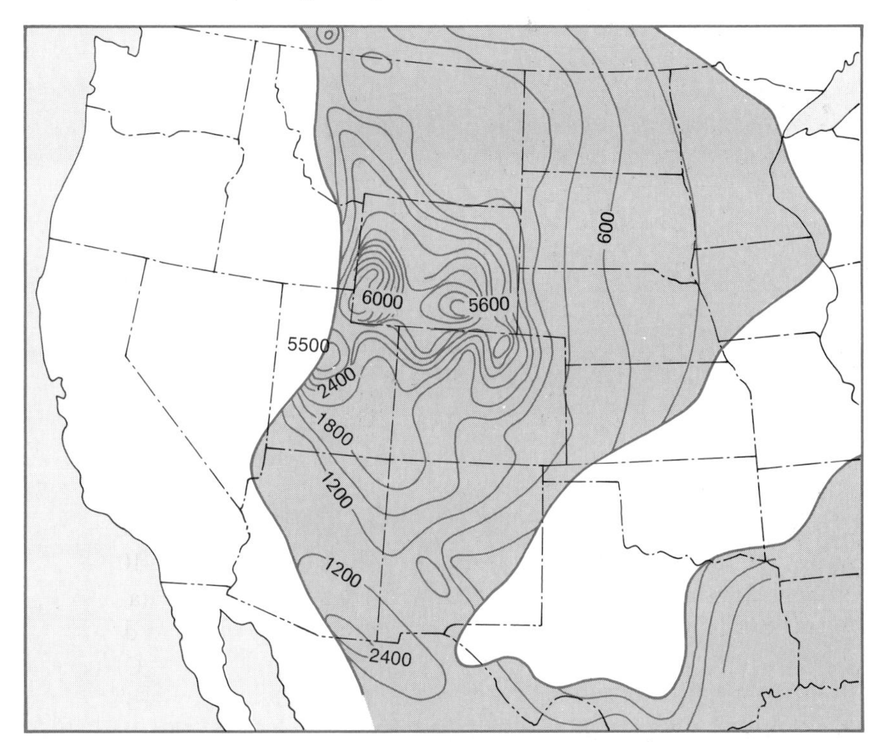

The voluminous quantity of sediment supplied by the Mesocordilleran highland contributed to the growth of extensive deltas along the western margin of the interior seaway. Study of large modern deltas like that of the Mississippi River has given us insight into the processes that must have taken place in ancient deltas as well. Marine deltas form where rivers flow into the sea and drop their sedimentary load. They persist and grow if the rate of influx of sediment exceeds the rate of removal by ocean waves and currents. As a delta grows seaward, some sediment is deposited on the delta plain, in streams, and in adjacent bays and marshes as **topset beds** (Figure 10.17). Most of the sediment is delivered to the delta margin, where the sand fraction is distributed parallel to the shoreline by ocean currents and deposited as bars and beaches. The finer silt and clay continue seaward, suspended in the plume of river water, which, being lighter than the ocean water below, flows outward as a surficial layer. The fine material finally settles beyond the river mouth on the gentle sloping delta front, where it forms **foreset beds** of silty clay. As the delta builds seaward, the shoreline sands and other topset beds build seaward over the top of the homogeneous marine foreset beds, which, in turn, overlie the **bottomset beds** of the sea floor (Figure 10.17). Large deltas, like those postulated for the western interior Cretaceous seaway, actually cause the crust beneath them to subside isostatically; they produce a much thicker accumulation of sediments than do the adjacent shoreline environments.

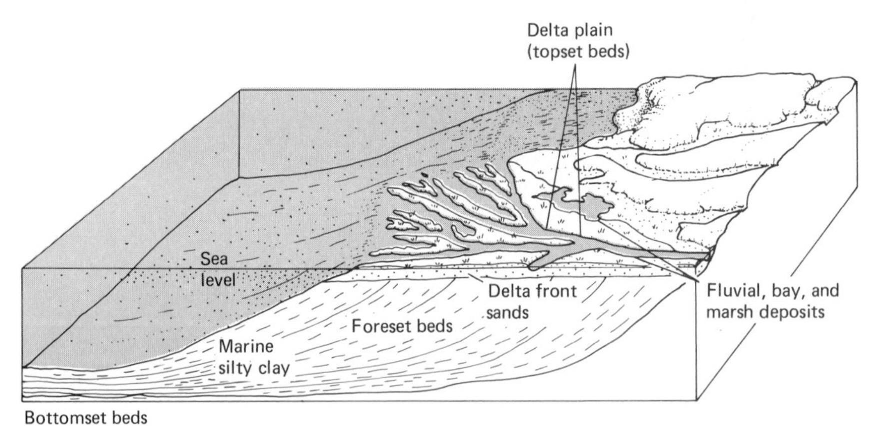

FIGURE 10.17 Cross section through a growing delta. Vertical scale exaggerated 50 times.

Shapes of modern deltas vary, depending on the extent of local tides and on wave and current energy (Figure 10.18). Regardless of shape, all deltas are crossed by **distributaries**, which branch off the main channel, radiate across the delta plain, and discharge at several localities on the delta margin. Between the distributaries are found marshes in which thick beds of peat may accumulate as the sediments there compact and sink. Many of the thick Cretaceous coals in the western interior probably formed in this way on extensive subsiding delta plains.

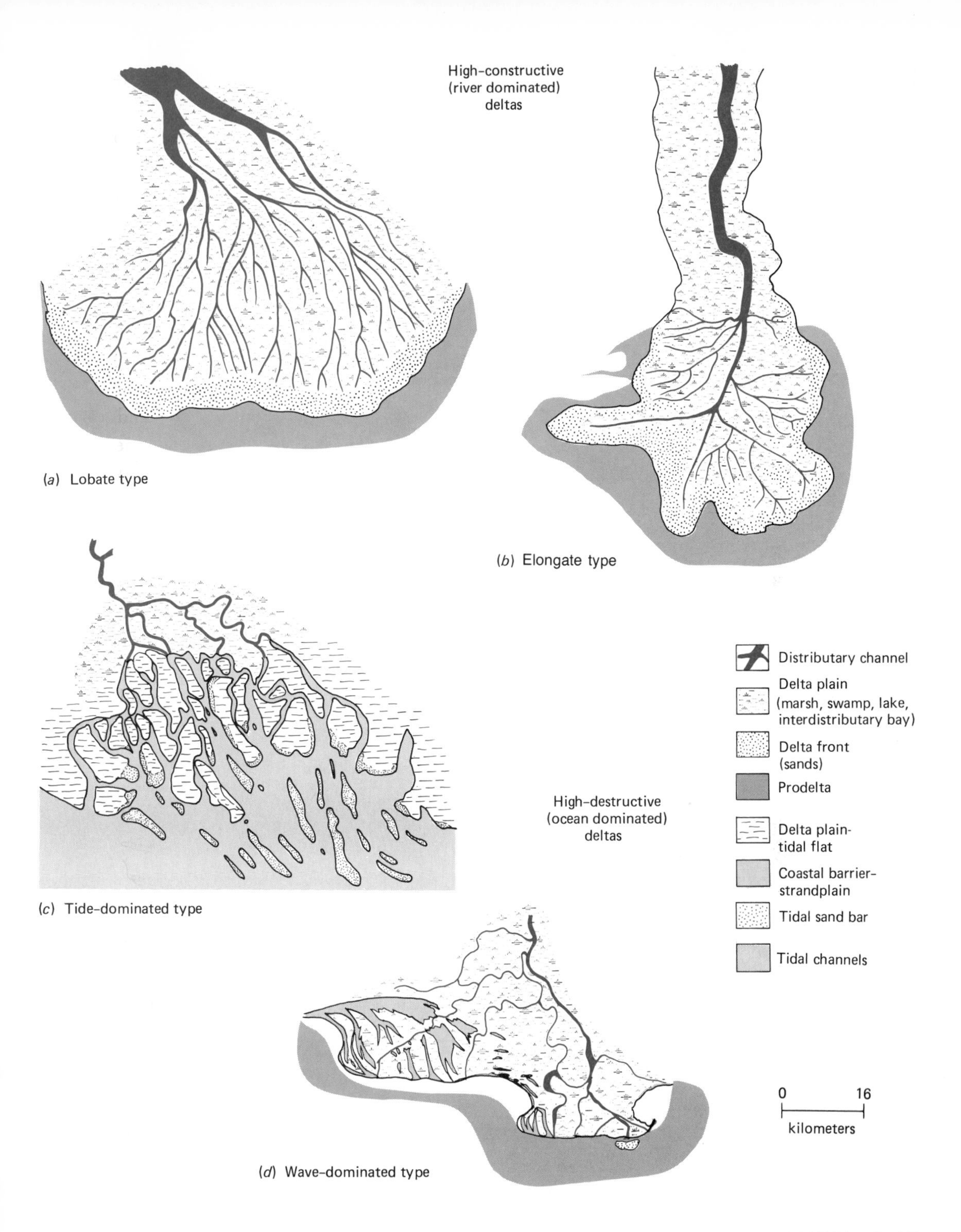

Four types of deltas: (*a*) Lafourche delta, forerunner of the modern Mississippi birdfoot delta; (*b*) modern Mississippi birdfoot delta; (*c*) delta of the Gulf of Papua, New Guinea, which is strongly influenced by tides that flood the lower parts of the distributaries and create tidal sand bars; (*d*) Rhone delta, shaped by strong waves. (*Fisher and others, 1969*)

FIGURE 10.18

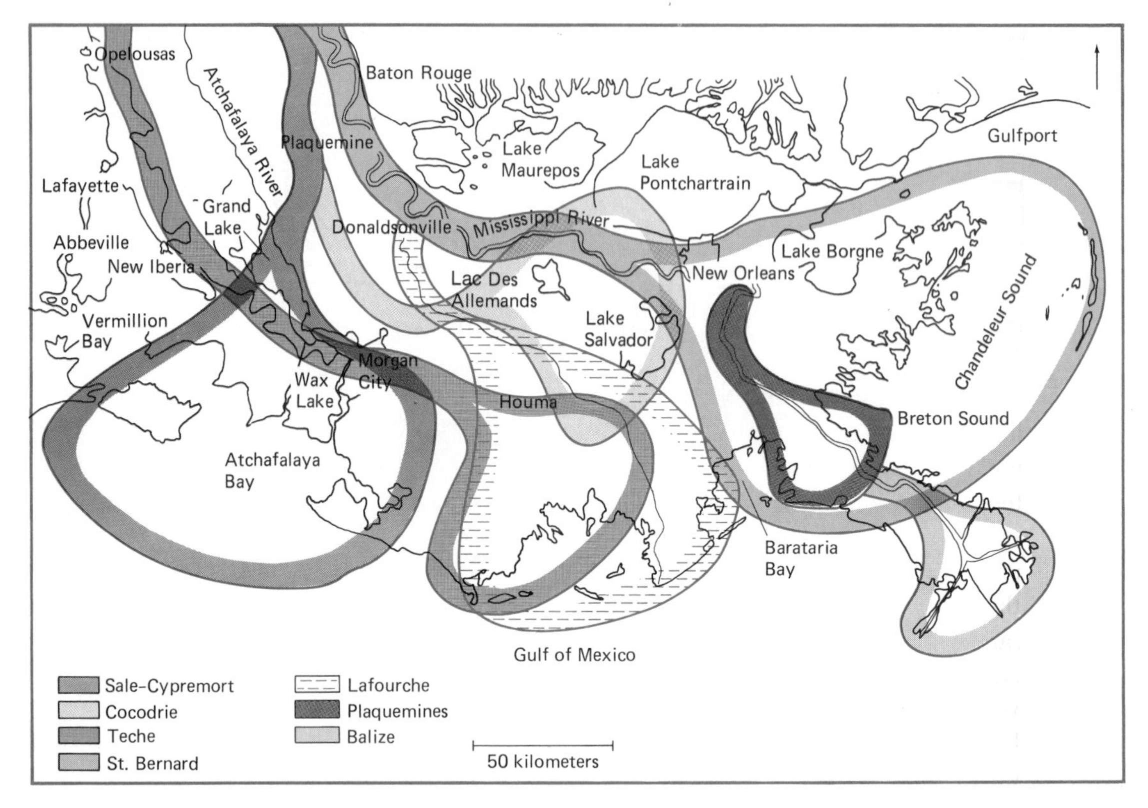

FIGURE 10.19 Overlapping subdelta lobes of the modern Mississippi delta complex during the last 5,000 years. The site of major deltaic sedimentation has shifted numerous times. (*Kolb and van Lopik, 1966*)

One way in which ancient deltas produced such coal beds is illustrated by the modern Mississippi Delta. As the delta grows seaward, the river must flow progressively farther across the delta plain to reach the sea, and thus its gradient across the delta plain decreases. Soon shorter, steeper courses to the sea become available. When the river overflows its banks in flood, it abruptly abandons the delta lobe that it built and takes a new course to the sea. Then a new delta lobe is begun adjacent to the old one, which sinks entirely below sea level as the area continues to subside. After the new lobe has extended seaward, the river again shifts its course, and a still newer delta lobe begins to build outward, in part over sunken portions of older delta lobes. Individual lobes build quickly. The modern Mississippi Delta consists of seven partially overlapping delta lobes that have been built and abandoned in succession during the last 5,000 years (Figure 10.19). Ancient delta complexes in the stratigraphic record similarly consist of many alternations of marine and marginal-marine strata. In addition to their coal beds, ancient deltas are sought for their petroleum-producing potential as well. Commonly, the well-sorted sandstones along the fronts of delta plains make excellent reservoir rocks.

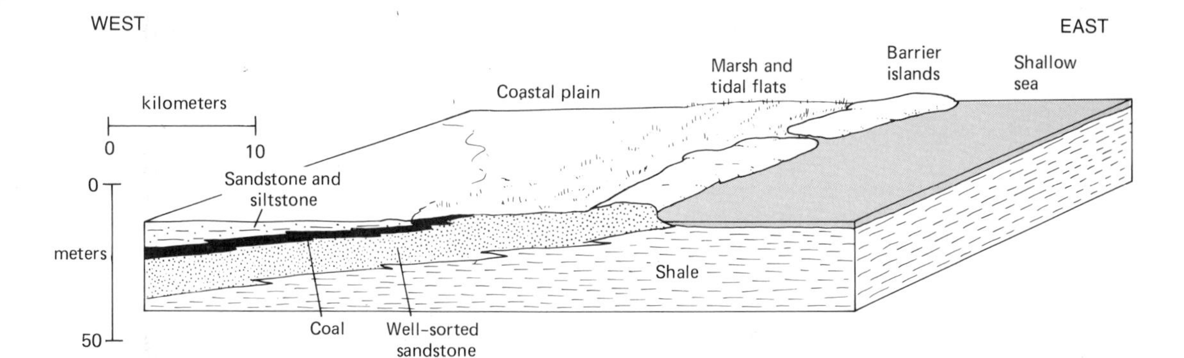

Barrier island-tidal marsh depositional model for Cretaceous coal deposits in the western interior of the United States. Vertical exaggeration 200 times.

FIGURE 10.20

Not all the sediment transported from the Mesocordilleran highland to the western interior Cretaceous seaway ended up in deltas. Much of it was transported laterally along the shoreline by waves and currents and finally deposited in elongate barrier islands (Figure 10.20). Seaward, east of the barrier islands, mud was deposited in the deeper waters of the seaway. To the west, between the barrier islands and the coastal plain, lay vast areas of marshes, tidal flats, and lagoons, which were protected from the open sea by the barrier islands themselves. Thick peat beds, which were later to become coal, formed in extensive marshes behind the barrier islands. Because the entire region was continually subsiding, the beds of peat grew very thick; the resulting coals commonly achieve thicknesses of several meters.

A given barrier-island coastline tended not to remain stationary, but to prograde seaward, the sand barriers building eastward over the former offshore muds. The protected marshes and associated enviornments behind the barriers also shifted progressively seaward and gradually covered the retreating barrier-island sands. As a result of **progradation**, the comparatively narrow band of near-shore sediments, including barrier-island sand and marsh peat, could form blanket deposits easily 100 kilometers wide (Figure 10.20). Finally, each eastward progradational episode ended when the shoreline again transgressed westward and buried the near-shore sediments with marine-transgressive deposits.

Repeated regressions and transgressions of the sea produced thick sequences of alternating marine and near-shore strata that contain many coal beds. The Cretaceous coals that were produced during these shoreline fluctuations in back-barrier environments and on adjacent delta plains are found today in a broad belt from the Arizona and New Mexico region northward into Canada. Together with coal beds of early Cenozoic age, these constitute the important coal region shown in Figure 10.21.

Lacustrine Oil Shales Following the Laramide Orogeny, the mountain blocks in the western interior of the United States began to shed fluvial sediments into the intervening basins. During the Eocene Epoch, three of these

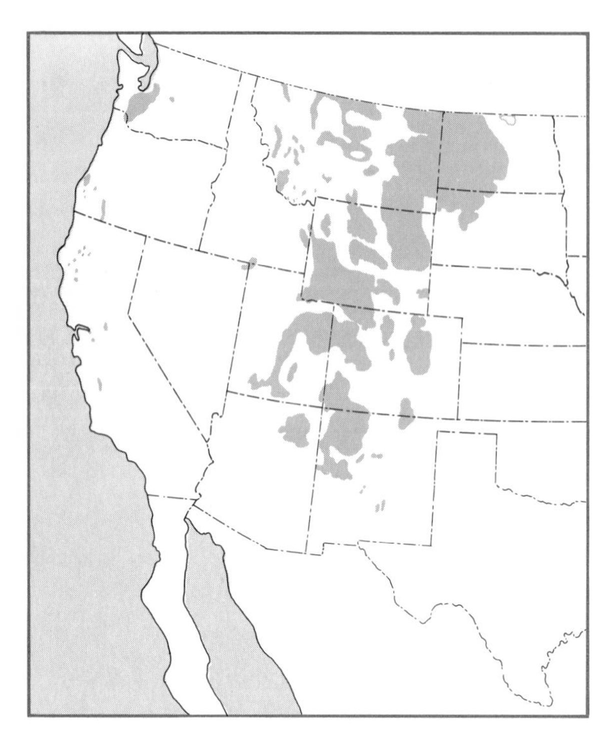

FIGURE 10.21 Cretaceous and early Cenozoic coal fields of the western United States. (*U.S. Bureau of Mines*)

basins in Utah, Colorado, and Wyoming became the sites of large inland lakes (Figure 10.22). These lake deposits, which make up the Green River Formation, constitute one of the world's largest single petroleum reserves.

The petroleum in the Green River Formation is in the form of **oil shale,** which is a dark gray or brown laminated rock composed chiefly of very fine grains of dolomite, calcite, organic material, and clay minerals. These rocks contain **kerogen,** a bituminous material that is solid at normal temperatures; upon heating, kerogen becomes liquid oil and flows from the rock. The richest oil shales yield up to 250 liters of oil per metric ton. Of the three distinct basins occupied by the Green River lakes, the richest and thickest deposits of oil shale occur in the central portions, where they are interbedded with evaporite deposits, chiefly sodium carbonate minerals and halite. These deposits indicate that the lakes were at times strongly saline, much like Great Salt Lake today.

The Green River Formation covers 100,000 square kilometers and attains a thickness of about 1,000 meters; as a result, it contains an enormous quantity of oil. In the Piceance [Pea′-ance] Basin of Colorado alone, those shales, which assay 50 or more liters of oil per metric ton, contain a total of 1,800 billion barrels of oil. To place this figure in perspective, the total quantity of conventional petroleum remaining in the United States is estimated to be only 140 billion barrels. One day the Green River Formation may be exploited on a large scale for its petroleum.

For many years geologists believed that the Green River Formation was deposited in one or two large, fairly deep permanent lakes. Shoals or bars were thought to have trapped brines in local segments of the lakes where evaporite minerals were precipitated. Today, however, many think that the oil shales and accompanying evaporites formed in a series of **playa** lakes, which were very shallow and occasionally even dry. The mountains surrounding the lakes must have received most of the regional precipitation, and the water flowed from the mountains into the basins (Figure 10.23). Early in their history the lakes were probably quite fresh. They were populated by fish and a number of fresh-water invertebrates, including crustaceans, snails, and clams. Later, as the climate became more arid, the Green River lakes may have become playas that alternately expanded and contracted, exposing a wide expanse of mud flats. Carbonate-rich waters flowed into the basins, and calcite and dolomite precipitated on the periodically exposed flats. During dry periods when the lakes shrank, mud cracks formed in the exposed mud flats. Simultaneously, brines rich in sodium carbonate and sodium chloride became concentrated in the shrinking playa lakes where evaporite minerals precipitated (Figure 10.23).

FIGURE 10.22

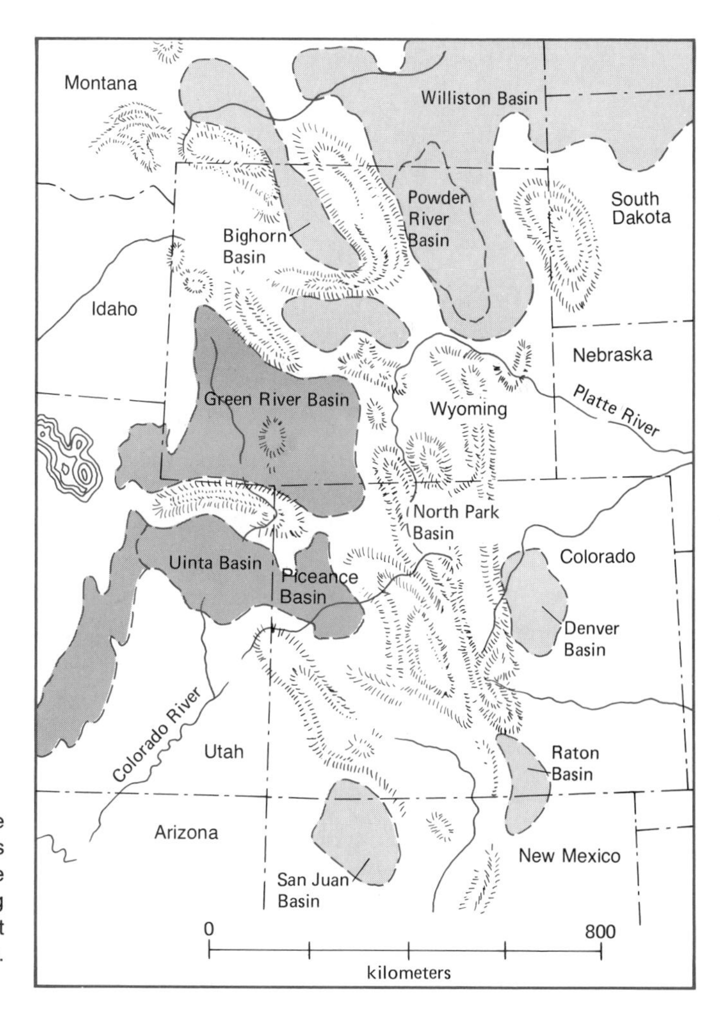

Relationship of the Green River lake basins to the other Rocky Mountain basins that contain Cenozoic deposits. These basins were produced by block faulting during the Laramide Orogeny in latest Cretaceous and earliest Cenozoic time. (*Stokes, 1973*)

FIGURE 10.23

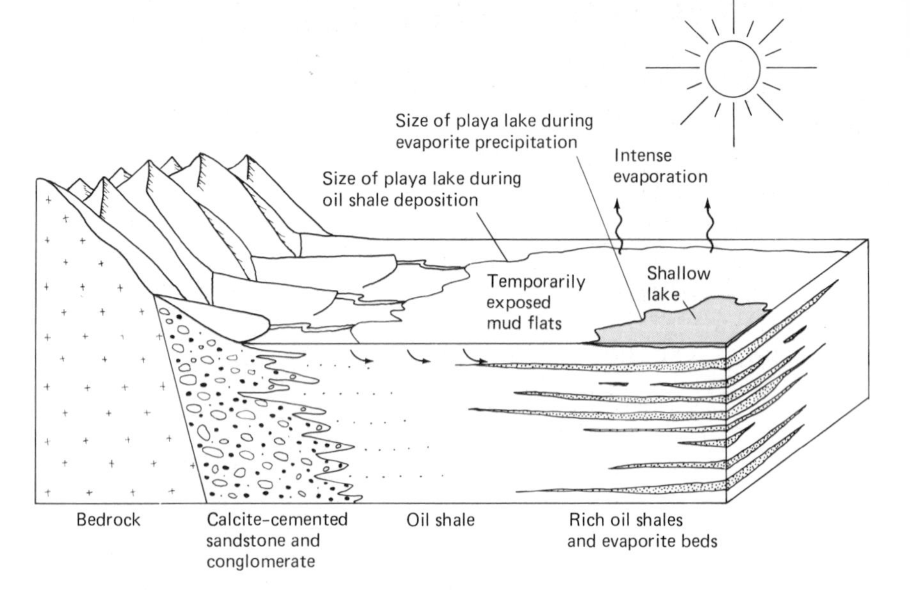

Playa lake model for deposition of the Green River oil shale and interbedded evaporites. Vertical scale greatly exaggerated. (*Eugster and Hardie, 1975*)

Oil shale formed during wet periods when the lakes expanded. At those times, the waters became fresh enough to support extremely rich blooms of algae (Figure 10.23). The blooms were periodic, and this periodicity produced very thin, alternating organic-rich and organic-poor laminae, which can be traced for many kilometers. Possibly the productivity fluctuations were seasonal. Most geologists consider the laminations to represent **varves,** that is, seasonally produced annual layers, which provide a built-in measure of the rate of sedimentation for the oil shale. The Green River lakes finally ceased to exist when the basins became filled with sediment and fluvial conditions returned to the region.

The Changing Oceans

Deep-Ocean Environments

During the Jurassic Period, shortly after the modern ocean basins had begun to form, tiny plants and animals that secrete calcium carbonate skeletons first appeared in the marine plankton. They quickly spread throughout the oceans and profoundly altered the worldwide pattern of carbonate sedimentation. Prior to the Jurassic, no important carbonate-secreting planktonic organisms existed. Virtually all the calcium carbonate removed from the oceans was precipitated by benthic organisms along continental margins and in epeiric seas. In these shallow settings, the resulting limestones were recycled frequently by episodes of uplift and erosion. Then, in the Jurassic, the evolution of microscopic planktonic algae called **coccoliths** (Figure 10.24*a*) and planktonic **foraminifera** (Figure 10.24*b*) initiated the deposition of large quantities of carbonate in the deep sea as **calcareous ooze.**

Calcareous ooze today covers almost half the ocean floor and is by far the most widespread kind of sediment on earth. Deep-sea drilling has revealed that during the late Mesozoic and Cenozoic it was equally widespread. The rate at which calcareous ooze accumulates is very slow, around ten meters in one million years, but in very deep water it does not accumulate at all, because calcium carbonate dissolves much more rapidly in very deep water than in water of moderate depth. The boundary below which calcium carbonate is absent is surprisingly sharp; it is known as the **calcite compensation depth**, or simply **CCD**. The CCD corresponds to the depth at which the rate of solution of calcium carbonate is *equal* to the rate at which the skeletons of calcareous plankton are supplied from above. At depths greater than the CCD, the rate of solution *exceeds* the rate of supply, and at depths shallower than the CCD, solution is *less* than supply.

The CCD has been compared to a snow line on land, above which snow persists. Like a snow line on land, the CCD is not at the same elevation everywhere. Solubility of calcium carbonate increases with an increase in the quantity of dissolved carbon dioxide, and cold water has a greater capacity for carbon dioxide than warm water. The cold bottom water that forms a thick, deep, bottom layer in all oceans strongly influences the depth of the CCD. In most oceans the CCD lies between 4,000 and 5,000 meters below sea level.

FIGURE 10.24

(*a*) A modern coccolith, *Coccolithus pelagicus* (X 10,000). (*Courtesy of C. Howard Ellis*) (*b*) A modern planktonic foraminifer, *Globorotalia cultrata* (X 800). (*Courtesy of Peter Roth*)

(*a*)

(*b*)

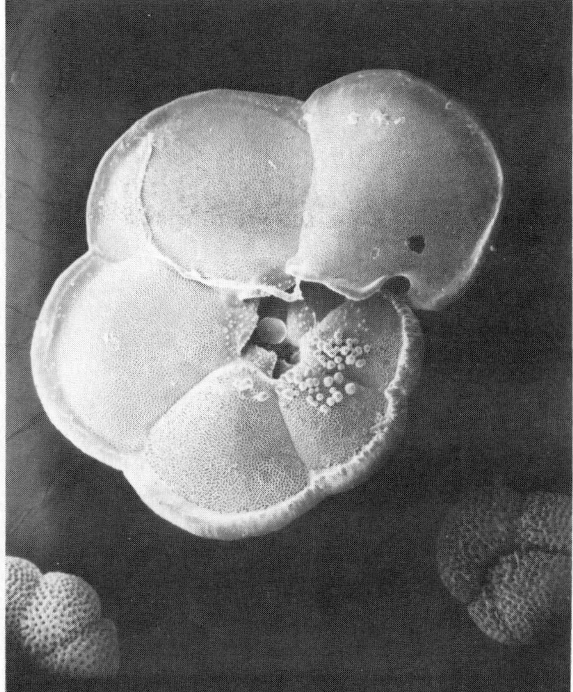

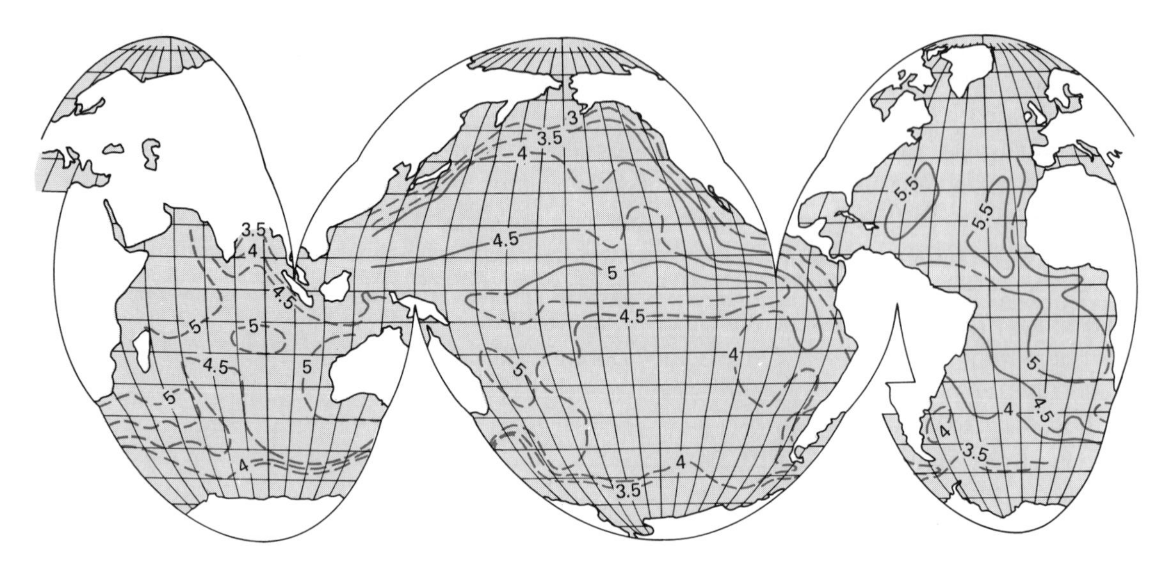

FIGURE 10.25 Depth in kilometers below sea level of the calcite compensation depth (CCD) in the world's oceans. (*Berger and Winterer, 1974*)

FIGURE 10.26 (*a*) Diatomaceous sediment. The elongate specimen is *Nitzschia sp.* (X 2,000). (*Courtesy of Peter Roth*) (*b*) A spumellarian radiolarian (X 800). (*Courtesy of Peter Roth*)

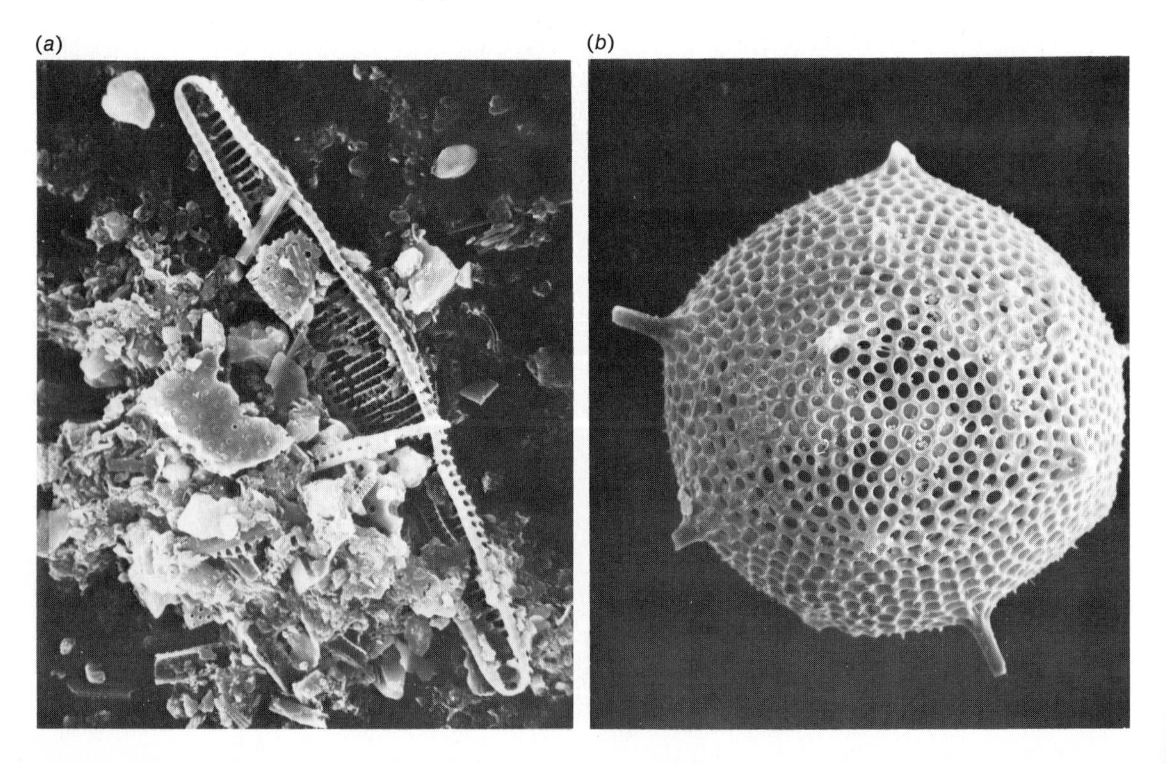

However, at high latitudes where the cold bottom water reaches nearly to the surface, as around Antarctica and in the northern Atlantic, the CCD is comparatively shallow, as Figure 10.25 shows.

Below the CCD, the most widespread sediment is **abyssal clay,** which is commonly red to brown in color. Much of the material that makes up abyssal clay comes from dust, some of it cosmic or volcanic in origin, and some of it probably derived directly from the continents. The clay particles are very fine and accumulate at a very slow rate—only about one meter in one million years. A significant component of abyssal clay is represented by zeolite minerals, which are **authigenic;** that is, they form in place in the marine sediment by the chemical interaction of ocean water with clay minerals.

The other common sediment type below the CCD is **siliceous ooze,** which contains more skeletons of **diatoms** (siliceous phytoplankton) or **radiolarians** (siliceous protozoans) than it does skeletons of calcium carbonate-secreting organisms (Figure 10.26). Siliceous ooze tends to accumulate where large masses of plankton are produced, chiefly at polar latitudes and in regions of upwelling, such as the equatorial Pacific. In those regions, siliceous skeletons are also supplied to the bottom where it is above the CCD, but they are overwhelmingly masked by calcareous skeletons. Fourteen percent of the deep-sea floor is covered by siliceous ooze. Through diagenesis, that is, compaction and cementation, siliceous oozes become beds of chert, which are common in ancient deep-sea deposits.

Local areas of high productivity may depress the depth of the CCD on the sea floor below. For example, in the equatorial Pacific, where productivity is much greater than in adjacent waters to the north and to the south, the CCD is much deeper because much more organic calcium carbonate is produced and contributed to the sea floor (see Figure 10.25).

Oceanic crust originates at ridges, as we have seen. The sea floor subsides gradually as it is transported away from the ridges during spreading. Hence, all oceanic crust now at abyssal depth below the CCD originated and spent its early history well above the present CCD. Typically, the stratigraphic sequence on an abyssal plain consists of calcareous ooze below and abyssal clay above (Figure 10.27*a*).

In Cenozoic sediments of the western part of the North Pacific Ocean, however, this sequence is repeated twice. A basal layer of calcareous ooze is overlain by abyssal clay, and this, in turn, is overlain by another layer of calcareous ooze, and finally more abyssal clay at the top. The ocean crust in the North Pacific where the four layers are found is now believed to have been produced *south* of the equator. As this crust was transported down the ridge flank, it first received calcareous ooze and then, as it passed below the CCD, abyssal clay. However, this portion of the ocean floor was transported beneath the equator as a result of the northward movement of the Pacific plate (see Figure 9.12). Here it encountered the equatorial belt of high productivity where the CCD is greatly depressed, and it again began to receive calcareous

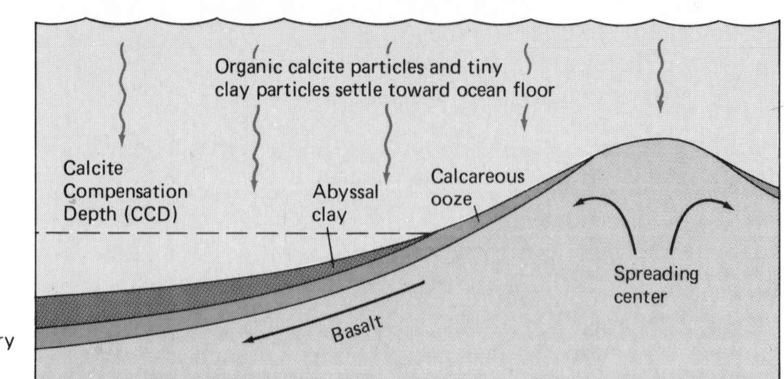

(*a*) Normal 2-layered sedimentary section

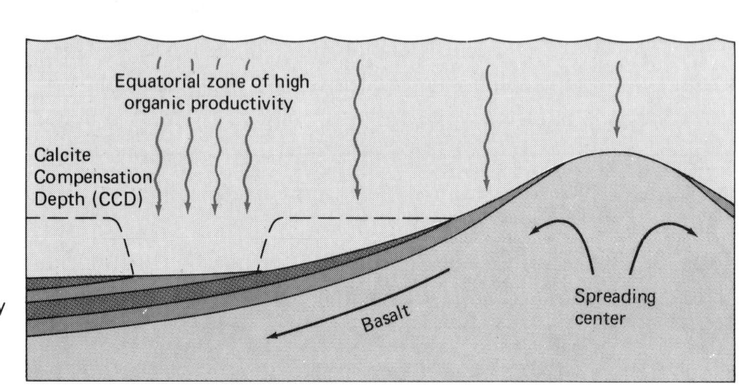

(*b*) 4-layered sedimentary section

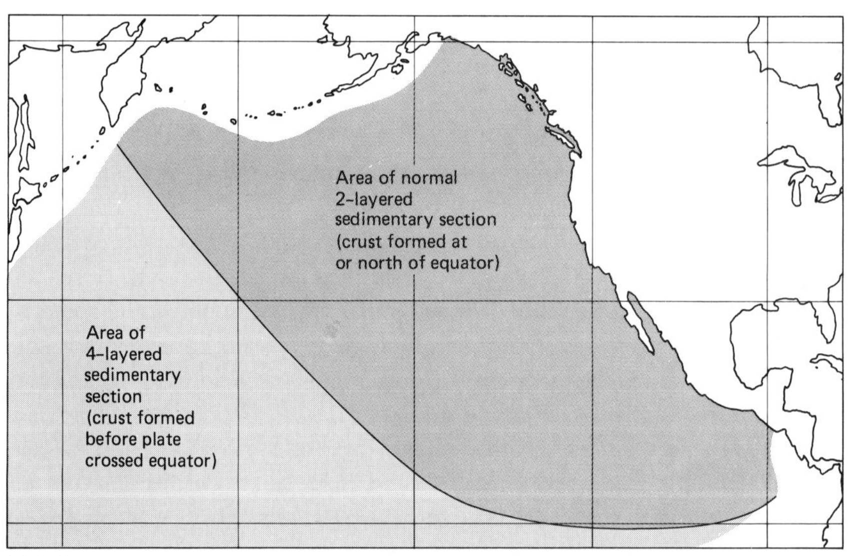

(*c*) Index map

FIGURE 10.27

(*a*) The northeastern North Pacific Ocean floor originated within or north of the equatorial zone of high productivity and received only the normal two-fold sequence of calcareous ooze below and an uninterrupted abyssal clay sequence above. (*b*) A large area of the western North Pacific originated south of the equator and traveled northward through the equatorial zone of high productivity, which produced a second layer of calcareous ooze within the abyssal clay sequence. (*c*) Index map of the area. (*Heezan and MacGregor, 1973*)

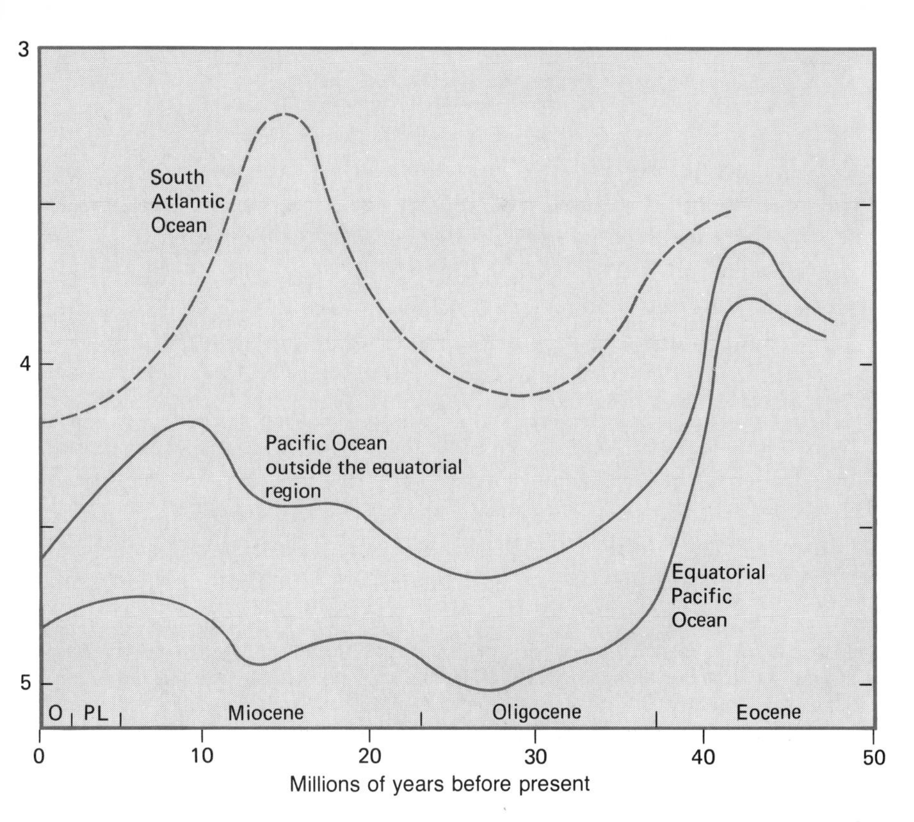

FIGURE 10.28

Fluctuations in the level of the calcite compensation depth (CCD) during the Cenozoic for three different regions of the world's oceans. (*Berger and Roth, 1975*)

ooze. With continued northward transport into the North Pacific, this portion of the ocean floor left the area of high productivity and received only abyssal clay. This very special transport path through the greatly depressed CCD at the equator accounts for the second layer of calcareous ooze in the middle of the abyssal clay sequence throughout a large region in the central North Pacific (Figure 10.27*b*).

Just as the CCD is not at a constant depth everywhere in today's ocean, neither has it remained constant throughout geologic time in any region. The deep-sea stratigraphic record indicates that the CCD fluctuated widely during the Cenozoic (Figure 10.28). These fluctuations were produced by major changes in oceanic circulation, by major transgressions and regressions of the sea, and by global temperature changes. For example, during the Pleistocene glacial episodes the CCD was deeper at the equator and shallower at high latitudes than during interglacial episodes.

Epeiric Seas

Epeiric seas were rarely stable for long. Most of the time their shorelines were moving either very gradually landward or very gradually seaward in marine transgression or regression. Marine transgression is caused by a relative rise of sea level. Whether sea level actually rises or a portion of the crust subsides tectonically, the shoreward movement of the shoreline and the sedimentary record produced are the same. If only a local area is affected by the transgression, the cause is almost certainly local subsidence. A local regression, however, is not necessarily caused by a local uplift. Instead, the local retreat of the shoreline may result from rapid deposition along the margin of the sea. In this case, the shoreline progrades as the sea is actually displaced by infilling marginal sediments.

Major transgressions and regressions of the Jurassic and Cretaceous epeiric seas, like those graphed in Figure 10.2, occurred simultaneously on several continents. These fluctuations must have resulted from the rising and lowering of sea level everywhere. Changes in the actual level of the sea, which affect the entire world ocean and all of its shorelines, are called **eustatic changes.** Eustatic changes in sea level may be caused by: (1) changing the quantity of water in the oceans, or (2) changing the capacity of the ocean basins to hold the water they have. In recent geologic history, eustatic sea-level changes resulted from the first cause. When ice accumulated periodically on the conti-

FIGURE 10.29

Simplified drawing of marine transgression and regression caused by temporary increase in the rate of sea-floor spreading. (*a*) Low spreading rate: narrow, steep-sided ridges, low sea level. Continent stands high relative to sea level. (*b*) Rapid spreading rate: broad ridge, sea level rising, marine transgression onto continent. (*c*) Again low spreading rate: narrow ridge, sea level falling, marine regression from continent.

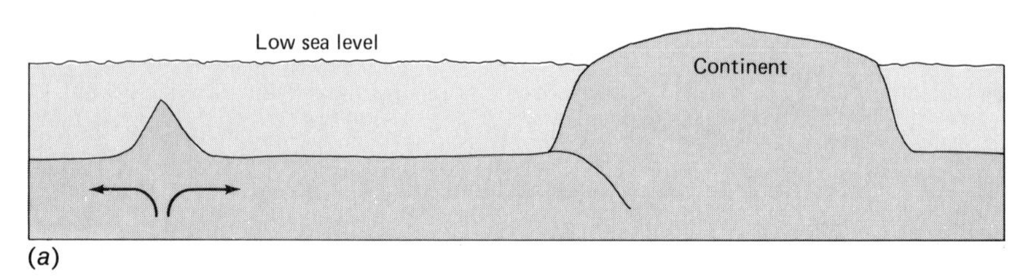

(*a*)

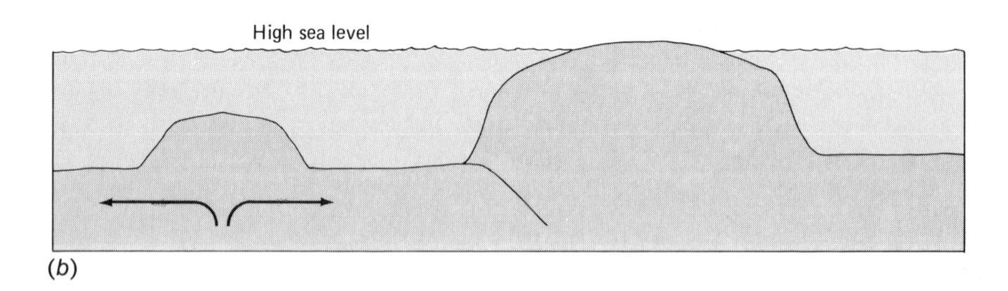

(*b*)

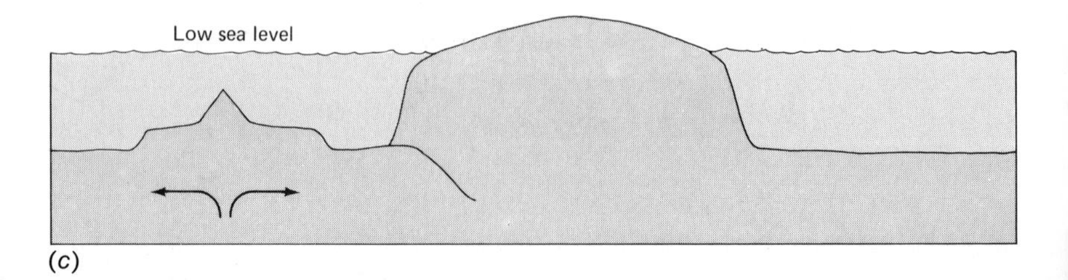

(*c*)

nents during the Pleistocene, sea level dropped by 200 meters or so due to withdrawal of water. During interglacial times, sea level rose again. Shorelines on broad shelves regressed and transgressed tens, and in many cases hundreds, of kilometers. However, this mechanism is not adequate to account for the substantially greater eustatic sea-level changes that occurred in the Jurassic and Cretaceous. Besides, there is no evidence for continental glaciation during these times.

The enormous eustatic changes in the Jurassic and Cretaceous seas probably reflect changes in the actual capacity of the ocean basins. We don't know the cause of these changes, but one possibility is that they were related to variations in the rate of sea-floor spreading. Ocean floor that is created at an oceanic ridge begins to subside continuously as it is transported away from the ridge as part of a diverging lithospheric plate. The subsidence of the new ocean floor is a function of time, not distance from the ridge. Accordingly, the actual slope of a given ridge flank is comparatively steep where the spreading rate is slow, and gentle where the spreading rate is fast.

Figure 10.29 shows how the shape of the ocean floor reflects differences in spreading rates. Slow spreading produces narrow ridges that occupy comparatively little space in the ocean basins. An increase in the spreading rate produces wider ridges that displace a great deal of water. Sea level then rises, and the oceans spill onto the continents in marine transgression. A decrease in spreading rate would similarly make more room for water in the sea and cause a regression. This mechanism may have caused the worldwide Jurassic and Cretaceous transgressive-regressive marine cycles, but whether the Jurassic and Cretaceous transgressions in fact occurred at times of abnormally rapid sea-floor spreading is not yet known.

Jurassic Ironstones: Unusual Sedimentary Rocks The stratigraphic record of the epeiric seas contains many rock types that are difficult to explain, largely because there seem to be no modern counterparts. One such rock that occurs in the Jurassic of Europe and in several other parts of the geologic record is **ironstone.** The Jurassic ironstones are bluish-green, iron-rich sedimentary rocks that weather dark brown. Typically they consist of **ooliths** of iron silicate or iron oxide in a matrix of iron carbonate (siderite), and they contain numerous marine fossils. Cross-bedding, scour surfaces, and intraformational conglomerate, together with the oolites, indicate that the water was probably shallow, with considerable current activity. Although ironstones occur widely in the European Jurassic, individual deposits are generally only a few meters thick and extend over an area of only a few tens of square kilometers before passing into iron-poor lateral equivalents. Some ironstones contain up to 40 percent iron. Historically they have been important sources of iron ore in central England and in France.

The iron minerals in the Jurassic ironstones are believed to be authigenic and to have formed in the unconsolidated marine sediment by chemical reaction between the sediment and seawater. Similar iron minerals are known from marine muds in some regions today, but they are rare and widely disseminated. Jurassic ironstones appear to have resulted in part from very slow deposition. Jurassic lands were low, and when they were partially flooded by the epeiric seas, they furnished much less sand and clay to the seas than would comparable lands of today. The iron was probably transported to the sea as a colloidal suspension; once in the sea, it could be carried far from shore.

Cretaceous Chalk The distinctive rock **chalk** consists chiefly of weakly cemented skeletons of coccoliths and planktonic foraminifera; it is lithified calcareous ooze. Most chalk deposits are in the deep sea, where they make up a large part of the sedimentary strata that overlie the basaltic crust. Chalks that are found on the continents are mainly of Cretaceous age and were deposited in epeiric seas that were probably never more than a few hundred meters deep. Cretaceous chalks are widely exposed in Europe, the Middle East, Australia, and North America.

Nowhere today is foraminiferal and coccolith ooze being deposited on or near the continental margins. This is not because coccoliths and planktonic foraminifera do not live there, but because the rate of accumulation of terrigenous sediment is so great that the skeletons of calcareous plankton never make up a significant proportion of the sediment. For Cretaceous chalks to accumulate in relatively shallow seas, there must have been a very low input of terrigeneous sediment, similar to that found today only in the deep sea. Hence, the Cretaceous chalk seas must have been vast, and the surrounding lands low and far removed.

The Drying Up of the Mediterranean Sea

By the late Miocene, the Mediterranean Sea looked very much as it does today. In the latest Miocene time—about 6 million years ago—the narrow connection with the Atlantic Ocean through the Strait of Gilbraltar closed, probably as a result of tectonic uplift, aided by falling sea level. The result was truly spectacular, for the Mediterranean Sea, which is 3,000 meters deep, quickly dried up (Figure 10.30). This surprising discovery was made in 1971 by deep drilling in the bottom of the Mediterranean. Here, beneath Pleistocene and Pliocene clays, widespread beds of anhydrite and halite reach local thicknesses of 1,500 meters.

Andhydrite and halite are evaporites; that is, they precipitate from seawater when it is concentrated by evaporation. The evaporites of the Mediterranean Sea could have been produced either in deep water or in shallow water. In the deep-water model, the inlet to the sedimentary basin is severely restricted, so that seawater comes in steadily but little escapes, except by evaporation; as a result, the salts that are left behind become increasingly concentrated. Although sea level in the basin is maintained, the water soon becomes saturated; salts then precipitate from the surface layer and accumulate on the bottom. In the shallow-water model, marine water continues to flow into the basin but at

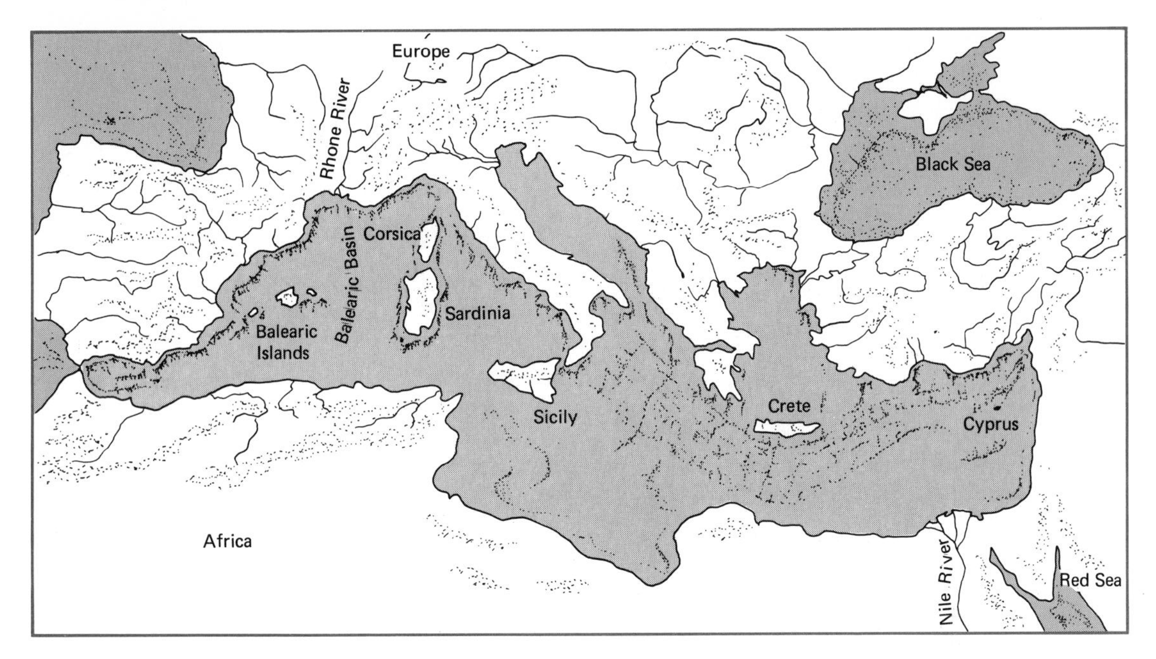

FIGURE 10.30

The dried-out Mediterranean Sea is represented by this panoramic drawing of the modern submarine topography of the Mediterranean basin. Approximately 6 million years ago, this area was a great interior desert lying 3,000 meters below sea level. The Balearic Basin was then a salt lake where evaporite minerals, including rock salt, were precipitated. Gravels and silts were deposited around the edge of the basin at the foot of the steep slope. (*Hsü, 1972*)

such a slow rate that it cannot replenish the quantity evaporated. Hence, sea level drops progressively, and probably sporadically. During periods when the basin evaporates nearly to dryness, evaporites are produced. At times evaporation is complete, and the Mediterranean becomes a deep playa covered with salts.

The Late Miocene deposits in the Mediterranean do not appear to be of deep-water origin because they contain stromatolites, which are formed by blue-green algae. Inasmuch as blue-green algae require sunlight, they must grow in shallow waters. Most living stromatolites, as we have seen in Chapter 3, occur in near-shore supratidal environments. The Late Miocene Mediterranean deposits also contain nodular anhydrite, which today forms only at temperatures above 35°C. Such high temperatures are unlikely in water depths of 3,000 meters, but they would be easily achieved on a desert playa, especially one that is 3,000 meters below sea level!

Supporting evidence that the sea disappeared from the Mediterranean basin comes from bordering coasts in adjacent North Africa and southern Europe, where river gorges are incised several hundred meters below present sea level (Figure 10.31). The rivers that cut these gorges in latest Miocene time necessarily flowed well below the present water-covered Mediterranean shelf, and even below the upper part of the continental slope. This could have occurred only if the Mediterranean had shrunk drastically, lowering base level for the streams flowing into the basin.

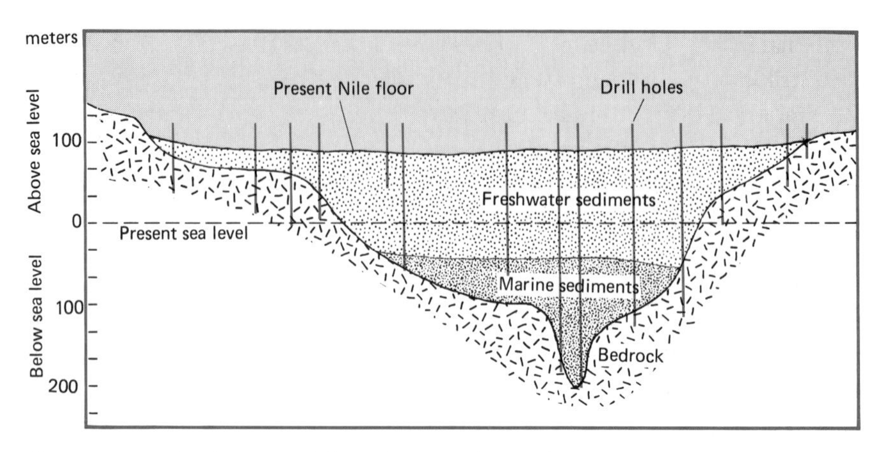

FIGURE 10.31 Deep gorge under the upper Nile valley near Aswan, Egypt was cut about 200 meters below the present level in latest Miocene time when the Mediterranean was dry. When the Mediterranean filled again at the beginning of the Pliocene, the sea also invaded the Nile, and marine sediments were deposited there. (*Hsü, 1972*)

FIGURE 10.32 Distribution of the evaporites in the Balearic Basin shows halite in a central portion, surrounded by anhydrite and then carbonates. This suggests shallow-water deposition in playas because halite would be precipitated last from a shrinking salt lake. (*Hsü, 1972*)

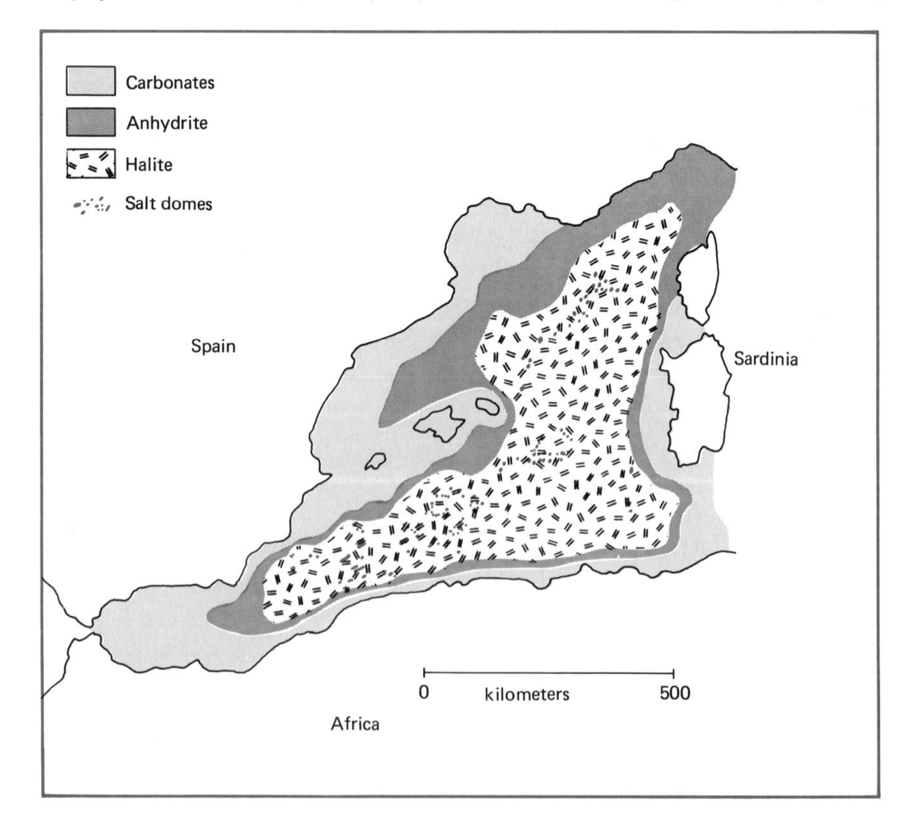

A final line of evidence for the drying up of the Mediterranean comes from the distribution on the deep Mediterranean floor of the evaporites themselves. The evaporitic facies in the Balearic Basin, one of the deep depressions on the bottom of the Mediterranean, forms a bull's-eye pattern, with the most soluble minerals in the center (Figure 10.32). This pattern would be expected only from a shrinking saline lake as it approached dryness.

It is not difficult to imagine how the Mediterranean might have dried up. At present, about 4,000 cubic kilometers of water evaporates from the Mediterranean every year. Only 10 percent of this loss is replenished by rain and by inflowing rivers. The other 90 percent flows into the Mediterranean from the Atlantic through the Strait of Gibraltar, which is 13 kilometers wide and 300 meters deep. If the Strait were totally closed today, so that the evaporation loss was not replaced, the Mediterranean would dry up completely in about 1,000 years. Evaporation of all the water in the present Mediterranean would produce an evaporite deposit only a few tens of meters thick. To produce 1,500 meters of evaporite deposits on the bottom of the Mediterranean during the late Miocene, the Strait of Gibraltar must have closed slowly; in the process, seawater must have poured into the basin as a partial replacement of that which was evaporating. In fact, normal marine sediments interbedded with the evaporites indicate that the basin dried up and refilled several times.

The quantity of salt contained in the Mediterranean Miocene evaporites is equivalent to about 6 percent of the total amount of salt dissolved in present-day oceans. Not only did precipitation of that much salt in the Mediterranean during the latest Miocene lower the salinity of the world's oceans directly, but the water that evaporated from the Mediterranean must have fallen elsewhere as additional precipitation, so that the surface salinity of the oceans may have been lowered still further.

Since lowered salinity raises the freezing point of water, this probably facilitated the formation of sea ice at high latitudes. In areas where large quantities of ice form, the **albedo** of the region increases dramatically, producing further cooling and promoting glaciation. Thus the drying up of the Mediterranean could possibly have been a factor in enhancing the glaciation taking place in the Arctic and Antarctic regions during the latest Miocene.

At the end of the Miocene—about 5.5 million years ago—a rise in sea level, accompanied by tectonic lowering of the Gibraltar sill, allowed the Atlantic Ocean to refill the Mediterranean basin for the last time, an event that must have been spectacular! The seawater must have flowed through the Gibraltar Straits as a gigantic waterfall. Had the water merely trickled in, it would have evaporated before it could fill the basin. In order to produce the near-normal salinities indicated for the sediments that immediately overlie the evaporites, the influx had to exceed the evaporative loss by a factor of ten. Therefore, some 40,000 cubic kilometers of seawater roared over the falls each year. This volume is 1,000 times greater than that of Niagara Falls. What a tourist attraction it would have been—had anyone been around to see it!

Changing Life on Land

The Mesozoic modernization of the earth's landscapes and oceans was accompanied by some equally profound changes in the living world. In the seas, plants, invertebrate animals, and fishes all went through several drastic episodes of extinction and evolutionary radiation. The changes in oceanic life did not lead to dramatically new and different forms; land-dwelling life, on the other hand, was to change almost beyond recognition. In the plant world, primitive Triassic cone-bearing species progressively gave way to the host of flowering plants that now dominate the land surface. Dinosaurs and other bizarre reptiles ruled the continents during Jurassic and Cretaceous time, only to be replaced, during the Cenozoic Era, by a steady progression of mammals and birds.

Flowering Plants

With the decline of seedless Paleozoic plant groups (Figure 8.23), seed-bearing plants of the Subphylum Spermopsida came to dominate the land. The Spermopsida are divided into five classes: The first four (seed ferns, cycads, ginkgoes, and conifers) flourished in late Paleozoic and early Mesozoic time. They were largely replaced during the Cretaceous Period by the more advanced flowering plants of the fifth class, the Angiospermophyta.

Today, flowering plants overwhelmingly dominate the land. Of the approximately 260,000 living species of vascular plants, about 250,000, or 96 percent, are angiosperms. The other 10,000 species are mostly ferns. Conifers and their relatives, despite wide distribution and local importance, have only 700 surviving species.

The great evolutionary success of flowering plants is partly due to the development of the flower and its enclosed seed. The flowerless conifers had to rely on the wind to carry pollen from plant to plant—a relatively wasteful and inefficient means of fertilization. Flowers are specialized reproductive structures that normally bear both the seed-producing and pollen-producing organs, surrounded by modified colored leaves (the petals and related structures) that serve principally to attract insects or birds to transport the pollen.

The Age of Reptiles

The Mesozoic changes in land plants that led to the present dominance of flowering forms were paralleled by equally dramatic changes in the dominant land-dwelling vertebrate animals—all of which ultimately depend on land plants for their nourishment. In Chapter 8 we traced the history of the earliest land vertebrates—primitive amphibians that arose from lobe-finned fish. There we saw that amphibians have never fully adapted to land life because they lack a means of protecting the embryo from drying out, and thus they must return to the water to lay their eggs. This problem was overcome in the **reptiles**, a group that arose from the early amphibians. Reptiles, in turn, were the ancestors of the two most successful classes of land vertebrates, the birds and mammals.

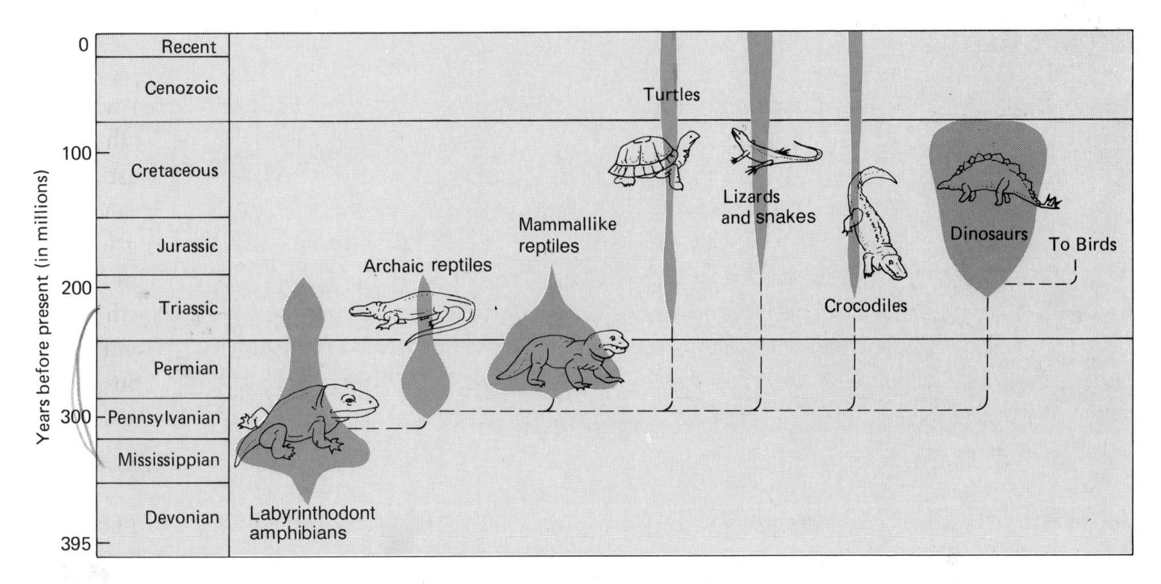

FIGURE 10.33

Evolutionary history of the reptiles. The dashed lines show the most probable evolutionary relations of the groups. The width of the vertical areas indicates the approximate abundance of each group. During Permian through Cretaceous time, the dominant land animals were, first, mammallike reptiles, then, dinosaurs. Surviving modern reptile groups (turtles, lizards, crocodiles) arose in Triassic time but were overshadowed by their dinosaur contemporaries.

Early Reptiles Reptiles overcame the problem of reproduction on land with the shelled egg, a device that allows the embryo to grow in its own self-contained liquid environment. After development is completed, the offspring breaks out of the shell as a small, but otherwise fully formed, animal. The reptilian egg freed land vertebrates from the necessity of living near bodies of water and ultimately permitted them to wander freely over the land surface in search of food and favorable habitats.

Although it is impossible to determine directly from skeletal features whether a given fossil vertebrate layed shelled eggs, there are some minor skeletal differences, principally in the patterns of skull bone and vertebral construction, that distinguish amphibians from reptiles. These features are useful in separating fossil representatives of the two groups. Such evidence shows that, by late in the Pennsylvanian Period, the first archaic reptiles had evolved from closely similar amphibian ancestors (Figure 10.33). During the succeeding Permian Period, reptiles went through an explosive evolutionary radiation that began their long domination of the land, a dominance that was to last until the end of the Mesozoic Era. During Permian and much of Triassic time, the dominant reptiles were the mammallike forms, an abundant and diverse group that was to give rise to the **mammals.** In Late Triassic time, mammallike reptiles were replaced by a more familiar group—the **dinosaurs**—that were to rule the land throughout the Jurassic and Cretaceous Periods.

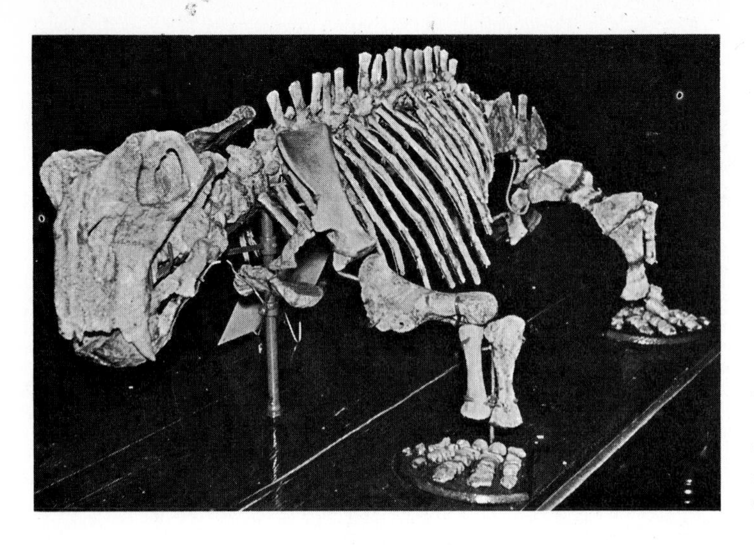

FIGURE 10.34

Skeleton of the aquatic, herbivorous mammallike reptile *Lystrosaurus,* about three feet long, from Early Triassic rocks. The sprawling posture characterized all mammallike reptiles and even early mammals. (*Courtesy of A. W. Crompton*)

In spite of their name, most mammallike reptiles did not look very much like mammals. Most were large-headed, bulky animals, from 1 to 5 meters long, that must have moved rather awkwardly, for their legs were not directly beneath the body, as in most dinosaurs and mammals, but instead extended outward from the side of the body (Figure 10.34). Their skulls and teeth show that both herbivores and predatory carnivores were common. During the Triassic Period, several different lines developed progressively more mammalian skull features; one of these lines gave rise to the mammals in Late Triassic time.

Dinosaurs The most dramatic land animals ever to evolve were the many dinosaurs that dominated Jurassic and Cretaceous landscapes. *Dinosaur* is a popular term for two distantly related groups of large land reptiles, the saurischians and ornithischians, that made up the second of the three major groups of dominant land-dwelling vertebrates: mammallike reptiles during the Permian and Triassic Periods, dinosaurs during the Jurassic and Cretaceous Periods, and mammals since the close of the Cretaceous Period. All three groups have tended to evolve many specialized herbivores and relatively few carnivores, which prey upon the herbivores. This trend is particularly well illustrated by the dinosaurs.

There were six principal kinds of dinosaurs (Figure 10.35). Studies of fossil skulls and teeth show that five of them were herbivorous: the *ornithopods, stegosaurs, ankylosaurs, horned dinosaurs,* and *sauropods.* The sixth group, the *theropods,* were carnivorous. Each of the six lines shows an evolutionary development from the relatively small, unspecialized forms found in Upper Triassic and Lower Jurassic rocks to the familiar huge and impressive animals of Late Jurassic and Cretaceous time. In the five herbivorous lines, the evolutionary trends appear to have been directly related to defense against the theropods, which, in the familiar *Tyrannosaurus,* developed the most awesome flesheater to evolve. Each of the five herbivorous groups developed means of protection from the theropods: some by becoming rapid runners (ornithopods), others by developing protective armor and weapons for defense (stegosaurs, ankylosaurs, horned dinosaurs), and still others by becoming too large for attack (sauropods).

The six principal dinosaur groups did not all originate at the same time. The sauropods, theropods, ornithopods, and stegosaurs evolved in Late Triassic or Early Jurassic time, whereas the ankylosaurs and horned dinosaurs developed in the Late Jurassic and Cretaceous Periods. Almost all the dinosaur lines persisted until late in the Cretaceous Period, when they abruptly became extinct. The only exception were the stegosaurs, which died out in Early Cretaceous time.

The cause of the demise of the dinosaurs is uncertain. Apparently, some environmental factor or factors changed so rapidly that dinosaur evolution could not keep pace. Whatever the cause, the extinction of the dinosaurs left vacant much of the land surface and paved the way for the great evolutionary expansion of the mammals, which were to be the dominant land vertebrates throughout the Cenozoic Era.

Birds In their egg-laying reproduction and general anatomy, birds can accurately be characterized as "feathered reptiles." Among the most remarkable fossils ever found are several skeletons from Jurassic limestones of southern Germany that are clearly transitional between birds and their reptilian ancestors (Figure 10.36). These skeletons have teeth (all modern birds are toothless) and other features that are so much like those of certain small dinosaurs that, had the skeletons alone been preserved, they probably would have been described as reptiles. Fortunately, the fine-grained limestone in which the skeletons are preserved also shows clear impressions of flight feathers on the tail and on the elongated front limbs. Feathers, found only in birds, indicate

FIGURE 10.35

Evolutionary history of the dinosaurs. The width of the vertical areas indicates the approximate abundance of each group.

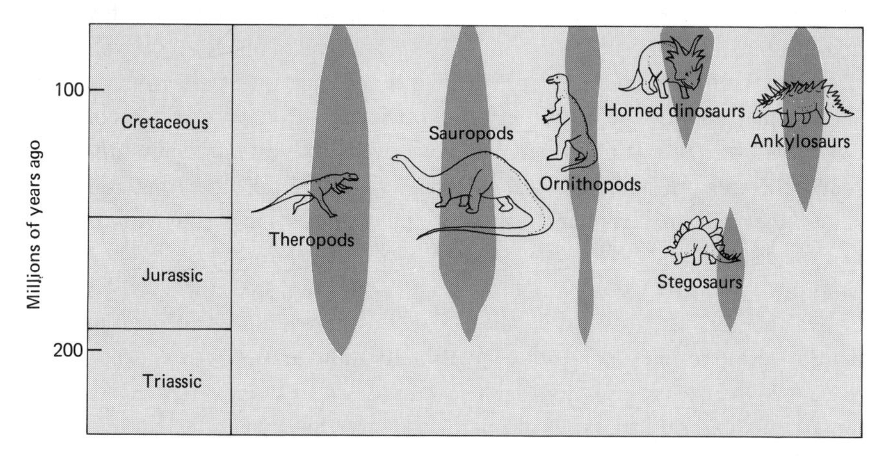

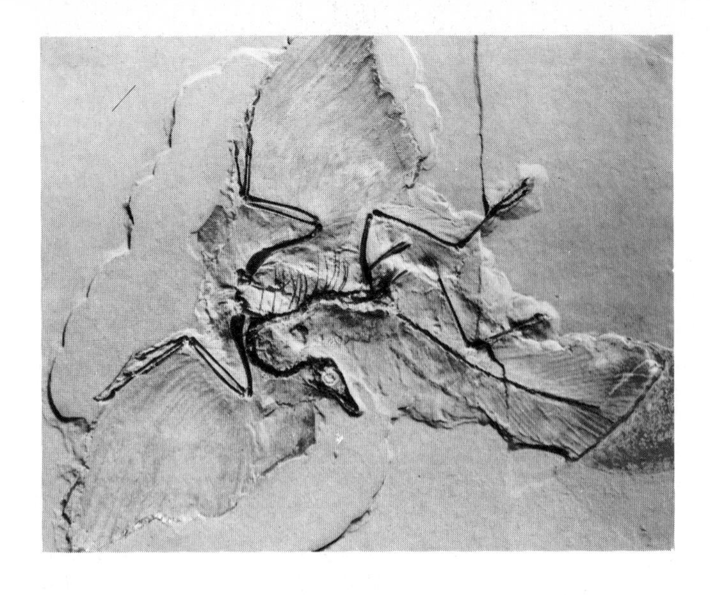

A nearly complete fossil of the transitional Jurassic bird *Achaeopteryx* showing the reptilelike skeleton as well as the impressions of wing and tail feathers. The skeleton is about 30 centimeters long. (*Courtesy of American Museum of Natural History*)

FIGURE 10.36

that this strange animal was indeed a primitive bird that still retained most of the skeletal features of its reptilian forebears. Regrettably, relatively few comparable bird skeletons are known from younger rocks, and these give a mere glimpse of the evolutionary history of the class.

The Rise of Mammals

Mammals maintain a constant body temperature, whereas the body temperature of present-day reptiles is determined by the surrounding air temperature. This adaptation permits mammals to lead an active and diversified life; they can survive in cold regions and can search for food in all seasons, during the cool of the night and the warmth of the day.

Mammals also differ from reptiles in their manner of reproduction. The shelled reptilian egg was an important advance over the amphibians, yet it had the disadvantage of being easily destroyed by predators. The mammal overcame this danger by internalizing the egg. The egg is retained within the body of the female, where the embryo can develop, protected, before being born alive. A few fish and reptiles hold their eggs within the body cavity until the young hatch, but only in mammals has this become an almost universal adaptation.

Along with internal embryonic development, mammals developed another special feature that permits the young animals to become still more fully developed before beginning an independent life—the milk-producing mammary glands from which the class takes its name. The strengthening nourishment of milk and the maternal care that goes with it help insure that the young mammal will reach maturity without becoming a meal for some predator.

In Triassic time, several separate lines of mammallike reptiles independently developed more and more mammalian jaw and tooth patterns. From these evolved the oldest unmistakable mammalian fossils, which are the teeth, jaw fragments, and rare skulls of small, shrewlike forms found in Late Triassic rocks (Figure 10.37). Mammals originated at about the same time as the

FIGURE 10.37

Fossil jawbone of a Jurassic mammal. Mesozoic mammals were small, inconspicuous contemporaries of the dinosaurs. They are known mainly from fragmentary fossil remains, such as the jawbone shown here. The specimen is about 3 centimeters long. (*Courtesy of Her Majesty's Geological Survey Photographs, Crown Copyright*)

dinosaurs, but throughout the Mesozoic Era they remained small, inconspicuous animals that are relatively uncommon fossils. Although not closely related to modern rodents, they were generally mouselike animals that fed on seeds and insects. The largest was about the size of a house cat. Most early mammals were evolutionary dead ends that failed to lead to more advanced groups. However, during the Cretaceous Period two groups arose—the **marsupials** and **placental mammals**—that were to dominate the land after the extinction of the dinosaurs (Figure 10.38).

Marsupials were the less successful of the two, for they make up only about five percent of all Cenozoic mammals. They differ from the more successful placentals principally in mode of reproduction. In marsupials the young are born alive, but at a very early stage of development. These tiny, immature young then crawl into a pouch on the mother's abdomen, attach themselves to a mammary nipple, and complete their development in this external pouch. This pattern is still found in all living marsupial mammals, including the familiar pouched kangaroos and opossums, and was almost certainly characteristic of fossil marsupials as well.

In contrast, placental mammals develop inside the uterus of the female with a special structure, the **placenta**, that allows the developing young to be nourished directly by the mother's body fluids. This structure permits a long period of development within the body of the female so that birth does not occur until the young are relatively mature and independent.

The ancestral placental mammals were small, superficially mouselike forms called insectivores. This group includes the modern shrew, moles, and hedgehogs, all of which show primitive anatomical features. In early Cenozoic time, insectivores went through an explosive evolutionary expansion that ultimately led to such diverse mammals as whales, bats, horses, and elephants. The earliest representatives of most groups of placental mammals did not closely resemble their modern counterparts, but were generally smaller and less specialized. For example, ancestral carnivores were no larger than a rab-

FIGURE 10.38

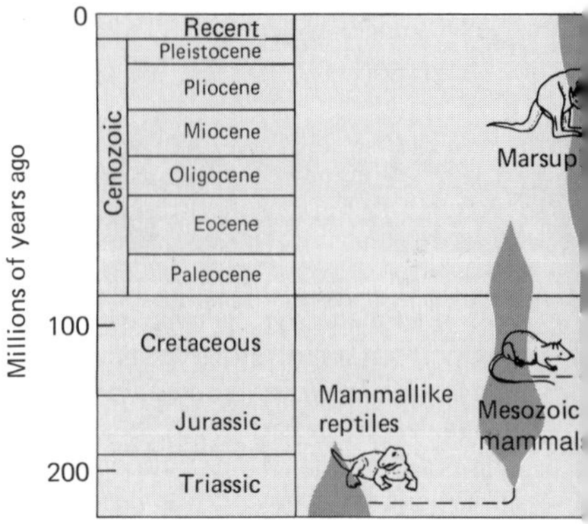

Evolutionary history of the mammals. The dashed lines show the most probable evolutionary relations of the groups. The width of the vertical areas indicates the approximate abundance of each group.

bit, and early horses were four-toed animals about the size of a fox terrier. Throughout the Cenozoic, some members of each order tended to increase in size and to diversify into increasingly specialized types. Some groups, such as the horses and elephants, reached their evolutionary climax earlier in the era and are reduced in numbers and diversity today. Others, such as rodents and primates, appear to have steadily increased in diversity until modern times.

Chapter Summary

Evolving Landscapes

Tectonic patterns: Plate motions that split open Pangaea led to two worldwide zones of mountain-building deformation that have shaped the modern landscape.

The Alps and Himalayas were formed as the former Tethys Sea was closed by the convergence of the African, Eurasian, and Indian continental masses.

The Cordilleran region, a second zone of mountain-building deformation, appeared along the western margins of North and South America as they periodically moved against plates underlying the Pacific Ocean.

Tectonic control of sedimentary environments: Mountain-building forces form highlands that shed sediment into adjacent lowlands, or basins; sedimentary patterns can reveal details of the deformations that produce them.

Flysch and molasse: First recognized in the Alps, flysch are deep-water marine sediments that pass upward into nonmarine molasse deposits that record the final stages of uplift in the adjacent mountain ranges.

Delta and barrier-island systems can reach great thicknesses near active mountain-building; many Cretaceous rocks of the western United States formed in this way.

Lacustrine oil shales accumulated in basins of the western United States that were occupied by inland lakes during Eocene time; these areas contain enormous quantities of solid petroleum particles that can be removed by heating.

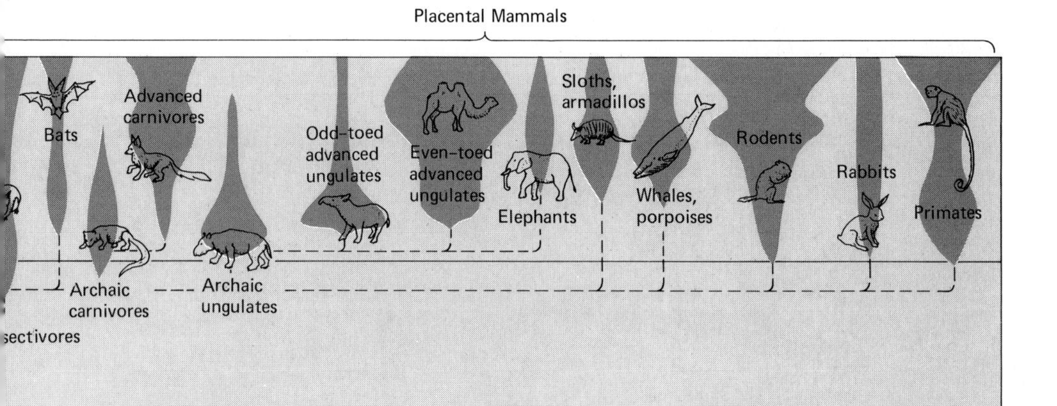

The Changing Oceans

Deep-ocean environments changed with the Jurassic evolution of new planktonic life with calcium carbonate shells; these shells accumulate in enormous quantities on the ocean floor; in the deepest parts of the ocean the shells are dissolved, and thus they accumulate only in intermediate depths.

Epeiric seas repeatedly spread over much of the surface of the continents in Jurassic and Cretaceous time; these variations probably reflect changes in the volume of the ocean basins caused by plate motions.

Jurassic ironstones, unusual sedimentary rocks, consist of tiny spheres of iron silicate or iron oxide that accumulated in shallow-marine settings; they appear to have no modern counterpart.

Cretaceous chalk consists of vast accumulations of tiny planktonic shells deposited in shallow seas; such deposits are forming today only in deep-ocean environments.

The drying up of the Mediterranean Sea occurred in Miocene time and produced great thicknesses of evaporite deposits.

Changing Life on Land

Flowering plants expanded to dominance in Cretaceous time; reproductive advances made the group far more diverse than its flowerless predecessors.

The age of reptiles:

Early reptiles developed the shelled egg as a means of reproducing far from bodies of water; in Permian and Triassic time, mammallike reptiles were most diverse and abundant.

Dinosaurs replaced the mammallike reptiles during Jurassic and Cretaceous time; the several herbivorous groups had varying defenses against the carnivorous theropods.

Birds arose from small dinosaurs in Jurassic time; birds are seldom preserved as fossils.

The rise of mammals: Mammals arose from mammallike reptiles in Triassic time but expanded to dominance only after the latest Cretaceous extinction of the dinosaurs.

Important Terms

abyssal clay
albedo
authigenic mineral
barrier island
bottomset bed
calcareous ooze
calcite compensation depth (CCD)
Cascadian Orogeny
chalk
coccolith
delta
diatom
dinosaur
distributary
eustatic change
flysch
foraminifera
foreset bed
ironstone
kerogen
Laramide Orogeny
mammal
marsupial
Mesocordilleran highland
microcontinent
molasse
Nevadan Orogeny
oil shale
oolith
placenta
placental mammal
playa
progradation
radiolarian
reptile
sedimentary basin
Sevier Orogeny
siliceous ooze
sole markings
topset bed
varve

Review Questions

1. What are the principal mountain belts that formed from the breakup of Pangaea?
2. How have the tectonic events related to the breakup of Pangaea influenced patterns of sediment accumulation?
3. Summarize the changing patterns of shallow seas on the continents since Jurassic time.
4. What are some possible explanations for the origin of Jurassic ironstones? Cretaceous chalk?
5. Outline the evolutionary history of reptiles. What advantage did reptiles have over their amphibian ancestors?
6. From what reptile group did mammals arise? What was the Mesozoic history of mammals?

Additional Readings

Hsü, K. J. "When the Mediterranean Dried Up," *Scientific American,* vol. 227, no. 6, pp. 27–36, 1972. *Reviews the unexpected discovery of the Sea's complex history.*

Kurtén, B. *The Age of Dinosaurs,* McGraw-Hill, New York, 1968. *A nontechnical summary.*

Kurtén, B. *The Age of Mammals,* Columbia University Press, New York, 1972. *A nontechnical survey.*

Lajoie, J. (ed.) *Flysch Sedimentology in North America,* Geological Association of Canada, Waterloo, Ontario, 1970. *Technical papers on these important sediments.*

Molnar, P. and P. Tapponnier "The Collision Between India and Eurasia," *Scientific American,* vol. 236, no. 4, pp. 30–40, 1977. *Summarizes the events leading to the formation of the Himalayas.*

Reineck, H. E. and I. B. Singh *Depositional Sedimentary Environments,* Springer-Verlag, New York, 1973. *Excellent reviews of environments of sand and mud deposition, both terrestrial and marine.*

Roberts, R. J. "The Evolution of the Cordilleran Fold Belt," *Geological Society of America Bulletin,* vol. 83, pp. 1989–2003, 1972. *An advanced review.*

THE PLEISTOCENE EARTH AND MAN

11

From the perspective of 4 billion years of the earth's past, the events of the last two million years, known as the Pleistocene Epoch of the Neogene Period, represent a mere moment of time. Yet this most recent interval of earth history is of extraordinary interest for two reasons. First, it is a time of unusual climatic events, as great ice sheets alternately advanced and retreated over much of the land surface. From evidence preserved in ancient rocks, similar intervals of widespread glaciation can be seen to have taken place in the more distant past, but these occurrences were rare, indicating that Pleistocene climates were exceptional, when viewed over the long course of earth history.

Still more important to us as humans is the fact that much of the crucial history leading to our own species, *Homo sapiens*, took place in the two million years of Pleistocene time.

Glacial Landscapes

The discovery that much of the land surface of the northern hemisphere was recently covered by thick sheets of ice required some clever scientific detective work. In the early 1800s, geologists were puzzled by boulder-filled clay deposits that cover much of northern and central Europe. They observed that these sediments, which form random mounds and ridges, were quite unlike ordinary water-lain deposits, and they referred to them as **drift.** Some cobbles and boulders in the drift could not have been derived from any bedrock in the immediate neighborhood, but could be matched with distant rocks. It was at first suggested that the drift had been deposited by icebergs or by the waters of the Biblical flood, but neither of these mechanisms could explain many of the features of the drift.

The idea that the drift had been deposited directly by glaciers was proposed very early, but it was not proven to the satisfaction of most scientists until near the middle of the nineteenth century. At that time, Alexander Agassiz, a young Swiss zoologist, argued that the widespread erosional and depositional features associated with the drift were identical to those being produced by the still-active valley glaciers in the Alps. He showed that the former glaciers had not been confined to valleys, but were spread widely over northern Europe and much of North America as thick, continuous sheets of ice. Agassiz concluded that the glaciation of these vast regions had occurred in the recent geologic past during a "great ice age." His idea is now accepted fully; in fact, it is now clear that the ice age is not over!

Pleistocene Geography

Today, ice covers about 10 percent of the earth's land surface, mostly in Antarctica, Greenland, Iceland, and in scattered mountain ranges throughout the world. At their maximum extent in the Pleistocene Epoch, ice sheets covered more than three times this area, or about 30 percent of the earth's land surface (Figure 11.1). In North America, most of Canada and much of the

Ice sheets of the present day (dark color) and of about 20,000 years ago (light color) in the northern hemisphere. (*Bloom, 1969*)

FIGURE 11.1

northern United States were covered by ice. Greenland also lay beneath a great mass of ice, as it does now. In Europe, an ice sheet spread southward from Scandinavia across the Baltic Sea into Germany and Poland; the Alps and the British Isles supported their own ice caps. Continental glaciers extended throughout the northern plains of Russia and large sections of Siberia. They covered the Kamchatka Peninsula and the high mountains and plateaus of Central Asia, where some of the ice caps were larger than those of the Alps. In North America, the southernmost extent of the continental ice sheets corresponded closely to the present courses of the Missouri and Ohio Rivers. The high mountains of the western United States were heavily glaciated by small ice caps and valley glaciers. In the southern hemisphere, Antarctica was covered by ice, as it is now. New Zealand, Tasmania, and southern South America were all heavily glaciated. Even in tropical latitudes, high mountains were glaciated, as in Hawaii and New Guinea.

Glaciers are powerful agents of erosion whose effects are much different from those of running water. As glaciers form in a previously unglaciated area, the streams of ice at first follow preexisting water-worn valleys; but as the ice sheets combine and thicken, the contour of the underlying land exerts less and less effect. Fully formed continental ice sheets may be more than 3,000 meters thick. Their direction of flow is determined by the slope of their *upper* surfaces, not by the nature of the rock surface below. Thus, the area that receives the greatest accumulation of snow eventually becomes the central point of outward flow for the entire sheet.

Ice sheets, with their load of rock fragments, push down on the land surface with tremendous force. The prolonged advance of a thick ice sheet levels and smooths the underlying surface, as the rocks carried along at the bottom of the ice are dragged across the buried landscape. Glaciated surfaces show numerous elongate depressions and scratches, which indicate the direction in which the ice moved (Figure 11.2). Local areas of hard rock may be left in relief and softer areas may be deeply excavated, but, on the whole, relief becomes considerably smoothed and subdued. The material thus removed is deposited far away, at the melting edge of the ice. In this fashion, particles the size of a house may be carried hundreds of kilometers.

FIGURE 11.2

This smooth and striated rock surface in Alaska indicates that it has been overridden by a glacier. (*U.S. Geological Survey*)

Glacial Lakes

At the melting fringes of ice sheets, the meltwater produces large streams that carry great quantities of glacially provided gravel and sand as **outwash**. On the Columbia River in Canada, for example, immense terraces of coarse sand and gravel more than 30 meters high attest to huge volumes of meltwater in the recent geologic past. In many areas of northern Europe and North America, the advancing glaciers themselves greatly modified drainage patterns and produced ephemeral lakes. Preexisting stream valleys that were gouged by the ice commonly became elongate depressions. Those that subsequently filled with fresh water became **finger lakes** (the elongate lakes of upper New York State are examples), and those that filled with the sea became **fjords.** The Great Lakes basins had a similar origin, and they underwent a complex history of excavation during advances and retreats of the ice. Figure 11.3 shows the Great Lakes region about 10,000 years ago, as the most recent ice sheet retreated. In addition to the present-day Great Lakes, note that additional large lakes, now drained, were then present in the same region.

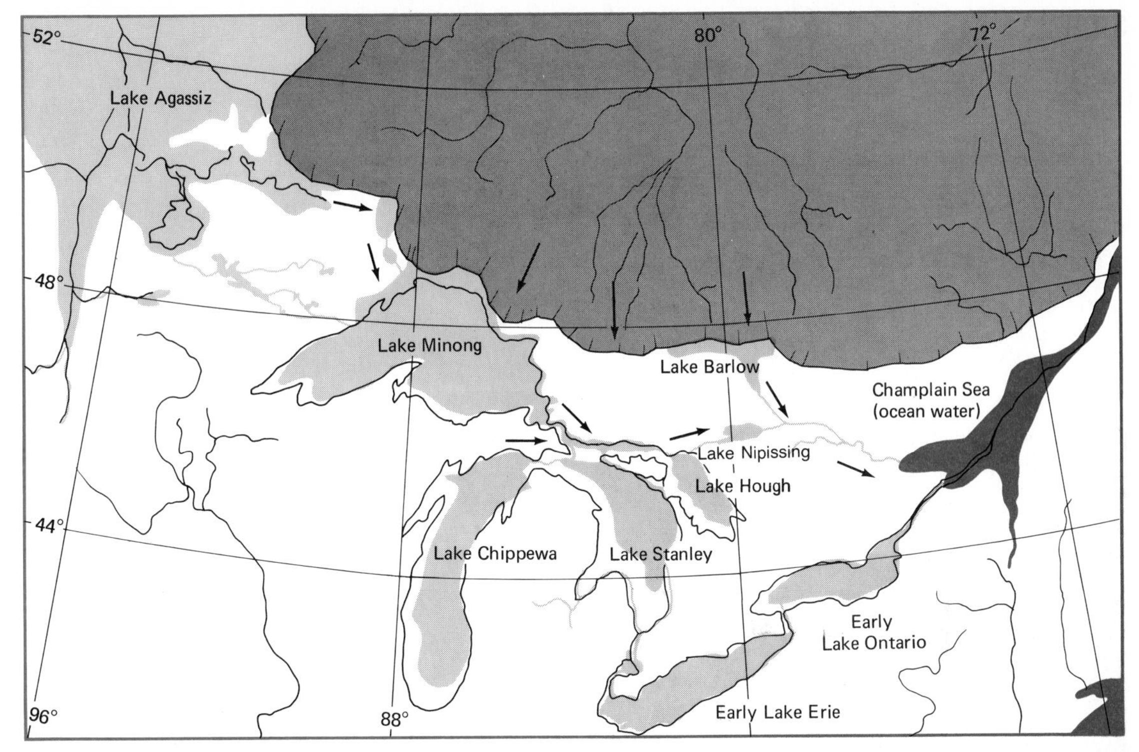

The edge of the North American ice sheet and the adjacent lakes of the present Great Lakes region as they appeared about 10,000 years ago. (*Prest, 1969*)

FIGURE 11.3

The channeled scablands of eastern Washington were produced about 22,000 years ago by a catastrophic flood that was caused by the breakup of the ice dam that contained glacial Lake Missoula. (*Baker, 1973*)

FIGURE 11.4

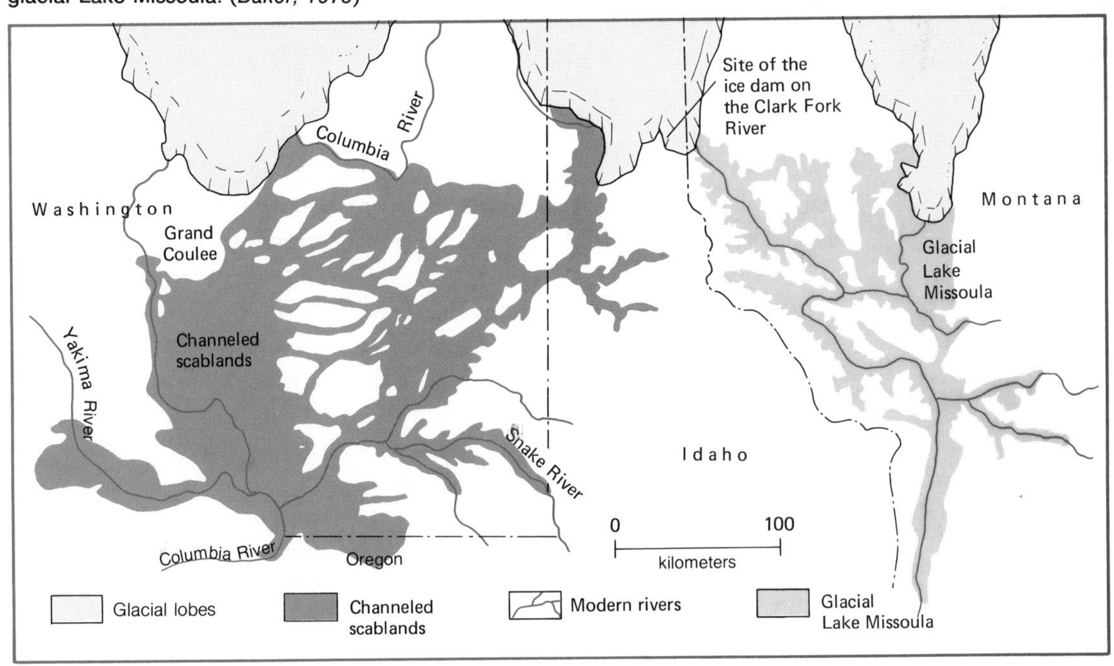

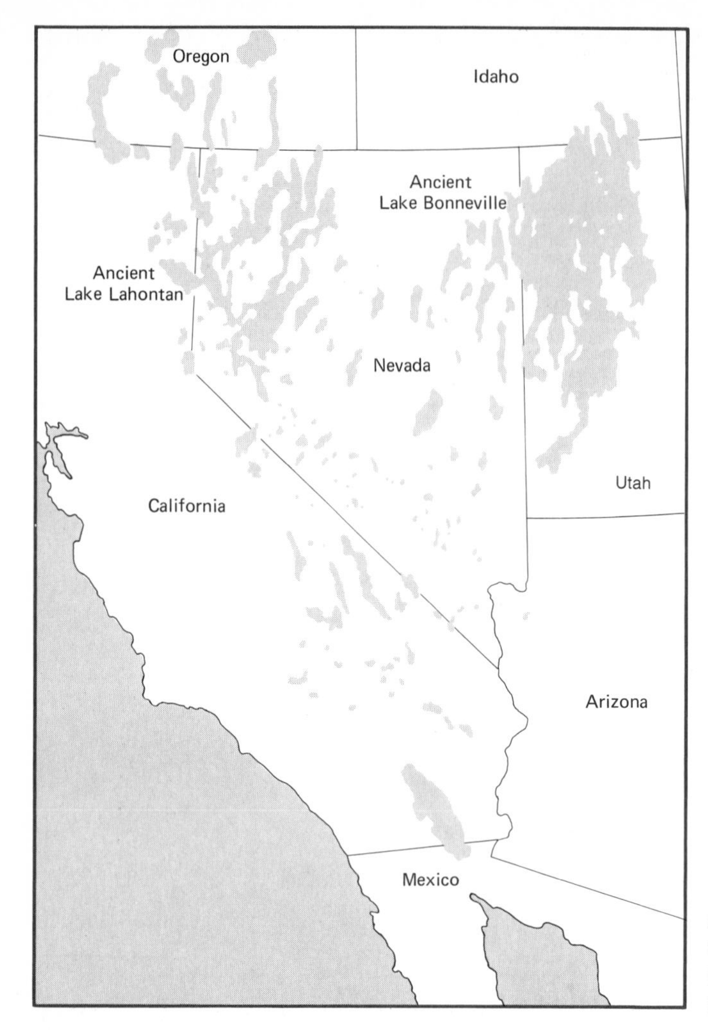

About 15,000 to 20,000 years ago, a much more moist climate produced some huge, deep lakes in the now-arid Basin and Range province of the western United States.

FIGURE 11.5

Similar large lakes, now mostly gone, formed in the northern Great Plains and the northern Rocky Mountains. Most were dammed by glacial deposits, but some were formed by the margins of the ice sheet itself and were filled with glacial meltwater. On several occasions, the dams made of glacial material broke, causing catastrophic flooding. In western Montana, for example, a large basin filled with water to form Lake Missoula (Figure 11.4). About 22,000 years ago, its dam of glacial ice and sediment broke abruptly, sending a wall of water rushing with incredible speed across northern Idaho and eastern Washington. This enormous flood scoured channels and deposited immense gravel bars over a large part of the Columbia Plateau, thus creating a unique landscape that has become known as the **channeled scablands.**

Glacially induced climatic changes also produced huge lakes in the now-arid Basin and Range region of the western United States, far from the margin of the ice-sheet. The two largest of these were Lake Bonneville, of which Great Salt Lake is a shrunken remnant, and Lake Lahontan in western Nevada (Figure 11.5).

Sea-Level Changes

The water that was locked up in the ice of Pleistocene glaciers originally came from the oceans. It was evaporated, transported landward by winds, precipitated as snow, and compacted into ice. At their maximum extent, the Pleistocene ice sheets had a volume of about 77 million cubic kilometers. As a result of this conversion of seawater to ice, sea level then was about 120 meters lower than it is now. At the present time, about 26 million cubic kilometers of ice still remain on the continents. If all the present glaciers were to melt, the water produced would raise sea level an additional 65 meters, greatly changing the outline of the earth's land areas and completely submerging most of the large coastal cities of the world. Long Island, most of New Jersey, the Delaware-Maryland-Virginia peninsula, and most of Florida would be flooded. It is clear that the expanding and contracting Pleistocene ice sheets caused profound, worldwide changes in the distribution of land and sea.

Isostasy

Closely related to Pleistocene sea-level changes is an additional phenomenon caused by the enormous weight of the thick ice sheets. The crust of the earth sinks gently when it accumulates a sufficiently heavy burden, whether of sediment, water, or ice; conversely, it rises slowly when such loads are removed. The overloaded area of crust actually sinks into a layer of higher density, which buoys up the heavy loads. The condition of balance thus achieved is called *isostasy.*

FIGURE 11.6

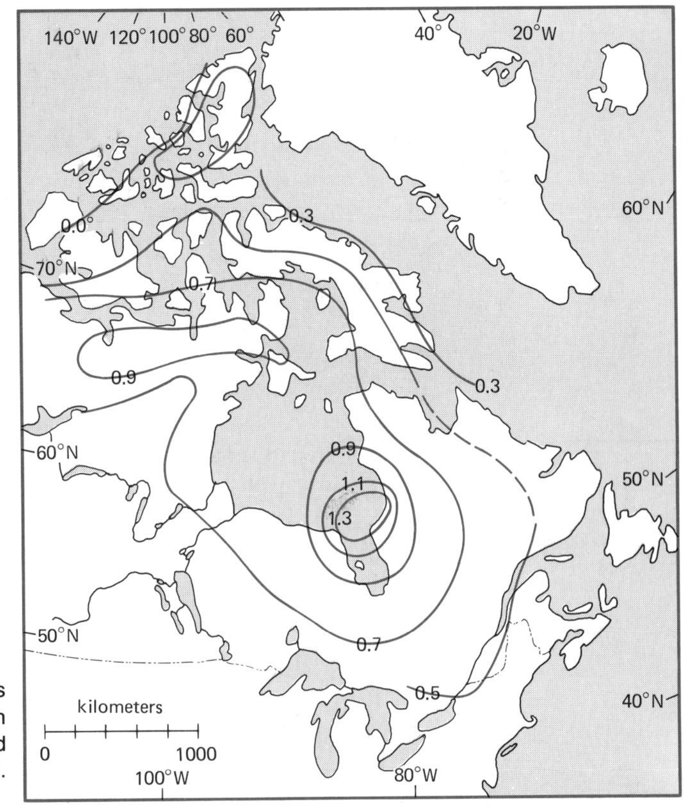

Present rate of uplift in meters per 100 years in northeastern North America. The southern Hudson Bay area already has rebounded about 240 meters since the ice sheet melted. (*Andrews, 1970*)

The concept of isostasy helps to explain the behavior of areas that have supported thick ice sheets. The gradual ice accumulation depresses the underlying crust until the actual land surface may lie well below sea level. For example, in the central part of Greenland, the present ice sheet is more than 3,000 meters thick, and its base is about 350 meters below sea level. If the ice were taken away, the area would temporarily become a shallow sea; not for several thousand years would the crust rise to a new position of equilibrium. Hudson Bay in Canada is an example of such an ice-depressed area that has not yet fully rebounded. This shallow sea occupies the site that was formerly covered by the thickest portion of the North American ice sheet. Raised marine sediments show that about 240 meters of rebound has so far occurred since the ice melted (Figure 11.6).

Pleistocene Chronology

A fundamental issue in Pleistocene history is the exact magnitude and timing of the many expansions and contractions of the ice sheets. Early in this century, studies of ice-deposited sediments, both in Europe and in the American midwest, seemed to indicate four major episodes of glaciation, separated from each other by an equal number of warmer interglacial intervals. Later work appeared to confirm these results, and the idea of four glaciations became firmly established.

A major difficulty with such chronology based on glacial deposits is that each advance of an ice sheet tends to destroy the sediments left by the preceding advance, thus obscuring earlier chronological evidence. For this reason, sediments deposited by the most recent ice sheet, which persisted until about 10,000 years ago, are abundant and easily studied, whereas those of earlier ice sheets are known from only a few areas where, for various reasons, they escaped destruction by the most recent advance.

Deep-Sea Cores

Beginning in the 1960s, an entirely new kind of evidence began to suggest that the traditional four-glaciation idea had greatly oversimplified the complexities of Pleistocene history. This newer approach to glacial chronology is based on the fact that the fluctuations of Pleistocene climates caused changes in the temperature and chemistry of the oceans. These changes were reflected in the shells of certain microscopic animals (mostly foraminifera) that float near the ocean surface. After death these shells accumulate on the sea floor, where they can be sampled by collecting **cores,** long cylinders of sediment obtained by dropping weighted hollow pipes into the ocean bottom. Because these sediments accumulate very slowly, a single core can contain a layered sequence of tiny shells that rained down from above over long intervals of Pleistocene time.

Two principal techniques are used to determine climatic events from these shells. The first depends on the fact that many species live only in waters of a

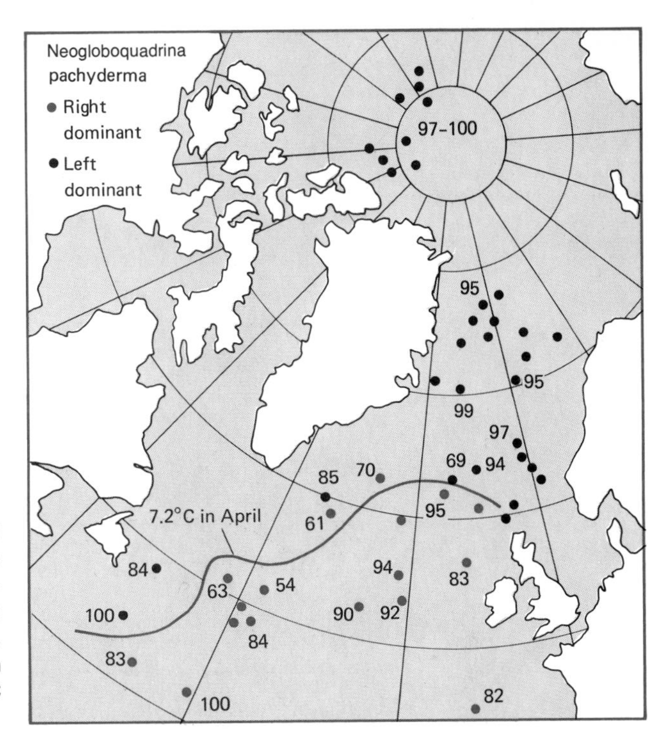

FIGURE 11.7

Correlation between the coiling direction of *Neogloboquadrina pachyderma* and surface water temperature. The figures shown are percentages. This forminiferan coils to the left in colder waters and to the right in warmer waters. The dividing line marks the boundary between water temperatures below and above 7.2°C in April. (*Ericson and Wollin, 1964*)

particular temperature. For example, the foraminifer *Neogloboquadrina pachyderma* occurs principally in cold waters, while the related form, *Globorotalia cultrata*, occurs principally in warm waters. In addition, some individual species may coil in one direction in cold water and in another direction in warm water (Figure 11.7). Changes in the distribution or coiling of such species through a core can thus provide a continuous document of intervals of colder and warmer ocean water, which, in turn, reflect the expansion and contraction of ice sheets on land.

A second method of determining climatic change depends on analysis of the oxygen contained in the shells. Oxygen dissolved in ocean water consists of two isotopes, oxygen-18 and oxygen-16. The exact ratio of the two fluctuates, however, as different amounts of ocean water are stored on land in ice sheets. This is because water evaporated from the ocean is slightly enriched in oxygen-16, the lighter isotope. When this "light" water is precipitated on land as snow and then becomes part of an ice sheet that persists for tens of thousands of years, it causes the ocean to become slightly depleted in oxygen-16. This change is recorded in the oxygen contained in the calcium carbonate shells. Variations in the oxygen-isotope ratios through a core thus indicate the extent to which ice sheets covered the land and are, therefore, an indirect index of Pleistocene climates.

Studies of deep-ocean cores began in the 1930s but they have been a principal means of understanding Pleistocene history only since the 1960s. At first the influence of the traditional idea of four glaciations was so great that the deep-sea evidence was interpreted in the same way; indications of more than four cold intervals were discounted as minor fluctuations, or were assumed to

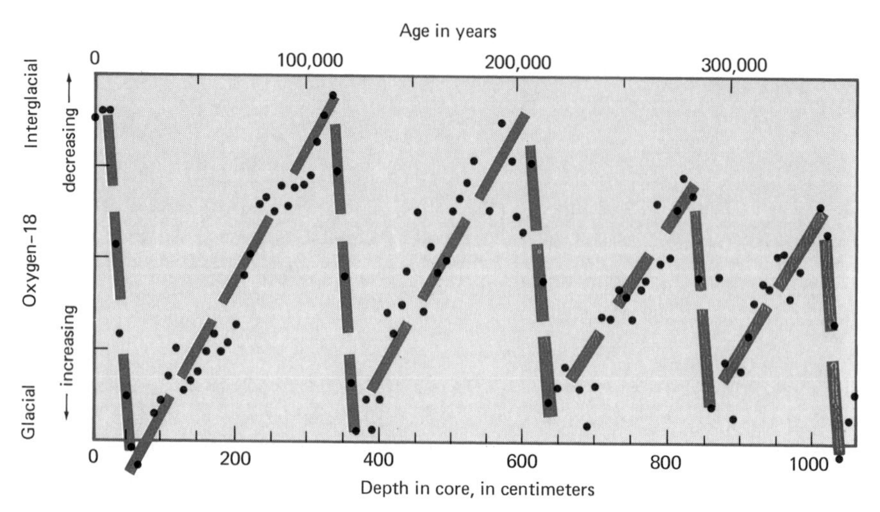

FIGURE 11.8 Oxygen isotope analysis of a 10-meter deep-sea core from the Carribean region. Note the sawtooth pattern of slow glacial buildup over about 90,000 years, followed by rapid deglaciation over about 10,000 years. (*Broeker and Van Donk, 1970*)

FIGURE 11.9 The earth at the maximum extent of the last glacial interval, about 18,000 years ago. Note the expanded land area due to the 120-meter drop in sea level. (*Climap Project members, 1976*)

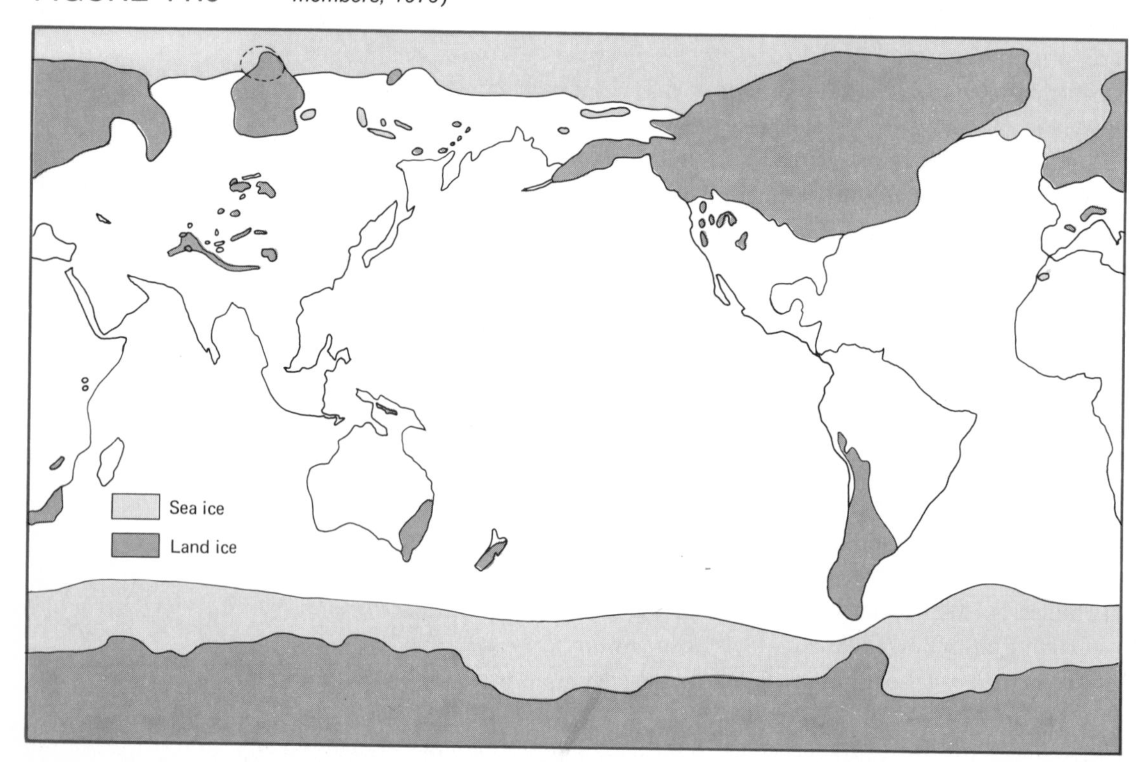

be older than the "true" Pleistocene. More refined studies during the 1970s have shown, however, that at least *eight* episodes of glaciation have occurred during the last 700,000 years, which is only about one-third of Pleistocene time. Many earlier glaciations also occurred. Consequently, an entirely new chronology of Pleistocene history is necessary, the details of which are just beginning to be worked out. Even so, several important facts have already become clear.

In the first place, the deep-sea oxygen-isotope data show that the cool glacial intervals lasted much longer than the warmer interglacial periods (Figure 11.8). Typically, ice sheets built up to their maximum extent over about 90,000 years and then dissipated rather quickly, to be followed in 10,000 to 20,000 years by the onset of still another glacial interval. Following this pattern, the most recent glacial interval, which is also by far the best understood, began around 100,000 years ago and reached its maximum extent about 18,000 years ago, when thick ice sheets covered much of northern Europe, Asia, and North America. Figure 11.9 is a reconstruction of worldwide geography at that time. Note particularly the expanded land area as a result of the 120-meter drop in sea level.

Beginning about 14,000 years ago, a worldwide warming trend brought this long glacial interval to an end. Most of the ice was gone from North America and Eurasia 9,000 years ago, although Greenland and Antarctica retained thick ice sheets, as they still do today. Studies of the sequence of ancient ice and sediment preserved in these surviving ice sheets, combined with studies of fossil plant pollen and ice-deposited sediments in other regions, show that about 6,000 years ago world climates were several degrees warmer than they have been since. That time most likely marked the peak of the present interglacial interval. The cooling trend of the past 6,000 years, although marked by climatic oscillations, may be the early stages of yet another glacial interval, with still further cooling likely over the next few thousand years.

Causes of Glaciations

These dramatic changes raise the broader question of the ultimate cause of changing climates, particularly those severe enough to cause widespread continental glaciation. In Chapter 9 we saw that the motions of crustal plates since Mesozoic time have generally moved the northern hemisphere continents into more northerly, and thus cooler, positions. This pattern undoubtedly accounts for much of the observed climatic cooling of Cenozoic time. Yet it seems insufficient to explain the glaciations of the past two million years, because the continents were already in approximately their present positions many millions of years earlier. Numerous additional theories have been advanced to explain the Pleistocene glaciations.

One theory suggests that glacial cycles are triggered by cyclic changes in the shape of the earth's orbit and in the inclination of its axis, which affect the amount of solar radiation (**insolation**) received at various latitudes for different times of the year. Cyclic changes in orbital geometry are well known; variations in summertime insolation for northern hemisphere latitudes may be calculated from them. When these are plotted against time, they show a fairly close correlation with glacial intervals based on oxygen-isotope studies (Figure 11.8). This relationship suggests that slight variations in solar insolation may well have been responsible for important climatic changes once glaciations got started. However, these same astronomical effects have operated throughout geologic history, whereas major glaciations are rare, occurring at only a few widely spaced intervals of geologic time. For this reason, many geologists believe that the astronomical effects are too small to *cause* glaciations, but they may operate only when other factors have reduced the average temperature at high latitudes sufficiently so that slight changes sway the balance toward expansion and contraction of ice sheets.

Variation in the carbon dioxide content of the atmosphere is another suggested cause of glaciations. Increases in the carbon-dioxide level might result at first in warmer, and then in colder, climates. Nearly half of the incoming solar radiation consists of medium-wavelength visible light, which can readily pass through the atmosphere. But most radiation reflected *from* the earth is in the long-wavelength part of the spectrum, and these wavelengths do not pass easily through the atmosphere. Instead, water vapor and carbon dioxide molecules absorb much of this long-wave energy, causing them to heat up and thereby warm the atmosphere. This added heat, in turn, enhances evaporation and thus provides moisture for additional clouds and glacier-forming precipitation. This cloud cover would decrease summer surface temperatures, even though the overall temperature of the atmosphere increased. It has been estimated that an increase of only 2.5 percent in the earth's average annual cloud cover could reduce summer temperatures to levels necessary for a typical glacial episode. Unfortunately, we cannot accurately estimate changes in past carbon dioxide content of the atmosphere, so the theory remains highly speculative.

Still another theory proposes that the late Cenozoic glaciations in the northern hemisphere were made possible by the tectonic movement of the Arctic Ocean Basin to a polar position. Because the Arctic Ocean has little circulation with the rest of the world's oceans, it is thermally isolated; as a result, it and the surrounding lands may have cooled independently to begin high-latitude glaciations. The increased reflectiveness of perpetually snow-covered lands then accelerated the cooling and the growth of the ice sheets. Finally, the Arctic Ocean cooled enough to freeze over, and the adjacent North Atlantic and North Pacific Oceans cooled substantially. This cooling greatly decreased ocean evaporation, causing precipitation to decrease on the nearby lands. Thereafter, the continental glaciers began to shrink. The rise in sea level due to this glacial melting enhanced the exchange of water between the Atlantic and the Arctic Oceans through the Greenland Straits. Eventually this exchange and the resulting warming of the northern hemisphere caused the Arctic Ocean to become ice free.

Increased precipitation from the newly open Arctic led to the accumulation of snow on the northern land masses, just as evaporation from the open Greenland Sea now feeds precipitation on the Greenland ice cap. As glaciers formed anew, temperatures decreased and sea level fell, eventually restricting flow between the Arctic and Atlantic. The cycle then repeated itself, with the freezing over of the Arctic Ocean. When precipitation was again drastically reduced, the glaciers again began to waste away. According to this theory, the Pleistocene glacial cycles will continue in this fashion until the north pole is no longer centered in the isolated Arctic Basin.

The Coming of Man

Whatever their cause, the dramatic climatic changes of Pleistocene time served as a background for what is, to us as humans, an extraordinarily significant event—the development of our own species, *Homo sapiens.*

Human Origins

Homo sapiens is a member of the mammalian Order Primates, which also includes monkeys, apes, and related forms. The first primates of early Cenozoic time diverged from their less specialized, mouselike ancestors by developing adaptations for living in trees, rather than on the ground. Among these adaptations were flexible fingers and an opposable thumb for securely grasping tree limbs. In addition, a shift in the position of the eyes toward the front of the face made possible precise, stereoscopic vision—a prime necessity if a tree dweller is to avoid fatal falls (Figure 11.10). These two key adaptations are characteristic of all primates, including humans. They provided the basis for the later evolutionary success of human beings on the ground, for the grasping hand and precise vision proved ideal for shaping and using tools.

FIGURE 11.10

Two unique primate adaptations, the grasping hand and stereoscopic vision, originated as specializations for life in trees. The primitive insectivore's fan-shaped arrangement of digits (*a*) can only grab by digging with claws, but the primate's opposable thumb (*b*) allows a stronger grip. The overlap of right and left eye vision is narrow in the insectivore (*c*) and broad in the primate (*d*) because of the different placement of the eyes in the head.

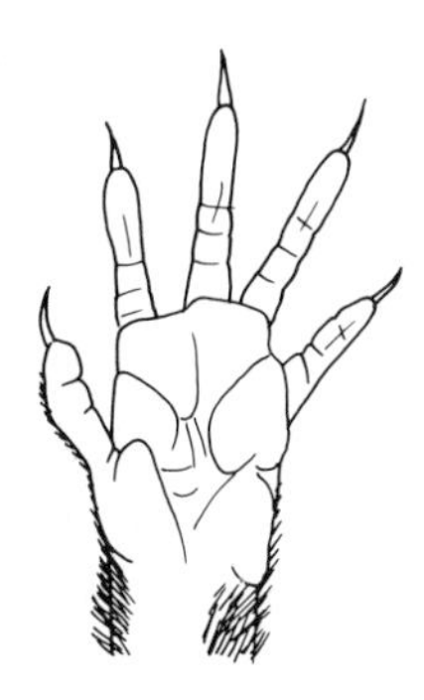
(*a*)

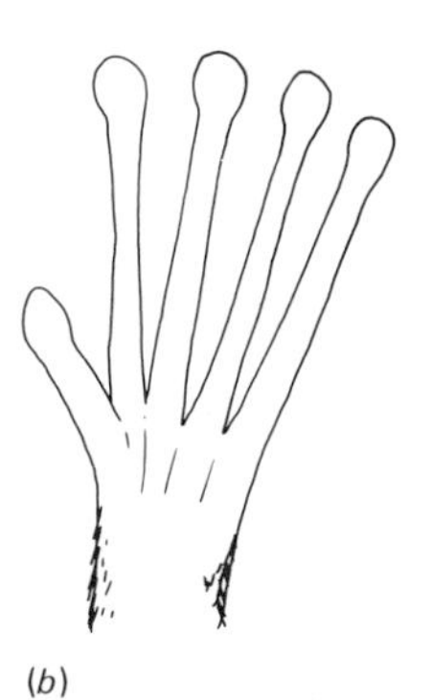
(*b*)

(*c*)

(*d*)

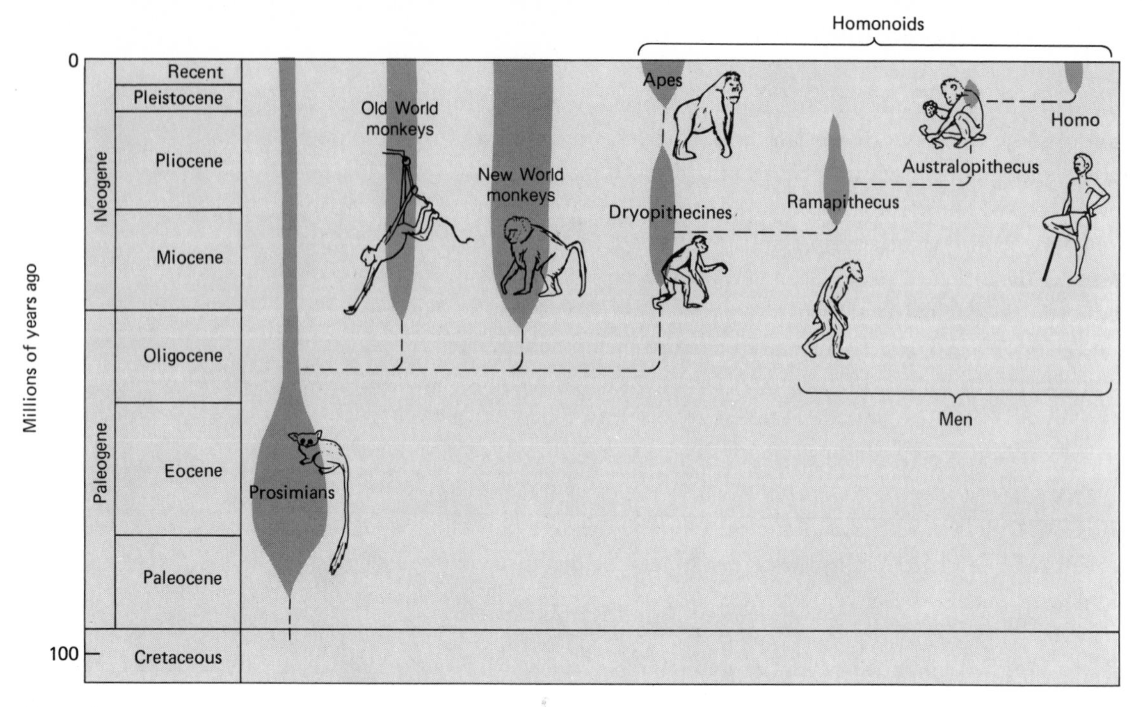

FIGURE 11.11 The geologic record of primates. The width of the vertical areas indicates the approximate abundance of each group. The larger modern apes and man are believed to have arisen from tailless, upright Miocene primates called dryopithecines. The three principal stages in human evolution are *Ramapithecus, Australopithecus,* and *Homo.*

During early Cenozoic time, the only primates were several relatively small, primitive types known as *prosimians* (pre-monkeys) (Figure 11.11). Fortunately, we understand a great deal about these earliest primates because some of their relatively unchanged descendants still survive in parts of Africa and Southeast Asia. In mid-Cenozoic time, three goups of larger and more advanced primates arose from these prosimian ancestors (Figure 11.11). One of these groups, the "New World monkeys," includes the monkeys of South America, which developed independently on that isolated continent. A second group, the "Old World monkeys," developed at the same time in Africa and were widely distributed in both Africa and southern Eurasia. The third and final group that arose in mid-Cenozoic time, the *hominoids,* are of greatest interest to us, for this group includes our own species, along with the four surviving kinds of apes (chimpanzees, gorillas, orangutans, and gibbons), as well as several extinct species of apes and men.

The earliest hominoids of mid-Cenozoic time, called *dryopithecines,* were relatively small animals, about 50 to 75 centimeters high. They differed from their monkey relatives in lacking a tail and in having a more upright posture. Both of these features were probably adaptations for spending more time on the ground than did their tree-dwelling monkey contemporaries. The dryopithecines gave rise to two main groups of descendants: one led to the larger modern apes; the other to the earliest close relatives of man.

Human Fossils

The oldest fossil remains that are clearly humanlike are found in sedimentary rocks of Miocene age, about 10 to 15 million years old. Most of these, unfortunately, are only teeth or fragmentary pieces of jaws and skulls containing teeth. (Teeth, being the hardest and the most resistant part of the vertebrate skeleton, are the most common vertebrate fossil.) Even these scattered fragments, which have been given the name *Ramapithecus,* are sufficient to show that close relatives of modern human beings have existed for at least 15 million years.

Fortunately, Upper Pliocene and Pleistocene sediments contain well-preserved bones, skulls, and occasionally even complete skeletons of ancient men that give a more complete record of their later evolutionary history. These show that three principal stages occurred in human evolution: *Australopithecus* and primitive *Homo* in the Late Pliocene and Early Pleistocene, *Homo erectus* in the Middle Pleistocene, and *Homo sapiens* in the latest Pleistocene (Figure 11.12).

Australopithecus The oldest well-understood human fossils are assigned to the genus *Australopithecus,* which is one of only three genera now recognized in the human family, the Hominidae. (The other two are the older *Ramapithecus* fragments mentioned above and our own genus, *Homo.*) Australopithecines were first discovered in South Africa in 1924, when a single well-preserved skull was found. Since then, many additional skulls, jaws, and limb bones have been collected in various parts of Africa. Fragmentary but probably identical specimens have also been discovered as far away as Java, suggesting that australopithecines were widely distributed in the Old World.

The oldest australopithecine specimen is about three million years, and the youngest about one million years in age. They appear to represent two species that were generally similar, but differed in size. The larger, *Australopithecus robustus,* was about the size of a gorilla; the smaller, *Australopithecus africanus,* was about the size of a chimpanzee. Both species were ground dwellers with upright posture, as indicated by limb and pelvic bones (Figure 11.13). The brain volume was, however, less than half that of modern humans. In shape, if not in size, the skulls, like the limb bones, are remarkably similar to ours. In short, australopithecines resembled later humans in every feature except their smaller brains.

The evolution of the upright human posture apparently preceded, and probably paved the way for, a later expansion in brain size. The upright stance completely freed the hands from use in locomotion; instead, they could be used for such tasks as toolmaking and weapon-throwing. These tasks put a premium on increased intelligence and awareness, and thus led to the evolution of a larger brain. In this regard, it is most significant that several sites bearing australopithecine fossils also contain crude stone tools and the fractured skulls of baboons and other animals, indicating that our early relatives were already toolmakers and weapon-users, in spite of their relatively small brains.

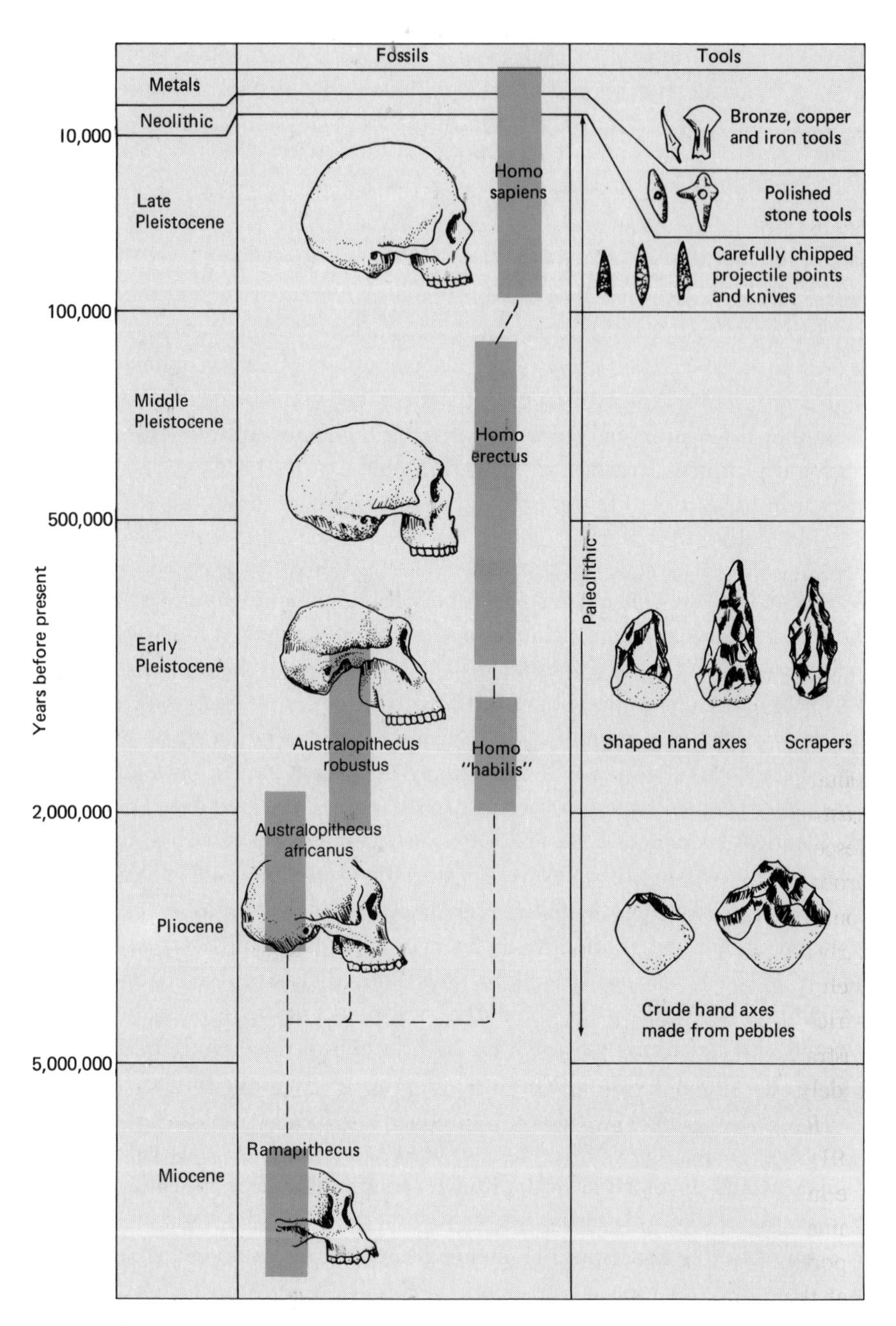

FIGURE 11.12

Summary of the evolutionary history of man. The vertical areas show the known fossil record and the dashed lines the most probable evolutionary relations of the groups.

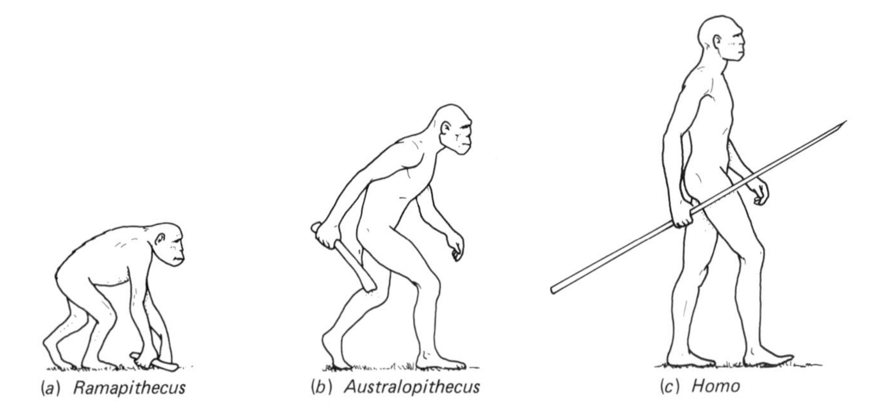

FIGURE 11.13 Inferred posture and locomotion in fossil men. (*Washburn, 1978*)

Homo Until rather recently, all Early Pleistocene fossils were thought to be australopithecines. They were succeeded in Middle Pleistocene time by a widely distributed and larger-brained species, *Homo erectus,* that is so like modern man that it clearly belongs in the same genus. In 1973, however, an Early Pleistocene skull was found in northern Kenya that is far more humanlike than its australopithecine contemporaries. Since then, additional discoveries in East Africa have shown that both *Homo erectus* and some still earlier fossils, tentatively called *Homo habilis,* occur in lower Pleistocene deposits along with *Australopithecus.* These discoveries suggest that *Homo* may not be a direct descendant of the advanced australopithecines. Instead, it appears more probable that both *Homo* and advanced *Australopithecus* diverged from a common ancestor in Pliocene time. This ancestral form was most probably an early and less specialized species of *Australopithecus.* Specimens thought to be such a species have recently been reported from Late Pliocene sites in east Africa. The fossil record makes it clear that by Middle Pleistocene time the australopithecines were becoming extinct, while *Homo erectus* was becoming widely distributed throughout the Old World.

Homo erectus became known when part of a skull was found in Java in 1891; this was one of the first human ancestors discovered. Thirty years after the initial discovery of this "Java Man," additional specimens were found in China; since that time, many additional skulls, jaws, and limb bones have been reported from Europe and Africa, as well as from Asia. Modern study shows that these are best interpreted as a single species that was somewhat larger than the australopithecines and had a brain volume only slightly smaller than that of modern man. *Homo erectus* used more complex stone tools than did *Australopithecus.* These were principally large hand axes made from pebbles of flint that were sharpened on one side by chipping. These tools show a progressive advance in design and workmanship through Middle Pleistocene time, indicating that man was becoming more adept at toolmaking as his brain size increased.

Specimens of *Homo erectus* have been found in sediments that range in age from 1,500,000 to about 200,000 years. This species may have been the direct ancestor of our own species, *Homo sapiens,* which first occurs in the fossil record about 100,000 years ago.

Human Culture

The origin of *Homo sapiens* begins the final act in the drama of human evolution. This event set the stage for the rise of those complex cultural systems whose study occupies the disciplines of anthropology and ancient history. We have already noted that stone tools occur along with fossil human bones and teeth throughout the two million years of Pleistocene time and are even found in Pliocene deposits. By Middle Pleistocene time, these stone tools are far more abundant and widespread than are human fossils. For this reason, much of our understanding of the development of *Homo sapiens,* and of *Homo erectus* before him, is based on analyses of man-made objects that also provide valuable clues to early patterns of human culture.

Anthropologists recognize three principal stages in this cultural evolution: the **Paleolithic,** or Old Stone Age, characterized by chipped stone tools; the **Neolithic,** or New Stone Age, characterized by ground and polished stone; and, finally, the Age of Metals, characterized by objects made first of copper and bronze, later of iron (Figure 11.12). Most of the several million years of human history fall within the Paleolithic Old Stone Age. Neolithic cultures began only about 10,000 years ago and were followed about 5,000 years ago by metal-using cultures.

The entire cultural history of *Australopithecus* and *Homo erectus,* as well as most of the existence of *Homo sapiens,* is represented by Paleolithic chipped stone tools. As we have noted, Paleolithic tools show progressive improvement in toolmaking techniques through Pleistocene time. The earliest were crude hand axes used by *Australopithecus,*and the most advanced are carefully styled projectile points and knives made by the Late Paleolithic *Homo sapiens,* beginning about 40,000 years ago. Late Paleolithic men and women also used bone for making fine tools, such as needles, and had a highly developed artistic sense, as is evident from elaborately carved bone objects and from the well-known Late Paleolithic cave paintings of France and Spain (Figure 11.14).

Between ten and fifteen thousand years ago, the chipped stone tools of Late Paleolithic time gave way to more advanced stone tools made by grinding and polishing. The Neolithic peoples who made these tools developed still more important cultural advances: They learned to make pottery for storing food and water and, most important of all, they began to cultivate plants and domesticate animals. With this step, for the first time in his long history, man no longer had to depend on hunting and the gathering of wild plants, but could grow his own food.

The development of agriculture by Neolithic man was one of the most significant events in all of human history, for it not only permitted the development of permanent communities, some of which were later to become the first cities, but it also allowed a division of labor—some individuals provided the food, while others became artisans, priests, tradesmen, and scholars. After the development of agriculture in this Neolithic Revolution, human cultures evolved rapidly. About 5,000 years ago, metals were first used for tools, and at about the same time the development of writing in Egypt and Mesopotamia led to the beginnings of recorded human history.

These striking advances in cultural evolution beginning about 40,000 years ago are also reflected in human physical evolution. At about that time, an earlier and now extinct subgroup of large-brained humans, *Homo sapiens neanderthalensis* (Neanderthal Man), was replaced by our own subgroup, *Homo sapiens sapiens*. Our group differs from the earlier neanderthals in having somewhat lighter bone structure and in other relatively minor features of skull and skeletal anatomy. Although it is not revealed by direct skeletal evidence, some anthropologists now believe that this rise of *Homo sapiens sapiens* reflects the beginnings of man's final and most distinctive adaptive feature—the ability to communicate by complex spoken language.

The origin of speech permitted elaborate ideas to be rapidly communicated within human societies and also meant that accumulated knowledge, legends, and traditions could be passed from generation to generation. This unique advantage paved the way for the enormous cultural progress of modern civilization, progress that culminated about 5,000 years ago in the development of writing and the end of the long "prehistoric" phase of human history.

The oldest art object known, a horse carved from mammoth ivory, found in Germany. It is about 30,000 years old. (*Courtesy of Alexander Marshack: © 1972*)

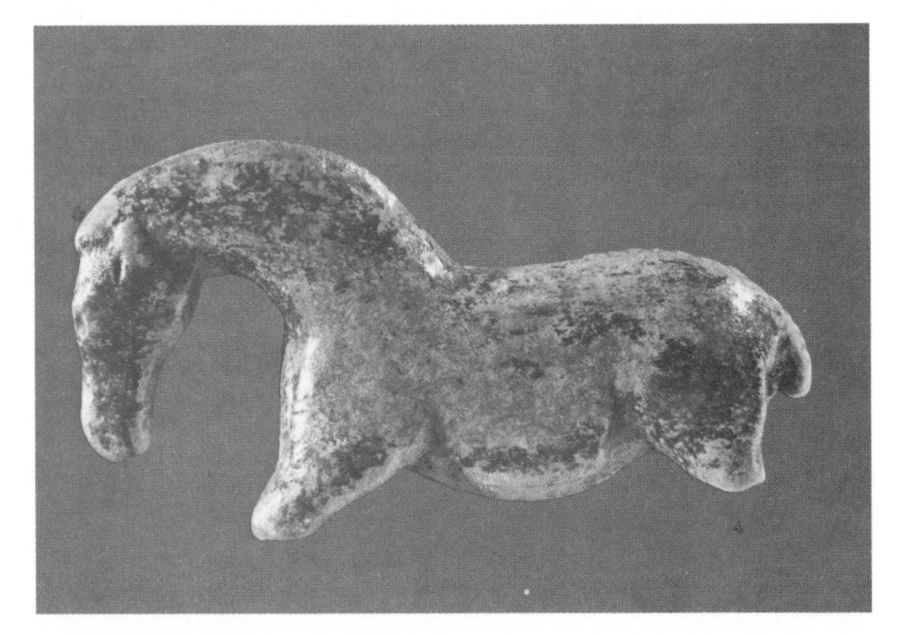

FIGURE 11.14

Chapter Summary

Glacial Landscapes

Pleistocene geography: Ice sheets, which today cover 10 percent of the land surface, have repeatedly expanded to cover as much as 30 percent during the past 2 million years; these ice fluctuations have modified drainage patterns and shaped much of the land surface, particularly in the northern hemisphere.

Glacial lakes were produced by erosion and sediment deposition and the subsequent damming of meltwater near the margins of the ice sheets; among their remnants are the Great Lakes and Finger Lakes of New York.

Sea-level changes: Ice accumulation on land removed water from the oceans and led to a drop in sea level of about 120 meters during ice sheet maxima.

Isostasy is the tendency of massive parts of the earth's crust to "float" on the more dense materials below; the weight of the ice sheets depresses the underlying crust, which slowly rises after the ice retreats.

Pleistocene Chronology

Deep-sea cores show that ice sheets expand slowly over about 90,000 years; they then dissipate quickly, to begin the cycle again after a 10,000-to-20,000-year interglacial interval.

Causes of glaciations are uncertain, but are most probably related to the geometry of the earth's spin and orbit, combined with longer-term changes in the position of the continents and oceans in polar regions.

The Coming of Man

Human origins: Men and modern apes arose from a less specialized group of ground-dwelling mid-Cenozoic apes called dryopithecines.

Human fossils belong to three genera: *Ramapithecus,* the earliest and least understood, *Australopithecus,* and *Homo;* the latter two are found together in Late Pliocene and Early Pleistocene deposits. *Australopithecus* became extinct in Middle Pleistocene time.

Human culture: Stone tools made by early man record a long sequence of cultural advances throughout Late Pliocene and Pleistocene time.

Important Terms

channeled scabland
core (deep-sea)
drift, glacial
finger lake
fjord
insolation
Neolithic Age
outwash
Paleolithic Age

Review Questions

1. Describe the maximum geographic distribution of Pleistocene ice sheets.
2. What effect did large continental ice sheets have on sea level? On land-drainage patterns? On the level of the underlying crust?
3. What contribution has the study of deep-sea cores made to our understanding of Pleistocene history?
4. Summarize some suggested causes for the repeated ice-sheet expansions and contractions of Pleistocene time.
5. Outline the evolutionary events leading to the genus *Homo*.
6. What are the stages in human evolution that are recorded by man's stone tools?

Additional Readings

Flint, R. F. *Glacial and Quaternary Geology,* John Wiley, New York, 1971. *A standard advanced text and reference.*

Match, C. L. *North America and the Great Ice Age,* McGraw-Hill, New York, 1976. *A contemporary summary of Pleistocene history.*

Rosen, S. I. *Introduction to the Primates,* Prentice-Hall, Inc. Englewood Cliffs, New Jersey, 1974. *An excellent popular summary.*

Walker, A. and R. E. F. Leakey "The Hominids at East Turkana," *Scientific American,* vol. 239, no. 2, pp. 54–66, 1978. *Summarizes important new discoveries in East Africa.*

Washburn, S. L. "The Evolution of Man," *Scientific American,* vol. 239, no. 3, pp. 194–208, 1978. *A readable, up-to-date review.*

EPILOGUE
The History of Earth History

12

The "prehistoric" phase of earth history and human history ended about 5,000 years ago. We know the rest, or at least much of the rest, because it is written down. One facet of this recorded history that is especially pertinent to our story concerns the way in which the pioneer geologists of the eighteenth and nineteenth centuries came to understand the long sequence of earth history that we have been summarizing. By way of epilogue, we shall here briefly outline some of the high points of this quest.

In Chapter 2 we saw that one of the first insights into the earth's past came in 1669, when Nicolaus Steno recognized that sedimentary layers were originally deposited horizontally in continuous sheets, one on top of the other. Working in the same region of western Italy a century later, Giovanni Arduino (1760) divided the rocks of the area into three categories: **Primary, Secondary,** and **Tertiary.** The Primary contained old crystalline rocks with metallic ores; the Secondary, hard sedimentary rocks with fossils; and the overlying Tertiary, weakly consolidated sediments, commonly containing marine shells. At about the same time, in what is now southeastern Germany, Johann Gottlob Lehmann (1756) was recognizing a similar threefold grouping of rocks, which he characterized as crystalline rocks, stratified rocks, and alluvial material. Within a few years, the rocks of many areas of Europe were divided into these three major units, each containing numerous subunits. This simple framework became the first stratigraphic classification.

These early workers all made one profound mistake: They incorrectly equated *rock types* with particular *times* in earth history. The stratified series of Lehmann, for example, was generally believed to be the result of a single event—the Biblical flood. In their search for principles of rock classification, these early workers assumed that even the numerous smaller rock units that made up the three major groupings were distributed very widely, and that each of them represented a discrete, worldwide episode of earth history.

Late in the eighteenth century, this idea was carried to its ultimate extreme by Abraham Gottlob Werner of Freiburg, Saxony (now Germany), who believed that at any one time only one kind of rock was being deposited, and that a particular type of *rock* was then characteristic of a particular *time*. By the end of the eighteenth century, this point of view had become widely accepted. Werner thought that all rocks—limestones, granites, basalts, other igneous rocks, and metamorphic rocks as well—were deposits of a primeval ocean that once covered the entire earth, even the highest mountains. This ocean was full of minerals that were deposited successively in one great rock-making episode. Werner's **Primitive Series,** deposited first from the ocean, included granite, gneiss, schist, and other crystalline rocks. The **Transition Series** included hard slates and limestones. Both of these series consisted, Werner believed, of numerous **universal formations** that covered the entire earth like layers of an onion. The overlying **Stratified Series,** consisting of sandstone, limestone, salt, gypsum, coal, and basalt, and the **Alluvial Series,** consisting chiefly of sand, clay, and gravel, were deposited as the giant ocean subsided below the level of the highest mountain tops, leaving the earth that we see today.

Werner's simplistic scheme eventually gave way to the views of James Hutton, a Scot, who maintained that there was no vast, primeval ocean, and that many rocks were of igneous, not sedimentary, origin. Instead of undergoing a single tumultuous saga of creation during which conditions on the earth's surface were very different from what they are now, the earth, in Hutton's view, was a nearly eternal machine in which internal forces created stresses that, over the course of time, elevated new lands from the ocean bed, even as other exposed regions were being eroded. Hutton saw no evidence of an ocean that once covered mountains, or of the universal flood that was championed, in one form or another, by most naturalists of the time. He saw only signs of slow subsidence of the earth's surface in some places and its renewed uplift in others. Said Hutton:

> *From the top of the mountain to the shore of the sea . . . everything is in a state of change.* [and] *We have a chain of facts which clearly demonstrates . . . that the materials of the wasted mountains have traveled through the rivers* [and] *There is not one step in all this progress . . . that is not to be actually perceived.* [and finally] *What more can we require? Nothing but time.**

Indeed, Hutton's view of slow geologic changes appeared to be the product of almost limitless time. Man, Hutton contended, has before him today all of the principles "from whence he may reason back into the boundless mass of time already elapsed." Hutton's point of view soon came to be called **uniformitarianism.** This uniformitarian view, that the earth has always responded to the same laws that govern it today, became widely accepted following its popularization by the Englishman Charles Lyell, beginning in 1830.

Another cornerstone of the Wernerian scheme—that ages of rocks may be ascertained by their composition—also failed to stand up under field observations. It was replaced as a means of dating rocks in the early decades of the nineteenth century by the much more fruitful principle of **faunal succession** discovered by William Smith. Between the years 1793 and 1815, Smith, an English surveyor, examined rock strata throughout much of Great Britain, where he found that each sedimentary rock unit contained its own distinguishing assemblage of fossils. These permitted him to recognize specific time units over large areas where differing types of contemporaneous rocks occurred. Utilizing his new principle, Smith produced the first modern geologic map of England and Wales in 1815, a map that represented a landmark in the understanding of earth history (Figure 12.1).

Smith's methods permitted Georges Cuvier, a French zoologist, to work out the detailed stratigraphic sequence of fossils and rocks in the Cenozoic strata of the Paris Basin. In 1812 he showed that many fossil vertebrates had no

*James Hutton, *Theory of the Earth* (Edinburgh, 1795).

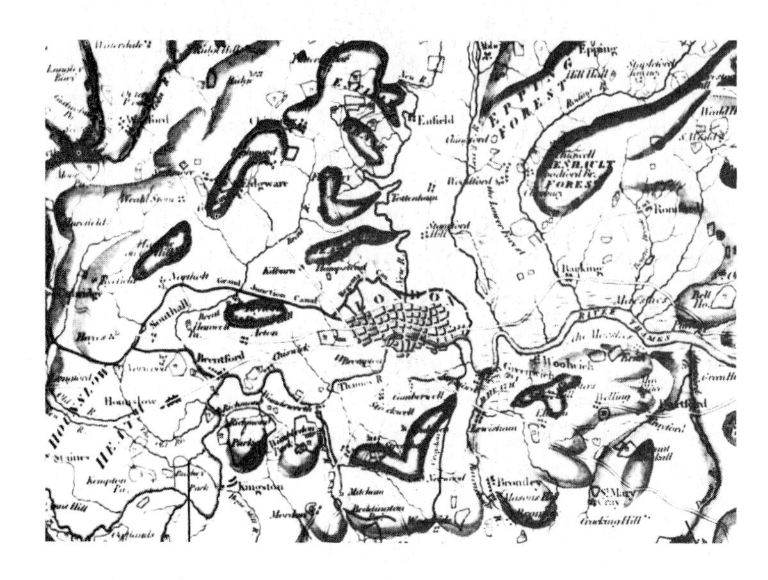

FIGURE 12.1 A portion of William Smith's landmark geologic map, published in 1815.

known counterparts living today. The reality that many once-thriving species had become extinct was at last becoming clear.

The discoveries of Smith and Cuvier made possible the recognition of rocks of the same age in widely separated areas—even on far distant continents, and even if the rock types were different. It then became feasible to define the geologic systems on the basis of sedimentary rocks and, using their distinctive fossils, to distinguish their time counterparts in remote regions. Thus the way was clear for a true scale of geologic time, and in the three decades following these discoveries, most of the geologic systems in use today were defined. The Phanerozoic time scale continued to evolve until the start of the twentieth century and has undergone only minor changes since then.

Most of the presently recognized geologic systems were defined in Europe. The actual type areas where they were first studied are shown in Figure 12.2. We shall briefly review the founding of these systems in stratigraphic order, beginning with the Cambrian.

In 1835 British geologists Adam Sedgwick and Roderick Murchison named the Cambrian and Silurian Systems from their studies of the geologically complex region of western Wales. Both Sedgwick and Murchison attempted to recognize breaks in the stratigraphic record as boundaries of the systems. Murchison began with the top of the sequence in the southeast, and Sedgwick began at the base in the northwest, and they worked toward one another. Murchison carefully documented the abundant fossils of his Silurian strata; but Sedgwick's strata were poorly fossiliferous, and his subdivisions of his Cambrian System were essentially based on rock type. When it became clear that their systems overlapped, a bitter quarrel ensued. The controversy was not resolved until 1879, when Charles Lapworth proposed the name Ordovician System, taken, like the names Cambrian and Silurian, from that of an ancient Welsh tribe. The Ordovician included the disputed interval between the

Cambrian and Silurian and expressed a threefold paleontologic division of early Paleozoic strata in Europe that had by then become apparent. Boundaries of Lapworth's Ordovician System were based solely on fossil occurrences.

The Devonian System was proposed jointly by Murchison and Sedgwick in 1840 for the rocks of Devonshire in southern England, an area they had studied prior to their misunderstanding over the contact between the Cambrian and Silurian. Devonshire is a poor type area because the rocks are intensely deformed and the base of the system is not exposed. Nevertheless, the rocks do contain fossils and it was these distinctive faunas—intermediate between those of the Silurian below and the Carboniferous above—that led to their identification as a separate system. Murchison and Sedgwick showed that fossils could be used to recognize the Devonian System in the German Rhineland, where it is much better exposed and much more fossiliferous.

Two British geologists, William Conybeare and William Phillips, proposed Carboniferous in 1822 for the strata in north central England that contained coal beds. The Carboniferous System was one of the first to be proposed following recognition of the value of fossils for correlation. The name, which means *coal-bearing*, is descriptive, but Conybeare and Phillips expected that the Carboniferous System would be widely recognizable by its distinctive fossils, rather than its lithology. The term is used throughout the world, including the United States, where it is more commonly subdivided into two full-scale systems, the Mississippian and the Pennsylvanian.

Summary of the naming of geologic systems in Europe.

FIGURE 12.2

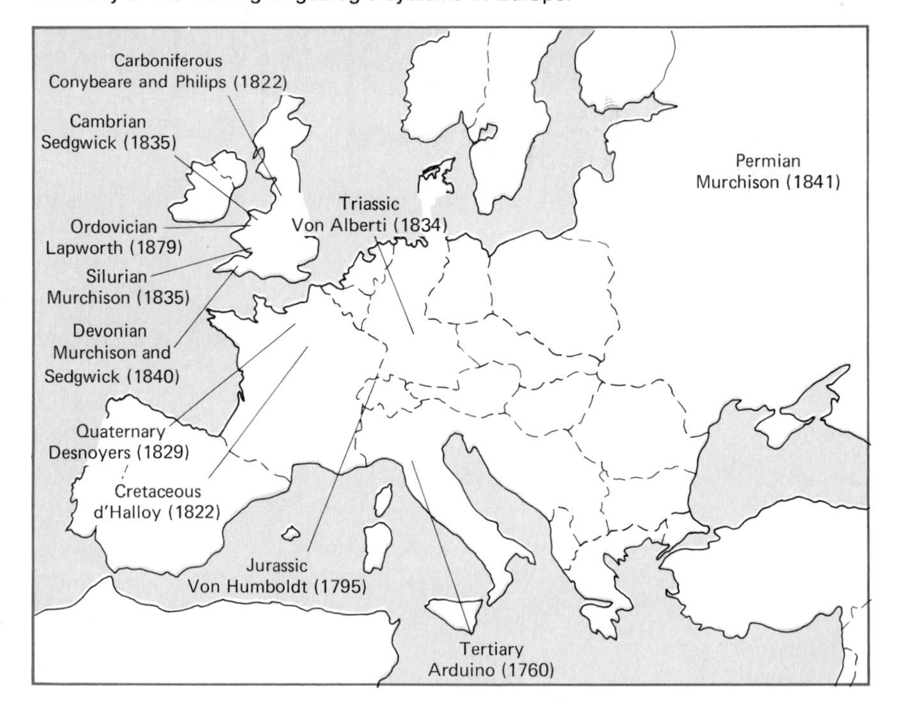

Alexander Winchell introduced Mississippian into American stratigraphic terminology in 1870 for the well-exposed lower Carboniferous strata of the Mississippi Valley. In 1891 Henry Shaler Williams coined the name Pennsylvanian, from the state of Pennsylvania, as an upper Carboniferous counterpart to Winchell's Mississippian strata. T. C. Chamberlain and R. D. Salisbury elevated both terms to system status in their influential geology textbook of 1906; they justified this division largely on the basis of the widespread unconformity that separated the two. Neither of these names has found use outside North America.

In 1841 Murchison named the Permian System from the province of Perm in Russia, where it consists of a great thickness of limestones overlying Carboniferous strata. Murchison recognized that the fossils differed from those found in older and younger strata, and, before he formally named this system, he ascertained that Permian strata could be identified in widespread areas by their distinctive fossils.

Triassic, meaning *three-part,* was coined in 1834 by Friedrich von Alberti, an official in the German salt-mining industry, for a sequence of rocks with a striking threefold subdivision. The term is essentially descriptive. In the type area of southern Germany, the strata are widely traceable but poorly fossiliferous. In the Alps to the south, a complete sequence of marine faunas today provides the standard of reference for worldwide correlation.

Alexander von Humboldt, a pioneer German geologist, used Jurassic in 1795 for the strata of the Jura Mountains in northern Switzerland, but at that early date he considered it only as another formation in Werner's Stratified Series. Leopold Von Buch in 1839 redefined the Jurassic as a system in its own right.

Cretaceous was proposed by a Belgian geologist, Omalius d'Halloy, in 1822 for strata encircling the Paris Basin. The term derives from the Latin word for chalk and is thus descriptive. As used today, the Cretaceous System includes more than just chalk beds, even in its type area, where the lower portion contains chiefly sandstone and shale. In other parts of the world, the system may be thousands of meters thick and include no chalk beds at all.

Tertiary is the only one of Arduino's (1760) terms still in general use. Originally defined in Italy, its constituent series—Paleocene, Eocene, Oligocene, Miocene, and Pliocene—have their type sections in both Italy and in France. The Eocene, Miocene, and Pliocene Series were defined by Charles Lyell (1833) on the basis, not of lithology, but of the relative proportions of the living and extinct fossils each contained: Eocene contained 3 percent living species, Miocene 17 percent, and Pliocene 50 to 65 percent. August von Beyrich (1854) later added the Oligocene, and Wilhelm Schimper (1874) added the Paleocene Series. In place of Tertiary and Quaternary, many geologists now employ the terms Paleogene and Neogene, using the top of the Oligocene as the dividing boundary between the systems.

In 1829 a French geologist, Jules Desnoyers, proposed the term Quaternary for the very young strata in the Paris Basin. Today it includes the Pleistocene Series (proposed by Lyell in 1839), which constitutes deposits formed during the most recent glacial ages, and the Holocene, or Recent Series, which is a rather poorly defined term for any postglacial deposits.

In summary, it is clear that the geologic systems arose without benefit of any grand design. Some were established on distinctive lithologies, some on distinctive faunal content, and some on major breaks in the stratigraphic record. Regardless of how the systems were conceived, however, most have been subsequently used on every continent on the basis of their distinctive fossils. These geologic systems have thus proven to be remarkably convenient units, and that is why they remain in use today.

We have now come full circle and have carried our story of the earth into the present century. Having gained an understanding of the earth's *past,* we should certainly be able to look *ahead.* This is what the thousands of people working as geologists spend much of their time doing. They are involved in a great search. Much of that search, of course, is focused on looking for mineral deposits and fossil fuels for future exploitation, both in rocks on the land and in the shelf seas surrounding the continents. Other geologists are concerned with predicting natural hazards. These hazards include not only earthquakes, landslides, and tidal waves, but also volcanic activity and long-term sea-level changes.

Geologists today are also becoming extensively involved in disposing of the growing volume of dangerous chemical and radioactive wastes, which pose the threat of poisoning or irradiating future generations. Where is it safest to store these wastes? In the oceans? In desert sediments? In deep caverns? All will undergo change. The problem is to determine which are the most stable, with changes that are the most predictable.

Many workers today are studying past climates in an attempt to predict future climatic change. Will climates remain stable? Not likely! But will they become warmer or colder? Wetter or dryer? Prediction of future climates rests heavily on a study of the past. Most of this study concerns the ocean floor, where, we have seen, climatic signals are extensively recorded. What do climatic changes portend for life in the sea and for food from the land? Which way is our civilization's current high output of carbon dioxide likely to change the climate, and how will it affect the oceans? No one yet knows.

Today geologists and other scientists in government, in large corporations, and in small specialized businesses are actively working on these and other problems, problems that we, as a civilization, urgently need to solve. Toward this end, the knowledge we have gained of the earth's past becomes a valuable basis for future planning. The scientists who will deal with these problems most successfully must maintain a sense of humility and avoid the trap of saying, "There was a time when they all thought the earth worked in *that* way, but now we know that it works *this* way." If the past is any guide, we are still in for some surprises!

Important Terms

Alluvial Series
faunal succession
Primary rocks
Primitive Series
Secondary rocks
Stratified Series
Tertiary rocks
Transition Series
uniformitarianism
universal formations

Questions

1. Who invented the first stratigraphic classification and where was it applied?
2. Some historians have termed Werner's concept of earth history "Neptunism." Why?
3. Briefly explain the uniformitarian philosophy of James Hutton.
4. How did the discovery of the principle of faunal succession help to elucidate earth history?
5. On what bases were the geologic systems originally distinguished, and how are they recognized today?

Additional Readings

Adams, F. D. *The Birth and Development of the Geological Sciences,* New York, Dover Publications, 1938. *An excellent book which begins with classical times and follows up the development of key ideas.*

Berry, W. B. N. *Growth of a Prehistoric Time Scale,* San Francisco, W. H. Freeman & Co., 1968. *A concise history of the development of the geologic time scale.*

Geikie, Archibald *The Founders of Geology,* London, Macmillan and Co., 1897. *An insightful, well-written book about the individuals who molded geology into a science.*

Schneer, C. J., editor *Toward a History of Geology,* Cambridge, Mass., The M. I. T. Press, 1969. *An excellent collection of papers on the history of geologic thought, many of which discuss historical controversies.*

Schneer, C. J., editor *Two Hundred Years of Geology in America,* Hanover, N.H., University Press of New England, 1978. *A modern, authoritative collection of papers on the growth of geologic understanding in the United States.*

von Zittel, K. A. *Geology and Paleontology,* London, Walter Scott, 1901. *A classical treatment of the history of geology to the end of the nineteenth century.*

GLOSSARY

abyssal clay a major kind of sediment in the deep ocean basins; typically brown or red in color.

Acadian Orogeny mountain-building episode that affected the eastern United States and Canada (from which it was named) late in the Devonian Period.

age geologic time unit that is a subdivision of an epoch.

albedo the percentage of total illumination received by a planet or satellite that is reflected from its surface.

algal mat a coherent surface layer of sediment bound together by filimentous blue-green algae.

Alluvial Series the youngest unit in Werner's eighteenth-century stratigraphic scheme, consisting of unconsolidated material.

amino acid an organic acid carrying an amino group ($-NH_2$); the building-block compound of proteins.

angular momentum the product of mass, speed, and radius of motion for a body moving in a circle; i.e. the quantity of rotational motion a mass possesses.

angular unconformity a surface of erosion in the stratigraphic record, the strata above which lie at an angle to, and truncate, the strata below.

Appalachian Orogeny the late Paleozoic episode of mountain building that produced the Appalachian Mountain system in the eastern United States and Canada simultaneously with the Hercynian Orogeny in Europe.

Archean Eon the oldest subdivision of Precambrian time, generally referring to rocks formed more than 2.5 billion years ago.

asexual reproduction production of offspring without the sharing of genetic material from separate parents.

asthenosphere zone of hot, plastic material in the mantle at depths of about 100 to 400 kilometers; the rigid lithosphere above moves on the soft asthenosphere.

atomic number the number of protons in the nucleus of an atom. This constitutes the positive charge of the nucleus and determines the element that the atom belongs to.

aulacogen a rift approximately perpendicular to a continental margin that contains a thick sequence of sedimentary rocks; aulacogens form at the time the continental margin forms by rifting at a spreading center.

authigenic mineral generated on the spot; the term applies to those constituents that form during or after the formation of the rock of which they constitute a part, such as cement in sedimentary rock.

autotroph organism capable of manufacturing organic nutrients from inorganic raw materials.

Baltica an early Paleozoic continent that included present-day Europe and eastern Asia.

banded iron formation cherts with alternating iron-rich and iron-poor laminae, commonly a centimeter or less in thickness, that constitute the bulk of the world's iron ore reserves; all are of Precambrian age, the majority around 2 billion years.

barrier island a large, elongated island of sand that parallels the coastline and is separated from it by a lagoon; commonly barrier islands are covered in part by dunes and by some vegetation.

bed load sediment of sand size and larger transported on the bed of a stream by rolling and saltation, as opposed to suspended load.

benthos animals and plants that live on the sea floor.

bentonite clay layer formed by the alteration of volcanic ash.

big bang theory the idea that the present expanding universe originated from a compact, extremely dense concentration of matter that exploded violently.

biofacies the biological aspects of a mass of sediments or sedimentary rocks, particularly with respect to their differences from those of adjacent sediments or sedimentary rocks.

biogeographic province geographic area showing similar associations of animal and plant species.

biostratigraphic unit stratigraphic unit defined only on the basis of characteristic fossils.

biostratigraphy the science of dating sedimentary rocks by means of fossils.

block mountain a mountain produced by near-vertical faulting in which tensional forces are dominant, as opposed to fold mountains.

blueschist a bluish metamorphic rock formed under conditions of high pressure and relatively low temperature.

bottomset beds the basal beds in a deltaic sequence. Strata deposited on the basin floor beyond the delta over which the delta builds later.

brachiopod a member of the Phylum Brachiopoda; superficially clamlike in having two valves, but unrelated to the clams.

calcareous ooze a deep-sea sediment containing more than 30 percent calcium carbonate; modern calcareous oozes consist largely of the skeletons of planktonic foraminifera and coccoliths.

calcite compensation depth (CCD) the depth at any point in the ocean below which calcium carbonate is not preserved because the rate of its dissolution exceeds the rate of supply; in low latitudes today this is generally between 4,000 and 5,000 meters.

Caledonian Orogeny a Silurian mountain-building episode in northwestern Europe; named after the Latin name for Scotland.

carbonate mud silt and clay-sized particles composed of calcium carbonate (either calcite or aragonite).

carbonate sand sand grains made out of calcium carbonate (the minerals calcite or aragonite).

carnivore-scavenger an animal whose food consists of other animals or their remains.

Cascadian Orogeny a Cenozoic episode of volcanism and tectonism on the west coast of the United States and Canada during which the Cascade Mountains were formed.

chalk a very soft porous limestone which easily abrades to powder; most chalks are lithified calcareous ooze and consist chiefly of planktonic foraminifera and coccoliths.

channeled scabland area of deep channels in central Washington state; produced by catastrophic floods of glacial meltwater.

chemical sediment sediment that is produced subaqueously by organisms or by evaporation from material dissolved in the water, as opposed to detrital sediment, which is transported into the environment as particulate material.

chromosome a filament-like structure in the cell nucleus of eucaryotes, along which the genes are located.

class category next in rank below the phylum and above the order in biologic classification.

climate-sensitive rock a term used to refer to particular kinds of sedimentary rocks, such as evaporites or tillites, whose deposition generally occurs under particular climatic regimes.

coccolith a microscopic calcite plate from a marine planktonic alga belonging to the Phylum Coccolithiphoridae.

compressional force in structural geology a pushing together of a portion of the earth's crust, typically producing folds and thrust faults.

concordia a curving line on a graph of uranium-238: lead-206 and uranium-235: lead-207 at any point on which the age determinations of the two uranium: lead ratios are identical.

concurrent range zone zone based on the overlapping ranges of fossil species.

continental drift horizontal displacement of continents relative to one another.

continental rise gently sloping prism of sediments at the base of the continental slope which it joins with the deeper ocean basins.

coral a sessile invertebrate belonging to the Phylum Coelenterata that secretes calcareous skeletons; some corals are solitary; others are colonial.

core (deep-sea) samples of sediments obtained by dropping or drilling a cylindrical tube from shipboard into the ocean floor.

core (of the earth) the inner portion of the earth, having a radius of 3,473 kilometers and believed to consist mostly of largely molten nickel and iron.

cosmic-ray induced nuclide a radioactive nuclide that is kept in constant supply by being produced in the upper atmosphere from nonradioactive nuclides as a result of their bombardment by cosmic rays from outer space.

craton the long-stable region of a continent whose Precambrian rocks are either at the surface or under a relatively thin cover of younger sedimentary rocks.

crinoid a kind of echinoderm that consists of a cup or calyx and plated arms that serve to filter food particles from the ocean water. They are particularly important in Paleozoic rocks; most grew on long stalks fastened to the sea floor.

cross beds beds that are inclined relative to the main planes of stratification of the strata concerned.

cross-cutting relationship a means of dating rocks that are cut by intrusive rock bodies or by faults; the rock that is cut must be older than the intrusive body or the fault.

crossopterygian a primitive group of fishes, characterized by lobe-shaped fins with internal skeletons.

crust the outer portion of the lithosphere, about 10 kilometers thick beneath the oceans and 35 kilometers thick beneath the continents, in which relatively low velocities of earthquake waves prevail.

Curie point the temperature, usually well below the melting point, at which minerals lose their permanent magnetization.

cyclothem repetitive sequence of coal-bearing strata; originally the term was applied to the Pennsylvanian cyclic coal-bearing sequences in the central and eastern United States.

daughter nuclide a nuclide that is produced by the decay of a radiometric nuclide.

decay constant the rate of decay of a radioactive nuclide, expressed as a proportion per unit of time.

delta a deposit of sediment formed at the mouth of a river that results in progradation of the shoreline.

detrital sediment sediment consisting of particles of material (detritus) transported from an external source into an area of accumulation.

diagenesis postdepositional chemical reactions, such as cementation and replacement, that take place between groundwater and mineral grains in a sediment or sedimentary rock.

diapir an upward injection into overlying strata of weak material, commonly salt, as a result of its plastic flow; diapirs typically produce domes in the strata they intrude.

diatom one of a group of siliceous algae that builds intricate microscopic shells; diatoms occur in both marine and fresh-water environments.

dinosaur a group of diverse extinct reptiles belonging to the Orders Saurischia or Ornithischia; the dinosaurs were confined to the Mesozoic Era.

disconformity a surface of erosion or nondeposition between sedimentary strata that are parallel and not angularly discordant.

distributary one of several channels into which a stream divides on its delta plain.

DNA (deoxiribonucleic acid) chain-like organic acid thought to be the carrier of genetic information within the cell.

dolomitization the diagenetic replacement of some of the calcium ions in carbonate sediment by magnesium ions, thus producing dolomite from calcite or aragonite.

Doppler shift the change in the frequency of light waves or sound waves caused by movement of the source relative to the observer.

drift, glacial rock material deposited by glacial ice or meltwater.

echo sounder shipboard device for determining the depth of the ocean, and thus the topography of the ocean floor, by the travel time of reflected sound waves.

encounter theory the idea that the solar system was formed from material torn from the sun's surface by the gravity of a passing star that narrowly missed the sun.

epeiric sea a widespread, shallow sea on a continent; a synonym of epicontinental sea.

epicontinental sea a shallow sea that temporarily invades the interior portion of a continent; such seas were common at times in the geologic past, but there are few examples today.
epifaunal animal one that lives exposed on the sea floor and not within the sediment.
epoch geologic time unit represented by the time-stratigraphic unit called a series.
eucaryote organism whose cells all have a membrane-bounded nucleus; characteristic of all organisms except bacteria and blue-green algae.
eugeosyncline the oceanward side of a geosyncline, characterized by thick deposits of chert, poorly sorted sandstones, and volcanic rocks.
eustatic change simultaneous, worldwide change in sea level.
evolutionary radiation rapid divergence of ancestral organisms into new adaptive types.

facies the appearance or the mineralogical or fossil content of a mass of sediments or sedimentary rocks that differs in one or more characteristics from laterally adjacent sediments or sedimentary rocks.
family category next in rank below the order and above the genus in biologic classification.
Farallon plate a lithospheric plate in the eastern Pacific Ocean that was largely consumed at the subducting margin of the North American continent; the small Juan de Fuca plate is a modern remnant.
fault-block basin a downdropped segment of the crust produced during large-scale block faulting.
faunal succession the principle, discovered by William Smith in the early nineteenth century, that successive groups of fossils succeed one another in the same sequence in sedimentary strata everywhere.
felsic igneous rocks light colored igneous rocks, rich in silica and typically made up largely of quartz, feldspar, and muscovite.
filter-feeder a sessile invertebrate that feeds by filtering suspended particulate matter from water by means of a specialized gill system.
finger lake a long, narrow lake occupying a formerly glaciated, steep-sided valley.
fjord a long, narrow arm of the sea occupying a formerly glaciated, steep-sided valley; most common along shorelines recently depressed by ice sheets.
flysch widespread deposits of sandstone and shale, including some conglomerate and some marl beds; the term was applied originally to such deposits on the northern and southern borders of the Alps.
foraminifera an important order of microscopic marine protozoans that have left an extensive fossil record in rocks of Ordovician and younger age; both benthic and planktonic in habit, their small shells are biostratigraphically important in late Paleozoic and younger strata.
foreset bed a stratum deposited on the sloping delta front as it progrades into a lake or sea; foreset beds are overlain by topset beds of the delta plain.
formation a unit of rocks that is mappable at ordinary mapping scales, distinguishable from adjacent units in the field, and consisting of two names, the first, geographic and the second lithologic or simply "Formation."

fossil the remains or traces of an ancient organism.

fractional crystallization separation of a magma into two phases, crystals and liquid, which may be followed by a physical separation through settling of the crystals or migration of the liquid.

gamete a reproductive cell that must usually fuse with another before development begins; an egg or sperm.

gene a unit of inheritance.

genus (plural: genera) category next in rank below the family and above the species in biologic classification.

geosyncline initially conceived as a great, elongate downfold in the earth's crust with dimensions in the order of hundreds of kilometers, but recognized always by the thickness of accumulated rocks, on the order of 10,000 to 15,000 meters; the so-called downfolds are now believed to be simply subsiding continental margins.

geotherm the change in temperature of the earth's interior with depth.

Gondwana (or Gondwanaland) a huge Paleozoic southern hemisphere continent that consisted of present-day Africa, Antarctica, India, South America, and Australia.

graded bedding beds of sandstone or conglomerate in which the coarser particles are concentrated near the bottom and which become progressively finer upward.

grapestone sand-sized grains of carbonate that are aggregates of silt-sized particles.

graptolite a Paleozoic group of organic-walled, colonial planktonic organisms.

greenstone belt thick, slightly metamorphosed sequences of early Precambrian sedimentary and volcanic rocks that occur in great elongated downwarps several tens of kilometers long.

greywacke a poorly sorted sandstone with considerable clay matrix, generally considered to be "immature" in the sense of not being reworked or winnowed.

group rock unit made up of two or more formations.

guide fossil fossil of value in identifying the age of the rocks in which it occurs.

guyot flat-topped submerged mountain rising from the ocean floor.

gymnosperm a seed-bearing plant in which the seeds are exposed on structures such as cones.

half-life the length of time required for one-half of an initial quantity of radioactive nuclide to decay.

Hercynian Orogeny a late Paleozoic episode of mountain building in Europe that coincided with the Appalachian Orogeny in eastern north America.

heterotroph organism incapable of manufacturing organic compounds from inorganic raw materials, therefore requiring organic nutrients from its environment.

homogeneous accretion the idea that the earth formed as a homogeneous body from the simultaneous accretion of both light and heavy materials, and that it became differentiated later in geologic history into crust, mantle, and core.

hot spot a persistent source of volcanism that remains stationary as a moving lithospheric plate passes over it, commonly resulting in a volcanic chain.
hydrothermal a mineral deposit produced by very hot waters.

inclination (magnetic) the plunge or lines of force of the earth's magnetic field at a locality.
index fossil see guide fossil.
infaunal animal one that lives within the sediment in a subaqueous environment.
inhomogeneous accretion the idea that the earth formed as a differentiated body from the accretion of, first, iron and nickel that make up the present core, followed by heavy ferromagnesian silicates that make up the mantle, and finally by lighter silicates that are concentrated in the crust.
insolation solar radiation received by the earth or other planets.
interstitial water water contained in the minute pores or spaces between grains or in fractures in rock.
intertidal flat a depositional environment within the normal tidal range and hence normally exposed at low tide and covered by water at high tide.
ironstone a sedimentary rock, usually shale or limestone, rich in iron carbonate or iron oxide; much of the iron produced in France and Britain comes from such rocks.
isochron a straight line on a graph of radioactive parent and/or radiogenic daughter nuclides that changes its slope as a function of time.
isopach line of equal thickness on a map of a stratigraphic unit.
isostasy the principle that a given area of the earth's surface has the same amount of rock mass beneath it as any other area of the same size; thus mountains are compensated by having a mass deficiency beneath them; depressed areas have a mass excess.
isotherm a surface of equal temperature.
isotope an atom of a given element that has a particular atomic weight; isotopes of an element differ in atomic weight as a result of containing differing numbers of neutrons in the nucleus.

Kazakhstania a Paleozoic continent, now a part of central Asia; named for a state in the south-central Soviet Union.
kerogen the solid bituminous matter in oil shale that yields oil when the shale is heated and undergoes destructive distillation.
key bed distinctive bed that permits correlation within a rock sequence.

labyrinthodont a Paleozoic amphibian with specialized teeth that possessed a folded enamel.
lagoon a bay extending roughly parallel to the coast and separated from the open ocean by barrier islands.
Laramide Orogeny a mountain-building episode in Colorado, Wyoming, and adjacent areas beginning at the end of the Cretaceous Period that produced much of the structure of the present-day Rocky Mountains.
Laurasia portion of Pangaea north of the Tethys seaway that contained the present-day northern hemisphere continents.

Laurentia an early Paleozoic continent consisting of much of present-day North America, not including a part of New England, the Canadian maritime provinces, and the southeastern United States.

Laurussia a late Paleozoic continent made up of the preexisting continents of Laurentia and Baltica.

lithofacies the lithologic or mineralogic aspects of a mass of sediments or sedimentary rocks, particularly with respect to their differences from those of adjacent sediments or sedimentary rocks.

lithosphere the rigid outer shell of the earth; about 100 kilometers thick, it includes both the crust and the uppermost part of the mantle.

lycopsids club mosses; dominant in the late Paleozoic; rare at present.

mafic igneous rocks rocks containing between 45 and 75 percent mafic minerals, i.e. the ferromagnesian rock-forming silicates.

magnetic anomaly an aberrant area on a map of regional magnetic field strength.

magnetic polarity epoch extended intervals during which the polarity of the earth's magnetic field was dominantly either normal or reversed; shorter-term reversals within these epochs are called magnetic polarity events.

magnetic reversal abrupt change in the polarity of the earth's magnetic field, the south pole becoming the north pole, and vice versa.

magnetometer instrument for measuring the intensity of the earth's magnetic field.

mammal a class of vertebrates including man, quadripeds, seals, and so on, whose females have mammae to nourish their young.

mantle the interior of the earth between the crust above and the core below, having a thickness of about 2,900 kilometers and composed of rocks more dense than those in the crust in which relatively high velocities of earthquake waves prevail.

marginal basin a small ocean basin that lies between a subduction zone and either a continent or another subduction zone; most marginal basins are in the western Pacific region.

maria from the latin word for "seas," referring to dark areas of the moon now known to be lava-covered lowlands.

marine regression movement of the shoreline seaward, a result of falling sea level or uplift of the land.

marine transgression movement of the shoreline landward, a result of rising sea level or subsidence of the land.

marsupial an order of mammals, including the opossum and kangaroo, whose females carry their undeveloped young in a pouch or marsupium.

mass number the sum of the number of protons and neutrons in the nucleus of an atom.

meiosis process of nuclear division within the cell in which the number of chromosomes is reduced by half.

member subdivision of a geologic formation.

Mesocordilleran highland an important sediment source in the Idaho-Nevada-Western Utah region in the Mesozoic Era; probably mountainous at times.

metaphyte any multicellular plant with differentiated cells.

metazoan any multicellular animal with differentiated cells.

microcontinent a fragment of continental crust belonging to a small lithospheric plate; much smaller than most continents.

microfacies the texture, sorting, and overall lithologic and biotic makeup of a rock as viewed in thin section under magnification.

miogeosyncline the continentward side of a geosyncline, characterized by thick deposits of well-sorted sandstones, limestones, and shales, and generally lacking in volcanics.

mitosis process of nuclear division within the cell that results in a new cell with the same number of chromosomes as in the original nucleus.

mobile belt an elongate region of the earth's crust characterized by tectonic instability, subsiding at times to receive extraordinarily thick sequences of sediments and rising at other times to become mountain chains.

molasse the terrigenous detritus eroded from elevated land masses during and immediately after a major mountain-building episode.

mold in paleontology, an impression of a fossil left in the surrounding rock after the fossil has been dissolved.

natural selection differential reproduction in nature leading to changes in gene frequency and, ultimately, to new species.

nebular theory the idea that the entire solar system, including the sun, the planets, and their satellites, accreted at the same time from a vast interstellar cloud of gas and dust particles (a nebula).

nekton pertains to aquatic animals that spend most of their lives actively swimming.

Neolithic Age changes in human culture around 10,000 years ago that led to polished stone tools, pottery, and agriculture.

neutron capture a process in which the nucleus of an atom absorbs a high-velocity neutron, thereby increasing its mass number; in some cases the captured neutron instantly decays to a proton, thus increasing the atomic number as well.

Nevadan Orogeny mountain-building and widespread igneous intrusive activity in the western portion of the Cordilleran geosyncline during the Jurassic and Early Cretaceous.

nonconformity a surface of erosion in the stratigraphic record in which stratified rocks above rest on unstratified crystalline rocks of igneous or metamorphic origin below.

nuclide a single kind of atom with a particular atomic number and mass number.

oceanic ridge-rise system world encircling system of linear mountain chains rising from the ocean floor.

oil shale fine-grained sedimentary rock containing sufficient quantities of hydrocarbons that it yields mineral oil on slow distillation.

oolite a sediment composed of spherical or nearly spherical grains, usually of calcium carbonate, that have a concentric structure produced during their growth in suspension in agitated water.

oolith an individual spherical or ovoid grain of carbonate sand, generally concentric in structure, of which an oolite rock is composed.

ophiolite associations of basalt, ultramafic igneous rocks, and deep-ocean sediments; thought to reflect the sites of former ocean basins.

order category next in rank below the class and above the family in biologic classification.

organic evolution change in the genetic makeup in a population through time, leading ultimately to new species.

original horizontality the concept that strata are deposited horizontally; dips observed in the field are the result of subsequent tilting.

original lateral continuity the idea that truncated edges of strata along, for example, opposite sides of a valley can be inferred to be erosional, the strata having been deposited across the area the valley now occupies before the region was eroded.

orogenic front the boundary of a folded and faulted region where it abuts a relatively undeformed region or another orogenic belt of different age.

orogeny the process of uplifting, folding, and faulting by which systems of mountains are formed.

Ouachita mobile belt late Paleozoic mountain belt in southern Oklahoma.

outgassing the emission of volatile substances from the interior of a planet on a large scale through volcanic activity.

outwash sediments deposited by glacial meltwater.

paleoecology the study of the interactions of ancient organisms with their environments.

paleogeographic map map showing a reconstruction of ancient geography during a specific interval of geologic time.

Paleolithic Age the "Old Stone Age" of human culture; characterized by progressively refined tools of chipped stone.

paleomagnetism magnetism imparted to minerals in the distant past and preserved to the present time.

Pangaea a single continent that, during the Triassic Period, contained virtually all the earth's land mass; Pangaea was built from small continents during the Paleozoic and is commonly divided into two subcontinents, Gondwana on the south and Laurasia on the north.

paraconformity a bedding surface representing a large time break but lacking field criteria for significant erosion or other evidence of the missing time; usually recognizable only by dating the strata by means of fossils.

parent nuclide a radioactive nuclide that decays to another nuclide, which is termed the daughter nuclide.

partial melting the melting of some of the minerals in a rock as the temperature is increased while minerals requiring higher temperatures before melting remain solid.

patch reef a small, roughly circular reef.

pelagic ooze open-marine calcareous or siliceous sediment made predominantly of mud-sized skeletons of zooplankton and phytoplankton.

pellet mud carbonate mud deposited under very low-energy conditions such that fecal pellets produced by deposit-feeding invertebrates remain intact.

period fundamental unit of geologic time represented by the time-stratigraphic unit called a system.

photosynthesis the process by which plants form carbohydrates from carbon dioxide, inorganic salts, and water through the agency of sunlight.

phylum (plural: phyla) a major category of biologic classification, second only to kingdom.

pillow lava basic lavas that consist of an agglomeration of rounded masses that fit closely upon one another, formed as a result of subaqueous deposition.

placenta in higher mammals the organ that connects the developing fetus to the womb, and through which the fetus is nourished prior to birth.

placental mammal one of a group of mammals in which the fetus is nourished by means of a placenta.

plankton marine plants and animals that do not swim actively but are transported by currents; they range in size from microorganisms to jellyfish.

plate tectonics the concept that the earth's rigid outer shell is divided into several huge structural plates that move to produce earthquakes, mountain chains, and other geologic features.

platform in stratigraphy, the area of comparatively thin sediments continentward from a geosynclinal sequence of thicker equivalent strata.

playa the shallow central basin of a desert plain which may at times be occupied by standing water.

polar wandering the apparent movement of the magnetic poles during geologic time.

polar-wandering curve points on a map that trace the path of the north or south magnetic poles from rocks of increasing age on the same continent.

potassium-argon method radioactive potassium-40 decays by electron capture to argon-40 with a half-life of 1.3 billion years, providing one of the most convenient and widely applicable methods of radiometric dating.

Precambrian shield the largely exposed Precambrian nucleus of a continent.

primary nuclide a long-lived radioactive nuclide that was part of the earth when it formed, and none of which has been produced subsequently.

Primary rocks the oldest of a threefold stratigraphic classification proposed by Nicolaus Steno in 1669.

Primitive series the basal unit of Werner's eighteenth-century stratigraphic scheme, consisting of granite, gneiss, schist, and other crystalline rocks.

procaryote organism whose cells lack a membrane-bounded nucleus; characteristic only of bacteria and blue-green algae.

progradation a seaward advance of the shoreline resulting from the near-shore deposition of sediments.

Protoatlantic Ocean the ocean that lay between the continents of Laurentia and Baltica during the early Paleozoic; this ocean closed completely in the Devonian Period.

quasar originally applied to starlike objects (quasi-stellar objects) that were sources of X-rays and that showed extreme red shifts; now the term includes some objects with extreme red shifts that are not X-ray sources.

radioactive element an element that undergoes radioactive decay.

radiolarian a microscopic marine protozoan that builds a delicate ornate shell of opaline silica.

radiometric dating process of determining the age of natural substances based on their content of radioactive parent and radiogenic daughter nuclides.

range zone all the strata that contain the fossil remains of a particular species.

redbed red or red-tinted sedimentary rocks, usually sandstones and shales, whose color is due to ferric iron compounds.

remanent magnetism weak reflection of the earth's magnetic field induced into some igneous and sedimentary minerals as they are incorporated into rocks.

reptile one of a class of cold-blooded, air-breathing vertebrates, including snakes, lizards, turtles, and a host of extinct groups, including the dinosaurs.

reservoir rock a rock that contains petroleum or natural gas and that is sufficiently permeable to yield them in commercial quantities.

rift valley an elongated valley formed by the depression of a block of the earth's crust between two approximately parallel fault zones.

rock cycle cycle in which rocks are uplifted and eroded to form sediments that lithify into sedimentary rocks that, in turn, may be eroded or buried to form new igneous or metamorphic rocks to begin the cycle anew.

rubidium-strontium method radioactive rubidium-87 decays by beta decay to strontium-87 with a half-life of 47,000 million years, providing a widely applicable method of radiometric dating.

sea-floor spreading the process through which new ocean-floor rocks are created at oceanic ridges and moved from the ridge by plate motions.

Secondary rocks the middle unit in Nicolaus Steno's 1669 stratigraphic classification.

section in stratigraphy, the sequence of stratified rocks at a given locality or representative of a given area.

sedimentary basin an area that receives sediments.

sediment-feeder an invertebrate that ingests mud systematically, digests the organic constituents, and expels the remainder as fecal material.

seismic discontinuity a boundary surface within the earth across which the velocity of earthquake waves changes abruptly.

seismic wave waves generated by an earthquake, which consist of primary (P) and secondary (S) waves that travel through the earth and of surface waves that are known as L waves because they have long periods.

seismograph an instrument that detects earthquake waves.

sequence stratigraphic unit bounded by widespread unconformities.

series time-stratigraphic unit, a subdivision of a system.

Sevier Orogeny a mountain-building episode in Idaho, eastern Nevada, and western Utah, in Early Cretaceous time.

sexual reproduction production of offspring through the sharing of genetic material from separate parents.

siliceous ooze a fine-grained pelagic deposit with more than 30 percent material of organic origin, a large percentage of which is siliceous skeletal material produced by planktonic plants and animals.

silicification diagenetic replacement by silica of another mineral such as calcium carbonate.

skeletal sand carbonate sand made of fragments of algae and invertebrate animals, such as corals, bivalves, bryozoans, and crinoids.

sole markings sedimentary structures on the bottom of a bed, usually of sandstone, such as groove casts, flute casts, load casts, and organic tracks.

source bed a sedimentary rock rich in organic material which, under burial and heat, furnishes hydrocarbons to adjacent reservoir rocks from which they are recoverable.

species a group of organisms capable of interbreeding to produce fertile offspring.

spectrograph an instrument for photographing the precise wavelengths of light emitted by a source.

sphenopsid horsetails; dominant in the Paleozoic; one genus, Equisetum, survives.

spreading center zone where new basaltic material is added to lithosphere plates; usually occurs along oceanic ridges.

stage time-stratigraphic unit, a subdivision of a series.

standing diversity the number of taxa in an area at a given time.

Stratified series a unit in Werner's eighteenth-century stratigraphic scheme consisting of sandstone, limestone, salt, gypsum, coal, and basalt.

stratigraphy the study of stratified rocks.

stromatolite a laminated deposit built up by blue-green algae. The layers are formed mostly of carbonate minerals but clay or even fine sand may be bound into the structure; forms of stromatolites range from flat to moundlike or fingerlike.

structural province a portion of a Precambrian shield area with distinctive radiometric ages and a distinctive structural style (that is, trend and intensity of deformation) relative to adjacent regions.

subduction zone a boundary between two converging lithospheric plates where one descends beneath the other.

supernova a violently exploding star whose brightness may, for a few days, rival that of an entire galaxy.

superposition a principle of organizing stratified rocks that stipulates that a given stratum is older than the one that overlies it and younger than the one that underlies it.

supratidal flat a depositional environment that lies above the normal high tide level and is covered by water only during big storms.

suspended load sediment of silt and clay size transported by a stream or other current of water and kept in suspension by turbulence.

suture zone zone of collision between two once-separated continental masses that are now united.

system fundamental time-stratigraphic unit; the basic subdivision of the geologic time scale.

Taconic Orogeny mountain-building episode that affected the eastern margin of the United States and Canada during Late Ordovician time named from easternmost New York.

tectonic melange rock in which a mixture of fragments of varying sizes and compositions is consolidated by deformational pressures.

tensional force pull-apart force; in structural geology, responsible for block faulting.

Tertiary rocks initially the youngest unit of Nicolaus Steno's 1669 stratigraphic scheme for western Italy; at present used for the oldest system of the Cenozoic Era; includes rocks formed between about 2 million and 65 million years ago.

tillite lithified glacial till.

topset bed the material laid down in horizontal layers on top of a delta.

tracheophyte vascular plants, many of which are hard-tissued and hence abundant as fossils.

transcurrent fault fault along which rocks move horizontally; large transcurrent faults bound some parts of most lithospheric plates.

Transition Series a stratigraphic unit in Werner's eighteenth-century scheme between the primitive series below and the stratified series above, consisting mainly of slate, greywacke, schist, and some limestone.

trilobite a Paleozoic class of marine arthropods that consist of a readily distinguishable head, body, and tail; hence the name.

triple junction a point at which three lithospheric plates are in contact.

turbidite detrital sediment deposited by the mechanism of turbidity currents; generally the beds are graded and the sediment poorly sorted.

turbidity current a mass of water made dense by its content of suspended sediment that flows downslope under gravity and deposits its material from suspension, coarse sediment first.

type area the area from which the name and definition of a stratigraphic unit is taken.

type section the actual rock outcrops used to define a stratigraphic unit.

ultramafic igneous rocks rocks containing more than 75 percent of dark, ferromagnesian silicate minerals.

unconformity a surface of erosion or nondeposition—usually the former—that separates younger strata from older rocks.

uniformitarianism the concept that the present is the key to the past; the term embodies the idea that natural laws are inviolate.

universal formations formations belonging to Werner's primitive and transition series in his eighteenth-century stratigraphic scheme, which were believed to represent individual worldwide depositional events.

uranium-lead method uranium-235 decays to lead-207 and uranium-238 decays to lead-208, both in a series of alpha and beta decay steps; method provides a cross-check in the radiometric dating of igneous rocks that improves accuracy.

varve a sedimentary bed or lamination that is deposited within one year's time.

volcanic arc arcuate chain of volcanic islands; usually formed above subduction zones.

volcaniclastic rock a broad term for any rock containing abundant clasts of volcanic origin, ranging from coarse breccia to fine-grained tuffs.

volcanic tuff a rock formed of volcanic fragments, generally smaller than 4 millimeters in size.

whole-rock method a method of radiometric dating, usually employing the potassium-argon or rubidium-strontium methods, in which particular minerals are not separated from the rock for analysis; instead the rock is analyzed as a single entity.

zone the principal biostratigraphic unit, defined solely on the basis of fossils.

REFERENCES

Alexander, E. C., Jr., Lewis, R. S., Reynolds, J. H., and Michel, M. C., 1971 Plutonium-244: Confirmation as an extinct radioactivity. *Science,* 172, 837–840.

Andrews, J. T., 1970 Present and postglacial rates of uplift for glaciated northern and eastern North America derived from postglacial uplift curves. *Can. Jour. Earth Sciences,* 7, 703–715.

Anhaeusser, C. R., Mason, R., Viljoen, M. J., and Viljoen, R. P., 1969 A reappraisal of some aspects of Precambrian shield geology. *Geol. Soc. America Bull.,* 80, 2175–2200.

Awramik, S. M., 1971 Precambrian columnar stromatolite diversity: Reflection of metazoan appearance. *Science,* 174, 825–827.

Baker, V. R., 1973 *Paleohydrology and sedimentology of Lake Missoula flooding in eastern Washington.* Geol. Soc. America Special Paper 144, 79 pp.

Barager, W. R. A., and McGlynn, F. C., 1976 *Early Archean basement in the Canadian Shield: A review of the evidence.* Geol. Survey Canada Paper 76–14.

Bassett, H. G., and Stout, J. G., 1967 Devonian of western Canada, in *International Symposium on the Devonian System,* D. H. Oswald, editor. Calgary, Alberta: Soc. Petrol. Geologists, pp. 717–752.

Brookins, D. G., Berdan, J. H., and Stewart, D. B., 1973 Isotopic and paleontologic evidence for correlating three volcanic sequences in the Maine coastal volcanic belt. *Geol. Soc. America Bull.,* 84, 1619–1628.

Berger, W. H., and Roth, P. H., 1975 Oceanic micropaleontology: Progress and prospect. *Reviews of Geophysics and Space Physics,* 13, no. 3, 561–585.

Berger, W. H., and Winterer, E. L., 1974 Plate stratigraphy and the fluctuating carbonate line, in *Pelagic sediments on land and under the sea,* Hsü and Jenkyns, eds. Oxford: International Assoc. of Sedimentologists, Blackwell Scientific Publications, Ltd., Special Publication 1, pp. 11–48.

Berry, W. B. N., 1973 Silurian–Early Devonian graptolites, in *Atlas of Paleobiogeography,* A. Hallam, ed. Amsterdam: Elsevier Sci. Publ. Co., pp. 81–87.

Bird, J. M., and Dewey, J. F., 1970 Lithosphere plate-continental margin tectonics and the evolution of the Appalachian Orogen. *Geol. Soc. of America Bull.,* 81, 1031–1060.

Blatt, H., Middleton, G., and Murray, R., 1972 *Origin of sedimentary rocks.* Englewood Cliffs, New Jersey: Prentice-Hall, Inc., 634 pp.

Bloom, A. L., 1969 *The surface of the earth.* Englewood Cliffs, N.J.: Prentice-Hall, Inc., 152 pp.

Broeker, W. S., and Van Donk, J., 1970 Insolation changes, ice volumes, and the O^{18} record in deep-sea cores. *Reviews of Geophysics and Space Physics,* 8, 169–198.

Burke, K., 1975 Atlantic evaporites formed by evaporation of water spilled from Pacific, Tethyan, and southern oceans. *Geology,* 3, 613–616.

Burke, K. C., and Wilson, J. T., 1976 Hot spots on the earth's surface. *Scientific American,* 235, no. 2, 46–57.

Carr, A., and Coleman, P. J., 1974 Sea-floor spreading theory and the odyssey of the green turtle. *Nature,* 249, 128–130.

Clark, S. P., Jr., 1971 *Structure of the earth.* Englewood Cliffs, N.J.: Prentice-Hall, Inc.

CLIMAP Project Members, 1976 The surface of the ice-age earth. *Science,* 191, no. 4232, 1131–1137.

Coates, A. G., 1973 Cretaceous Tethyan coral-rudist biogeography related to the evolution of the Atlantic Ocean, in *Organisms and continents through time,* N. F. Hughes, ed. London: Palaeontological Assn., Special Papers in Palaeontology 12, pp. 169–174.

Condie, K. C., 1976 *Plate tectonics and crustal evolution.* New York: Pergamon Press, Inc., 288 pp.

Dewey, J. F., Pitman, W. C., III, Ryan, W. B. F., and Bonnin, J., 1973 Plate tectonics and the evolution of the Alpine system. *Geol. Soc. America Bull.,* 84, 3137–3180.

Dietz, R. S., 1972 Geosynclines, mountains, and continent-building. *Scientific American,* 226, no. 3, 30–38.

Dietz, R. S., and Holden, J. C., 1970 Reconstruction of Pangaea: Break-up and dispersion of continents, Permian to Recent. *Jour. Geophysical Research,* 75, 4939–4956.

Donn, W. L., and Shaw, D. M., 1977 Model of climate evolution based on continental drift and polar wandering. *Geol. Soc. America Bull.,* 88, 390–396.

Ericson, D., and Wollin, G., 1964 *The deep and the past.* New York: Alfred A. Knopf, Inc., 292 pp.

Ernst, W. G., 1972 Occurrence and mineralogic evolution of blueschist belts with time. *Am. Jour. Science,* 272, 657–668.

Eugster, H. P., and Hardie, L. A., 1975 Sedimentation in an ancient playa-lake complex: the Wilkins Peak Member of the Green River Formation of Wyoming. *Geol. Soc. America Bull.,* 86, 319–334.

Fisher, W. L., Brown, L. F., Jr., Scott, A. J., and McGowen, J. H., 1969 *Delta systems in the exploration for oil and gas.* Austin, Texas: Bur. Econ. Geol., 78 pp.

Folk, R. L., and Robles, R., 1964 Carbonate sands of the Isla Perez, Alacran reef complex, Yucatan. *Jour. Geology,* 72, 255-292.

Fraser, J. A., Hoffman, P. F., Irvine, T. N. and Mursky, G., 1972 The Bear Province: in Price, R. A., and Douglas, R. J. W., *Variations in Tectonic Styles in Canada.* Geol. Assoc. Canada, Special Paper 11, pp. 453-503.

Gilluly, J., 1963 The tectonic evolution of the western United States. *Quart. Jour. Geol. Soc. London,* 119, 133-174.

Glikson, A. Y., 1972 Early Precambrian evidence of a primitive ocean crust and island nuclei of sodic granite. *Geol. Soc. America Bull.,* 83, 3323-3344.

Gordon, W. A., 1975 Distribution by latitude of Phanerozoic evaporite deposits. *Jour. Geology,* 83, 671-684.

Green, D. H., 1972 Archaean greenstone belts may include terrestrial equivalents of lunar maria. *Earth and Planetary Science Letters,* 15, 263-270.

Hedberg H. D., 1964 Earth history and the record in the rocks. *Proc. Am. Philosophical Soc.,* 109, no. 2, 99-104.

Heezen, B. C., and MacGregor, I. D., 1973 The evolution of the Pacific. *Scientific American,* 229, no. 5, 102-112.

Hey, R. N., Deffeyes, K. S., Johnson, G. L., and Lowrie, A., 1972 The Galapagos triple junction and plate motions in the East Pacific. *Nature,* 237, 20-22.

Hite, R. J., and Cater, F. W., 1972 Pennsylvanian rocks and salt anticlines, Paradox Basin, Utah and Colorado, in *Geologic Atlas of the Rocky Mountain Region.* Denver: Rocky Mountain Association of Geologists, pp. 133-138.

Hoffman, P., 1967 Algal stromatolites: Use in stratigraphic correlation and paleocurrent determination. *Science,* 157, 1043-1045.

Hoffman, P., Dewey, J. F., and Burke, J., 1974 Aulacogens and their genetic relation to geosynclines, with a Proterozoic example from Great Slave Lake, Canada, in Dott and Ahaver, eds., *Modern and Ancient Geosynclinal Sedimentation.* Society of Economic Paleontologists and Mineralogists, Special Publication 19, 38-55.

House, M. R., 1971 Devonian faunal distributions: in *Faunal provinces in space and time,* Middlemiss, Rawson and Newall, eds. Liverpool: Seel House Press, pp. 77-94.

Hsü, K. J., 1972 When the Mediterranean dried up. *Scientific American,* 227, no. 6, 26-36.

Hurley, P. M., and others, 1967 Test of continental drift by comparison of radiometric ages. *Science,* 157, 495-500.

Hurley, P. M., and Rand, J. R., 1969 Pre-drift continental nuclei. *Science,* 164, 1229-1242.

Irving, E., North, F. K., and Couillard, R., 1974 Oil, climate and tectonics. *Can. Jour. Earth Sci.,* 11, 1-17.

James, H. L., 1960 Problems of stratigraphy and correlation of Precambrian rocks with particular reference to the Lake Superior Region. *Am. Jour. Science,* 258-A, 104-114.

Johnson, J. G., 1971 Timing and coordination of orogenic, epeirogenic and eustatic events. *Geol. Soc. America Bull.,* 82, 3263-3298.

Karig, D. E., 1971 Origin and development of marginal basins in the western Pacific. *Jour. Geophysical Research,* 76, 2542-2561.

Karig, D. E., 1974 Evolution of arc systems in the western Pacific. *Ann. Rev. Earth Planet. Sci.,* 2, 51-75.

Kay, M., and Colbert, E. H., 1965 *Stratigraphy and life history.* New York: John Wiley & Sons, 736 pp.

King, P. B., 1948 Geology of the southern Guadalupe Mountains, Texas. *U.S. Geol. Survey,* Professional Paper 215, 183 pp.

Kolb, C. R., and Van Lopik, J. R., 1966 Depositional environments of the Mississippi River deltaic plain—southeastern Louisiana, in *Deltas in their geologic framework,* Shirley and Ragsdale, eds., Houston, Texas: Houston Geol. Society, pp. 17-62.

Kumar, N., and Gamboa, L. A. P., 1979 Evolution of the Sao Paulo Plateau (southeastern Brazilian margin) and implications for the early history of the South Atlantic. *Geol. Soc. America Bull.,* 90, 281-293.

Laporte, L. F., 1968 *Ancient environments.* Englewood Cliffs, N.J.: Prentice-Hall, Inc., 115 pp.

Larson, R. L., and Pitman, W. C., III, 1972 World-wide correlation of Mesozoic magnetic anomalies, and its implications. *Geol. Soc. America Bull.,* 83, 3645-3662.

Lipman, P. W., Protska, H. J., and Christiansen, R. L., 1971 Evolving subduction zones in the western United States, as interpreted from igneous rocks. *Science,* 174, 821-825.

Logan, B. W., Rezak, R., and Ginsburg, R. N., 1964 Classification and environmental significance of algal stromatolites. *Jour. Geol.* 72, 68-83.

Mallory, W. W., 1972 Pennsylvanian arkose and the ancestral Rocky Mountains, in *Geologic Atlas of the Rocky Mountain Region.* Denver: Rocky Mountain Assoc. Geologists, pp. 131-132.

McAlester, A. L., 1973 Phanerozoic biotic crises, *Proceedings of the International Conference on the Permian-Triassic.* Bulletin of Canadian Petroleum Geology Special Publication 2, 1-5.

McDougall, I., 1964 Potassium-Argon ages from lavas of the Hawaiian Islands, *Geol. Soc. America Bull.,* 75, 107-128.

McElhinny, M. W., 1973 *Paleomagnetism and plate tectonics.* Cambridge Univ. Press, 358 pp.

McKee, E. D., and others, 1956 Paleotectonic maps of the Jurassic System. *U.S. Geol. Survey,* Map I-175.

McKee, E. D., and others, 1975 Paleotectonic investigations of the Pennsylvanian System in the United States. *U.S. Geol. Survey,* Professional Paper 853.

McKenzie, D. P., 1972 Plate tectonics and sea-floor spreading. *Am. Scientist,* 60, 425-435.

McKerrow, W. S., ed., 1978 *The ecology of fossils.* Cambridge, Mass.: The MIT Press, 384 pp.

Molnar, P., and Tapponnier, P., 1977 The collision between India and Eurasia. *Scientific American,* 236, no. 4, 30–41.

Moore, R. C., 1970 Stability of the earth's crust. *Geol. Soc. America Bull.,* 81, 1285–1324.

Newell, N. D., 1972 The evolution of reefs: *Scientific American,* 226, no. 6, 54–64.

Opdyke, N. D., Glass, B., Hays, J. D., and Foster, J., 1966 Paleomagnetic study of Antarctic deep-sea cores, *Science,* 154, 349–357.

Palmer, A. R., 1974 Search for the Cambrian world. *Am. Scientist,* 62, no. 2, 216–224.

Parker, R. H., 1959 Macro-invertebrate assemblages of central Texas coastal bays and Laguna Madre. *Am. Assoc. Petrol. Geologists Bull.,* 43, 2100–2166.

Pitman, W. C., III, and Talwani, M., 1972 Sea-floor spreading in the North Atlantic, *Geol. Soc. America Bull.,* 83, 619–646.

Press, F., and Siever, R., 1978 *Earth,* 2nd ed. San Francisco: W. H. Freeman and Co., 649 pp.

Prest, V. K., 1969 Quaternary geology, in *Geology and Economic Minerals of Canada.* Geol. Survey of Canada, Econ. Geol. Report 1, pp. 675–764.

Pryor, W. A., and Amaral, E. J., 1971 Large-scale cross-stratification in the St. Peter Sandstone. *Geol. Soc. America Bull.,* 83, 239–244.

Reynolds, J. H., 1960 The age of the elements in the solar system. *Scientific American,* 203, no. 5, 171–182.

Robinson, P. L., 1973 Palaeoclimatology and continental drift, in Tarling and Runcorn, eds., *Implications of continental drift to the earth sciences,* vol. 1. London: Academic Press, pp. 451–476.

Ronov, A. B., 1964 Common tendencies in the chemical evolution of the earth's crust, ocean, and atmosphere. *Geochemistry,* 8, 715–743.

Schenck, H. G., and Graham, J. J., 1960 Subdividing a geologic section. *Science Reports,* Tohoku University, Sendai, Japan, 2nd Ser. (Geol.), Spec. Vol., no. 4, 92–99.

Schumm, S. A., 1968 Speculations concerning paleohydrologic controls of terrestrial sedimentation. *Geol. Soc. America Bull.,* 79, 1573–1588.

Scotese, C., Bambach, R. K., Barton, C., Van der Voo, R., and Zeigler, A., 1979 Paleozoic base maps. *Jour. Geology,* 87, no. 3, 217–277.

Semikhatov, M. A., 1976 Experience in stromatolite studies in the U.S.S.R., in *Stromatolites.* M. R. Walter, ed. New York: Elsevier, pp. 337–358.

Seyfert, C. K., and Sirkin, L. A., 1973 *Earth history and plate tectonics, an introduction to historical geology.* New York: Harper & Row, Pubs., 504 pp.

Sillitoe, R. H., 1972 A plate tectonic model for the origin of porphyry copper deposits. *Economic Geology,* 67, 184–197.

Sloss, L. L., 1963 Sequences in the cratonic interior of North America. *Geol. Soc. America Bull.,* 74, 93–114.

Stainforth, R. M., Lamb, J. L., Lutherbacher, H. P., Beard, J. H., and Jeffords, R. M., 1975 Cenozoic planktonic foraminiferal zonation and characteristics of index forms. *Univ. Kansas Paleontological Contribution,* Art. 62, 425 pp.

Stewart, J. H., 1972 Initial deposits in the Cordilleran Geosyncline: Evidence of a late Precambrian (<850 m.y.) continental separation. *Geol. Soc. America Bull.,* 83, 1345–1360.

Stockwell, C. H., McGlynn, J. C., Emslie, R. F., Sanford, B. V., Norris, A. W., Fahrig, W. F., and Currie, K. L., 1970 Geology of the Canadian Shield, in *Geology and Economic Minerals of Canada.* Geol. Survey of Canada, Econ. Geol. Report 1, pp. 45–150.

Sullivan, Walter, 1974 *Continents in motion: The new earth debate.* New York: McGraw-Hill, 399 pp.

Teichert, C., 1958 Some biostratigraphical concepts. *Geol. Soc. America Bull.,* 69, 99–120.

Textoris, D. A., 1968 Petrology of supratidal, intertidal, and shallow subtidal carbonates, Black River Group, Middle Ordovician, New York, U.S.A., *23rd Internat. Geol. Cong.,* Prague, pp. 227–248.

Valentine, J. W., 1969 Patterns of taxonomic and ecological structure of the shelf benthos during Phanerozoic time. *Palaeontology,* 12, pt. 4, 684–709.

Walker, T. R., 1967 Formation of red beds in modern and ancient deserts. *Geol. Soc. America Bull.,* 78, 353–368.

Washburn, S. L., 1978 The evolution of man. *Scientific American,* 239, no. 3, 194–208.

Welland, M. J. P., and Mitchell, A. H. G., 1977 Emplacement of the Oman ophiolite: Mechanism related to subduction and collision. *Geol. Soc. America Bull.,* 88, 1081–1088.

Wensink, H., 1973 The Indo-Pakistan subcontinent and the Gondwana reconstructions based on palaeomagnetic results, in Tarling and Runcorn, eds., *Implications of continental drift to the earth sciences.* London: Academic Press, pp. 103–116.

Wetherill, G. W., 1971 Of time and the moon. *Science,* 173, 383–392.

Wickham, J., Roeder, D., and Briggs, G., 1976 Plate tectonics models for the Ouachita foldbelt. *Geology,* 4, 174–176.

INDEX

D

E

F

N

O

P

Q

R

S

T

U

V

THE TIME SCALE OF EARTH HISTORY

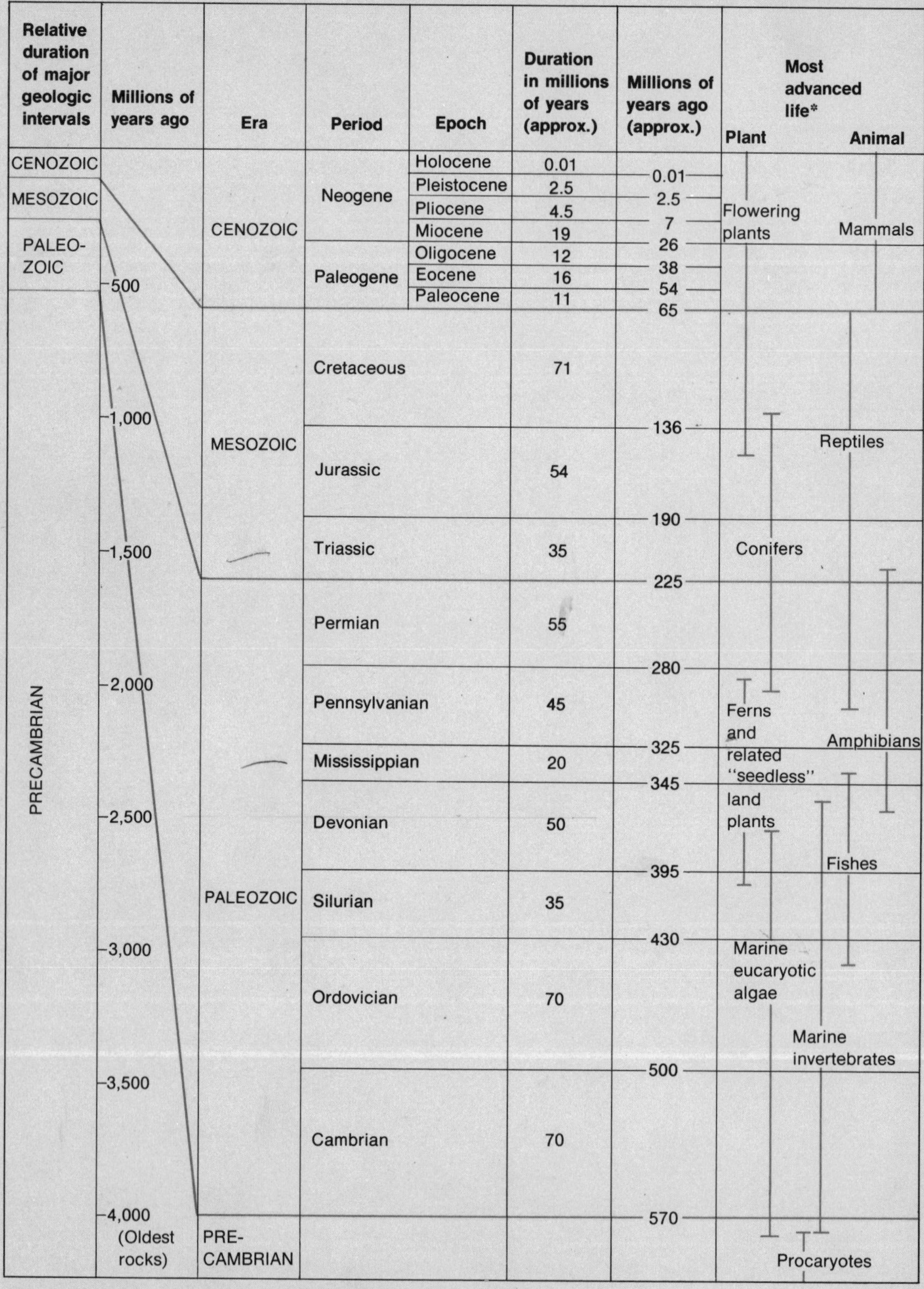

Era	Period	Epoch	Duration in millions of years (approx.)	Millions of years ago (approx.)
CENOZOIC	Neogene	Holocene	0.01	0.01
		Pleistocene	2.5	2.5
		Pliocene	4.5	7
		Miocene	19	26
	Paleogene	Oligocene	12	38
		Eocene	16	54
		Paleocene	11	65
MESOZOIC	Cretaceous		71	136
	Jurassic		54	190
	Triassic		35	225
PALEOZOIC	Permian		55	280
	Pennsylvanian		45	325
	Mississippian		20	345
	Devonian		50	395
	Silurian		35	430
	Ordovician		70	500
	Cambrian		70	570
PRE-CAMBRIAN				

*The vertical lines in this column show only the intervals of dominant importance, not total ranges. All groups shown persist to the present day.